Genome Mapping and Genomics in Animals

Volume 5

Series editor
Chittaranjan Kole, Mohanpur, India

For further volumes:
http://www.springer.com/series/7518

Preface

In recent years, there has been phenomenal progress in the understanding of the genetic architecture of normal and disease-related complex phenotypes. The progress has been fueled by an explosion of research activities related to the Human Genome Project and subsequent sequencing projects, and the nonhuman primate comprehensive sequencing projects. Advances in molecular genetics, statistical genetics, medical genetics, and bioinformatics have accompanied this progress.

Tracing its roots back to the laws of inheritance established by Mendel, which continue to be the basic tenets underlying modern genetics, the field of genetics has expanded tremendously and has richly diversified over the years. Gene mapping efforts and genomic research on humans and nonhuman primates have generated an enormous amount of information relevant for studies of evolution, phylogenetics, human genetics, anthropological genetics, and for biomedical research. Elucidation of gene function, expression, and regulation and of genetic variation and conservation among primate species has exciting potential for informing research in the areas of biology, evolution, population genetics, anthropological genetics, and biomedicine. The huge increase in the amount of available genomic information and advances in the tools available to analyze that data have already had a tremendous impact on disciplines such as evolutionary biology, bioinformatics, genetic epidemiology, medicine, pharmacogenetics, pharmacogenomics, and anthropology.

This volume is an attempt to provide researchers and academicians with a review of advanced methodologies and applications in gene mapping and genomics of humans and nonhuman primates, with an emphasis on genetics of complex phenotypes and diseases. As a part of the "Genome Mapping and Genomics in Animals" series (Dr. C. Kole, Editor), this volume is designed to illustrate ongoing research activities related to gene mapping and genomics in human and nonhuman primates. The topic of this volume is broad and a full coverage of such a huge area of research would be impossible. Therefore, we limited the volume to 16 chapters that illustrate the amazing changes in genomic studies that have occurred since the Human Genome Project. From the initiation and expansion of the Human Genome Project to revolutionary next generation sequencing approaches, we have seen dramatic improvement in the understanding of the genetic architecture of complex phenotypes in human and nonhuman primates.

This volume constitutes an overview of the impact of the genomic revolution on research related to human and nonhuman primate populations. It also reviews the state-of-the-science with respect to the molecular, statistical genetics, and genetic epidemiologic techniques that are used to dissect the genetic architecture of normal and disease-related complex phenotypes using data from human and nonhuman primates. We present examples of successful applications of genomic methods to traits of particular interest in biomedical research and evolutionary biology, and provide discussions of future directions in human and nonhuman primate genomics.

Since genetic investigation of complex phenotypes is by nature multidisciplinary, efforts were made to provide readers with review papers which illustrate the full range of methodological and analytical approaches being applied to human and nonhuman primate population data sets. Examples and applications were drawn from diverse areas including evolutionary genetics, population structure, genetic epidemiology, transcriptomics, copy number variation, molecular ecology, comparative genomics, and gene mapping for phenotypes related to behavior, skeletal biology, and cardio-metabolic disease in human and nonhuman primate populations.

We are in the midst of an exciting scientific era with constantly changing technology revolutionizing genomic research approaches over and over again. The advances will ensure continued interest in explorations of genomics and other "omics" approaches as they relate to normal variation and disease-related traits in human and nonhuman primate populations. Progress in gene mapping and genomic sequencing will add further momentum to progress in comparative genomics, evolutionary genomics, and biomedical research as it corresponds to disease prevention and treatment, pharmacogenomics, and personalized medicine.

We are grateful to the contributors to this volume who have prepared comprehensive and informative reviews of advanced, complex genomics-related topics. We thank Drs. Vidya S. Farook, Sobha Puppala, Geetha Chittoor, and Laura Cox for reviewing one or more chapters of this volume. The editors also express their gratitude to Ms. Maria Messenger whose expert skills in proofreading and formatting greatly improved the quality of this volume.

Ravindranath Duggirala
Laura Almasy
Sarah Williams-Blangero
Solomon F.D. Paul
Chittaranjan Kole

Contents

1 **The Utility of Genomics for Studying Primate Biology** . . . 1
Sarah Williams-Blangero and John Blangero

2 **The Human Genome Project: Where Are We Now and Where Are We Going?** 7
Satish Kumar, Christopher Kingsley, and Johanna K. DiStefano

3 **Linkage Mapping: Localizing the Genes That Shape Human Variation** 33
Laura Almasy, Mark Zlojutro Kos, and John Blangero

4 **Association Studies to Map Genes for Disease-Related Traits in Humans** 53
Robert L. Hanson and Alka Malhotra

5 **Gene Expression Studies and Complex Diseases** 67
Harald H.H. Göring

6 **Copy Number Variations and Chronic Diseases** 85
August N. Blackburn and Donna M. Lehman

7 **Applications of Genomic Methods to Studies of Wild Primate Populations** 103
Mary A. Kelaita

8 **Comparative Genomics: Tools for Study of Complex Diseases** 113
Laura A. Cox

9 **Genetic Structure and Its Implications for Genetic Epidemiology: Aleutian Island Populations**. 129
Michael H. Crawford

10 Mapping Genes in Isolated Populations: Lessons from the Old Order Amish 141
Braxton D. Mitchell, Alejandro A. Schäffer, Toni I. Pollin, Elizabeth A. Streeten, Richard B. Horenstein, Nanette I. Steinle, Laura Yerges-Armstrong, Alan R. Shuldiner, and Jeffrey R. O'Connell

11 Genetics of Cardiovascular Disease in Minority Populations 155
Jean W. MacCluer, John Blangero, Anthony G. Comuzzie, Sven O.E. Ebbesson, Barbara V. Howard, and Shelley A. Cole

12 Mapping of Susceptibility Genes for Obesity, Type 2 Diabetes, and the Metabolic Syndrome in Human Populations 181
Rector Arya, Sobha Puppala, Vidya S. Farook, Geetha Chittoor, Christopher P. Jenkinson, John Blangero, Daniel E. Hale, Ravindranath Duggirala, and Laura Almasy

13 Genetic Influence on the Human Brain 247
D. Reese McKay, Anderson M. Winkler, Peter Kochunov, Emma E.M. Knowles, Emma Sprooten, Peter T. Fox, John Blangero, and David C. Glahn

14 Variation, Genetics, and Evolution of the Primate Craniofacial Complex 259
Richard J. Sherwood and Dana L. Duren

15 Genetic Influences on Behavior in Nonhuman Primates . . . 277
Julia N. Bailey, Christopher Patterson, and Lynn A. Fairbanks

16 Genomic Studies of Human Populations: Resequencing Approaches to the Identification of Human Quantitative Loci 289
Joanne E. Curran, Claire Bellis, Laura Almasy, and John Blangero

Index 301

The Utility of Genomics for Studying Primate Biology

1

Sarah Williams-Blangero and John Blangero

1.1 Genomics of Primate Populations Writ Large

This volume was organized with the intent to review progress in primate genomics and, in particular, to show the value of studying primate genomics for understanding the determinants of risk for disease in human populations. Humans, our hominid ancestors, and nonhuman primates (and their ancestors) share most of their genetic material. The evolutionary proximity of nonhuman primates to humans provides us with a particularly valuable set of tools to make inferences about the causes of human phenotypic variation using experimental techniques applied to our close animal relatives. There is a remarkable amount of anatomical and physiological similarity across all primates that justifies the use of nonhuman primate models rather than more phylogenetically remote animal models, such as the mouse, for many types of studies. However, the use of nonhuman primates for modeling the basis of human phenotypic variation is associated with considerable costs due to the comparatively large size of the animals, their relatively long generation times, and the general expense of working with nonhuman primates.

The utility of considering primate biology writ large as a major source for inference about human biology stems from the close genetic relationship between humans and nonhuman primates. With the advent of large-scale genome sequencing, we now know precisely the extent of genetic similarity among the phylogenetically most proximate relatives.

The chimpanzee genome was the first nonhuman primate genome to be sequenced and was completed in 2005 (Chimpanzee Sequencing and Analysis Consortium 2005). From these data, we know that chimpanzees and humans diverged about 6 M years ago and share ~98 % sequence identity. About 29 % of orthologous proteins are identical between human and chimpanzees with most proteins differing by an average of only two amino acids (Chimpanzee Sequencing and Analysis Consortium 2005). This protein similarity greatly facilitates cross-inference of biological mechanism between the humans and chimpanzees species.

Even in the presence of substantial protein similarity, the genetic differences between the primate species clearly lead to striking phenotypic divergence. For example, comparative quantitative proteomic and metabolomics studies using chimpanzee and human biomaterials are finding fascinating and unexpected differences that may be of utility for understanding the substantial differences in brain and muscle

S. Williams-Blangero (✉) · J. Blangero
South Texas Diabetes and Obesity Institute, Regional Academic Health Center, University of Texas Health Science Center at San Antonio, 2102 Treasure Hills Blvd, Harlingen, TX 78550, USA
e-mail: WilliamsBlan@uthscsa.edu

J. Blangero
e-mail: blangero@uthscsa.edu

R. Duggirala et al. (eds.), *Genome Mapping and Genomics in Human and Non-Human Primates*,
Genome Mapping and Genomics in Animals 5, DOI 10.1007/978-3-662-46306-2_1,

function between the species (Bozek et al. 2014). Similarly, advanced neurophenotyping methods have revealed profound differences in synaptic phenotypes such as synaptic density between the humans and chimpanzees that appears to correlate with brain function (Liu et al. 2012). For many reasons, including smaller numbers of available colony-managed animals, greater expense, and extreme regulatory burden, chimpanzees are now little used in biomedical research despite being the most potentially useful of all primate species for making inferences about human health.

The rhesus macaque is the most widely used nonhuman primate model for human biology and its genomic sequence was first obtained in 2007 (Rhesus Monkey Sequencing and Analysis Consortium 2007). The sequence data shows that humans and rhesus monkeys have ~93 % total sequence identity, the reduction over that with chimpanzees correlating with the earlier divergence time of about 25 M years ago.

The baboon also is well utilized in biological research designed to be informative for human health (as evidenced in the chapters by Cox and Sherwood and Duren in this volume), as is the vervet (with examples provided by Bailey et al. in this volume). The genomes of the baboon and the vervet are in the process of being sequenced.

1.2 What Are We Trying to Explain?

In this volume, almost every chapter ultimately focuses on trying to explain the causal sources of human quantitative phenotypic variation. In many cases, the phenotypes under consideration are related to complex disease risk. In general, we would argue that the principle role of modern human genetics is to identify the causal sequence variants responsible for quantitative phenotypic variation. Most of the phenotypes in which we are interested exhibit complex causal pathways unlike those seen for simple monogenic traits. An overriding challenge in the analysis of complex traits as compared to the analysis of simple monogenic traits is that multiple loci may be contributing to the phenotype and, as a result, the effect size of any one locus is likely to be relatively small. In addition, there may be multiple types of sequence variation in play, ranging from substitutions of single nucleotides to rearrangements of chromosomal structure (sequence deletion, duplication, or inversion), which may not be equally detectable by any one analytical technique. Finally, even if the majority of a genetic effect was confined to a single locus, this could be either due to a single variable site or multiple rare alleles segregating in the population (s) under study.

1.3 Measuring Genetic Variation in Quantitative Traits

Many of the chapters in this volume at least implicitly involve characterizing how much of the observed phenotypic variation in primates is due to the action of genes. *Heritability* is the proportion of the total variance of a phenotype that is attributable to the additive effects of alleles. It represents an estimate of the relative extent of genetic variation in a given phenotype. Thus, heritability provides us with a single measure of how important a role genes likely play in the causal determination of a variable human trait. In a classical variance-components-based approach to quantitative genetic analysis, heritability is readily estimated by decomposing the phenotypic covariance between pairs of individuals based on their relatedness:

$$\begin{aligned} \mathrm{Cov}(i,j) &= 2\phi_{ij}\sigma_g^2 \\ \mathrm{Cov}(i,i) &= \sigma_g^2 + \sigma_e^2 \\ h^2 &= \sigma_g^2/\sigma_P^2 \end{aligned} \tag{1.1}$$

where $\mathrm{Cov}(i,j)$ is the covariance between different individuals i and j, and $\mathrm{Cov}(i,i)$ represents the variance for the ith individual, ϕ_{ij} is the kinship coefficient between i and j, σ_g^2 is the additive genetic variance, σ_e^2 is the error (sometimes termed environmental) variance, σ_P^2 is the total phenotypic variance of the trait given by $\sigma_P^2 = \sigma_g^2 + \sigma_e^2$, and h^2 is the additive genetic heritability which measures the relative contribution of additive genetic factors to the overall observed phenotypic variance.

Twice the 'kinship' coefficient in this context refers to the expected proportion of alleles shared *identical by descent* (IBD) by two individuals given their degree of relatedness: siblings share half their alleles IBD, half-siblings one-fourth of their alleles, and so on. Classically, we estimate this relatedness by knowledge of pedigree records. However, with the advent of high density assays of genetic variants (such as whole genome sequencing), we can now empirically estimate genetic relatedness in the absence of knowledge of the pedigree relationships among individuals. This latter development opens up vast opportunities for the study of wild primate populations. The genetic variance is a cumulative variance; it represents the summation over all additive genetic factors for the phenotype. Hence, depending upon the phenotype, it may represent the influence of a single genetic variant or that of many hundreds of genetic variants.

If variation in a phenotype were entirely due to genetic causes (and these could be clearly discerned), the heritability of the trait would, of course, be 1. Due to multifactorial causation and measurement error, a typical range of heritability for many quantitative traits is between 30 and 80 % of the total variance, and estimates may differ widely from one study to another due to sampling error. The heritability is a critical measure of the importance of within-population genetic variation. This single metric conveys whether or not the search for the individual contributing genes is merited for a given phenotype.

There have been thousands of studies of heritability of human phenotypes but relatively few of nonhuman primate phenotypes. The lack of nonhuman primate studies presumably is due to the paucity of pedigreed populations. However, the studies that have been conducted show that a substantial amount of genetic variation relevant for complex phenotype variation is segregating in nonhuman primate colonies.

Life history traits such as life span (Martin et al. 2002) and age at first birth (Williams-Blangero and Blangero 1995) show significant heritable components in pedigreed baboons. Standard hematological parameters in chimpanzees show significant heritable components (Williams-Blangero et al. 1993), as do many different lipid parameters in baboons (Blangero et al. 1990). Anatomical phenotypes, in particular, show high heritabilities (Mahaney et al. 1993; Rogers et al. 2007). Other even more complex phenotypes, such as the response of liver enzymes to experimental infection with hepatitis C virus in chimpanzees (Williams-Blangero et al. 1996) and longitudinal changes in fetal baboon morphometrics (Jaquish et al. 1997) also are significantly heritable in nonhuman primates. The evidence for substantial additive genetic variation in fundamental anatomical, physiological, and biochemical phenotypic dimensions in nonhuman primates parallels that seen in human studies and further confirms the utility of nonhuman primate models for studying human biological problems.

1.4 Functional Genetic Variation Determines Heritability

What is the biological source of heritability in humans? Ultimately, it comes from observable functional genetic variation at the sequence level. A functional variant is one that influences the focal phenotype via some molecular mechanism. Thus, functional variants can be considered to be phenotype-specific in this context. If a variant influences a quantitative trait (such as performance), we term it a *quantitative trait nucleotide* variant (QTN). The effect of a functional variant on the phenotype can be quantified by the QTN-specific variance which is given by $\sigma_q^2 = 2p(1-p)\alpha^2$, where p is the minor allele frequency of the QTN and α is one-half the difference between phenotypic means of the two homozygotes. Biologically, we expect α to be determined by biophysical molecular properties of the QTL and relatively constant across populations. The term, $2p(1-p)$, is also known as the expected heterozygosity of the underlying genotype and measures the variance of a trait that is scored as the number of minor alleles in the diploid genotype. The relative genetic signal intensity for this QTN is given by the QTN-specific heritability $h_q^2 = \sigma_q^2/\sigma_P^2$ where σ_P^2 is the total variance of the phenotype. The relative genetic signal for the QTL is determined by the sum of the QTN-specific heritabilities (although these must be corrected for possible linkage disequilibrium amongst variant sites) in the immediate region of the QTL and thus will be influenced by all of the relevant functional variants in the region. In algebraic form, the QTL-specific heritability is $h_q^2 = \frac{\sum 2p_i(1-p_i)\alpha_i}{\sigma_P^2} = \sum h_{qi}^2$ where the summation is over the functional variants in the regions of the QTL. Similarly, the total heritability of the phenotype is given by the sum of all of the QTL-specific heritabilities over the whole genome or $h^2 = \sum h_{qi}^2$.

1.5 Identifying Functional Sequence Variants Is the Critical Problem in Primate Biology

One of the main reasons that we study nonhuman primate species is to aid in the identification of the function of sequence variants. The four chapters that specifically utilize nonhuman primate genomics in this volume ultimately point to ways to better identify human genes and their sequence variants that influence human phenotypic variation.

In Chap. 7, Kelaita provides an overview of genomic methods applied to studies of wild primate populations. Besides the obvious utility of genomic methods for aiding our understanding of primate microevolution and primate population structure, she also suggests that phylogenetic inference can aid our interpretation of human adaptations (which are ultimately about human phenotypic variation).

Information on sequence variation across species can be used to make evolutionary inferences about genes likely to be under the influence of natural selection. Such selection only will occur for functionally relevant sequence variation (and nearby variants that are in linkage disequilibrium). Information on selection can be accumulated and used to aid studies of human sequence variation when attempting to determine which variants are most likely to be functional.

There is even more potential value likely to come from studies of wild nonhuman primate populations. Using our quantitative genetic model as described above, we would further suggest the great potential to better assess the importance of genetic variation for complex phenotypes observed in wild nonhuman primate populations. Now that accurate and direct molecular assay of genetic relatedness can be performed via sequencing technology, the potential to better understand the genetic basis of many complex phenotypes that are only observable in wild populations is greatly enhanced.

In Chap. 8, Cox directly shows how causal gene discovery of relevance for human disease risk can be directly performed in captive pedigreed nonhuman primate colonies. She highlights a clear benefit of using nonhuman primates for making inferences about human biology which is the access to tissues that are extremely difficult to obtain on a large scale in human studies. In studies of nonhuman primates, it is possible to safely obtain tissues such as liver or kidney that may be of critical value in understanding the biological mechanisms underlying functional sequence variation. Indeed, the potential for deep cellular phenotyping of many different nonhuman primate samples represents one of the major benefits of the primate animal model. Although the general paradigm Cox utilizes is that of complex phenotype gene discovery widely used in human populations, she shows that the ability to manipulate the other main component of the causal players, the environment, is possible in such experimental situations. By carefully controlling the environment, it is feasible to truly test for such complexities such as genotype-by-environment interaction. In her case, she focuses on genotype-by-diet interaction effects on lipid variation. This type of experiment involving the rigorous control of diet is extremely difficult to directly perform in humans and thus the benefit of doing such in nonhuman primates is obvious.

In Chap. 14, Sherwood and Duren examine the genetic determinants of variation in the primate craniofacial complex. This phenotypic dimension also is of obvious utility for understanding a large number of human disorders. Again, they employ standard gene discovery approaches to captive primate colonies. The ability to deeply phenotype animal models becomes a substantial benefit when dealing with potentially high dimensional imaging-derived phenotypes.

Finally, in Chap. 15 Bailey and colleagues apply similar approaches to nonhuman primate behavioral phenotypes. These are the most complex of all phenotypes and also among the most difficult to study. They review a number of studies including work on pedigreed vervets that show a consistent heritable component for complex primate behaviors. Other groups working on vervets also have used advanced genomic and transcriptomic methods to identify likely genes involved in complex phenotypes (Jasinska et al. 2012).

Advanced phenotyping relevant for behavior and psychiatric disease risks such as brain imaging are possible in large numbers of related primates to act as potential endophenotypes of relevance for the focal behaviors/disease risks. For example, an advanced brain imaging study that identified substantial genetic variation in the neural basis of anxious temperament in pedigreed rhesus macaques undergoing brain PET was performed successfully (Oler et al. 2010). Other clear benefits to working with nonhuman primates for psychiatric disease studies include the ability to get cerebrospinal fluid samples from large numbers of animals, a tissue that is exceedingly hard to justify in human studies of normal variation (Rogers et al. 2004).

1.6 Where Are We Going?

Most of the nonhuman primate applications in this volume still focus on causal gene discovery. Like others (Aitman et al. 2011), we believe that the tremendous advances in studies of human genetics will soon eliminate this focus. The field of human complex disease genetics is currently transitioning away from the study of common sequence variants of small effect (such as those that have been found typically in genome wide association studies) to the study of rare variants that are difficult to capture in sufficient numbers for testing except in extended families. These human studies often point to genes that require additional biological investigation in more controlled experimental circumstances.

Given that we can now sequence very large numbers of humans to directly search for likely functional variants, the utility of nonhuman primate genomic studies should transition to our second major question, that of aiding the functional characterizations of sequence variants. One of the most obvious ways to utilize nonhuman primates in this context is the exploitation of existing sequence variation through experimental breeding. For example, in the near future, all nonhuman primate colonies can easily be sequenced and all rare coding variation identified. For a given gene of interest, we can then identify those animals harboring the most likely consequential functional variation and design a breeding plan that will generate sufficient numbers of copies of variant animals for deep phenotyping studies. The ability to get at critical tissues, such as neuronal tissues of relevance for many psychiatric diseases, will greatly facilitate our ability to decide what the most important genes and variants are. Indeed, advances in genome editing and gene therapy are further likely to give us the tools to study the functional consequences of human sequence variants in nonhuman primates. This should help us identify and prioritize causal genes that may be of greatest importance for human disease pathways with the main advantage of providing an in vivo biological context that is extremely similar to that which we observed in humans.

References

Aitman TJ, Boone C, Churchill GA, Hengartner MO, Mackay TF, Stemple DL (2011) The future of model organisms in human disease research. Nat Rev Genet 12:575–582

Blangero J, MacCluer JW, Kammerer CM, Mott GE, Dyer TD, McGill HC Jr (1990) Genetic analysis of apolipoprotein A-I in two dietary environments. Am J Hum Genet 47:414–428

Bozek K, Wei Y, Yan Z, Liu X, Xiong J, Sugimoto M, Tomita M, Pääbo S, Pieszek R, Sherwood CC, Hof PR, Ely JJ, Steinhauser D, Willmitzer L, Bangsbo J, Hansson O, Call J, Giavalisco P, Khaitovich P (2014) Exceptional evolutionary divergence of human muscle and brain metabolomes parallels human cognitive and physical uniqueness. PLoS Biol 12(5):e1001871

Chimpanzee Sequencing and Analysis Consortium (2005) Initial sequence of the chimpanzee genome and comparison with the human genome. Nature 437:69–87

Jaquish CE, Leland MM, Dyer T, Towne B, Blangero J (1997) Ontogenetic changes in genetic regulation of fetal morphometrics in baboons (*Papio hamadryas* subspp.). Hum Biol 69:831–848

Jasinska AJ, Lin MK, Service S, Choi OW, DeYoung J, Grujic O, Kong SY, Jung Y, Jorgensen MJ, Fairbanks LA, Turner T, Cantor RM, Wasserscheid J, Dewar K, Warren W, Wilson RK, Weinstock G, Jentsch JD, Freimer NB (2012) A non-human primate system for large-scale genetic studies of complex traits. Hum Mol Genet 21:3307–3316

Liu X, Somel M, Tang L, Yan Z, Jiang X, Guo S, Yuan Y, He L, Oleksiak A, Zhang Y, Li N, Hu Y, Chen W, Qiu Z, Pääbo S, Khaitovich P (2012) Extension of cortical synaptic development distinguishes humans from chimpanzees and macaques. Genome Res 22:611–622

Mahaney MC, Williams-Blangero S, Blangero J, Leland MM (1993) Quantitative genetics of relative organ weight variation in captive baboons. Hum Biol 65:991–1003

Martin LJ, Mahaney MC, Bronikowski AM, Carey KD, Dyke B, Comuzzie AG (2002) Lifespan in captive baboons is heritable. Mech Ageing Dev 123:1461–1467

Oler JA, Fox AS, Shelton SE, Rogers J, Dyer TD, Davidson RJ, Shelledy W, Oakes TR, Blangero J, Kalin NH (2010) Amygdalar and hippocampal substrates of anxious temperament differ in their heritability. Nature 466:864–868

Rhesus Macaque Sequencing and Analysis Consortium (2007) The rhesus macaque genome. Science 316:235–237

Rogers J, Kochunov P, Lancaster J, Shelledy W, Glahn D, Blangero J, Fox P (2007) Heritability of brain volume, surface area and shape: an MRI study in an extended pedigree of baboons. Hum Brain Mapp 28:576–583

Rogers J, Martin LJ, Comuzzie AG, Mann JJ, Manuck SB, Leland M, Kaplan JR (2004) Genetics of monoamine metabolites in baboons: overlapping sets of genes influence levels of 5-hydroxyindolacetic acid, 3-hydroxy-4-methoxyphenylglycol, and homovanillic acid. Biol Psychiatry 55:739–744

Williams-Blangero S, Blangero J (1995) Heritability of age of first birth in captive olive baboons. Am J Primat 37:233–239

Williams-Blangero S, Blangero J, Murthy KK, Lanford RE (1996) Genetic analysis of serum alanine transaminase activity in normal and hepatitis C virus infected chimpanzees: an application of research-oriented genetic management. Lab Anim Sci 46:26–30

Williams-Blangero S, Brasky K, Butler T, Dyke B (1993) Genetic analysis of hematological traits in chimpanzees (Pan troglodytes). Hum Biol 65:1013–1024

2 The Human Genome Project: Where Are We Now and Where Are We Going?

Satish Kumar, Christopher Kingsley, and Johanna K. DiStefano

2.1 The Human Genome Project: Where Have We Been?

An explosion in our understanding of genetics and biochemistry, which began in the 1970s, led to the rapid development of diverse laboratory techniques such as restriction enzymes, cloning vectors, nucleic acid hybridization, and DNA sequencing. Together these methods revolutionized research in molecular biology. It was here, in this fertile atmosphere, that the seeds of genome sequencing were sown. The progressive spirit pervading research in the life sciences at this time consequently helped to fuel the conception of the Human Genome Project (HGP), whose primary aims were to determine the identity of the three billion nucleotides comprising the human genome and characterize the full repertoire of genes encoded therein.

S. Kumar
Department of Genetics, Texas Biomedical Research Institute, San Antonio, TX 78245, USA
e-mail: skumar@txbiomedgenetics.org

C. Kingsley · J.K. DiStefano (✉)
Diabetes, Cardiovascular & Metabolic Diseases Division, Translational Genomics Research Institute, 445 North Fifth Street, Phoenix, AZ 85004, USA
e-mail: jdistefano@tgen.org

C. Kingsley
e-mail: ckingsley@tgen.org

2.1.1 Historical Background of the HGP

The HGP is considered one of the most ambitious and successful international research collaborations in the history of biology. Those individuals and organizations responsible for bringing the HGP to fruition were both visionary and innovative, considering that the technological and computational tools commonplace today were unheard of 20 years ago when the idea of sequencing the human genome was germinated. Because thorough and engaging accounts of the conception, implementation, and completion of the HGP have already been presented elsewhere (Roberts 2001; Choudhuri 2003), we will provide only a brief synopsis of its history here.

The idea of sequencing the human genome was first discussed in 1984 at a meeting in Salt Lake City, Utah, hosted by the Department of Energy (DOE) and the Internal Commission for Protection Against Environmental Mutagens and Carcinogens. Although the purpose of this meeting was focused on mutation detection, the value of a human genome reference sequence was acknowledged, albeit in an oblique manner (Cook-Deegan 1989). The actual merit of sequencing the human genome was brought forward as a focus topic for the first time in 1985 during a conference at the University of California, Santa Cruz. Meeting participants generally supported the idea of such a project, but largely agreed that the endeavor laid outside the then current realms of feasibility and/or practicality.

R. Duggirala et al. (eds.), *Genome Mapping and Genomics in Human and Non-Human Primates*, Genome Mapping and Genomics in Animals 5, DOI 10.1007/978-3-662-46306-2_2,

Enthusiasm for the initiative quickly mounted during the following year at meetings held consecutively at Los Alamos National Laboratory and Cold Spring Harbor Laboratory (Roberts 2001). Debate about the value, expense, and potential consequences of the initiative continued until 1988, when the National Research Council panel officially endorsed the HGP. At that time, the panel refined the initiative, recommending that physical maps of each chromosome be constructed, and genomes of simple organisms be investigated prior to the full-scale sequencing of the human genome. In addition to sequencing the entire human genome, the HGP also aimed to identify all genes in the human genome, store sequence information in publicly available databases, develop and/or improve tools for analyzing sequence data, help transfer technologies resulting from the HGP to the private sector, and address relevant ethical, legal, and social issues (http://www.ornl.gov).

The HGP was officially launched on October 01, 1990, following the initiation of large-scale sequencing trials on *Mycoplasma capricolum*, *Escherichia coli*, *Caenorhabiditis elegans*, and *Saccharomyces cerevisiae*. The International Human Genome Sequencing Consortium (IHGSC), comprised of the National Institutes of Health (NIH), the DOE, and a collaborative of investigators from the United Kingdom, France, Germany, Japan, and China, was formed to implement the goals of the HGP. In 1998 this effort was joined by Celera Genomics, a privately funded venture formed jointly by Dr. J. Craig Venter, from The Institute for Genomic Research (TIGR), and the Perkin-Elmer Corporation. Venter proposed to sequence the human genome in a shorter period of time and at less cost than the publicly funded effort, using the relatively novel technique of whole genome shotgun sequencing. In early 2001, both IHGSC (Lander et al. 2001) and Celera Genomics (Venter et al. 2001) published working draft sequences of the human genome. Although these drafts covered only ~90 % of the euchromatic genome, was interrupted by ~150,000 gaps, had many misassembled segments and errors in the nucleotide sequence, the accomplishment of such a tremendous effort was generally applauded among the scientific community.

Following the publication of these rough draft versions of the genome, the IHGSC initiated efforts to finish sequencing the euchromatic genome and resolve areas containing gaps and misalignments. Results of these efforts were published in 2004 (International Human Genome Sequencing Consortium 2004). This updated version of the human genome covered 2.85 billion nucleotides, corresponding to ~99 % of the euchromatic genome. The near-complete draft was highly accurate: the error rate of the new genome sequence was reduced to <1 event/100,000 bases, a figure that surpassed the original acceptable estimate of the project (International Human Genome Sequencing Consortium 2004). The number of gaps was likewise decreased from ~150,000 to only 341, and most of these remaining gaps were associated with segmental duplications that are not amenable to current methods of sequencing. With the release of the near-complete human genome sequence, the original goals of the HGP were largely achieved (International Human Genome Sequencing Consortium 2004). Despite the incompleteness of this "finished" version, the availability of these sequence data has already had an irrevocable impact on the study of human disease.

2.2 Impact of the Human Genome Project: Where Are We Now?

Completion of the Human Genome Project has provided us with a greatly enhanced understanding of human genetics, including a greater appreciation of how DNA shapes species development and evolution, biology, and disease susceptibility. The HGP has also affected the development and/or maturation of research disciplines such as genome annotation, knowledge of genome evolution and segmental duplication, and comparative genomics, among others. Below we discuss the areas in which completion of the HGP has influenced our basic understanding of genetics, while subsequent sections will address the impact of the HGP on the manner in which we approach disease risk and development of treatment strategies based on genetic predisposition.

2.2.1 Enhanced Understanding of Human Genetics

2.2.1.1 Genome Annotation

The sequencing portion of the HGP was a significant technological feat, and provided the scientific community with a comprehensive accounting of the working material of the genome. However, acquisition of DNA sequence was only the first step toward the ultimate aim of understanding how the human genome functions at the molecular level. Necessary next steps toward this goal include the systematic identification and characterization of the functional units of the genome. This process of genome annotation is currently a multidisciplinary field, integrating the results of many different analytical approaches, both experimental and computational, to build our understanding of the functional underpinnings of the human genome (Table 2.1).

Prior to the completion of the HGP, the field of genome annotation was largely focused on the comprehensive identification of protein-coding genes, which was primarily achieved through the use of large-scale sequencing of cDNA libraries derived from reverse-transcribed mRNA transcripts. The resulting expressed sequence tags (ESTs) were grouped together based on sequence similarity using multiple sequence alignment algorithms. It was generally held that if the starting material was comprised of a mixture of mRNAs purified from numerous tissue types, then the number of groups produced by this process would provide a rough estimate of the total number of protein-coding genes expressed throughout the body. Prior to the publication of the human genome sequence, estimates on the total number of genes varied widely, from 35,000 to 150,000 (Pennisi 2007).

While cDNA sequencing approaches were fairly open ended in nature, the HGP produced a finite database of sequence information that could be easily searched for the presence of protein-coding genes. Yet, due to the low proportion of coding sequence in the human genome, the large number of exons per genes, and the relatively small exon size, gene annotation presented a much more difficult proposition in

Table 2.1 Experimental and computational methods of genome annotation

Genomic feature	Experimental/computational approach
Gene identification	cDNA and peptide sequencing
	Computational prediction
	Comparative genomics
Transcript identification	Tiling microarray
	cDNA sequencing
	Computational prediction
	Comparative genomics
Regulatory sequence identification	Chromatin Immunoprecipitation and tiling microarray (ChIP-Chip)
	Computational prediction of factor binding sites
	Promoter/enhancer assays
Sequence variation	DNA resequencing
	Copy number microarray
Chromatin structure	*DNaseI* sensitivity assay
	Tiling microarray

A number of methods are currently employed to identify functional regions of the genome. The first column lists several genomic features that are commonly annotated, and the second column lists the experimental or computational approaches that can be used to identify those features in genome sequence assemblies humans compared to previously sequenced organisms, such as *Drosophila melangaster*, *C. elegans*, or various prokaryotes. Because of this fact, a hybrid approach was taken that incorporated multiple lines of evidence, including homology of genome sequence to ESTs, similarity to other known genes or proteins, and statistical strategies that took into account splice site structure, amino acid coding bias, and known distributions of intron and exon lengths. Using these approaches with the newly available human genome sequence, a surprisingly low estimate of only 30,000–40,000 protein-coding genes was obtained, but the estimate involved considerable guesswork owing to the imperfections of the draft sequence and the inherent difficulty of gene identification (Lander et al. 2001; Venter et al. 2001). In the years following these initial estimates, it was discovered that many open reading frames (ORFs) that occur at random in transcripts are actually nonfunctional, and the total number of protein-coding genes has been steadily revised downward since. Currently, the human genome is estimated to contain approximately 20,000–21,000 protein-coding genes (Clamp et al. 2007; Pennisi 2007). Recent RNA-Seq projects have confirmed the gene catalog, while illuminating alternative splicing, which seems to occur at >90 % of protein-coding genes and results in many more proteins than genes. At this time, the proteome is now known to be similar across placental mammals, with about two-thirds of protein-coding genes having 1:1 orthologues across species and most of the rest belonging to gene families that undergo regular duplication and divergence—the de novo creation of fundamentally new proteins is considered a rare phenomenon (Lander 2011).

The human genome also gives rise to a large number of noncoding RNAs (Kapranov et al. 2007). Oligonucleotide-based tiling microarrays that interrogate every base pair of genome sequence over expansive regions have revealed that a much larger percentage of the human genome is transcribed compared to what was originally presumed (Cheng et al. 2005). While only 1–2 % of the human genome codes for proteins, approximately 15 % of all interrogated bases were able to detect RNA molecules from a single cell line, indicating that the vast majority of transcription from the human genome produces noncoding RNA products. The novel RNA transcripts are often transcribed from both strands, and transcription of coding sequences from the antisense strand is particularly common (Cheng et al. 2005). While the function of most of these products is not yet known, some noncoding RNAs exert regulatory effects on coding transcripts through complementary nucleotide base pairing. This hybridization decreases transcript stability by targeting it for degradation or translational repression (Kim and Nam 2006).

One of the surprising discoveries about the human genome was that the majority of the functional sequence does not encode proteins. Inferring these non-neutral, conserved noncoding elements in humans was a challenge before the HGP. Soon after the first draft the comparative analysis of the human and mouse genomes showed a substantial excess of conserved sequence, relative to the neutral rate in ancestral repeat elements (Mouse Genome Sequencing Consortium 2002).

Research groups working independently of one another have performed most of the approaches applied toward annotating the human genome (Table 2.1). The National Human Genome Research Institute (NHGRI) launched a public research consortium named ENCODE, the **ENC**yclopedia **O**f **D**NA **E**lements, in September 2003, to systematically integrate the genome annotation efforts in identifying all functional elements in the human genome sequence. (ENCODE Project Consortium 2004). The project started with two components—a pilot phase and a technology development phase. The pilot phase of the ENCODE project tested and compared the existing arsenal of annotation approaches on a series of 44 genomic regions comprising approximately 30 Mb, or roughly 1 % of the human genome. About half of the targets were chosen to contain extensively characterized genes or functional regions, while the other half were randomly selected (ENCODE Project Consortium 2004). The findings of the pilot project were published in June 2007

(ENCODE Project Consortium 2007) and scores of important information highlighted includes:

- There is abundant transcription beyond the known protein-coding genes both intragenic and intergenic transcription, including noncoding RNA and transcribed pseudogenes. While this has been previously observed in other studies, the ENCODE pilot phase confirmed this phenomenon on a global level.
- At the same time, known protein-coding genes revealed unexpected complexity in distal, untranslated regions (UTRs), exons located as far as 200 kb away, overlapping or interleaved loci, and antisense transcription. Together, these findings challenged the conventional definition of a "gene".
- Patterns of histone modifications and DNase sensitivity revealed domains of packed or accessible chromatin. These accessibility patterns correlate well with rates of transcriptions, DNA replication, and regulatory protein factors binding to the DNA. These results served to underscore the regulatory importance of epigenetic factors.

Combined, the ENCODE findings changed our conceptual framework of the organization and functional aspects of the genome. Two additional goals of the pilot ENCODE Project were to develop and advance technologies for annotating the human genome, with the combined aims of achieving higher accuracy, completeness, and cost-effective throughput and establishing a paradigm for sharing functional genomics data.

In 2007, the ENCODE Project was expanded to study the entire human genome, capitalizing on experimental and computational technology developments during the pilot project period. The genome-wide ENCODE phase is currently in progress focusing on the completion of two major classes of annotations—genes (both protein-coding and noncoding) and their RNA transcripts and transcriptional regulatory regions.

Gene Annotation. A major goal of ENCODE is to annotate all protein-coding genes, pseudogenes, and noncoding transcribed loci in the human genome and to catalog the products of transcription, including splice isoforms. Although the human genome contains 20,000 protein-coding genes (International Human Genome Sequencing Consortium 2004), accurate identification of all protein-coding transcripts has not been straightforward. Annotation of pseudogenes and noncoding transcripts also remains a considerable challenge. While automatic gene annotation algorithms have been developed, manual curation remains the approach that delivers the highest level of accuracy, completeness, and stability (Guigo et al. 2006). This annotation process involves consolidation of all evidence of transcripts (cDNA, EST sequences) and proteins from public databases, followed by building gene structures based on supporting experimental data (Harrow et al. 2006). More than 50 % of annotated transcripts have no predicted coding potential and are classified by ENCODE into different transcript categories. A classification that summarizes the certainty and types of the annotated structures is provided for each transcript. Pseudogenes are identified primarily by a combination of similarity to other protein-coding genes and an obvious functional disablement such as an in-frame stop codon. Ultimately, each gene or transcript model is assigned one of the three confidence levels. Level 1 includes genes validated by RT-PCR and sequencing, plus consensus pseudogenes. Level 2 includes manually annotated coding and long noncoding loci that have transcriptional evidence in EMBL/GenBank. Level 3 includes Ensembl gene predictions in regions not yet manually annotated or for which there is new transcriptional evidence. The result of ENCODE gene annotation "GENCODE" is a comprehensive catalog of transcripts and genemodels. ENCODE gene and transcript annotations are updated bimonthly and are available through the UCSC ENCODE browser, Distributed Annotation Servers (DAS), and the Ensembl Browser (Flicek et al. 2010; ENCODE Project Consortium 2011, 2012).

RNA Transcripts. The work on comprehensive genome-wide catalog of transcribed loci that characterizes the size, polyadenylation status, and subcellular compartmentalization of all transcripts is also ongoing at ENCODE, with transcript data generated from high-density

(5 bp) tiling DNA microarrays (Kampa et al. 2004) and massively parallel DNA sequencing methods (Mortazavi et al. 2008; Wold and Myer 2008; Wang et al. 2009). Because subcellular compartmentalization of RNAs is important in RNA processing and function, such as nuclear retention of unspliced coding transcripts (Schmid and Jensen 2010) or small nucleolar RNA (snoRNA) activity in the nucleolus (Bachellerie et al. 2002), ENCODE is analyzing not only total whole cell RNAs but also those concentrated in the nucleus and other subcellular compartments, providing catalogs of potential microRNAs (miRNAs), snoRNA, promoter-associated short RNAs (PASRs) (Kapranov et al. 2007), and other short cellular RNAs. These analyses revealed that the human genome encodes a diverse array of transcripts. Additional transcript annotations include exonic regions and splice junctions, transcription start sites (TSSs), transcript 39 ends, spliced RNA length, locations of polyadenylation sites, and locations with direct evidence of protein expression (ENCODE Project Consortium 2011, 2012).

Transcriptional Regulatory Regions. Transcriptional regulatory regions include diverse functional elements such as promoters, enhancers, silencers, and insulators, which collectively modulate the magnitude, timing, and cell specificity of gene expression (Maston et al. 2006). The ENCODE Project is using multiple approaches to identify *cis*-regulatory regions, including localizing their characteristic chromatin signatures and identifying sites of occupancy of sequence-specific transcription factors. These approaches are being combined to create a comprehensive map of human *cis*-regulatory regions.

Chromatin Structure and Modification. Chromatin accessibility and histone modifications provide independent and complementary annotations of human regulatory DNA, and massively parallel, high-throughput DNA sequencing methods are being used by ENCODE to map these features on a genome-wide scale. Deoxyribonuclease I (DNaseI) hypersensitive sites (DHSs) and an expanding panel of histone modifications are also being mapped (Barski et al. 2007; Johnson et al. 2007; Mikkelsen et al. 2007; Robertson et al. 2007). ENCODE chromatin annotation data such as chromatin accessibility, DNase I hypersensitive sites, and selected histone modifications are available through the UCSC browser (http://genome.ucsc.edu/).

Transcription Factor and RNA Polymerase Occupancy. Much of human gene regulation is determined by the binding of transcriptional regulatory proteins to their cognate sequence element in *cis*-regulatory region. To create an atlas of regulatory factor (i.e., transcription factors, RNA polymerase 2, both initiating and elongating, and RNA polymerase 3) binding, ENCODE is applying chromatin immunoprecipitation and DNA sequencing (ChIP-seq) technology, which enables genome-wide mapping of transcription factors occupancy pattern in vivo (Barski et al. 2007; Johnson et al. 2007; Robertson et al. 2007). Alternative technologies, such as epitope tagging of transcription factors in their native genomic context using recombineering (Poser et al. 2008; Hua et al. 2009), are also being explored.

ENCODE Additional Data. ENCODE is also generating additional data types to complement gene and regulatory region annotations and that includes data on DNA methylation, DNase I footprinting, long-range chromatin interaction, protein–RNA interaction, and genetic and structural variation in the cell types used in ENCODE production phase. The key features of the production phase include use of several cell types for the main data collections efforts and the use of these cell types by all project teams to maintain consistency. The cell types are organized into tiers to prioritize experimental investigations. These features are expected to enable better coordination of studies and interpretation of results.

2.2.1.2 Segmental Duplications

The HGP has also extended our understanding of segmental duplications (SDs). Eukaryotic organisms have evolved a complex, highly regulated cellular machinery to insure the proper replication, condensation, and segregation of chromosomes during cell division (Hirano 2000). However, errors in the distribution of genetic material during cell division occasionally occur, leading to daughter cells that receive more or less than the usual complement of genomic DNA following cell division. If such an alteration in DNA copy number occurs in the germ cell lineage of a multicellular organism, then the progeny of that organism can inherit the change in DNA copy number. Over many generations, copy number changes that occur in a single individual can spread through a population, leading to a situation in which the copy number status of a chromosomal region can be considered a type of genetic polymorphism, typically referred to as a copy number polymorphism (CNP) or copy number variation (CNV) (Bailey et al. 2002; Sebat et al. 2004).

The human genome is enriched for SDs that vary extensively in copy number (Bailey et al. 2002; Iafrate et al. 2004; Redon et al. 2006; Kidd et al. 2008). There are about 25,000–30,000 SDs with ≥90 % sequence identity and ≥1 kb length have been identified in the human genome, which cover about 5–6 % of the total genome (Bailey et al. 2002). It has also been reported that SDs are highly enriched with genes and pseudogenes in the human genome (i.e., SDs comprise ~5 % of the genome and contain ~17.8 % of human genes and ~36.8 % of human pseudogenes) (Bailey et al. 2002; Zheng 2008).

When a SD contains a functional gene, the new sequence may contain a paralog performing the same function as the original gene or a new function. Duplicated pseudogenes are formed when the new sequence undergoes mutations that result in the loss of original function (Fig. 2.1). The process of SD such as retrotransposition events may also result in the loss of function (LOF) of the duplicated gene; such genes are referred as processed pseudogenes (Mighell et al. 2000; Harrison and Gerstein 2002). Processed pseudogenes usually lack promoter sequences, and hence are considered dead on arrival.

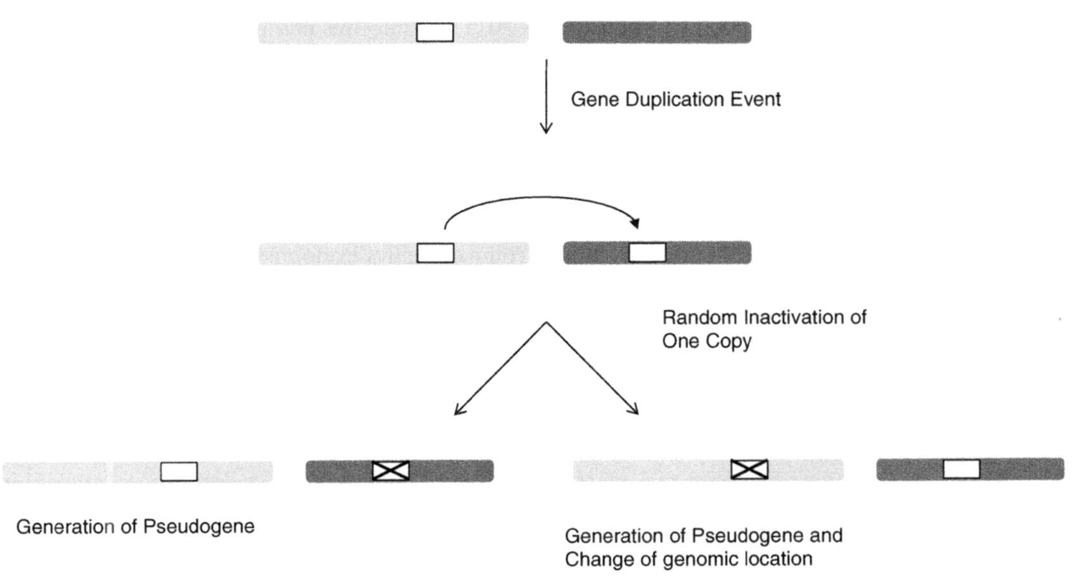

Fig. 2.1 Pseudogene generation by gene duplication and random inactivation. The creation of a novel pseudogene is initiated by a gene duplication event in which a sequence containing a functional gene (*white box*) is duplicated and inserted into a separate site in the genome (shown here as a duplication from one chromosome to another). In most cases of gene duplication, one of the two copies will be randomly silenced and inactivated by mutations, leading to the creation of a pseudogene (checked *white box*). Depending on which of the two copies is inactivated during this process, the genomic position of the original gene can change

Although pseudogenes are assumed to have lost the original coding functions of their parent genes due to the presence of disablements such as premature stop codons or frameshift mutations, recent studies indicate that they might have some regulatory roles (Sasidharan and Gerstein 2008). Automated methods of annotating genomic DNA sequences have identified more than 20,000 pseudogenes (International Human Genome Sequencing Consortium 2004).

Although studies have begun to define the important roles of SDs in generating novel genes through adaptive evolution, gene fusion, or exon exaptation (Lynch and Conery 2000; Taylor and Raes 2004; Bailey and Eichler 2006), it remains a mystery how duplicated copies have evolved from an initial state of complete redundancy (immediately after duplications) to a stable state where both copies are maintained by natural selection. Some glimpse into this important evolutionary process comes from the investigations of duplicated protein-coding genes or gene families showing that duplicated genes can evolve different expression patterns, leading to increased diversity and complexity of gene regulation, which in turn can facilitate an organism's adaptation to environmental change (Gu et al. 2004, 2005; Hittinger and Carroll 2007; Louis 2007). Furthermore, the studies of histone modification in human SDs have also demonstrated that parental and duplicated copies are not functionally identical even though they share ≥90 % identity in their primary sequences, suggesting that descendants in a new genomic environment are more likely the candidates for sequence degeneration or functional innovation (Zhao et al. 2007; Zheng 2008).

Despite recent technological advances in copy number detection, a global assessment of genetic variation of these regions has remained elusive. Commercial single nucleotide polymorphism (SNP) microarrays frequently bias against probe selection within these regions (Estivill et al. 2002; Locke et al. 2006; Cooper et al. 2008; Pinto et al. 2011). Array comparative genomic hybridization (array CGH) approaches have limited power to discern copy number differences, especially as the underlying number of duplicated genes increases and the difference in copy number with respect to a reference genome becomes vanishingly small (Locke et al. 2003; Sharp et al. 2005; Redon et al. 2006; Pinto et al. 2011). Even sequence-based strategies such as paired-end mapping (Tuzun et al. 2005; Korbel et al. 2007) frequently cannot unambiguously assign end sequences in duplicated regions, making it impossible to distinguish allelic and paralogous variation. Consequently, duplicated regions have been largely refractory to standard human genetic analyses (Conrad et al. 2010; Sudmant et al. 2010).

However, a great deal of interest has developed around the role of CNPs/CNVs in inherited diseases, since Lupski et al. (1991) showed for the first time, that a duplicated region on chromosome 17 caused an inherited form of Charcot–Marie–Tooth disease. Since that initial finding, numerous CNPs have been shown to be associated with several human diseases such as psoriasis, Crohn's disease, lupus, rheumatoid arthritis, Parkinson's, Alzheimer's, autism, neuroblastoma, obesity, coronary heart disease, and type 2 diabetes (Cohen 2007; Girirajan et al. 2011). While the number of such cases is still relatively small compared to the number of inherited diseases shown to be caused by point mutations in protein-coding sequences, the importance of CNPs/CNVs in human disease has become increasingly apparent over the past few years. It is now known that at least 15 % of human neurodevelopmental diseases are due to rare and large copy number changes that result in local dosage imbalance for dozens of genes (Giriraj et al. 2011). Other large CNVs, both inherited and de novo, have been implicated in the etiology of autism, schizophrenia, kidney dysfunction, and congenital heart disease. Surprisingly, studies of the general population suggest that although such alleles are rare, collectively they are quite common and under strong purifying selection. These features mean that a significant fraction of the human population carries an unbalanced genome. Such individuals may be sensitized for the effect of another variant that could potentially interact with these CNVs in a digenic manner. The co-occurrence of multiple, rare CNVs has been used to explain the comorbidity and variable expressivity associated with particular variants in cases of severe developmental delay. There is circumstantial evidence that the full complement of both CNVs and SNPs may be important for understanding genetic diseases more broadly (O'Roak et al. 2011).

2.2.1.3 Comparative Genomics and Genome Evolution

Comparative genomics is the study of relationships among genome sequences of different species. Although a relatively young discipline, comparative genomics has been used to refine our understanding of a number of phenomena, including the evolutionary relationship between species, and the content and function of genomes. From an evolutionary perspective, the similarities and differences between genomic sequences can serve to infer phylogenetic relationships between species based upon molecular criteria in the same fashion that morphological and physiological criteria were used to distinguish species in the past. Identification of conserved regions may also help to elucidate functionally important sequences such as genes, regulatory sites, and structural elements.

Before the availability of whole genome assemblies, comparative genomic analyses were performed using a small number of homologous sequences that were individually isolated from different organisms and sequenced (Murphy et al. 2001). As crucial as these studies were for establishing broad phylogenetic relationships between and among species, the relatively small fraction of genomic sequence used for such analyses was a significant limitation. The recent explosion in the field of comparative genomics results directly from the efforts of numerous sequencing projects and the widespread availability of whole genome assemblies from a variety of different species. The Genomes Online Database (GOLD), which is a World Wide Web resource for comprehensive access to information regarding genome and metagenome sequencing projects, and their associated metadata, documented 11,472 ongoing and completed genome projects by September 2011. These comprise 8,473 bacterial, 329 archaeal, and 2,204 eukaryal genomes. Additionally, 340 metagenomic projects are tracked with a total of 1,927 samples associated with them. GOLD also tracks well over 1,000 proprietary projects, currently not available to the public, whose metadata will be accessible once the principal investigators of these projects give consent for their public release. In terms of status, 1914 different organisms are completely sequenced and their final sequence has been released from GenBank. From those, 1,644 are bacterial, 117 are archaeal, and 153 are eukaryal. A constantly increasing number of sequencing projects are completed at the level of a draft genome and their final sequences are submitted in GenBank. These projects are identified as "Permanent Draft" genomes. There are currently 989 genomes at this stage (28 archaeal, 949 bacterial, and 12 eukaryal). As of September 2011, the total number of complete genomes is 2,907, which is the sum of the finished and the permanent draft genomes (Pagani et al. 2012).

With the availability of genomes representing multiple species, comprehensive comparisons have produced results that have been both informative and unexpected. Primarily, our understanding of the functional contents of the human genome has been substantially enhanced by comparisons with the genomes of other species. For example, comparison of the human genome with distantly related organisms (e.g., the fruit fly) has been critical for determining the core set of genes necessary for the development and function of multicellular eukaryotes. Similarly, comparison of genomes from humans and vertebrate species of intermediate evolutionary distance (e.g., the mouse) can identify both coding and noncoding sequences that are likely to be functional based on strong evolutionary conservation (Fig. 2.2). Finally, comparison of genomes from humans and closely related primates will help identify the small percentage of divergent sequence that is responsible for specifically human traits. The following paragraphs touch briefly on each of these kinds of comparisons.

The divergence of humans and fruit flies (*D. melanogaster*) from a common ancestor is

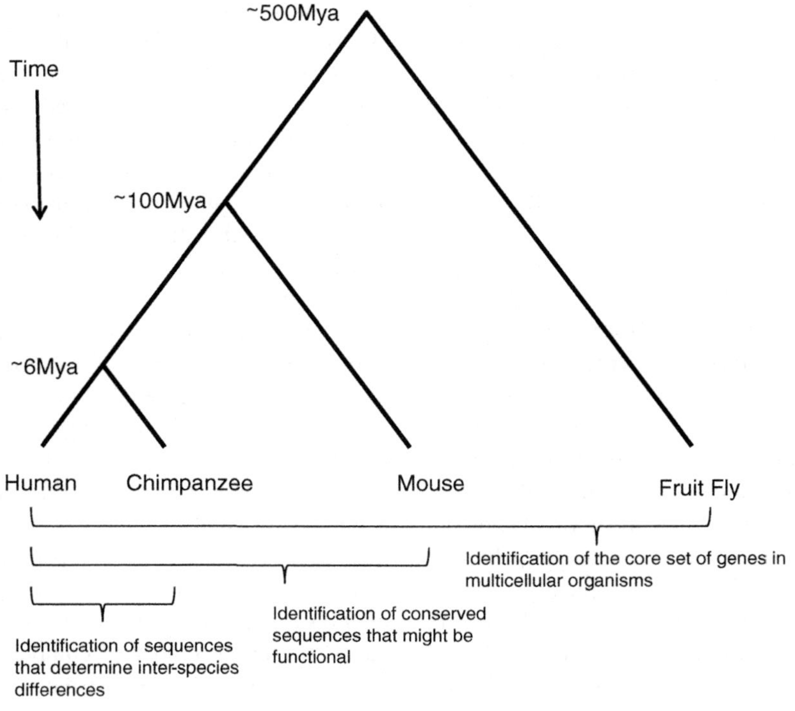

Fig. 2.2 Comparative genomics of species at different evolutionary distances. Genomic comparison of two species can yield different conclusions depending on the degree of genetic difference between them. The evolutionary tree shows the estimated time (in millions of years) from the divergence of human, chimp, mouse, and fruit fly from their common ancestor. The text at *bottom* indicates the information that can be inferred from comparing the human genome to that of a closely related species (chimp), a species of intermediate evolutionary distance (mouse), or a species of great evolutionary distance (fruit fly)

estimated to have occurred over half a billion years ago. The obvious morphological differences between the two species are reflected in the substantial differences at the level of the genome, with the most apparent discrepancies being genome size and gene content (Adams et al. 2000). The human genome spans ~3.1 billion base pairs compared to the 180 million base pairs comprising the *drosophila* genome (Adams et al. 2000), yet contains less than twice as many genes compared to the fly. This size–content disparity is generally consistent with the large expansion of nongenic sequence present in the human lineage, resulting mostly from simple repetitive elements, which are not present in the *drosophila* genome. Despite relatively comparable content, human genes undergo vastly greater amounts of alternative transcription and splicing events, which lead to a much greater diversity of protein products. For example, the ~20,000 genes comprising the human genome give rise to more than 100,000 proteins. Further comparison of protein-coding sequences from the genomes of both species reveals that many genes involved in basic cellular functions such as metabolism, DNA replication and repair, core transcriptional regulation, and cell cycle regulation are conserved. In contrast, human-specific gene expansions are observed for many different functional groups, several of which would be expected given the anatomical and physiological differences between the two species. In general, these expansions occur mainly in gene families involved in adaptive immunity (a vertebrate-specific process), neuronal function, hemostasis, and programmed cell death (Venter et al. 2001).

The first large-scale comparison of two mammalian genome assemblies was performed between human and mouse (*Mus musculus*), two species separated by 75–100 million years of evolution (Mouse Genome Sequencing Consortium 2002; Mural et al. 2002). The human and mouse genomes share ~80–90 % of the same genes, while the remaining unshared genes represent mostly species-specific expansions of functional groups including olfaction, immunology, reproduction, and detoxification (Mouse Genome Sequencing Consortium 2002). One of the most significant and unexpected findings of the human/mouse genome comparison was the large fraction of highly conserved sequences that are neither protein-encoding nor related to known genes (Mural et al. 2002). While ~5 % of the human genome is significantly conserved with that of the mouse (>70 % identity over 100 bp or more), only ~1.5 % of each genome was found to correspond to protein-coding sequence (Dermitzakis et al. 2003). This finding suggests that conserved nonprotein coding sequence is almost twice as abundant as conserved coding sequence. Further, the degree of conservation is estimated to be even greater for noncoding than coding sequences, implying a substantial degree of selective pressure on noncoding sequences (Dermitzakis et al. 2003). Recent comparisons of vertebrate genome assemblies from organisms as diverse as human, rat, mouse, dog, and chicken have provided additional support for this relationship by identifying hundreds of "ultra-conserved" elements, in which an extremely high level of conservation is present among sequences (>95 % over 200 bp or more), and with most of the conserved regions occurring outside of known genes (Bejerano et al. 2004). Although a substantial portion of this conserved sequence is posited to serve a regulatory function (Pennacchio et al. 2006; Prabhakar et al. 2006; Xie et al. 2007), and a very weak selection could also maintain the sequence conservation of ultraconserved elements in noncoding regions (Kryukov et al. 2005; Chen et al. 2007), the reason for this extremely high level of conservation in noncoding regions over millions of years remains unknown.

The completion of genomic assemblies from closely related primates has enabled focus on more recent events in the molecular evolution, molecular adaptation, and genome structure of *Homo sapiens* (Fig. 2.3). Currently, the genome sequences of 13 nonhuman primates are available and at least 11 are approved sequencing targets (Enard 2012). These genomic assemblies together with future sequencing will reveal basic insights into evolutionary processes of mutation, selection and recombination (Marques-Bonet et al. 2009), will be essential tools for primate model organisms (Sasaki et al. 2009), and will also be directly informative for medically relevant questions (Enard 2012). Among the first completed after human are chimpanzee (*Pan troglodytes*) (Chimpanzee Sequencing and Analysis Consortium 2005) and rhesus macaque (*Macaca mulatta*) (Rhesus Macaque Genome Sequencing and Analysis Consortium 2007), which diverged from humans ~6 and 25 million years ago, respectively. Genome-wide comparative analyses of the human, macaque, and chimpanzee genomes have revealed some important features and general principles of primate genome evolution. The alignment of the majority of genomic sequence from closely related primates is relatively trivial (Ebersberger et al. 2002; Thomas et al. 2003) and shows a neutral pattern of single nucleotide variation consistent with the primate phylogeny, although the rate of single nucleotide variation has varied by a factor of threefold within different lineages (Li and Tanimura 1987; Steiper et al. 2004; Elango et al. 2006). Notably, the pattern of single nucleotide variation also varies as a function of chromosome structure and organization (Chimpanzee Sequencing and Analysis Consortium 2005; Rhesus Macaque Genome Sequencing and Analysis Consortium 2007). On average, 10 % of the genomic sequence has proven more elusive in terms of orthologous alignment. This includes SDs, subtelomeric regions, pericentromeric regions, and lineage specific repeats.

Comparative sequence data highlight the value of genomic sequence from nonhuman primates to determine the ancestral and derived status of human alleles (Chen and Li 2001;

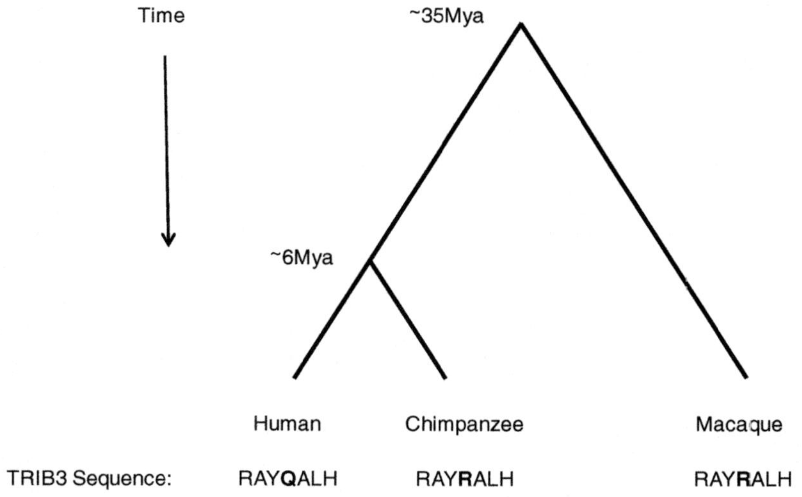

Fig. 2.3 Ternary analysis of closely related primate species. Evolutionary triangulation can identify the lineage in which a sequence variant evolved. The evolutionary tree shows the estimated time (in millions of years) from the divergence of human, chimp, and macaque from their common ancestor. As an example, the protein sequence shown at bottom is derived from a portion of the *TRIB3* gene from each species. Since the sequence variant in the *TRIB3* gene is common to chimp and macaque, it likely occurred in the human lineage within the last 6 million years. Interestingly, the ancestral *TRIB3* allele observed in the chimp and macaque is associated with insulin resistance when present in humans

Kaessmann et al. 2001). There have been some surprises. Phylogenetic analysis of resequenced regions among humans and the great apes reveal that as many as 18 % of genomic regions are inconsistent with the Homo-Pan clade, and, rather, support a Homo-Gorilla clade (Chen and Li 2001). This has been taken as evidence of lineage-sorting and/or an ancestral hominid population size greater than five times that of the effective human population size (n = 10,000). Another surprise has been the identification of ancestral allelic variants that now occur as disease alleles within the human population, i.e., phenylketonuria, macular dystrophy, and cystic fibrosis pyrin and familial Mediterranean fever (Schaner et al. 2001; Rhesus Macaque Genome Sequencing and Analysis Consortium 2007). Such findings suggest that the functional and selective effects of mutations change over time, perhaps as a result of environmental changes or compensatory genetic mutations.

Despite the ease at which genomic sequences can be aligned among primate genomes, the number of genes that can be assigned to 1:1:1 orthologous group has changed only slightly with the first two nonhuman primate genomes sequenced. A three-way comparison involving chimp–human–mouse identified 7,645 orthologues (Clark et al. 2003) as compared to 10,376 by human–chimp–macaque (Rhesus Macaque Genome Sequencing and Analysis Consortium 2007) over the total estimated 20,000 genes in the human genome, suggesting that a large fraction of human genes are yet to be subjected to orthologous comparisons and the pattern of selection operating on these genes is yet to be adequately interrogated. Among the primate order of mammals, comparative genomic studies have advanced more rapidly for taxa closely related to humans, chimpanzees, macaques, and baboons. As complete genome sequencing projects advance for other primate families, including the New

World monkeys (Cebidae) and strepsirhine primates (lemurs, lorises, aye-aye, pottos, and galagos), new insights are anticipated as, particularly for a lemur genome project, new information about primate adaptations and evolution can be anticipated (Horvath and Willard 2007).

However, identification of the most recent events in the speciation of *H. sapiens* will require comparative analyses between the genomes of humans and other members of the genus *Homo*. While genetic material for such species has been available for years, the reliable amplification and sequencing of DNA extracted from ancient bone samples has not been tenable until recently. Careful collection procedures, performed under exceedingly pristine conditions, have enabled 1.3x coverage from three Neanderthal individuals (Green et al. 2006; Noonan et al. 2006) and the 1.9x coverage from a small finger bone found in the Denisova cave in Siberia (Reich et al. 2010). These genomes are on average slightly more related to each other than to modern human genomes, but most genomic regions still fall within the variation of modern humans (Reich et al. 2010). Interestingly, those regions where this is not the case, i.e., where all modern humans are closely related to each other than to Denisovans or Neanderthals, are enriched for regions that have been positively selected after the population split some 270,000–440,000 years ago (Green et al. 2006). While a comprehensive comparison of human and Neanderthal DNA sequence has the potential to identify the relatively small number of genetic changes that occurred over the span of time in which *H. sapiens* evolved into a distinct species. Further data and the identification of additional fossils will lead to considerably better assemblies of these ancient genomes and 30x coverage data for Denisovans was recently made available (Meyer et al. 2012). Although it is unlikely that endogenous DNA sequences can be obtained from much older hominin fossils, the unexpected finding of Denisovans allows optimism that genomes from more hominins can be discovered and will improve our understanding of human evolution and even some aspects of human disease.

2.2.2 Genetic Studies of Complex Traits

Perhaps the greatest impact of the HGP has been on the manner in which researchers investigate the causes of complex human diseases. Unlike monogenic diseases, which arise due to a single genetic aberration, complex diseases result from a complicated interaction of multiple genetic and environmental determinants, none of which are amenable to identification and characterization using the traditional approaches to monogenic disease gene discovery. Completion of the HGP gave rise to the development of efforts and technology to characterize genetic variation on a genome-wide scale, including the genotyping of common variants, which has led directly to the application of whole genome association studies to identify common alleles which contribute to complex disease risk, or the very recent whole genome sequencing efforts to identify low-frequency and rare variants in diverse populations. Each of these areas is discussed in the following sections.

2.2.2.1 The International HapMap Project

The sequence data resulting from the HGP paved the way for the development of an effort lead by the International HapMap Consortium to characterize all common variation within the human genome (International HapMap Consortium 2005). The most common type of genetic variant is the SNP, which occurs with the presence of two or more different alleles at the same nucleotide position. In humans, polymorphisms occur at a rate of approximately one variant every kilobase (Wang et al. 1998; Lander et al. 2001), and the presence of 11 million SNP sites with a minimal minor allele frequency of 1 % that constitute ~90 % of the variation in the world's population has been estimated (Kruglyak and Nickerson 2001).

The HapMap Project, currently completed phase III, was officially launched in 2002 to create a public, genome-wide database of common human sequence variation, providing information needed as a guide to genetic studies of clinical phenotypes and consists of collaborators from the United States, Canada, the United Kingdom, China, Nigeria, and Japan (International HapMap Consortium 2003).

The Phase I of the HapMap Project contains high-quality genotype data on more than 1 million SNPs, genotyped on 270 samples from 90 individuals (30 parent–parent–offspring trios) of European descent from Utah (CEU), 90 Yoruba individuals (30 trios) from Ibadan, Nigeria (YRI), 45 unrelated Japanese from Tokyo (JPT), and 45 unrelated Han Chinese from Beijing (CHB). Although the goal of Phase I was to genotype at least one common SNP (minor allele frequency ≥0.05) every 5 kb across the genome and SNP selection was agnostic to functional annotation, 11, 500 nonsynonymous SNPs are prioritized in choosing SNPs for each 5 kb region (International HapMap Consortium 2005).

The Phase I HapMap Project data had a central role in the development of methods for the design and analysis of Genome-Wide Association (GWA) studies. For example, the HapMap resource provides critical information regarding the extent of linkage disequilibrium among SNPs in each of the four distinct populations represented in the project. In this way, knowledge of a particular SNP allele at one site can predict specific alleles at nearby sites (allele combinations along a chromosome are known as haplotypes). Approximately, 50–75 % of all SNPs in the HapMap database are highly correlated with other genotyped markers and >90 % are associated with nearby SNPs at levels of statistical significance (International HapMap Consortium 2005). These advances, alongside the release of commercial platforms for performing economically viable genome-wide genotyping, have led to a new phase in human medical genetics.

Large-scale GWA studies have identified novel loci involved in multiple complex diseases (Altshuler and Daly 2007; Bowcock, 2007). In addition, the HapMap data have led to novel insights into the distribution and causes of recombination hotspots (International HapMap Consortium 2005, Myers et al. 2005), the prevalence of structural variation (Conrad et al. 2006; McCarroll et al. 2006), and the identity of genes that have experienced recent adaptive evolution (International HapMap Consortium 2005; Voight et al. 2006).

In Phase II of the HapMap project an additional 2.1 million SNPs were genotyped on the same individuals from Phase I. The resulting HapMap Phase I and II datasets (3.1 million SNPs) constitute ~25–30 % of the 9–10 million estimated common SNPs (minor allele frequency ≥0.05) in the assembled human genome. The Phase II HapMap differs from the Phase I not only in SNP spacing, but also in minor allele frequency (MAF) distribution and patterns of linkage disequilibrium. Because the criteria for choosing additional SNPs did not include consideration of SNP spacing or preferential selection for high MAF, the SNPs added in Phase II are, on average, more clustered and have lower MAF than the Phase I SNPs. One notable consequence is that the Phase II HapMap includes a better representation of rare variation than the Phase I HapMap (International HapMap Consortium 2007). The HapMap dataset and other resources such as public catalog of variant sites (dbSNP) and databases of structural variants (SVs) have driven disease gene discovery in the first generation of GWA studies, wherein genotypes at several hundred thousand variant sites, combined with the knowledge of LD structure, allowed the vast majority of common variants (MAF ≥ 0.05) to be tested for association with disease (International HapMap Consortium 2007). Over 6–7 years, GWA studies have identified more than a thousand genomic regions associated with disease susceptibility and other common traits (Hindorff et al. 2012). Genome-wide collections of both common and rare SVs have similarly been tested for association with disease (Wellcome Trust Case Control Consortium 2010). Despite successes, these studies raise many questions, such as why the identified variants have low-associated risks and account for so little heritability (Goldstein 2009). Explanations for this apparent gap are being sought. It is possible that these studies were limited with respect to variant type, frequency, and population diversity. Only common DNA variants (MAF ≥ 0.05) have been well studied, even though the contributions of rare variants, which were not captured by GWA studies; SVs, which were poorly captured, and other forms of genomic variation; or interactions between genes or between genes and environmental factors may be important (Manolio et al. 2009). Furthermore, despite their value in locating the vicinity of genomic variants that may be related to the susceptibility to disease, few of the SNPs identified in GWA studies have clear functional implications that are relevant to mechanisms of disease (Hindorff et al. 2009). Narrowing an implicated locus to a single variant with direct functional consequences has proven challenging. Together, these findings suggest that additional work will be necessary to achieve a deep understanding of the genetic contribution to human phenotypes and diseases (Manolio et al. 2009).

Once a region has been identified as harboring a risk locus, a detailed study of all genetic variants in the locus is required to discover the causal variant(s), to quantify their contribution to disease susceptibility, and to elucidate their roles in functional pathways. A much more complete catalog of human DNA variation is a prerequisite to fully understanding the role of common and low-frequency variants in human phenotypic variation. The efforts aimed at illuminating the gaps in the first generation of databases that contain mostly common variant sites were made. The HapMap project was expanded into Phase III to perform genome-wide SNP genotyping and CNP detection, as well as polymerase chain reaction (PCR) resequencing in selected genomic regions on a larger set of 1,184 samples from 11 populations (International HapMap3 Consortium 2010). Also during the same time another consortium project called "1,000 Genomes" aimed to discover additional genotypes and to provide accurate haplotype information on all forms of human DNA polymorphism in multiple human populations by next generation sequencing, was initiated (1000 Genomes Project Consortium 2010).

The HapMap Phase III. Despite great progress in identifying genetic variants influencing human diseases, most inherited risk remains unexplained. A more comprehensive strategy that fully examines the low-frequency and rare variants in populations of diverse ancestry is required to understand the genetic architecture of human diseases. Accordingly, the HapMap Phase I and II resources were expanded by genotyping 1.6 million SNPs and CNP detection in 1,184 samples from 11 populations. These included all Phase I and II samples, along with additional samples from the same four populations (i.e., samples from 165 individuals (trios) of European descent from Utah (CEU), 167 Yoruba individuals (trios) from Ibadan, Nigeria (YRI), 86 unrelated Japanese from Tokyo (JPT), and 84 unrelated Han Chinese from Beijing (CHB)), and an additional 682 samples from seven new populations (i.e., 83 individuals (trios) of African ancestry from southwestern USA (ASW); 85 unrelated Chinese individuals from metropolitan Denver, Colorado, USA (CHD); 88 unrelated Gujarati Indian individuals from Houston, Texas, USA (GIH); 90 unrelated Luhya individuals from Webuye, Kenya (LWK); 171 Maasai individuals (trios + unrelated) from Kinyawa, Kenya (MKK); 77 unrelated individuals of Mexican ancestry from Los Angeles, California, USA (MXL); and 88 unrelated Tuscan individuals from Italy (Toscani in Italia, TSI). The new populations were included to provide further variation data from each of the three continental regions, as well as data from some admixed populations. Unlike Phase I and II, a much larger sample size of 692 unrelated individuals from ten populations (i.e., ASW, CEU, CHB, CHD, GIH, JPT, LWK, MXL, TSI, and YRI) were sequenced for 100 kb each of the ten ENCODE regions (see International HapMap 3 Consortium 2010 publication for details) by direct PCR-Sanger capillary sequencing in the Phase III. This direct sequencing of the selected regions, unlike SNPs genotyped using microarray platforms, which are intentionally biased toward high frequency by the discovery and selection process, the SNPs discovered by sequencing provide a direct estimate of the underlying allele frequency spectrum in each population. As in previous phases, common (MAF ≥ 0.05) and low-frequency (MAF = 0.005–0.05) variants account for the vast majority of the heterozygosity in each sample, but a large number of rare (MAF = 0.0005–0.005) and private (singletons and MAF < 0.0005) variants were also observed. Each population had 42–66 % of sites with a MAF < 0.05, compared to 10–13 % in the genotyping data; 37 % of SNPs with a MAF < 0.005 were observed in only one population. In total, 77 % of the discovered SNPs were new (that was, not in the SNP database (dbSNP) build 129) and 99 % of those had a MAF < 0.05 (International HapMap 3 Consortium 2010). The HapMap Phase III results underscored the need to characterize population-specific parameters, and for each stratum of allele frequency. As expected, lower frequency variation is less shared across populations, even closely related ones, highlighting the importance of sequencing and sampling widely to achieve a comprehensive understanding of human variation. With improvement in sequencing technology, whole genome sequencing is becoming increasingly accessible. This revolution will no doubt expand our ability to identify rare and private variations along with common variations to better understand the genetic architecture of human diseases.

2.2.2.2 The 1000 Genomes Project

Launched in 2008, the 1000 Genomes Project involving researchers from more than 75 institutions and companies in the United States, the United Kingdom, China, and Germany, set its sights on characterizing over 95 % of variants that have allele frequency of 1 %, or higher (MAF ≥ 0.01) in the five major population groups-West African, European, North American, and East and South Asian. The coding region of the genome was cataloged for variants of even lower allele frequencies (i.e., MAF ≥ 0.001) because coding regions will more often have variants with functional consequences, which may also have low allele frequency (1000 Genomes Project Consortium 2010; Patterson 2011).

The pilot phase of the project aimed at developing and comparing genome-wide sequencing strategies, sequenced three sets of samples at three different levels of sequencing coverage.

- Family trios: high coverage (average 42x) whole genome sequencing of two HapMap family trios (i.e., one YRI and one CEU).
- Low coverage: low coverage (2–6x) whole genome sequencing of 179 unrelated individuals from four HapMap populations (i.e., 59 from YRI, 60 from CEU, 30 from CHB, and 30 from JPT).
- Exon sequencing: targeted capture of the exons from nearly1,000 randomly selected protein-coding genes (total 1.4 Mb) followed by sequencing at high coverage (average > 50 x) in 697 individuals from 7 HapMap populations (i.e., YRI, LWK, CEU, TSI, CHB, JPT, and CHD).

The pilot project identified 15 million SNPs, 1 million short insertions and deletions of DNA, and 20,000 large SVs. Populations of African ancestry contributed the largest number of variants to the data, including the biggest portion of novel variants (1000 Genomes Project Consortium 2010). The pilot project data also showed that more than half of the genetic variants that were found were previously unknown. It has also been observed that an individual's genome contains many variants of functional consequence (10,000–11,000 nonsynonymous sites and 10,000–12,000 synonymous sites per genome that differs from reference). However, the number of variants with greater functional impact is much smaller (overall 340–400 premature stop codons, splice site disruptions, and frame shifts, affecting 250–300 genes per genome, as putative LOF variants). In addition, 50–100 of the variants had previously been associated with an inherited disease (1000 Genomes Project Consortium 2010).

The success of the pilot project paved the way for the production phase of the full 1000 Genomes Project, which aims to sequence 2,500 genomes from 27 populations worldwide. The data on genomes of 1,092 individuals from

14 populations from Europe, East Asia, sub-Saharan Africa, and the Americas, sequenced using combination of whole genome low coverage sequencing (2–6 x) and targeted deep sequencing (50–100 x) of the exome have been published recently (1000 Genomes Project Consortium 2012). The dataset provides a detailed view of variations across several populations. Individuals from different populations carry different profiles of rare and common variants, and low-frequency variants show substantial geographic differentiation, which is further increased by the action of purifying selection. Most common variants (94 % with MAF $\geq$ 0.05) were previously known and their haplotype structures were also mapped through earlier projects (International HapMap Consortium 2007; 1000 Genomes Project Consortium 2010). In contrast, only 62 % of variants with the MAF range 0.005–0.05 and 13 % with MAF $\leq$ 0.005 had been described previously. A validated haplotype map of 38 million SNPs, 1.4 million short indels, and more than 14,000 larger deletions has been developed using this phase dataset. The Phase I data also show that at the most highly conserved coding sites, 85 % of the nonsynonymous variants and more than 90 % of stop-gain and splice disrupting variants are below 0.5 % in frequency, compared with 65 % of synonymous variants (1000 Genomes Project Consortium 2012).

2.3 Future Impact of the HGP: Where Are We Going?

2.3.1 Pharmacogenetics

Response to pharmacological interventions is variable and in most cases, difficult to predict. For instance, only about 66 % of individuals treated with beta blockers actually respond in the intended way with a reduction in blood pressure (Abbott 2003). In general, individuals can respond to drug treatment in one of three ways: favorably (i.e., as expected), unfavorably (i.e., adversely or with a blunted response), or not at all. Many factors, including age, ethnic background, gender, diet, interactions with other pharmaceuticals, and clearance function, influence the manner in which an individual will respond to a drug. In addition to these determinants, genetic factors are also known to impact the degree to which an individual will respond to a drug. The prediction of drug response based upon genetic variation has evolved into the field of pharmacogenetics. The closely related discipline of pharmacogenomics encompasses pharmacogenetics, but incorporates analysis of gene expression to understand genotype-drug interaction; thus, the main differences between the two disciplines lie mainly in the underlying technologies and the level at which a given gene is investigated. Because the focus of this chapter is the genome, we will address the intersection between genotype and drug response from a perspective favoring the field of pharmacogenetics.

Although the contribution of genetic variation on drug response has been recognized for decades, the availability of human genome reference sequence and a catalog of common genetic variation in the human genome has expanded the field tremendously (Collins et al. 2003). Indeed, it is in the field of pharmacogenetics that the clinical applicability of the HGP and HapMap resources has had the most impact. A number of genetic variants have been identified that at least partially predict drug response, including associations between *HLA-B* alleles and hypersensitivity to the anti-HIV therapeutic, abacavir (Hetherington et al. 2002). Among these patients, 46 % of individuals who had previously suffered an adverse immunological reaction to abacavir possessed the *HLA-B57* variant, compared to only 4 % of individuals who were not hypersensitive to the drug (Hetherington et al. 2002).

A specific haplotype within the vitamin K epoxide reductase gene (*VKORC1*) has been found to predict 21–25 % of required warfarin dose. When *VKORC1* haplotype is combined with genotypes in the cytochrome P450, subfamily IIC, polypeptide 9 gene (*CYP2C9*), 31 % of warfarin dose can be predicted (Rieder et al. 2005). This finding is particularly significant because warfarin, the most commonly prescribed anticoagulant, has a narrow therapeutic index and requires careful and regular monitoring. Dosing above the required concentration produces potentially life-threatening side effects, while dosing below delays therapeutic benefit. The use of *VKORC1* and *CYP2C9* genotypes, combined with age, sex, body, and surface area, can predict up to 60 % of warfarin dose, thereby better ensuring achievement of optimal therapeutic dose (Rieder et al. 2005; Marsh and McLeod 2006).

Clopidogrel therapy improves cardiovascular outcomes in patients with acute coronary syndromes and following percutaneous coronary intervention by inhibiting adenosine diphosphate (ADP)-dependent platelet activation. However, nonresponsiveness to the drug is widely recognized and is related to recurrent ischemic events. The cytochrome P450 2C19 (*CYP2C19*) and *ABCB1* genotypes were found to be associated with platelet response to clopidogrel treatment and in the prediction of major cardiovascular events beyond stent thrombosis in coronary patients treated with clopidogrel (Shuldiner et al. 2009; Mega et al. 2010). Similarly, it has been shown that in patients with diabetes, vitamin E significantly increases HDL function in haptoglobin 2-2 but significantly decreases HDL function in haptoglobin 2-1. Thus, vitamin E therapy provides cardiovascular protection to individuals with the haptoglobin 2-2 genotype, but appears to increase cardiovascular risk in individuals with the haptoglobin 2-1 genotype. This pharmacogenetic interaction was paralleled by similar nonsignificant trends in HDL-associated lipid peroxides, glutathione peroxidase, and inflammatory cargo (Farbstein et al. 2011).

Pharmacogenetics is a rising concern in clinical oncology, because the therapeutic window of most anticancer drugs is narrow and patients with impaired ability to detoxify drugs will undergo life-threatening toxicities. In particular, genetic deregulations affecting genes coding for DPD, UGT1A1, TPMT, CDA, and CYP2D6 are now considered as critical issues for patients treated with 5-FU/capecitabine, irinotecan, mercaptopurine/azathioprine/thiopurine, gemcitabine/capecitabine/AraC, and tamoxifen, respectively (Evans 2004; Marques and Ikediobi 2010; Yang et al. 2011; O'Donnell and Ratain 2012). Examples like this serve to underscore the reality that the real clinical impact of pharmacogenetics will be in identifying those patients who are most likely to experience the desired therapeutic effect from the drug under consideration. For these individuals, quicker control of disease symptoms, reduced likelihood of adverse events, and better disease management will be provided by pharmacogenetics. Together, these factors will also impact public health by decreasing healthcare costs.

2.3.2 Nutrigenetics and Nutrigenomics

Nutrigenetics is the study of the relationship between genetic variation and metabolic, biochemical, or physiological response to foods. The related field of nutrigenomics comprises nutrient impact at the levels of gene expression, transcript stability, and posttranslational modifications (Young 2002; Ghosh et al. 2007). Completion of the HGP and availability of sequence variants have significantly fueled the development of these complementary disciplines; similar to the promise of pharmacogenetics, both nutrigenetics and nutrigenomics have the potential to influence the development of "personalized" nutrition by delineating dietary composition based upon specific genotype.

Several variants have been found to impact upon the metabolism of various dietary components (Ghosh et al. 2007; Raqib and Cravioto 2009). For example, individuals with phenylketonuria, an autosomal recessive disorder characterized by a deficiency in phenylalanine hydroxylase, are unable to metabolize phenylalanine and in the presence of foods high in this amino acid, such as meats, nuts, cheese, and the artificial sweetener aspartame, develop severe neurological disorders, including mental retardation. Simple avoidance of such foods prevents significant medical problems for patients with this genetic susceptibility. Likewise, variants in *HLA DQ2* and *DQ8* have been linked with gluten in the development of celiac disease; more than 95 % of celiac patients are positive for either DQ2 or DQ8 (Sollid and Lie 2005). For individuals with these risk allelels, a gluten-free diet is recommended for disease management.

Considerable evidence also suggests that epigenetic abnormalities induced by diet are also among the most important factors affecting cancer risk. At least four distinct processes are involved with epigenetics: DNA methylation, histone modifications, microRNAs as well as other noncoding regulatory RNA, and chromatin modeling (Ross 2007). Some of the strongest data linking diet to epigenetic events come from studies with the agouti mouse model. Adding dietary factors (i.e., choline, betaine, or folic acid), which enhance methylation, to the maternal diet of pregnant agouti dams leads to a change in the phenotype of some of the offspring (Dolinoy 2008). Interestingly, adding genistein, which does not provide methyl groups, also leads to a change in the phenotype from a yellow to more agouti offspring (Dolinoy et al. 2006). Most importantly, these shifts in coat color are accompanied by a reduction in the risk of cancer, diabetes, and obesity. The shift in obesity in these animals is noteworthy because of the worldwide obesity epidemic. Such findings should serve as justification for additional attention to bioenergetic-epigenetic interrelationships, especially those that are modified by dietary factors.

Myzak and Dashwood (2006) have demonstrated that sulphoraphane, butyrate, and allyl sulfur are effective inhibitors of histone deacetylase (HDAC). HDAC inhibition was associated with global increases in histone acetylation, enhanced interactions of acetylated histones with the promoter regions of the *P21* and *BAX* genes, and elevated expression of p21Cip1/Waf1 and BAX proteins. Importantly, sulphoraphane has been reported to reduce HDAC activity in humans (Myzak et al. 2006). Future research likely needs to relate HDAC changes in humans to a change in cancer-related processes. Furthermore, since acetylation is only one method to regulate histone homeostasis (Ross 2007), greater attention needs to be given to how nutrition might influence the other types of histone modifications (Fenech et al. 2011).

In addition to the development of nutrient-related diseases, genetic variants can also interact with dietary components to produce subtle effects on metabolism. For example, a dose-dependent interaction between variants in the *APOA5* gene and dietary fat intake was found to increase risk for obesity in participants of the Framingham Heart Study (Corella et al. 2007). Similarly, individuals with the AA genotype at the G(-6)A marker in the angiotensinogen gene, which is associated with both higher circulating levels of angiotensinogen and elevated blood pressure, were more responsive to the effects of a diet high in fruits and vegetables and low in fat compared to individuals with the GG genotype (Svetkey et al. 2001). Other studies have found relationships between specific genetic variants and responsiveness to dietary components, and provide support for a role of dietary shifts in shaping human evolution. Perry et al. (2007) reported that individuals from populations with a typically high-starch diet (i.e., European Americans, Japanese, and Hadza hunter-gatherers) have more copies of the salivary amylase gene, which breaks down starch, compared to those from populations with a low-starch diet (i.e., Biaka, Mbuti, Datog pastoralists, and the Yakut). This finding is one of the first examples of positive selection on copy number variant, and further supports the idea that individuals may respond quite differently to the same diet given their respective genetic backgrounds.

2.4 Conclusions

The completion of HGP represents one of the momentous projects of modern scientific research. Delineation of the human genome sequence has consequently led to a greater understanding of human genetics and fueled the development of such diverse disciplines as comparative genomics, pharmacogenetics, and nutrigenomics. The fruits of the HGP directly contributed to the creation of the HapMap and the 1000 Genomes projects, which has since provided the basis for WGA studies. Results from these investigations will be instrumental in the elucidation of the genetic variants that contribute to the development of complex diseases such as cancer, diabetes, autoimmune syndromes, and neurological disorders. Thus, the HGP has produced a significant impact upon a variety of different areas, and in completely unexpected ways.

References

1000 Genomes Project Consortium, Abecasis GR, Altshuler D, Auton A, Brooks LD, Durbin RM, Gibbs RA, Hur es ME, McVean GA (2010) A map of human genome variation from population-scale sequencing. Nature 467:1061–1073

1000 Genomes Project Consortium, Abecasis GR, Auton A, Brooks LD, DePristo MA, Durbin RM, Handsaker RE, Kang HM, Marth GT, McVean GA (2012) An integrated map of genetic variation from 1,092 human genomes. Nature 491: 56–65

Abbott A (2003) With your genes? Take one of these, three times a day. Nature 425:760–762

Adams MD, Celniker SE, Holt RA et al (2000) The genome sequence of Drosophila melanogaster. Science 287:2185–2195

Altshuler D, Daly M (2007) Guilt beyond a reasonable doubt. Nat Genet 39:813–815

Bachellerie JP, Cavaille J, Huttenhofer A (2002) The expanding snoRNA world. Biochimie 84:775–790

Bailey JA, Gu Z, Clark RA, Reinert K, Samonte RV, Schwartz S, Adams MD, Myers EW, Li PW, Eichler EE (2002) Recent segmental duplications in the human genome. Science 297:1003–1007

Bailey JA, Eichler EE (2006) Primate segmental duplications: crucibles of evolution, diversity and disease. Nat Rev Genet 7:552–564

Barski A, Cuddapah S, Cui K, Roh TY, Schones DE, Wang Z, Wei G, Chepelev I, Zhao K (2007) High-resolution profiling of histone methylations in the human genome. Cell 129:823–837

Bejerano G, Pheasant M, Makunin I, Stephen S, Kent WJ, Mattick JS, Haussler D (2004) Ultraconserved elements in the human genome. Science 304:1321–1325

Bowcock AM (2007) Genomics: guilt by association. Nature 447:645–646

Chen FC, Li WH (2001) Genomic divergences between humans and other hominoids and the effective population size of the common ancestor of humans and chimpanzees. Am J Hum Genet 68:444–456

Chen CT, Wang JC, Cohen BA (2007) The strength of selection on ultraconserved elements in the human genome. Am J Hum Genet 80:692–704

Cheng J, Kapranov P, Drenkow J, Dike S, Brubaker S, Patel S, Long J, Stern D, Tammana H, Helt G, Sementchenko V, Piccolboni A, Bekiranov S, Bailey DK, Ganesh M, Ghosh S, Bell I, Gerhard DS, Gingeras TR (2005) Transcriptional maps of 10 human chromosomes at 5-nucleotide resolution. Science 308:1149–1154

Chimpanzee Sequencing and Analysis Consortium (2005) Initial sequence of the chimpanzee genome and comparison with the human genome. Nature 437:69–87

Choudhuri S (2003) The path from nuclein to human genome: a brief history of DNA with a note on human genome sequencing and its impact on future research in biology. Bull Sci Technol Soc 23:360–367

Clamp M, Fry B, Kamal M, Xie X, Cuff J, Lin MF, Kellis M, Lindblad-Toh K, Lander ES (2007) Distinguishing protein-coding and noncoding genes in the human genome. Proc Natl Acad Sci USA 104:19428–19433

Clark AG, Glanowski S, Nielsen R, Thomas PD, Kejariwal A, Todd MA, Tanenbaum DM, Civello D, Lu F, Murphy B, Ferriera S, Wang G, Zheng X, White TJ, Sninsky JJ, Adams MD, Cargill M (2003) Inferring nonneutral evolution from human-chimp-mouse orthologous gene trios. Science 302:1960–1963

Cohen J (2007) Genomics. DNA duplications and deletions help determine health. Science 317:1315–1317

Collins FS, Green ED, Guttmacher AE, Guyer MS, US National Human Genome Research Institute (2003) A vision for the future of genomics research. Nature 422:835–847

Conrad DF, Andrews TD, Carter NP, Hurles ME, Pritchard JK (2006) A high-resolution survey of deletion polymorphism in the human genome. Nat Genet 38:75–81

Conrad DF, Pinto D, Redon R, Feuk L, Gokcumen O, Zhang Y, Aerts J, Andrews TD, Barnes C, Campbell P, Fitzgerald T, Hu M, Ihm CH, Kristiansson K, Macarthur DG, Macdonald JR, Onyiah I, Pang AW, Robson S, Stirrups K, Valsesia A, Walter K, Wei J, Wellcome Trust Case Control Consortium, Tyler-Smith C, Carter NP, Lee C, Scherer SW, Hurles ME (2010) Origins and functional impact of copy number variation in the human genome. Nature 464:704–712

Cook-Deegan RM (1989) The Alta summit, December 1984. Genomics 5:661–663

Cooper GM, Zerr T, Kidd JM, Eichler EE, Nickerson DA (2008) Systematic assessment of copy number variant detection via genome-wide SNP genotyping. Nat Genet 40:1199–1203

Corella D, Lai CQ, Demissie S, Cupples LA, Manning AK, Tucker KL, Ordovas JM (2007) APOA5 gene variation modulates the effects of dietary fat intake on body mass index and obesity risk in the Framingham Heart Study. J Mol Med (Berl) 85:119–128

Dermitzakis ET, Reymond A, Scamuffa N, Ucla C, Kirkness E, Rossier C, Antonarakis SE (2003) Evolutionary discrimination of mammalian conserved non-genic sequences (CNGs). Science 302:1033–1035

Dolinoy DC, Weidman JR, Waterland RA, Jirtle RL (2006) Maternal genistein alters coat color and protects Avy mouse offspring from obesity by modifying the fetal epigenome. Environ Health Perspect 114:567–572

Dolinoy DC (2008) The agouti mouse model: an epigenetic biosensor for nutritional and environmental alterations on the fetal epigenome. Nutr Rev 66(Suppl 1):S7–S11

Ebersberger I, Metzler D, Schwarz C, Paabo S (2002) Genomewide comparison of DNA sequences between humans and chimpanzees. Am J Hum Genet 70:1490–1497

Elango N, Thomas JW, NISC Comparative Sequencing Program, Yi SV (2006) Variable molecular clocks in hominoids. Proc Natl Acad Sci USA 103:1370–1375

Enard W (2012) Functional primate genomics–leveraging the medical potential. J Mol Med 90:471–480

ENCODE Project Consortium (2004) The ENCODE (ENCyclopedia Of DNA Elements) project. Science 306:636–640

ENCODE Project Consortium (2007) Identification and analysis of functional elements in 1 % of the human genome by the ENCODE pilot project. Nature 447:799–816

ENCODE Project Consortium (2011) A user's guide to the encyclopedia of DNA elements (ENCODE). PLoS Biol 9:e1001046. doi:10.1371/journal.pbio.1001046

ENCODE Project Consortium, Dunham I, Kundaje A et al (2012) An integrated encyclopedia of DNA elements in the human genome. Nature 489:57–74

Estivill X, Cheung J, Pujana MA, Nakabayashi K, Scherer SW, Tsui LC (2002) Chromosomal regions containing high-density and ambiguously mapped putative single nucleotide polymorphisms (SNPs) correlate with segmental duplications in the human genome. Hum Mol Genet 11:1987–1995

Evans WE (2004) Pharmacogenetics of thiopurine S-methyltransferase and thiopurine therapy. Ther Drug Monit 26:186–191

Farbstein D, Blum S, Pollak M, Asaf R, Viener HL, Lache O, Asleh R, Miller-Lotan R, Barkay I, Star M, Schwartz A, Kalet-Littman S, Ozeri D, Vaya J, Tavori H, Vardi M, Laor A, Bucher SE, Anbinder Y, Moskovich D, Abbas N, Perry N, Levy Y, Levy AP (2011) Vitamin E therapy results in a reduction in HDL function in individuals with diabetes and the haptoglobin 2-1 genotype. Atherosclerosis 219:240–244

Fenech M, El-Sohemy A, Cahill L, Ferguson LR, French TA, Tai ES, Milner J, Koh WP, Xie L, Zucker M, Buckley M, Cosgrove L, Lockett T, Fung KY, Head R (2011) Nutrigenetics and nutrigenomics: viewpoints on the current status and applications in nutrition research and practice. J Nutr Nutr 4:69–89

Flicek P, Aken BL, Ballester B, Beal K, Bragin E, Brent S, Chen Y, Clapham P, Coates G, Fairley S, Fitzgerald S, Fernandez-Banet J, Gordon L, Graf S, Haider S, Hammond M, Howe K, Jenkinson A, Johnson N, Kahari A, Keefe D, Keenan S, Kinsella R, Kokocinski F, Koscielny G, Kulesha E, Lawson D, Longden I, Massingham T, McLaren W, Megy K, Overduin B, Pritchard B, Rios D, Ruffier M, Schuster M, Slater G, Smedley D, Spudich G, Tang YA, Trevanion S, Vilella A, Vogel J, White S, Wilder SP, Zadissa A, Birney E, Cunningham F, Dunham I, Durbin R, Fernandez-Suarez XM, Herrero J, Hubbard TJ, Parker A, Proctor G, Smith J, Searle SM (2010) Ensembl's 10th year. Nucleic Acids Res 38:D557–D562. doi:10.1093/nar/gkp972

Ghosh D, Skinner MA, Laing WA (2007) Pharmacogenomics and nutrigenomics: synergies and differences. Eur J Clin Nutr 61:567–574

Girirajan S, Campbell CD, Eichler EE (2011) Human copy number variation and complex genetic disease. Annu Rev Genet 45:203–226

Goldstein DB (2009) Common genetic variation and human traits. N Engl J Med 360:1696–1698

Green RE, Krause J, Ptak SE, Briggs AW, Ronan MT, Simons JF, Du L, Egholm M, Rothberg JM, Paunovic M, Paabo S (2006) Analysis of one million base pairs of Neanderthal DNA. Nature 444:330–336

Gu Z, Rifkin SA, White KP, Li WH (2004) Duplicate genes increase gene expression diversity within and between species. Nat Genet 36:577–579

Gu X, Zhang Z, Huang W (2005) Rapid evolution of expression and regulatory divergences after yeast gene duplication. Proc Natl Acad Sci USA 102:707–712

Guigo R, Flicek P, Abril JF, Reymond A, Lagarde J, Denoeud F, Antonarakis S, Ashburner M, Bajic VB, Birney E, Castelo R, Eyras E, Ucla C, Gingeras TR, Harrow J, Hubbard T, Lewis SE, Reese MG (2006) EGASP: the human ENCODE genome annotation assessment project. Genome Biol 7(Suppl 1:S2):1–31

Harrison PM, Gerstein M (2002) Studying genomes through the aeons: protein families, pseudogenes and proteome evolution. J Mol Biol 318:1155–1174

Harrow J, Denoeud F, Frankish A, Reymond A, Chen CK, Chrast J, Lagarde J, Gilbert JG, Storey R, Swarbreck D, Rossier C, Ucla C, Hubbard T, Antonarakis SE, Guigo R (2006) GENCODE: producing a reference annotation for ENCODE. Genome Biol 7 (Suppl 1:S4):1–9

Hetherington S, Hughes AR, Mosteller M, Shortino D, Baker KL, Spreen W, Lai E, Davies K, Handley A, Dow DJ, Fling ME, Stocum M, Bowman C, Thurmond LM, Roses AD (2002) Genetic variations in HLA-B region and hypersensitivity reactions to abacavir. Lancet 359:1121–1122

Hindorff LA, Sethupathy P, Junkins HA, Ramos EM, Mehta JP, Collins FS, Manolio TA (2009) Potential etiologic and functional implications of genome-wide association loci for human diseases and traits. Proc Natl Acad Sci USA 106:9362–9367

Hindorff LA, MacArthur J, Morales J, Junkins HA, Hall PN, Klemm AK, Manolio TA (2012) A catalog of published genome-wide association studies. www.genome.gov/gwastudies. Accessed (2012)

Hirano T (2000) Chromosome cohesion, condensation, and separation. Annu Rev Biochem 69:115–144

Hittinger CT, Carroll SB (2007) Gene duplication and the adaptive evolution of a classic genetic switch. Nature 449:677–681

Horvath JE, Willard HF (2007) Primate comparative genomics: lemur biology and evolution. Trends Genet TIG 23:173–182

Hua S, Kittler R, White KP (2009) Genomic antagonism between retinoic acid and estrogen signaling in breast cancer. Cell 137:1259–1271

Iafrate AJ, Feuk L, Rivera MN, Listewnik ML, Donahoe PK, Qi Y, Scherer SW, Lee C (2004) Detection of large-scale variation in the human genome. Nat Genet 36:949–951

International HapMap Consortium (2003) The International HapMap project. Nature 426:789–796

International HapMap Consortium (2005) A haplotype map of the human genome. Nature 437:1299–1320

International HapMap Consortium, Frazer KA, Ballinger DG, et al (2007) A second generation human haplotype map of over 3.1 million SNPs. Nature 449:851–861

International HapMap 3 Consortium, Altshuler DM, Gibbs RA, et al (2010) Integrating common and rare genetic variation in diverse human populations. Nature 467:52–58

International Human Genome Sequencing Consortium (2004) Finishing the euchromatic sequence of the human genome. Nature 431:931–945

Johnson DS, Mortazavi A, Myers RM, Wold B (2007) Genome-wide mapping of in vivo protein-DNA interactions. Science 316:1497–1502

Kaessmann H, Wiebe V, Weiss G, Paabo S (2001) Great ape DNA sequences reveal a reduced diversity and an expansion in humans. Nat Genet 27:155–156

Kampa D, Cheng J, Kapranov P, Yamanaka M, Brubaker S, Cawley S, Drenkow J, Piccolboni A, Bekiranov S, Helt G, Tammana H, Gingeras TR (2004) Novel RNAs identified from an in-depth analysis of the transcriptome of human chromosomes 21 and 22. Genome Res 14:331–342

Kapranov P, Willingham AT, Gingeras TR (2007) Genome-wide transcription and the implications for genomic organization. Nat Rev Genet 8:413–423

Kidd JM, Cooper GM, Donahue WF, Hayden HS, Sampas N, Graves T, Hansen N, Teague B, Alkan C, Antonacci F, Haugen E, Zerr T, Yamada NA, Tsang P, Newman TL, Tuzun E, Cheng Z, Ebling HM, Tusneem N, David R, Gillett W, Phelps KA, Weaver M, Saranga D, Brand A, Tao W, Gustafson E, McKernan K, Chen L, Malig M, Smith JD, Korn JM, McCarroll SA, Altshuler DA, Peiffer DA, Dorschner M, Stamatoyannopoulos J, Schwartz D, Nickerson DA, Mullikin JC, Wilson RK, Bruhn L, Olson MV, Kaul R, Smith DR, Eichler EE (2008) Mapping and sequencing of structural variation from eight human genomes. Nature 453:56–64

Kim VN, Nam JW (2006) Genomics of microRNA. Trends Genet 22:165–173

Korbel JO, Urban AE, Affourtit JP, Godwin B, Grubert F, Simons JF, Kim PM, Palejev D, Carriero NJ, Du L, Taillon BE, Chen Z, Tanzer A, Saunders AC, Chi J, Yang F, Carter NP, Hurles ME, Weissman SM, Harkins TT, Gerstein MB, Egholm M, Snyder M (2007) Paired-end mapping reveals extensive structural variation in the human genome. Science 318:420–426

Kruglyak L, Nickerson DA (2001) Variation is the spice of life. Nat Genet 27:234–236

Kryukov GV, Schmidt S, Sunyaev S (2005) Small fitness effect of mutations in highly conserved non-coding regions. Hum Mol Genet 14:2221–2229

Lander ES, Linton LM, Birren B et al (2001) Initial sequencing and analysis of the human genome. Nature 409:860–921

Lander ES (2011) Initial impact of the sequencing of the human genome. Nature 470:187–197

Li WH, Tanimura M (1987) The molecular clock runs more slowly in man than in apes and monkeys. Nature 326:93–96

Locke DP, Segraves R, Carbone L, Archidiacono N, Albertson DG, Pinkel D, Eichler EE (2003) Large-scale variation among human and great ape genomes determined by array comparative genomic hybridization. Genome Res 13:347–357

Locke DP, Sharp AJ, McCarroll SA, McGrath SD, Newman TL, Cheng Z, Schwartz S, Albertson DG, Pinkel D, Altshuler DM, Eichler EE (2006) Linkage disequilibrium and heritability of copy-number polymorphisms within duplicated regions of the human genome. Am J Hum Genet 79:275–290

Louis EJ (2007) Evolutionary genetics: making the most of redundancy. Nature 449:673–674

Lupski JR, de Oca-Luna RM, Slaugenhaupt S, Pentao L, Guzzetta V, Trask BJ, Saucedo-Cardenas O, Barker DF, Killian JM, Garcia CA, Chakravarti A, Patel PI (1991) DNA duplication associated with Charcot-Marie-Tooth disease type 1A. Cell 66:219–232

Lynch M, Conery JS (2000) The evolutionary fate and consequences of duplicate genes. Science 290:1151–1155

Manolio TA, Collins FS, Cox NJ, Goldstein DB, Hindorff LA, Hunter DJ, McCarthy MI, Ramos EM, Cardon LR, Chakravarti A, Cho JH, Guttmacher AE, Kong A, Kruglyak L, Mardis E, Rotimi CN, Slatkin M, Valle D, Whittemore AS, Boehnke M, Clark AG, Eichler EE, Gibson G, Haines JL, Mackay TF, McCarroll SA, Visscher PM (2009) Finding the missing heritability of complex diseases. Nature 461:747–753

Marques-Bonet T, Ryder OA, Eichler EE (2009) Sequencing primate genomes: what have we learned? Ann Rev Genomics Hum Genet 10:355–386

Marques SC, Ikediobi ON (2010) The clinical application of UGT1A1 pharmacogenetic testing: gene-environment interactions. Hum Genomics 4:238–249

Marsh S, McLeod, HL (2006) Pharmacogenomics: from bedside to clinical practice. Hum Mol Genet 15(Spec No 1):R89–R93

Maston GA, Evans SK, Green MR (2006) Transcriptional regulatory elements in the human genome. Annu Rev Genomics Hum Genet 7:29–59

McCarroll SA, Hadnott TN, Perry GH, Sabeti PC, Zody MC, Barrett JC, Dallaire S, Gabriel SB, Lee C, Daly MJ, Altshuler DM, International HapMap Consortium (2006) Common deletion polymorphisms in the human genome. Nat Genet 38:86–92

Mega JL, Close SL, Wiviott SD, Shen L, Walker JR, Simon T, Antman EM, Braunwald E, Sabatine MS (2010) Genetic variants in ABCB1 and CYP2C19 and cardiovascular outcomes after treatment with clopidogrel and prasugrel in the TRITON-TIMI 38 trial: a pharmacogenetic analysis. Lancet 376:1312–1319

Meyer M, Kircher M, Gansauge MT, Li H, Racimo F, Mallick S, Schraiber JG, Jay F, Prufer K, de Filippo C, Sudmant PH, Alkan C, Fu Q, Do R, Rohland N, Tandon A, Siebauer M, Green RE, Bryc K, Briggs AW, Stenzel U, Dabney J, Shendure J, Kitzman J, Hammer MF, Shunkov MV, Derevianko AP, Patterson N, Andres AM, Eichler EE, Slatkin M, Reich D, Kelso J, Paabo S (2012) A high-coverage genome sequence from an archaic Denisovan individual. Science 338:222–226

Mighell AJ, Smith NR, Robinson PA, Markham AF (2000) Vertebrate pseudogenes. FEBS Lett 468:109–114

Mikkelsen TS, Ku M, Jaffe DB, Issac B, Lieberman E, Giannoukos G, Alvarez P, Brockman W, Kim TK, Koche RP, Lee W, Mendenhall E, O'Donovan A, Presser A, Russ C, Xie X, Meissner A, Wernig M, Jaenisch R, Nusbaum C, Lander ES, Bernstein BE (2007) Genome-wide maps of chromatin state in pluripotent and lineage-committed cells. Nature 448:553–560

Mortazavi A, Williams BA, McCue K, Schaeffer L, Wold B (2008) Mapping and quantifying mammalian transcriptomes by RNA-seq. Nat Methods 5:621–628

Mouse Genome Sequencing Consortium, Waterston RH, Lindblad-Toh K et al (2002) Initial sequencing and comparative analysis of the mouse genome. Nature 420:520–562

Mural RJ, Adams MD, Myers EW et al (2002) A comparison of whole-genome shotgun-derived mouse chromosome 16 and the human genome. Science 296:1661–1671

Murphy WJ, Eizirik E, Johnson WE, Zhang YP, Ryder OA, O'Brien SJ (2001) Molecular phylogenetics and the origins of placental mammals. Nature 409:614–618

Myers S, Bottolo L, Freeman C, McVean G, Donnelly P (2005) A fine-scale map of recombination rates and hotspots across the human genome. Science 310:321–324

Myzak MC, Dashwood RH (2006) Histone deacetylases as targets for dietary cancer preventive agents: lessons learned with butyrate, diallyl disulfide, and sulforaphane. Curr Drug Targets 7:443–452

Myzak MC, Hardin K, Wang R, Dashwood RH, Ho E (2006) Sulforaphane inhibits histone deacetylase activity in BPH-1, LnCaP and PC-3 prostate epithelial cells. Carcinogenesis 27:811–819

Noonan JP, Coop G, Kudaravalli S, Smith D, Krause J, Alessi J, Chen F, Platt D, Paabo S, Pritchard JK, Rubin EM (2006) Sequencing and analysis of Neanderthal genomic DNA. Science 314:1113–1118

O'Donnell PH, Ratain MJ (2012) Germline pharmacogenomics in oncology: decoding the patient for targeting therapy. Mol Oncol 6:251–259

O'Roak BJ, Deriziotis P, Lee C, Vives L, Schwartz JJ, Girirajan S, Karakoc E, Mackenzie AP, Ng SB, Baker C, Rieder MJ, Nickerson DA, Bernier R, Fisher SE, Shendure J, Eichler EE (2011) Exome sequencing in sporadic autism spectrum disorders identifies severe de novo mutations. Nat Genet 43:585–589

Oak Ridge National Laboratory http://www.ornl.gov. Accessed 28 Mar 2013

Pagani I, Liolios K, Jansson J, Chen IM, Smirnova T, Nosrat B, Markowitz VM, Kyrpides NC (2012) The Genomes OnLine Database (GOLD) v. 4: status of genomic and metagenomic projects and their associated metadata. Nucleic Acids Res 40:D571–D5799. doi:10.1093/nar/gkr1100

Patterson K (2011) 1000 genomes: a world of variation. Circ Res 108:534–536

Pennacchio LA, Ahituv N, Moses AM, Prabhakar S, Nobrega MA, Shoukry M, Minovitsky S, Dubchak I, Holt A, Lewis KD, Plajzer-Frick I, Akiyama J, De Val S, Afzal V, Black BL, Couronne O, Eisen MB, Visel A, Rubin EM (2006) In vivo enhancer analysis of human conserved non-coding sequences. Nature 444:499–502

Pennisi E (2007) Genetics. Working the (gene count) numbers: finally, a firm answer? Science 316:1113

Perry GH, Dominy NJ, Claw KG, Lee AS, Fiegler H, Redon R, Werner J, Villanea FA, Mountain JL, Misra R, Carter NP, Lee C, Stone AC (2007) Diet and the evolution of human amylase gene copy number variation. Nat Genet 39:1256–1260

Pinto D, Darvishi K, Shi X, Rajan D, Rigler D, Fitzgerald T, Lionel AC, Thiruvahindrapuram B, Macdonald JR, Mills R, Prasad A, Noonan K, Gribble S, Prigmore E, Donahoe PK, Smith RS, Park JH, Hurles ME, Carter NP, Lee C, Scherer SW, Feuk L (2011) Comprehensive assessment of array-based platforms and calling algorithms for detection of copy number variants. Nat Biotechnol 29:512–520

Poser I, Sarov M, Hutchins JR, Heriche JK, Toyoda Y, Pozniakovsky A, Weigl D, Nitzsche A, Hegemann B, Bird AW, Pelletier L, Kittler R, Hua S, Naumann R, Augsburg M, Sykora MM, Hofemeister H, Zhang Y, Nasmyth K, White KP, Dietzel S, Mechtler K, Durbin R, Stewart AF, Peters JM, Buchholz F, Hyman AA (2008) BAC TransgeneOmics: a high-throughput method for exploration of protein function in mammals. Nat Methods 5:409–415

Prabhakar S, Noonan JP, Paabo S, Rubin EM (2006) Accelerated evolution of conserved noncoding sequences in humans. Science 314:786

Raqib R, Cravioto A (2009) Nutrition, immunology, and genetics: future perspectives. Nutr Rev 67(Suppl 2): S227–S236

Redon R, Ishikawa S, Fitch KR, Feuk L, Perry GH, Andrews TD, Fiegler H, Shapero MH, Carson AR, Chen W, Cho EK, Dallaire S, Freeman JL, Gonzalez JR, Gratacos M, Huang J, Kalaitzopoulos D, Komura D, MacDonald JR, Marshall CR, Mei R, Montgomery L, Nishimura K, Okamura K, Shen F, Somerville MJ, Tchinda J, Valsesia A, Woodwark C, Yang F, Zhang J, Zerjal T, Zhang J, Armengol L, Conrad DF, Estivill X, Tyler-Smith C, Carter NP, Aburatani H, Lee C, Jones KW, Scherer SW, Hurles ME (2006) Global variation in copy number in the human genome. Nature 444:444–454

Reich D, Green RE, Kircher M, Krause J, Patterson N, Durand EY, Viola B, Briggs AW, Stenzel U, Johnson PL, Maricic T, Good JM, Marques-Bonet T, Alkan C, Fu Q, Mallick S, Li H, Meyer M, Eichler EE, Stoneking M, Richards M, Talamo S, Shunkov MV, Derevianko AP, Hublin JJ, Kelso J, Slatkin M, Paabo S (2010) Genetic history of an archaic hominin group from Denisova Cave in Siberia. Nature 468:1053–1060

Rhesus Macaque Genome Sequencing and Analysis Consortium, Gibbs RA, Rogers J et al (2007) Evolutionary and biomedical insights from the rhesus macaque genome. Science 316:222–234

Rieder MJ, Reiner AP, Gage BF, Nickerson DA, Eby CS, McLeod HL, Blough DK, Thummel KE, Veenstra DL, Rettie AE (2005) Effect of VKORC1 haplotypes on transcriptional regulation and warfarin dose. N Engl J Med 352:2285–2293

Roberts L (2001) The human genome. Controversial from the start. Science 291:1182–1188

Robertson G, Hirst M, Bainbridge M, Bilenky M, Zhao Y, Zeng T, Euskirchen G, Bernier B, Varhol R, Delaney A, Thiessen N, Griffith OL, He A, Marra M, Snyder M, Jones S (2007) Genome-wide profiles of STAT1 DNA association using chromatin immunoprecipitation and massively parallel sequencing. Nat Methods 4:651–657

Ross SA (2007) Nutritional genomic approaches to cancer prevention research. Exp Oncol 29:250–256

Sasaki E, Suemizu H, Shimada A, Hanazawa K, Oiwa R, Kamioka M, Tomioka I, Sotomaru Y, Hirakawa R, Eto T, Shiozawa S, Maeda T, Ito M, Ito R, Kito C, Yagihashi C, Kawai K, Miyoshi H, Tanioka Y, Tamaoki N, Habu S, Okano H, Nomura T (2009) Generation of transgenic non-human primates with germline transmission. Nature 459:523–527

Sasidharan R, Gerstein M (2008) Genomics: protein fossils live on as RNA. Nature 453:729–731

Schaner P, Richards N, Wadhwa A, Aksentijevich I, Kastner D, Tucker P, Gumucio D (2001) Episodic evolution of pyrin in primates: human mutations recapitulate ancestral amino acid states. Nat Genet 27:318–321

Schmid M, Jensen TH (2010) Nuclear quality control of RNA polymerase II transcripts. Wiley Interdiscip Rev RNA 1:474–485

Sebat J, Lakshmi B, Troge J, Alexander J, Young J, Lundin P, Maner S, Massa H, Walker M, Chi M, Navin N, Lucito R, Healy J, Hicks J, Ye K, Reiner A, Gilliam TC, Trask B, Patterson N, Zetterberg A, Wigler M (2004) Large-scale copy number polymorphism in the human genome. Science 305:525–528

Sharp AJ, Locke DP, McGrath SD, Cheng Z, Bailey JA, Vallente RU, Pertz LM, Clark RA, Schwartz S, Segraves R, Oseroff VV, Albertson DG, Pinkel D, Eichler EE (2005) Segmental duplications and copy-number variation in the human genome. Am J Hum Genet 77:78–88

Shuldiner AR, O'Connell JR, Bliden KP, Gandhi A, Ryan K, Horenstein RB, Damcott CM, Pakyz R, Tantry US, Gibson Q, Pollin TI, Post W, Parsa A, Mitchell BD, Faraday N, Herzog W, Gurbel PA (2009) Association of cytochrome P450 2C19 genotype with the anti-platelet effect and clinical efficacy of clopidogrel therapy. JAMA 302:849–857

Sollid LM, Lie BA (2005) Celiac disease genetics: current concepts and practical applications. Clin Gastroenterol Hepatol 3:843–851

Steiper ME, Young NM, Sukarna TY (2004) Genomic data support the hominoid slowdown and an Early Oligocene estimate for the hominoid-cercopithecoid divergence. Proc Natl Acad Sci USA 101:17021–17026

Sudmant PH, Kitzman JO, Antonacci F, Alkan C, Malig M, Tsalenko A, Sampas N, Bruhn L, Shendure J, 1000 Genomes Project, Eichler EE (2010) Diversity of human copy number variation and multicopy genes. Science 330:641–646

Svetkey LP, Moore TJ, Simons-Morton DG, Appel LJ, Bray GA, Sacks FM, Ard JD, Mortensen RM, Mitchell SR, Conlin PR, Kesari M, DASH Collaborative Research Group (2001) Angiotensinogen genotype and blood pressure response in the Dietary Approaches to Stop Hypertension (DASH) study. J Hypertens 19:1949–1956

Taylor JS, Raes J (2004) Duplication and divergence: the evolution of new genes and old ideas. Annu Rev Genet 38:615–643

Thomas JW, Touchman JW, Blakesley RW et al (2003) Comparative analyses of multi-species sequences from targeted genomic regions. Nature 424:788–793

Tuzun E, Sharp AJ, Bailey JA, Kaul R, Morrison VA, Pertz LM, Haugen E, Hayden H, Albertson D, Pinkel D, Olson MV, Eichler EE (2005) Fine-scale structural variation of the human genome. Nat Genet 37:727–732

UCSC Genome Bioinformatics (2013) http://genome.ucsc.edu/ Accessed 28 Mar 2013

Venter JC, Adams MD, Myers EW et al (2001) The sequence of the human genome. Science 291:1304–1351

Voight BF, Kudaravalli S, Wen X, Pritchard JK (2006) A map of recent positive selection in the human genome. PLoS Biol 4:e72

Wang DG, Fan JB, Siao CJ, Berno A, Young P, Sapolsky R, Ghandour G, Perkins N, Winchester E, Spencer J, Kruglyak L, Stein L, Hsie L, Topaloglou T, Hubbell E, Robinson E, Mittmann M, Morris MS, Shen N, Kilburn D, Rioux J, Nusbaum C, Rozen S, Hudson TJ, Lipshutz R, Chee M, Lander ES (1998) Large-scale identification, mapping, and genotyping of single-nucleotide polymorphisms in the human genome. Science 280:1077–1082

Wang Z, Gerstein M, Snyder M (2009) RNA-seq: a revolutionary tool for transcriptomics. Nat Rev Genet 10:57–63

Wellcome Trust Case Control Consortium, Craddock N, Hurles ME, et al (2010) Genome-wide association study of CNVs in 16,000 cases of eight common diseases and 3,000 shared controls. Nature 464:713–720

Wold B, Myers RM (2008) Sequence census methods for functional genomics. Nat Methods 5:19–21

Xie X, Mikkelsen TS, Gnirke A, Lindblad-Toh K, Kellis M, Lander ES (2007) Systematic discovery of regulatory motifs in conserved regions of the human genome, including thousands of CTCF insulator sites. Proc Natl Acad Sci USA 104:7145–7150

Yang CG, Ciccolini J, Blesius A, Dahan L, Bagarry-Liegey D, Brunet C, Varoquaux A, Frances N, Marouani H, Giovanni A, Ferri-Dessens RM, Chefrour M, Favre R, Duffaud F, Seitz JF, Zanaret M, Lacarelle B, Mercier C (2011) DPD-based adaptive dosing of 5-FU in patients with head and neck cancer: impact on treatment efficacy and toxicity. Cancer Chemothe Pharmacol 67:49–56

Young VR (2002) 2001 W.O. Atwater memorial lecture and the 2001 ASNS president's lecture: human nutrient requirements: the challenge of the post-genome era. J Nutr 132:621–629

Zhao XD, Han X, Chew JL, Liu J, Chiu KP, Choo A, Orlov YL, Sung WK, Shahab A, Kuznetsov VA, Bourque G, Oh S, Ruan Y, Ng HH, Wei CL (2007) Whole-genome mapping of histone H3 Lys4 and 27 trimethylations reveals distinct genomic compartments in human embryonic stem cells. Cell Stem Cell 1:286–298

Zheng D (2008) Asymmetric histone modifications between the original and derived loci of human segmental duplications. Genome Biol 9:R105

Linkage Mapping: Localizing the Genes That Shape Human Variation

3

Laura Almasy, Mark Zlojutro Kos, and John Blangero

3.1 Conceptual Overview

Linkage analysis is a statistical technique used to localize one or more genes of interest against genotyped markers with known positions in the genome. This method is based on the co-segregation within families of loci that are located near each other on the same chromosome. Linkage between any two loci also can be measured and used to construct maps of relative genetic distances.

3.2 History of Linkage Analysis and Its Application in Human and Nonhuman Primate Genetics

To date genetic linkage analysis has proven to be a remarkably successful strategy for investigating inheritance patterns of variable traits and mapping of their underlying genes, including those for numerous human and nonhuman primate phenotypes. The approach was originally developed and applied in the early twentieth century by Thomas Hunt Morgan in his now famous "Fly Laboratory" at Columbia University. Studying the mechanisms of heredity and evolution in the fruit fly *Drosophila melanogaster*, Morgan (1910) was the first to report a connection between an observable trait and a specific chromosome, determining the locus responsible for eye color to be sex linked and located on the X chromosome based on Mendelian inheritance principles. Other sex-linked traits were discovered soon thereafter in Morgan's *Drosophila* colonies, displaying variable assortment with eye color in offspring transmissions. Morgan conceived from these results the notion of genetic linkage, with chromosomes representing linear assemblages of genes, and inferred the phenomenon of chromosome recombination, with the frequency of recombination being a function of the distance between genes on a chromosome. Based on these insights, Alfred Henry Sturtevant, then an undergraduate student working for

L. Almasy (✉)
South Texas Diabetes and Obesity Institute, University of Texas Health Science Center at San Antonio, 3463 Magic Drive, Suite 320, San Antonio, TX 78229, USA
e-mail: almasy@uthscsa.edu

M.Z. Kos
Department of Genetics, Texas Biomedical Research Institute, 7620 NW Loop 410, San Antonio, TX 78227, USA
e-mail: markz@txbiomedgenetics.org

J. Blangero
South Texas Diabetes and Obesity Institute, University of Texas at San Antonio Health Science Center Regional Academic Health Center, Harlingen, TX 78550, USA
e-mail: blangero@uthscsa.edu

R. Duggirala et al. (eds.), *Genome Mapping and Genomics in Human and Non-Human Primates*, Genome Mapping and Genomics in Animals 5, DOI 10.1007/978-3-662-46306-2_3,

Morgan, duly recognized that variations in the strength of linkage could be used as a means of mapping genes on chromosomes by determining relative spatial distances between genes. This principle was used to produce the first gene map (Sturtevant 1913), representing the relative locations of various genes on the *Drosophila* X chromosome, and would become the basis for classical gene mapping (Muller 1916), launching the field of experimental genetics.

Although linkage studies began very early in *Drosophila*, the paucity of genetic variation and poorly delineated chromosomes of mammals, including humans, limited gene mapping efforts in other species. The lone exception was the X chromosome, with its recognizable sex-linked pattern of inheritance. Hence, the first gene to be mapped in humans was for color blindness on the X chromosome by Wilson in (1911), which prompted more than 20 additional X-linked genes to be mapped in the ensuing 40 years, including those for hemophilia and Duchenne muscular dystrophy (Chakravarti and Lynn 1999). The relative locations of these genes on the X chromosome remained largely unknown, aside from those responsible for color blindness and hemophilia, which J.B.S. Haldane determined to be closely linked from their co-segregation in affected families (Bell and Haldane 1937). The first autosomal linkage in humans, on the other hand, would not be discovered until the early 1950s, between the loci for Lutheran blood groups and the ABH-secretor system using a sib-pair method (Mohr 1951, 1954). This was soon followed by a number of other identified autosomal linkages, many involving blood polymorphisms (Chalmers and Lawler 1953; Renwick and Lawler 1955), although with little knowledge of their chromosomal whereabouts. That is, it was known that these loci were linked, but not on which chromosome they resided. This situation would change markedly with the determination of the correct number of human chromosomes (Tijio and Levan 1956) and the corresponding advancement of cytogenetic techniques, revealing associations between chromosomal defects and human disease (Nowell and Hungerford 1960; Lejeune et al. 1963; Lele et al. 1963) and ultimately leading to the first unequivocal autosome assignment in 1968 between the Duffy blood group locus and a length polymorphism observed in the paracentric region of chromosome 1 based on pairwise linkage analysis (Donahue et al. 1968). However, despite these significant achievements, the assignment of traits and linkage groups remained at this time a largely opportunistic process due to a lack of detectable physical landmarks on human chromosomes.

By the 1970s, human gene mapping activity began to rapidly expand with the application of interspecies somatic cell hybrids to chromosome assignments (Weiss and Green 1967) and the advent of recombinant DNA technology, in particular the conceptualization of restriction fragment length polymorphisms (RFLPs) (Nathans and Smith 1975) and their use in mapping disease susceptibility loci (Botstein et al. 1980). Unlike classical genetic markers, such as blood group antigens and allozymes, RFLPs represented a source of highly polymorphic, anonymous DNA fragments from throughout the genome that exhibit Mendelian inheritance in pedigrees and thus are potentially informative for linkage studies and useful in the construction of comprehensive maps. Early application of these markers proved successful, revealing numerous polymorphisms around the human globin loci on chromosome 11, including ones linked to hemoglobinopathies such as sickle-cell anemia and β-thalassemia (Kan and Dozy 1978; Little et al. 1980). However, it was not until a seminal paper by Gusella et al. (1983) investigating the genetics of Huntington's disease (HD) that the value of DNA polymorphisms for linkage analysis and mapping of human disease genes was fully appreciated by molecular biologists. In that study, two large, multiplex families were typed for 12 RFLPs and tested for Mendelian model-based linkage with HD, with one marker, a probe hybridizing to *Hind*III-digested DNA, exhibiting a high likelihood of linkage to a disease locus (two-point LOD of 8.53), with no recombinants detected in either pedigree. The probe sequence was mapped to chromosome 4 using Southern blot analyses of human–mouse somatic cell hybrids, establishing for the first time the chromosomal location of the HD gene. This dramatic demonstration was closely followed by other studies successfully mapping Mendelian disorders, including cystic fibrosis on chromosome 7 (Knowlton et al. 1985; Wainwright et al. 1985; White et al. 1985) and polycystic kidney disease on chromosome 16 (Reeders et al. 1985), underscoring the clinical usefulness of linkage analysis involving RFLP markers.

It was not long after this that the first comprehensive human linkage map was constructed using 393 RFLPs and a small number of minisatellites (variable number tandem repeats) (Donis-Keller et al. 1987), spurring the development of increasingly detailed maps. With the later identification of microsatellite polymorphisms (Weber and May 1989) and the parallel improvement of molecular genetic procedures, most notably the polymerase chain reaction (PCR) (Saiki et al. 1985), genetic marker number, and polymorphic content increased substantially. This allowed for more extensive linkage scanning with high-throughput genotyping, which by the 1990s resulted in a series of genome-wide maps of approximately 1 cM resolution (Weissenbach 1993; Buetow et al. 1994; Gyapay et al. 1994; Murray et al. 1994). To better utilize the greater information content available from these tightly linked markers, multipoint methods (Lathrop et al. 1984) have been widely adopted for traditional model-dependent linkage analysis (O'Connell and Weeks 1995).

Although the model-dependent or "parametric" approach to maximum likelihood linkage analysis (Morton 1955) has had tremendous historical success in mapping genes underlying monogenic diseases with clear Mendelian inheritance patterns (as outlined above), many researchers questioned its usefulness for complex traits that do not follow a simple single-gene model. This genetic complexity may arise in a disease for a number of reasons, including heterogeneity in etiology, oligogenic inheritance, epistasis, and gene–environment interactions. Thus, the parametric method of assuming a known genetic model for linkage loses much of its power and may produce erroneous results (Risch et al. 1989; Risch and Guiffra 1992). In recognition of the need for linkage analysis that relies less completely on genetic model specification, "nonparametric" methods were developed that weakened one or more assumptions of the fully specified model, including parameters for the disease gene allele frequencies and penetrance functions, either by considering only affected individuals or by reparametrizing the genetic model. The first nonparametric method to be developed in humans was a sib-pair linkage approach, originally described in a seminal paper by Penrose in 1935 and later expanded by others (Risch 1990a, b; Haseman and Elston 1972; Weeks and Lange 1992; Kruglyak and Lander 1995). This approach has been increasingly applied in gene mapping efforts of complex disease in recent decades, with the first major breakthrough occurring for insulin-dependent (type 1) diabetes in 1994, producing evidence for three susceptibility loci that are in addition to previously identified candidate genes for HLA and insulin (Davies et al. 1994). This success was later extended to noninsulin-dependent (type 2) diabetes (Hanis et al. 1996), for which a susceptibility locus was identified in Mexican-American affected sibling pairs within the terminal portion of chromosome 2q (LOD of 3.2), designated *NIDDM1*. This large chromosomal region (10–20 cM) was later narrowed (7 cM) in follow-up linkage analysis (Cox et al. 1999), ultimately leading to the identification of *CALPAIN 10* as a potential susceptibility gene in the region based on both linkage and association evidence (Horikawa et al. 2000). Other successful applications of the relative pair approach to linkage analysis at this time included prostate cancer (Smith et al. 1996), end stage renal disease (Bowden et al. 1997), asthma (CSGA 1997), febrile convulsions (Johnson et al. 1998), and others.

More recently, however, pair-based linkage methods have been criticized by some as being less powerful in localizing genes than methods that utilize larger configurations of relatives (Williams et al. 1997; Alcais and Abel 2000; Blangero et al. 2000). A large number of papers on pedigree-based linkage methods that do not depend on a penetrance model have emerged (e.g., the variance components approach), in particular with relation to quantitative traits (Amos 1994; Amos et al. 1996; Almasy and Blangero 1998; de Andrade et al. 1999; Blangero et al. 2001). With the transition away from classical analysis of monogenic disorders to a new emphasis on the genetic basis of common complex disorders, researchers have become increasingly interested in measurable quantitative variation of intermediary phenotypes closely related to disease risk as a means of examining physiological pathways that are more proximate to gene action than the more complex (and statistically less informative) dichotomous disease outcome itself (Sing et al. 1996; Blangero et al. 2000). The first localization of a human quantitative trait locus (QTL) from a genome-wide scan was a linkage peak for obesity-related traits in Mexican-American families on chromosome 2p (Comuzzie et al. 1997), which has since been replicated in French (Hager et al. 1998) and African-American populations (Rotimi et al. 1999). Other well-replicated human QTL linkages include loci influencing variation in triglyceride levels on chromosome 15 (Arnett et al. 2004; Austin et al. 2003; Coon et al. 2001; Duggirala et al. 2000), body mass index on chromosome 3 (Kissebah et al. 2000; Wu et al. 2002; Luke et al. 2003), and reading disability on chromosome 6p (Cardon et al. 1994; Fisher et al. 1999; Grigorenko et al. 1997, 2000) to name just a few.

For some of the replicated QTL linkages, researchers have successfully identified specific associated genes under the linkage peaks. For instance, in a study by Duggirala et al. (1999), a QTL influencing type 2 diabetes and its age of onset was localized to chromosome 10q using variance component-based linkage analysis for a liability threshold model. This linkage peak was later confirmed (Reynisdottir et al. 2003) and eventually the gene *TCF7L2* from the linked region was identified as strongly associated with type 2 diabetes by several research groups (Grant et al. 2006; Lehman et al. 2007; Scott et al. 2007; Sladek et al. 2007; Tong et al. 2009. Another example are QTL signals obtained for brain oscillation measurements of alcohol-dependent individuals, with linkage peaks emerging on chromosomes 4p12 and 7q31-34 (Porjesz et al. 2002; Jones et al. 2004). From these two identified regions, significant associations have been found between neurotransmitter receptors GABRA2, GRM8, and CHRM2 and neuroelectrical measures (Porjesz et al. 2002; Jones et al. 2006; Chen et al. 2009), as well as associations with alcohol dependence in various study samples (Covault et al. 2004; Edenberg et al. 2004; Wang et al. 2004; Lappalainen et al. 2005; Luo et al. 2005; Drgon et al. 2006; Enoch et al. 2006; Fehr et al. 2006; Soyka et al. 2008; Chen et al. 2009), underscoring the effectiveness of intermediate, quantitative phenotypes (or *endophenotypes*) in dissecting the genetic underpinnings of complex clinical disorders (Almasy 2003; Gottesman and Gould 2003).

Although linkage analysis has played a long, important role in the genetic research of nonhuman species, especially *Drosophila* (Morgan 1910; Rubin and Lewis 2000) and laboratory mice (Haldane et al. 1915; Snell 1941; Lyon and Searle 1989), its application in nonhuman primates has been relatively limited in scale and impact, despite the obvious benefits that these species can have on biomedical research due to their close evolutionary relationship to humans. The first published study of genetic linkage in nonhuman primates was not until 1973, looking at polymorphisms in carbonic anhydrase genes of pig-tailed macaques, *Macaca nemestrina* (DeSimone et al. 1973). Most of the primate linkage studies that followed examined variation in immune responses and antibody reactions, mainly involving rhesus macaques (*Macaca mulatta*), establishing linkages to the major histocompatibility complex (MHC) gene cluster (Dorf et al. 1975; Maurer et al. 1979; Rogers et al. 2009), however, without assignment to specific chromosomes because of the lack of molecular genetic data. By the 1980s, genetic linkage analysis in nonhuman primates remained limited to classical polymorphisms, focusing on protein or isozyme variants and blood group or other immunological markers (e.g., Ferrell et al. 1985). Molecular markers began to be employed in primate linkage studies during the 1990s, with researchers using RFLPs and highly diverse microsatellites (Morin and Woodruff 1992; Inoue and Takenaka 1993; Rogers and Kidd 1993; Deka et al. 1994; Kayser et al. 1995; Rogers et al. 1995), leading to the construction of the first linkage map of a nonhuman primate, the baboon (*Papio hamadryas*), based on marker-to-marker microsatellite linkage analysis (Rogers et al. 2000). This linkage map has proven to be a valuable resource, allowing detailed analysis of locus order and recombination distances in baboon chromosomes and localization of QTLs that influence phenotypic variation related to human health and disease (Mahaney et al. 1999; Comuzzie et al. 2001; Martin et al. 2001; Kammerer et al. 2002; Rainwater et al. 2003; Havill et al. 2005). Since this major breakthrough, linkage maps of other nonhuman primates have been generated (Rogers et al. 2006; Jasinska et al. 2007), which will greatly benefit continued research of disease phenotypes in these species and provide important comparative mapping data as the whole genome sequences of rhesus macaque, chimpanzee, and other nonhuman primate species become available in the near future.

3.3 Marker-to-Marker Linkage

The most straightforward type of analysis used in linkage studies is that between genotyped markers. Before the development of high-resolution maps that enabled comprehensive, genome-wide scanning in the 1990s, linkage was used to find the chromosomal locations of newly identified genetic markers relative to the known positions of a limited number of genetic loci. Imagine that we have a family with 10 individuals (numbered 1–10) in which we have genotypes for two markers, a letter locus with alleles *A–H* and a number locus with alleles 1–8 (Fig. 3.1a). We can use what we observe about the transmission of alleles from the grandparents (individuals 1–4) to the parents (individuals 5 and 6) to "set phase" for the genotypic data. Setting phase essentially involves asking: If these markers are linked, which alleles are traveling together on the same chromosome in this family? Individual 5 got the *A* allele at the letter locus and the 1 allele at the number locus from his father and he received the *D* and 3 alleles from his mother. So if the loci are linked, in individual 5 the *A* is paired with the 1 and the *D* is paired with the 3 as transmitted from his parents. By the same logic, we can set phase in the mother (individual 6)—the *F* allele is with the 5 and the *G* with the 7. Based on the inferred phase, we can then test the hypothesis of linkage by examining the genotypes of the children. If the loci are in fact linked, then the two markers will be on the same chromosome, with the allelic combinations observed in the parents (*A* and 1, *D* and 3, *F* and 5, and *G* and 7) transmitted together to the offspring more often than expected by chance. If, on the other hand, the two markers are not linked and are thus located on different chromosomes (or far apart on the same chromosome), then based on Mendel's law of independent assortment, we would expect, for instance, the *A* allele at the letter locus to appear with equal frequency with the 1 and 3 alleles at the number locus among the children in this pedigree. Therefore, testing linkage is a matter of testing for violations in this expectation of independent assortment.

For loci that are linked, any new combination of alleles appearing in a child (e.g., *A* with 3, *D* with 1, *F* with 7, or *G* with 5) would be the result of genetic crossover, or recombination, between the loci in a parent during gametogenesis. The closer together two linked loci are, the lower the probability of a recombination event occurring. Thus, the frequency of recombination (θ) provides an indirect way of estimating the distance between two linked loci, as was originally deduced by A.H. Sturtevant.

Returning to Fig. 3.1a, children 7, 8, and 10 have genotypes that are consistent with the phase set in their parents and would require no recombinations. They have the *A* allele with the 1, *D* with 3, *F* with 5, and *G* with 7. Child 9, on the other hand, has inherited the *D* allele at the letter locus and the 1 allele at the number locus from his father. This particular arrangement of genotypes is not observed among the phased genotypes in the father and would require a recombination between the letter locus and the number locus. In Fig. 3.1b, the paternally

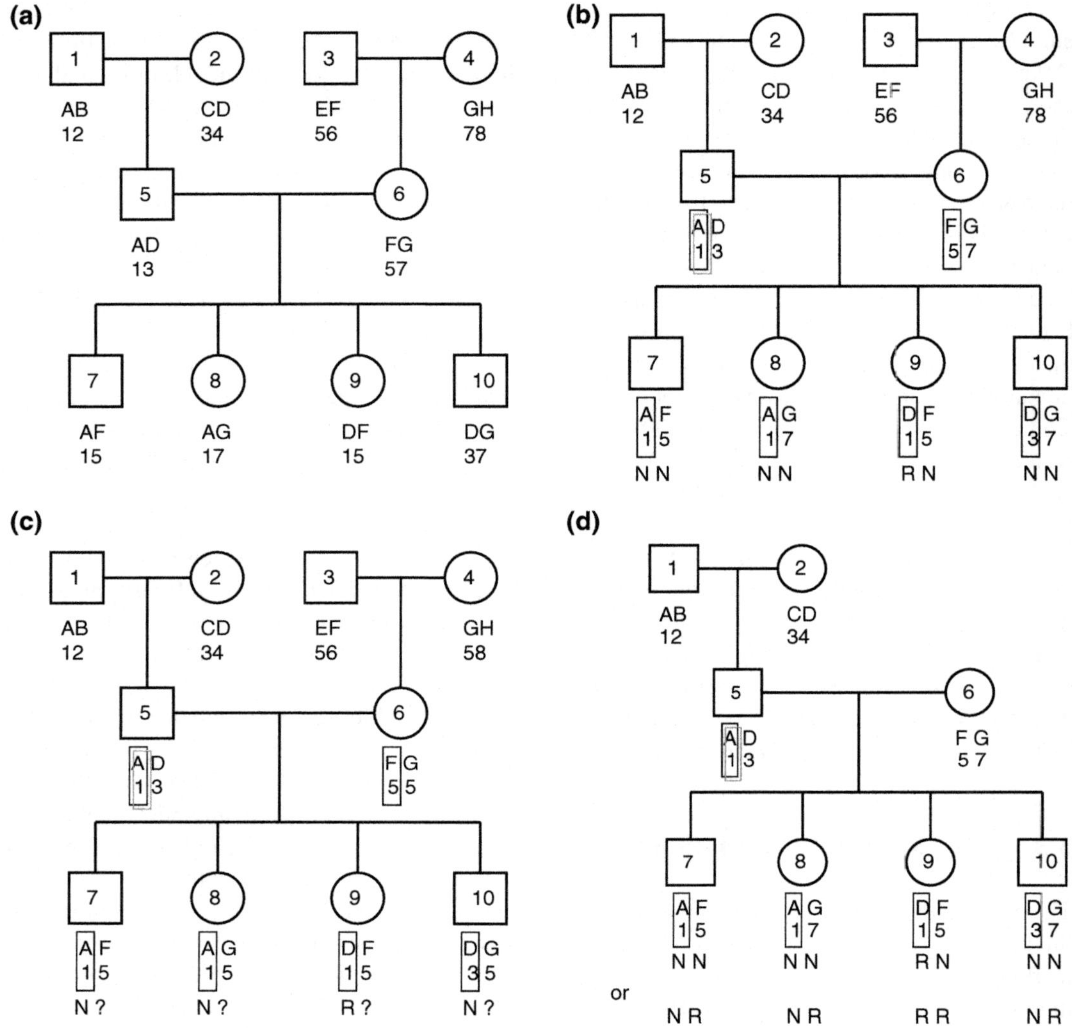

Fig. 3.1 Linkage between two genotyped marker loci. **a** Genotypes at two loci, a letter locus and a number locus. **b** Setting phase and counting recombinant (*R*) and nonrecombinant (*N*) transmissions. **c** Uninformative meioses due to a homozygous parent. **d** When phase cannot be set with certainty

inherited chromosomes are enclosed within boxes and the recombinant (*R*) and nonrecombinant (*N*) chromosomes are indicated for each of the children. The test for linkage and the estimation of genetic distance between the number locus and the letter locus is based on counting the number of recombinations that would have to occur to account for the observed arrangement of genotypes if the loci were linked. In this case, out of eight informative meioses (the transmissions from the father to each of the four children and from the mother to each of the four children), we infer only one recombination. If the two loci were unlinked and the alleles were assorting randomly, we would expect to observe four recombinations in eight meioses (i.e., 50 % recombination).

To assess the statistical support for linkage versus no linkage, a likelihood ratio test is used. The likelihood of observing *R* recombinations among *N* informative meioses given a certain value of θ is:

$$L(\theta) = \theta^R(1-\theta)^{N-R} \quad (3.1)$$

For the likelihood ratio test, we compare the likelihood for a value of θ we wish to test to the likelihood under random assortment, when $\theta = ½$. Traditionally, the odds ratio is expressed on a logarithmic scale (referred to as an LOD) such that a LOD score of 3 represents 1,000 times greater support for the hypothesis of linkage versus no linkage:

$$Z(\theta) = \log[L(\theta)/L(1/2)] \quad (3.2)$$

Of course, when $\theta = ½$, $1 - \theta$ is also ½ and Eq. 3.1 above simplifies to $½^N$. Putting Eqs. 3.1 and 3.2 together, we get:

$$Z(\theta) = \log[\theta^R(1-\theta)^{N-R}/½^N] \quad (3.3)$$

A positive LOD score indicates support for linkage and an LOD > 3 is generally considered significant in a genome-wide scan in humans. This threshold is based on the length of the human genome and calculations regarding the effective number of independent tests given that adjacent markers are correlated. Some have suggested that a threshold of 3.3 or 3.6 should be used in some studies (Lander and Kruglyak 1995); however, this is based on assuming an infinitely dense map of genotyped markers. In this type of model-based linkage, both marker-to-marker and for disease traits (described below), the LOD score may be negative, indicating greater support for the hypothesis of no linkage than for the hypothesis of linkage. This can be used for exclusion mapping, in which regions are "ruled out." A threshold of LOD < −2, indicating 100 times more support for the hypothesis of no linkage, is generally used for exclusion mapping in human studies.

The configuration of available individuals and the observed genotypic data within a pedigree can limit the linkage information used in analysis. For instance, what if the mother in our example had been homozygous for one of the genotyped markers, say the number locus? In this case (Fig. 3.1c), the meioses from the mother to the four children become uninformative. All of the children obligately must inherit the 5 allele from the mother, which prevents us from determining whether or not any recombinations occurred. We can still conduct our linkage analyses, but we can only score the four meioses from the father, reducing the information available and the power of this family for our linkage study.

Alternatively, what if some of the grandparents are unavailable and we cannot set phase in one of the parents? Then the linkage is conducted taking into account both possible phases (Fig. 3.1d). We cannot tell whether the mother inherited the *F* and the 5 from the same parent or the *F* and the 7. If the *F* and the 5 were inherited together, the recombination pattern is the same as before and we observe one recombinant and seven nonrecombinants in eight meioses. On the other hand, if the *F* and the 7 were inherited together, then the observed genotypes of all four children would require recombinations on the maternally transmitted chromosome and we would have a total of five recombinations and three nonrecombinations. To accommodate both scenarios, the numerator of the likelihood ratio test becomes a weighted average of the two possibilities. Without any information from the mother's parents or siblings, these two scenarios are equally likely, and we have:

$$L(\theta) = ½\theta^1(1-\theta)^7 + ½\theta^5(1-\theta)^3 \quad (3.4)$$

And lastly, what if there is more than one recombination in a given interval? If there has been an odd number of recombinations, we will observe a recombinant genotype in the offspring. But if there has been an even number, the resulting genotype cannot be distinguished from one that underwent zero recombinations.

3.3.1 Converting Recombination Fraction to Distance in cM

As noted above, the observed proportion of recombinations between two linked loci can be used to derive a genetic distance between them measured in centiMorgans (cM). Two widely used mapping functions for this were proposed by Kosambi and by Haldane (Ott 1999).

Haldane's formula

$$x = -\tfrac{1}{2}\ln(1 - 2\theta) \quad (3.5)$$

Kosambi's formula

$$x = \tfrac{1}{4}\ln\left((1 + 2\theta)/(1 - 2\theta)\right) \quad (3.6)$$

In each case, x represents the distance in morgans (1/100th the distance in cM). Using either formula, for small values of θ (i.e., short distances in which multiple recombinations in an interval are unlikely), $x = \theta$. Over longer distances, both formulas take into account the possibility of multiple recombinations between genetic markers. Additionally, Kosambi's formula allows for interference, the idea that crossing over of DNA between chromosomes takes up a certain amount of physical space such that it is unlikely or physically impossible for two recombinations to occur within a very short distance of each other.

Genetic measures of distance also have rough approximations on the scale of physical distances in base pairs along a chromosome. One cM is approximately equal to 1,000 kb. However, the rate of recombination is not constant across the genome (Yu et al. 2001). In regions with higher crossover rates, so-called recombination hotspots or "jungles", 1 cM will correspond to <1,000 kb.

3.3.2 Mendelian Model-Based Linkage

The same model of linkage described above for two genotyped markers has also been widely applied in analyses of simple Mendelian traits, beginning with T.H. Morgan's genetic research on *Drosophila* during the early twentieth century. However, the genotypes of one or both trait loci cannot be observed directly and must be inferred from the observed phenotypes of the family members. For fully penetrant traits with a clear pattern of inheritance, this is relatively simple. The terminology for this type of analysis comes largely from medical genetics, so let us label the two alleles at the trait locus *D* (representing the disease allele) and *d*. If the trait is autosomal dominant and fully penetrant, such as achondroplasia, then unaffected individuals are guaranteed to have the genotype dd and affected individuals carry at least one *D* allele. Because most simple Mendelian diseases are relatively rare, we assume the frequency of the *D* allele is low and thus it is statistically unlikely that an individual is homozygous DD. Under these assumptions, we can easily take the pedigree in Fig. 3.2a, where affected individuals are shaded and unaffected ones are not, and write in inferred genotypes at the trait locus as in Fig. 3.2b. Setting phase as we did above for marker-to-marker linkage analysis, we can then count recombinations between the loci (Fig. 3.2c). In this case, as in the example in Fig. 3.1c, only the transmissions from the father (individual 5) to his offspring can be scored because the mother (individual 6) is homozygous for the trait locus and therefore uninformative.

The same procedure can be used for a simple, fully penetrant autosomal recessive trait, such as cystic fibrosis. In this case, affected individuals obligately have genotype DD. If their parents are unaffected, they must have genotype Dd, as they have passed on a *D* allele to at least one of their offspring. Often in this case, we do not know which two of the four grandparents contributed the *D* alleles and we must allow for multiple possible phases, as we did in Fig. 3.1d. Genotypes at the trait locus can also be easily specified for highly penetrant X-linked traits such as hemophilia or Y-linked traits such as hairy ears.

Such model-based linkage becomes more difficult when the mode of inheritance of the trait is unclear, when the trait is not completely penetrant, or when there are phenocopies. Penetrance is defined as the probability of being affected, given one's genotype. For a simple dominant trait, the three penetrances for the three genotypes are f(DD) = 1, f(Dd) = 1, and f (dd) = 0, though again we would expect the genotype DD to be very rare. For a simple Mendelian recessive model, the penetrances are f (DD) = 1, f(Dd) = 0, and f(dd) = 0. Incomplete penetrance describes a situation where an individual has a disease-causing genotype, but is not affected. In this case, the penetrance of Dd, under a dominant model, or DD, under a recessive

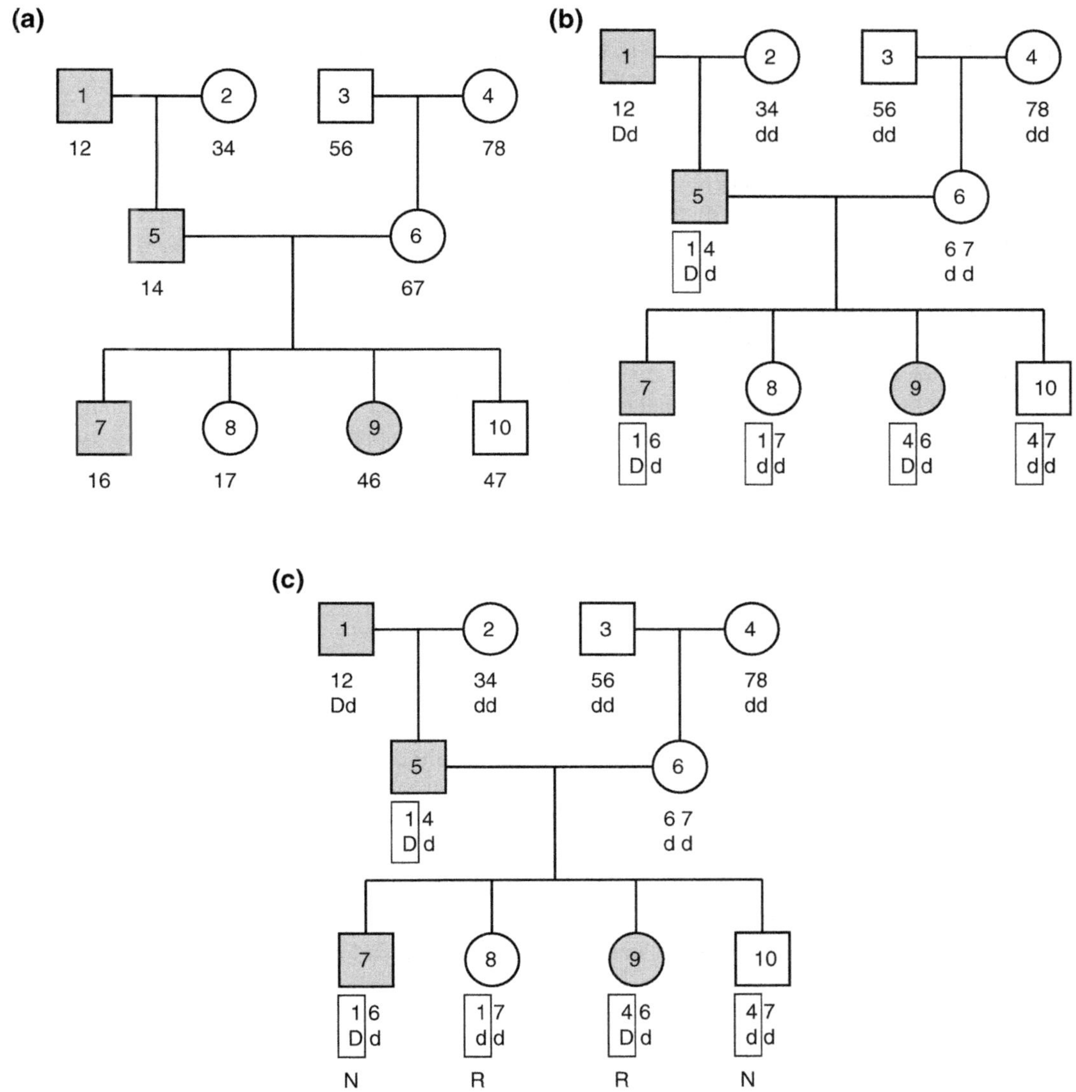

Fig. 3.2 Linkage between a trait locus and a genotyped marker. **a** Genotypes at a marker locus and disease status (*shaded* = affected, *unshaded* = unaffected). **b** Inferred genotypes at the trait locus and setting phase. **c** Counting recombinations

model, would be <1. A phenocopy is a case where an individual is affected, even though he or she does not have the risk genotype, meaning a penetrance >0 for genotype dd. When there is incomplete penetrance or phenocopies, we can no longer deduce a person's genotype at the trait locus given that person's trait phenotype, and the calculations must be conducted weighting across possible genotypes as well as possible phases. This quickly becomes more complex than is feasible to do by hand, but is easily accomplished with existing software packages, such as LINKAGE (Lathrop and Lalouel 1984; Lathrop et al. 1984) or GENEHUNTER (Kruglyak et al. 1996). The one key is that the user must be able to specify the presumed allele frequencies at the trait locus and the penetrances for each possible genotype at the trait locus. This is easily done for traits such as cystic fibrosis, hemophilia, or hairy ears but almost impossible for traits such as heart disease or schizophrenia or for quantitative traits such as height or head circumference. In theory, model-based linkage can be used for quantitative traits, but it requires specifying the mean trait values for each genotype, rather than the penetrance. Some model misspecification can be tolerated in linkage analysis (Ott 1977; Clerget-Darpoux et al. 1986), but a poor fit between the model parameters and the true mode of inheritance can lead to false negatives, even to the point of excluding linkage in regions that contain true loci (Risch and Giuffra 1992; MacLean et al. 1993). Power for penetrance model-based linkage analyses is dependent on the number of informative meioses and correct specification of the penetrance model. For an in-depth discussion of model-based linkage analyses, refer to the classic textbook by Ott (1999).

3.4 Identity by Descent

Because of the difficulties in specifying a simple genetic model for many traits, new methods of linkage analysis were developed that did not rely on specifying the properties of the locus being sought. These methods are commonly referred to as nonparametric or penetrance model-free and they rely on identity by descent sharing of alleles among relatives. For two alleles to be considered identical by descent (IBD) they must come from the same ancestral source and be copies of the same ancestral chromosome. Humans and nonhuman primates have two copies of each chromosome, one inherited from their mother and one from their father. They thus carry two alleles at any given locus and a parent may pass along either of their alleles to each offspring. If two siblings inherited copies of the same maternal allele, they are IBD for this allele. They may or may not also be IBD for the alleles they inherited from their father. So a pair of relatives may share zero, one, or two alleles IBD.

The expected IBD sharing for a relative pair is a function of their pedigree relationship. First degree relatives are expected to share half of their alleles, second degree relatives one-quarter, third degree relatives one–eighth, and so on, with the expected IBD sharing being $\frac{1}{2}^N$, where N is the degree of relationship. In general, the expected IBD sharing is also the proportion of their DNA that a relative pair would be expected to share on average. At any given marker, a sibling pair may have IBD of 0, 0.5, or 1, but their IBD averaged across many such markers across the genome will be approximately 0.5. Conversely at any given marker, while a particular sibling pair may have IBD of 0, 0.5, or 1, the average IBD at this marker across a large group of sibling pairs should be approximately 0.5. Many types of relative pairs cannot share both alleles IBD because they are related through only one parent, for example half siblings, aunts and uncles, nieces and nephews, or cousins. However, the same principle holds. They can share zero or one allele and on average across many markers or across many relative pairs, their expected IBD is as given in Table 3.1.

Two exceptions to this general pattern are parent–child pairs and identical twin pairs. In both of these cases, the IBD is invariant. Except when there is inbreeding, parents and children, by definition, share exactly one allele IBD and identical twins must share both alleles IBD. As such, samples with only these types of relative pairs are not informative for linkage. Their IBD will be the same at every marker genotyped.

By definition, a pair of individuals who are not related share no alleles IBD, even if they have identical genotypes at a locus. Unrelated individuals have no common ancestors (unless we are considering evolutionary time scales) and thus their alleles cannot be copies of the same ancestral chromosome.

IBD can be estimated for a particular genotyped marker or for a chromosomal location using multiple genotyped markers in the region. Such "multipoint" IBD estimation is more computationally intensive, but also more informative.

Support for the hypothesis of linkage in IBD-based methods can be expressed as an LOD score, in which case it is interpreted as described above, with an LOD > 3 being generally regarded as genome-wide significant for human

Table 3.1 Expected IBD sharing

Degree of relationship	Types of relative pairs	Pr (share 0)	Pr (share 1)	Pr (share 2)	E (IBD)
–	Identical twins	0	0	1	1
1	Parent–child	0	1	0	0.5
1	Siblings, including fraternal twins	0.25	0.50	0.25	0.5
2	Half-sibling; avuncular; grandparent–grandchild	0.5	0.5	0	0.25
3	First cousin; half avuncular	0.75	0.25	0	0.125
4	Half first cousin; first cousin once removed	0.875	0.125	0	0.0625

studies. However, it is not possible for the LOD score to be negative in IBD-based linkage analyses without assuming an effect size for the trait locus and IBD-based exclusion mapping is not generally employed. Linkage evidence from IBD-based approaches may also be given as an NPL (nonparametric linkage) score. This NPL score is on a slightly different scale, but can easily be converted using the following relationship (Abreu et al. 1999):

$$(\text{NPL})^2/2^*\ln(10) = \text{LOD} \qquad (3.7)$$

An NPL of 3.7 is roughly equivalent to an LOD score of 3.

3.5 Concordant and Discordant Sibling Pairs

This type of penetrance model-free linkage analysis is intuitively simple. In the region of a gene influencing the trait of interest, relatives who are phenotypically concordant should share more alleles IBD than expected and relatives who are discordant should share fewer alleles IBD than expected. This is true regardless of the underlying model of gene action. The simplest linkage test in this framework is assessing whether a set of concordant sibling pairs have a mean IBD > 0.5 at a given chromosomal location. Variations on this involve assessing the proportion of sibling pairs sharing 2 alleles IBD and the full distribution of the proportions sharing each of 0, 1, and 2 alleles IBD. The relative power of each of these tests depends on the underlying disease model, but the test of mean IBD > 0.5 optimizes power over the widest range of underlying models of gene action (Blackwelder and Elston 1985; Knapp et al. 1994). Maximum likelihood-based affected sibling pair analyses have also been developed (Risch 1990a, b).

In theory, similar analyses can be conducted with a sample of discordant relative pairs to identify regions of the genome where they share fewer alleles IBD than expected. In practice, few studies consist solely of discordant relative pairs. In cases where investigators have set out to collect a sample of discordant sibling pairs, it has been observed that this selection scheme enriches the sample for families with pedigree misspecifications (Neale et al. 2002). This illustrates the important point that linkage analyses are crucially dependent on correct specification of the pedigree structure. If, for example, the discordant siblings are in fact half siblings (sharing one parent) rather than full siblings (sharing both parents), their expected IBD is then 0.25. If a sample of discordant sibling pairs contains any substantial fraction of half siblings, the group will, on average, have an IBD < 0.5 at many locations in the genome that have nothing to do with the trait of interest, simply because the expected IBD against which we are comparing observed IBDs is based on false assumptions regarding the pedigree structures. This is one reason that standard practice for linkage studies includes verifying that the specified pedigree structure is consistent with observed genotypes prior to analysis. This can be done using a variety of programs, such as PREST (Sun et al. 2002). The specific cases of half siblings wrongly presumed to be full siblings in a discordant pair study and identical twins treated as regular full siblings in a concordant pair study are particularly dangerous because the pedigree misspecification introduces a consistent bias toward inflating linkage evidence. Studies using Mendelian penetrance-based models, those that include both concordant and discordant pairs, and quantitative trait studies discussed below are also dependent on correct specification of the pedigree relationships; however, misspecification in these cases generally does not introduce a consistent bias for or against the detection of linkage.

Concordant and discordant sibling pair analyses are implemented in programs such as SAGE (S.A.G.E. 2009), GENEHUNTER (Kruglyak and Lander 1995), ASPEX (Hinds and Risch 1996), and MERLIN (Abecasis et al. 2002). Power for concordant and discordant sibling pair linkage analyses depends on the genotype-specific relative risk conferred by the locus and the sample size. Note that the relative risk, the unit of effect size for affected and discordant pair analyses, is partly a function of the prevalence of a trait. Because of this, concordant and discordant pair analyses are more successful for rarer traits.

3.6 Quantitative Traits

Penetrance model-free quantitative trait linkage analysis relies on essentially the same idea as linkage analysis of concordant and discordant sibling pairs, but instead uses quantitative measurements of phenotypes rather than categorizing them into concordant or discordant groups, such as disease status. Essentially, the analysis tests whether relatives who are more alike phenotypically also share more alleles IBD in a particular chromosomal region. The simplest such test is the Haseman–Elston method, in which the squared difference between the trait values for pairs of siblings is regressed against the proportion of alleles shared IBD at each location being tested (Haseman and Elston 1972; Sun et al. 2002). In the presence of linkage, there should be a negative slope to the regression coefficient. Pairs with the smallest difference in trait values should share the most alleles IBD (Fig. 3.3a). In the absence of linkage, there should be no correlation between trait differences among sibling pairs and IBD allele sharing and the regression coefficient would not be different from zero (Fig. 3.3b). The Haseman–Elston linkage method, as well as revisions to it that are

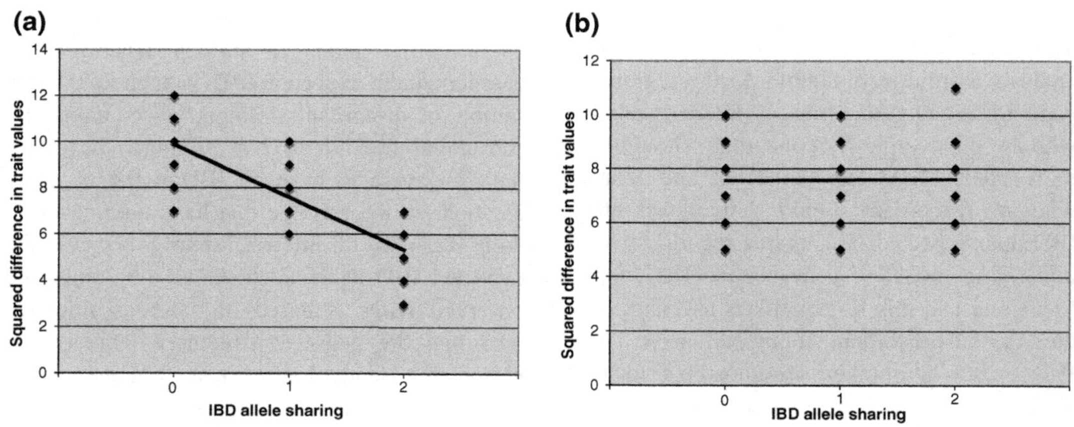

Fig. 3.3 The basic Haseman–Elston linkage approach. **a** In the presence of linkage between a locus influencing the trait and a genotyped marker, pairs with larger differences in trait values should share fewer alleles IBD at the marker, resulting in a negative slope to the regression line. **b** In the absence of linkage, the difference in trait values between siblings should not differ by their IBD at the genotyped marker

discussed below, are implemented in the program SAGE (S.A.G.E. 2009).

The variance component linkage approach operates on the same principal as the Haseman–Elston, but accommodates pedigrees of arbitrary size and complexity and typically uses maximum likelihood rather than regression. At its most basic, the variance component approach seeks to identify the factors that contribute to variation in a trait and quantify their contributions to the phenotypic variance observed among individuals. This is done by decomposing the observed variance in the trait into variance attributable to genes or QTLs (quantitative trait loci) in a region of linkage (σ_q^2), additive effects of unspecified genes elsewhere in the genome (σ_a^2), and environmental factors (σ_e^2). Rather than the difference between a pair of relatives as in the original Haseman–Elston, the variance component approach models the contrasts among individuals using a matrix of the phenotypic covariances among all pairs of relatives in a pedigree (Ω):

$$\Omega = \Pi\,\sigma_q^2 + 2\Phi\,\sigma_a^2 + \mathrm{I}\sigma_e^2 \qquad (3.8)$$

Each of the potential genetic and environmental components of variance (σ_q^2, σ_a^2, and σ_e^2) is structured by a matrix that describes the correlations among individuals that would be expected due to that component. In the simple linkage model in Eq. 3.8, Π is a matrix of observed IBD allele sharing in the chromosomal region where linkage is being assessed, Φ is a matrix of kinship coefficients among the pairs of individuals derived from the pedigree relationships, and I is an identity matrix, implying an unshared environmental component that is unique to each individual. Maximum likelihood methods are used to estimate the components of variance and linkage is tested by comparing the likelihood of a model where the QTL-specific variance, σ_q^2, is estimated to a model in which σ_q^2 is constrained to zero, to test whether the variance due to a QTL in this region is >0. This likelihood ratio test provides an LOD score that can be interpreted on the traditional scale, with an LOD of 3 representing 1,000 times more support for the hypothesis of linkage ($\sigma_q^2 > 0$) versus no linkage ($\sigma_q^2 = 0$). The simple model in Eq. 3.8 above is easily expanded to incorporate multiple loci, shared environmental factors, and gene–gene or gene–environment interaction (Blangero et al. 2000, 2001).

This same basic variance decomposition model also underlies the regression-based "revised Haseman-Elston" approach (Chen et al. 2004; Wang and Elston 2005) and has been implemented in Markov chain Monte Carlo frameworks where the number of QTLs genome-wide is estimated along with the variance attributable to each (Daw et al. 2003; Heath 1997). Another regression-based quantitative trait linkage approach reverses this basic model to estimate IBD allele sharing as a function of covariance in trait values among relatives (Sham et al. 2002). Variance component methods are implemented in SOLAR (Almasy and Blangero 1998), ACT (Amos et al. 1996), and GENEHUNTER (Kruglyak and Lander 1995). The power of IBD-based quantitative trait linkage analysis depends on the proportion of variance due to the QTL, the sample size, and the family configuration with larger families providing more power per person sampled. If sample size is held constant, power is maximized by concentrating these individuals into as few families as possible (Blangero et al. 2003).

3.7 Special Challenges for Linkage Analysis in Nonhuman Primates

Linkage studies in nonhuman primates face a number of extra challenges. First, linkage methods generally assume that the relationships among individuals are known and require that a pedigree structure be specified. Often, this information may not be available for nonhuman primate studies. Sometimes partial information is available, for example, when colony records make it possible to identify the mother of all individuals born in a facility. If there are a limited number of potential fathers and DNA is available for them, paternity testing may be performed to fill in the missing pedigree information. Or when it is known that individuals must be either full or half siblings because they share a mother, estimates of their IBD for many genotyped markers spread throughout the genome allow them to be correctly classified for sibling pair-based analyses. Complicating either of these is the possibility of inbreeding. Most pedigree checking programs are not capable of dealing with situations where the potential father shares more alleles than expected with a putative child because the father is genetically related to the mother or with situations where full or half siblings share more alleles than expected because their mother and father are related. In cases where there is substantial inbreeding or where pedigree relationships are unknown, one approach is to use IBD-based methods and estimate the null distribution of allele sharing empirically, averaging IBD across many genotyped markers, rather than relying on a kinship matrix derived from known pedigree relationships.

Any kind of genome scanning requires a minimum of hundreds of STR markers or thousands of SNPs of known chromosomal location and multipoint linkage analyses also require a genetic map where the order of the markers is specified along with the distances between them in cM. Such high-resolution maps and marker sets have been available for humans for over a decade, but exist for only a subset of the nonhuman primate species used in research. Linkage maps have been developed for the baboon (Cox et al. 2006; Rogers et al. 2000), rhesus macaque (Rogers et al. 2006), and vervet (Jasinska et al. 2007). Furthermore, a related issue is that the significance thresholds for genome-wide linkage screens discussed above are tailored to human studies. The cutoff of an LOD > 3 incorporates within it the expected number of independent tests in a genome-wide scan given the length of the human genome. The cutoff for genome-wide significance will differ for linkage studies of other species, generally being lower, as the total genome length in cM is shorter for the nonhuman primate species with available linkage maps.

3.8 Linkage and Linkage Disequilibrium are Different

Linkage disequilibrium (described in detail in the chapter on association analyses Hanson and Malhotra, this volume) is an association of two alleles at different loci that is present at a population level. This occurs when the frequency of a haplotype differs from the product of the two allele frequencies. Essentially, this requires not only that the loci be linked, but that the "phase" described above for marker-to-marker linkage be nonrandom. In other words, that one phase is more common across families. In Fig. 3.4, we can set phase for the transmissions from the heterozygous father (individual 3) to his offspring (individuals 5–8). In the family on the left, the *A* allele and the 1 allele are transmitted together as are the *D* allele and the 3 allele. In contrast, in the family on the right, the *A* allele is with the 2 and the *D* allele is with the 4. In other families, the *A* is inherited with the 3 or 4 and the *D* allele with the 1 or 2. There is no linkage disequilibrium between the letter locus and the number locus. The frequency of the A1 haplotype is merely the product of the frequency of the *A* allele and the frequency of the 1 allele. Knowing what allele someone has at the letter locus does not predict their allele at the number locus. However, there is still linkage information. We can still score these families for recombinant and nonrecombinant meioses, accounting for the differences between families by the fact that we set phase separately for each family, and it is still the case that individuals who are IBD at the letter locus will also be IBD at the number locus unless a recombination has occurred between the two loci. Linkage analyses do not require the presence of linkage disequilibrium between the trait locus and a genotyped marker.

3.9 Linkage Analysis in the WGS Era

With the advent of next generation sequencing technologies that are rapidly making whole genome sequence (WGS) studies practically and

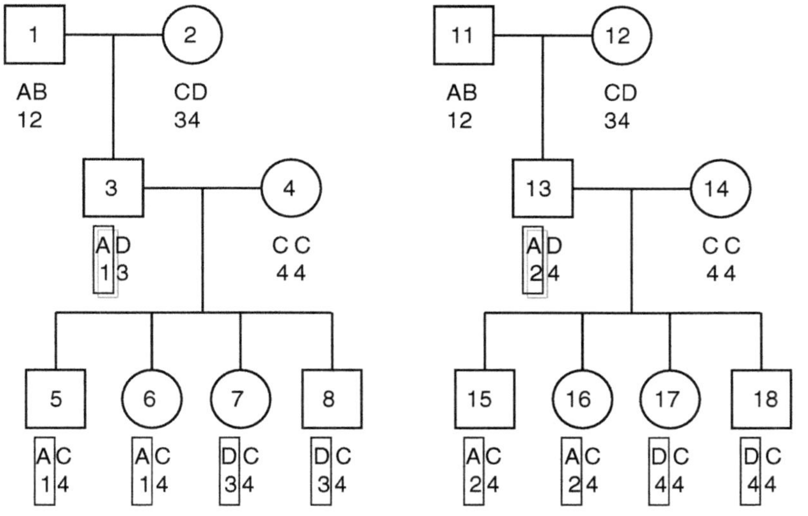

Fig. 3.4 Linkage does not require association or disequilibrium between loci because setting phase happens separately in each family

economically feasible, some have questioned the continued relevance of linkage analysis as a method for gene localization in the twenty-first century. When the variants influencing the focal phenotype are among the genotyped markers, as will be the case with complete WGS, association tests can be more powerful than linkage for gene localization. However, genome-wide association screening using WGS data necessarily entails millions of statistical tests and appropriately rigorous correction for multiple testing will again limit power. Given this, strategies will be needed for focused testing of subsets of variants drawn from WGS or for placing informative prior probabilities on subsets of variants. One such strategy will be to utilize the independent information from transmission within families, i.e., linkage, to limit association testing to targeted regions drawn from the WGS or to preferentially weight sequence variants in linkage regions. Importantly, the large pedigrees that provide the best power to detect linkage are also an ideal design for WGS studies as they have the potential to carry numerous copies of even rare alleles of interest when founders at the top of the pedigree structure have many descendants. Thus in many cases, samples selected for WGS will also be well optimized for linkage and linkage information can be utilized to augment analysis of WGS at no additional cost.

Acknowledgments This work was supported in part by NIH grants MH59490 and GM31575.

References

Abecasis GR, Cherny SS, Cookson WO, Cardon LR (2002) Merlin-rapid analysis of dense genetic maps using sparse gene flow trees. Nat Genet 30:97–101

Abreu PC, Greenberg DA, Hodge SE (1999) Direct power comparisons between simple LOD scores and NPL scores for linkage analysis in complex diseases. Am J Hum Genet 65:847–857

Alcais A, Abel L (2000) Linkage analysis of quantitative trait loci: sib pairs or sibships? Hum Hered 50:251–256

Almasy L (2003) Quantitative risk factors as indices of alcoholism susceptibility. Ann Med 35:337–343

Almasy L, Blangero J (1998) Multipoint quantitative-trait linkage analysis in general pedigrees. Am J Hum Genet 62:1198–1211

Amos CI (1994) Robust variance-components approach for assessing genetic linkage in pedigrees. Am J Hum Genet 54:535–543

Amos CI, Zhu DK, Boerwinkle E (1996) Assessing genetic linkage and association with robust components of variance approaches. Ann Hum Genet 60:43–160

Arnett DK, Miller MB, Coon H, Ellison RC, North KE, Province M, Leppert M, Eckfeldt JH (2004) Genome-wide linkage analysis replicates susceptibility locus for fasting plasma triglycerides: NHLBI Family Heart Study. Hum Genet 115:468–474

Austin MA, Edwards KL, Monks SA, Koprowicz KM, Brunzell JD, Motulsky AG, Mahaney MC, Hixson JE (2003) Genome-wide scan for quantitative trait loci influencing LDL size and plasma triglyceride in familial hypertriglyceridemia. J Lipid Res 44:2161–2168

Bell J, Haldane JBS (1937) The linkage between the genes for colour-blindness and haemophilia in man. Proc R Soc Lond B 123:119–150

Blackwelder WC, Elston RC (1985) A comparison of sib-pair linkage tests for disease susceptibility loci. Genet Epidemiol 2:85–97

Blangero J, Williams JT, Almasy L (2000) Quantitative trait locus mapping using human pedigrees. Hum Biol 72:35–62

Blangero J, Williams JT, Almasy L (2001) Variance component methods for detecting complex trait loci. Adv Genet 42:151–181

Blangero J, Williams JT, Almasy L (2003) Novel family-based approaches to genetic risk in thrombosis. J Thromb Haemost 1:1391–1397

Botstein D, White RL, Skolnick M, Davis RW (1980) Construction of a genetic linkage map in man using restriction fragment length polymorphisms. Am J Hum Genet 32:314–331

Bowden DW, Sale M, Howard TD, Qadri A, Spray BJ, Rothschild CB, Akots G, Rich SS, Freedman BI (1997) Linkage of genetic markers on human chromosomes 20 and 12 to NIDDM in Caucasian families with a history of diabetic nephropathy. Diabetes 46:882–886

Buetow KH, Weber JL, Ludwigsen S, Scherpbier-Heddema T, Duyk GM, Sheffield VC, Wang Z, Murray JC (1994) Integrated human genome-wide maps constructed using the CEPH reference panel. Nat Genet 6:391–393

Collaborative Study on the Genetics of Asthma (CSGA) (1997) A genome-wide search for asthma susceptibility loci in ethnically diverse populations. Nat Genet 15:389–392

Cardon LR, Smith SD, Fulker DW, Kimberling WJ, Pennington BF, DeFries JC (1994) Quantitative trait locus for reading disability on chromosome 6. Science 266:276–279

Chakravarti A, Lynn A (1999) Background, history, and current status of human genetic mapping. In: Birren B et al (eds) Genome analysis: a laboratory manual, vol 4. Cold Spring Harbor Laboratory Press, Cold Spring Harbor, New York

Chalmers JNM, Lawler SD (1953) Data on linkage in man: elliptocytosis and blood groups. I. Families 1 and 2. Ann Eugen 17:267–271

Chen WM, Broman KW, Liang KY (2004) Quantitative trait linkage analysis by generalized estimating equations: unification of variance components and Haseman-Elston regression. Genet Epidemiol 26:265272

Chen AC, Tang Y, Rangaswamy M, Wang JC, Almasy L, Foroud T, Edenberg HJ, Hesselbrock V, Nurnberger J Jr, Kuperman S, O'Connor SJ, Schuckit MA, Bauer LO, Tischfield J, Rice JP, Bierut L, Goate A, Porjesz B (2009) Association of single nucleotide polymorphisms in a glutamate receptor gene (GRM8) with theta power of event-related oscillations and alcohol dependence. Am J Med Genet B Neuropsychiatr Genet. 150B:59–368

Clerget-Darpoux F, Bonaiti-Pellie C, Hochez J (1986) Effects of misspecifying genetic parameters in lod score analysis. Biometrics 42:393–399

Comuzzie AG, Hixson JE, Almasy L, Mitchell BD, Mahaney MC, Dyer TD, Stern MP, MacCluer JW, Blangero J (1997) A major quantitative trait locus determining serum leptin levels and fat mass is located on human chromosome 2. Nat Genet 15:273–276

Commuzzie AG, Martin LJ, Cole SA, Rogers J, Mahaney MC, Blangero J, VandeBerg JL (2001) A quantitative trait locus for fat free mass in baboons localizes to a region homologous to human chromosome 6. Obes Res 9(Suppl):71S

Coon H, Leppert MF, Eckfeldt JH, Oberman A, Myers RH, Peacock JM, Province MA, Hopkins PN, Heiss G (2001) Genome-wide linkage analysis of lipids in the Hypertension Genetic Epidemiology Network (HyperGEN) blood pressure study. Arterioscler Thromb Vasc Biol 21:1969–1976

Covault J, Gelernter J, Hesselbrock V, Nellissery M, Kranzler HR (2004) Allelic and haplotypic association of GABRA2 with alcohol dependence. Am J Med Genet B Neuropsychiatr Genet 129:104–109

Cox NJ, Frigge M, Nicolae DL, Concannon P, Hanis CL, Bell GI, Kong A (1999) Loci on chromosome 2 (NIDDM1) and 15 interact to increase susceptibility to diabetes in Mexican Americans. Nat Genet 21:213–215

Cox LA, Mahaney MC, Vandeberg JL, Rogers J (2006) A second-generation genetic linkage map of the baboon (*Papio hamadryas*) genome. Genomics 88:274–281

Davies JL, Kawaguchi Y, Bennett ST, Copeman JB, Cordell HJ, Pritchard LE, Reed PW, Gough SC, Jenkins SC, Palmer SM, Karen M, Balfour KM, Beth R, Roweb R, Farrall M, Anthony H, Barnetta H, Bain SC, Todd AJ (1994) A genome-wide search for human type 1 diabetes susceptibility genes. Nature 371:130–136

Daw EW, Wijsman EM, Thompson EA (2003) A score for Bayesian genome screening. Genet Epidemiol 24:181–190

de Andrade M, Amos CI, Thiel TJ (1999) Methods to estimate genetic components of variance for quantitative traits in family studies. Genet Epidemiol 17:64–76

Deka R, Shriver MD, Yu LM, Jin L, Aston CE, Chakraborty R, Ferrell RE (1994) Conservation of human chromosome 13 polymorphic microsatellite (CA)n repeats in chimpanzees. Genomics 22:226–230

DeSimone J, Linde M, Tashian RE (1973) Evidence for linkage of carbonic anhydrase isozyme genes in the pig-tailed macaque, *Macaca nemestrina*. Nat New Biol 242:55–56

Donahue RP, Bias WB, Renwick JH, McKusick VA (1968) Probable assignment of the Duffy blood group locus to chromosome 1 in Man. Proc Natl Acad Sci USA 61:950–955

Donis-Keller H, Green P, Helms C, Cartinhour S, Weiffenbach B, Stephens K, Keith TP, Bowden DW, Smith DR, Lander ES, Botstein D, Akots G, Rediker KS, Gravius T, Brown VA, Rising MB, Parker C, Powers JA, Watt DE, Kauffman ER, Bricker A, Phipps P, Muller-Kahle H, Fulton TR, Ng S, Schumm JW, Braman JC, Knowlton RG, Barker DF, Crooks SM, Lincoln SE, Daly MJ, Muller-Kahle H,

Abrahamsont J (1987) A genetic linkage map of the human genome. Cell 51:319–337
Dorf ME, Balner H, Benacerraf B (1975) Mapping of the immune response genes in the major histocompatibility complex of the rhesus monkey. J Exp Med 142:673–693
Drgon T, D'Addario C, Uhl GR (2006) Linkage disequilibrium haplotype and association studies of a chromosome 4 GABA receptor gene cluster: candidate gene variants for addictions. Am J Med Genet B Neuropsychiatr Genet 141:854–860
Duggirala R, Blangero J, Almasy L, Dyer TD, Williams KL, Leach RJ, O'Connell P, Stern MP (1999) Linkage of type 2 diabetes mellitus and of age at onset to a genetic location on chromosome 10q in Mexican Americans. Am J Hum Genet 64:1127–1140
Duggirala R, Blangero J, Almasy L, Dyer TD, Williams KL, Leach RJ, O'Connell P, Stern MP (2000) A major susceptibility locus influencing plasma triglyceride concentrations is located on chromosome 15q in Mexican Americans. Am J Hum Genet 66:1237–1245
Edenberg HJ, Dick DM, Xuei X, Tian H, Almasy L, Bauer LO, Crowe RR, Goate A, Hesselbrock V, Jones K, Kwon J, Li TK, Nurnberger JI Jr, O'Connor SJ, Reich T, Rice J, Schuckit MA, Porjesz B, Foroud T, Begleiter H (2004) Variations in GABRA2, encoding the alpha 2 subunit of the GABA(A) receptor, are associated with alcohol dependence and with brain oscillations. Am J Hum Genet 74:705–714
Enoch MA, Schwartz L, Albaugh B, Virkkunen M, Goldman D (2006) Dimensional anxiety mediates linkage of GABRA2 haplotypes with alcoholism. Am J Med Genet B Neuropsychiatr Genet 141:599–607
Fehr C, Sander T, Tadic A, Lenzen KP, Anghelescu I, Klawe C, Dahmen N, Schmidt LG, Szegedi A (2006) Confirmation of association of the GABRA2 gene with alcohol dependence by subtype-specific analysis. Psychiatr Genet 16:9–17
Ferrell RE, Majumder PP, Smith DG (1985) A linkage study of protein-coding loci in *Macaca mulatta* and *Macaca fascicularis*. Am J Phys Anthropol 68:315–320
Fisher SE, Marlow AJ, Lamb J, Maestrini E, Williams DF, Richardson AJ, Weeks DE, Stein JF, Monaco AP (1999) A quantitative-trait locus on chromosome 6p influences different aspects of developmental dyslexia. Am J Hum Genet 64:146–156
Gottesman II, Gould TD (2003) The endophenotype concept in psychiatry: etymology and strategic considerations. Am J Psychiatry 160:636–645
Grant SF, Thorleifsson G, Reynisdottir I, Benediktsson R, Manolescu A, Sainz J, Helgason A, Stefansson H, Emilsson V, Helgadottir A, Styrkarsdottir U, Magnusson KP, Walters GB, Palsdottir E, Jonsdottir T, Gudmundsdottir T, Gylfason A, Saemundsdottir J, Wilensky RL, Reilly MP, Rader DJ, Bagger Y, Christiansen C, Gudnason V, Sigurdsson G, Thorsteinsdottir U, Gulcher JR, Kong A, Stefansson K (2006) Variant of transcription factor 7-like 2 (TCF7L2) gene confers risk of type 2 diabetes. Nat Genet 38:320–323
Grigorenko EL, Wood FB, Meyer MS, Hart LA, Speed WC, Shuster A, Pauls DL (1997) Susceptibility loci for distinct components of developmental dyslexia on chromosomes 6 and 15. Am J Hum Genet 60:27–39
Grigorenko EL, Wood FB, Meyer MS, Pauls DL (2000) Chromosome 6p influences on different dyslexia-related processes: further confirmation. Am J Hum Genet 66:715–723
Gusella JF, Wexler NS, Conneally PM, Naylor SL, Anderson MA, Tanzi RE, Watkins PC, Ottina K, Wallace MR, Sakaguchi AY, Young AB, Shoulson I, Bonilla E, Martin JB (1983) A polymorphic DNA marker genetically linked to Huntington's Disease. Nature 306:234–238
Gyapay G, Morissette J, Vignal A, Dib C, Fizames C, Millasseau P, Marc S, Bernardi G, Lathrop M, Weissenbach J (1994) The 1993–1994 Genethon human genetic linkage map. Nat Genet 7:246–339
Hager J, Dina C, Francke S, Dubois S, Houari M, Vatin V, Vaillant E, Lorentz N, Basdevant A, Clement K, Guy-Grand B, Froguel P (1998) A genome-wide scan for human obesity genes reveals a major susceptibility locus on chromosome 10. Nat Genet 20:304–308
Haldane JBS, Sprunt AD, Haldane NM (1915) Reduplication in mice. J Genet 5:133–135
Hanis CL, Boerwinkle E, Chakraborty R, Ellsworth DL, Concannon P, Stirling B, Morrison VA, Wapelhorst B, Spielman RS, Gogolina-Ewens KJ, Shephard JM, Williams SR, Risch N, Hinds D, Iwasaki N, Ogata M, Omori Y, Petzold C, Rietzsch H, Schröder HE, Schulze J, Cox NJ, Menzel S, Boriraj VV, Chen X, Lim LR, Lindner T, Mereu LE, Wang YQ, Xiang K, Yamagata K, Yang Y, Bell GI (1996) A genome-wide search for human non-insulin-dependent diabetes genes reveals a major susceptibility locus on chromosome 2. Nat Genet 13:161–166
Haseman JK, Elston RC (1972) The investigation of linkage between a quantitative trait and a marker locus. Behav Genet 2:3–19
Havill LM, Mahaney MC, Cox LA, Morin PA, Joslyn G, Rogers J (2005) A quantitative trait locus for normal variation in forearm bone mineral density in pedigreed baboons maps to the ortholog of human chromosome 11q. J Clin Endocrinol Metab 90:3638–3645
Heath SC (1997) Markov chain Monte Carlo segregation and linkage analysis for oligogenic models. Am J Hum Genet 61(3):748–760
Hinds DA, Risch N (1996) The ASPEX package: affected sib-pair exclusion mapping. http://aspex.sourceforge.net/. Accessed 05 Oct 2012
Horikawa Y, Oda N, Cox NJ, Li X, Orho-Melander M, Hara M, Hinokio Y, Lindner TH, Mashima H, Schwarz PEH, Bosque-Plata L, Horikawa Y, Oda Y, Yoshiuchi I, Colilla S, Polonsky KS, Wei S, Concannon P, Iwasaki N, Schulze J, Baier LJ, Bogardus C, Groop L, Boerwinkle E, Hanis CL, Bell GI (2000) Genetic variation in the gene encoding calpain-10 is associated with type 2 diabetes mellitus. Nat Genet 26:163–175
Inoue M, Takenaka O (1993) Japanese macaque microsatellite PCR primers for paternity testing. Primates 34:37–45

Jasinska AJ, Service S, Levinson M, Slaten E, Lee O, Sobel E, Fairbanks LA, Bailey JN, Jorgensen MJ, Breidenthal SE, Dewar K, Hudson TJ, Palmour R, Freimer NB, Ophoff RA (2007) A genetic linkage map of the vervet monkey (*Chlorocebus aethiops sabaeus*). Mamm Genome 18:347–360

Johnson EW, Dubovsky J, Rich SS, O'Donovan CA, Orr HT, Anderson VE, Gil-Nagel A, Ahmann P, Dokken CG, Schneider DT, Weber JL (1998) Evidence for a novel gene for familial febrile convulsions, FEB2, linked to chromosome 19p in an extended family from the Midwest. Hum Mol Genet 7:63–67

Jones KA, Porjesz B, Almasy L, Bierut L, Goate A, Wang JC, Dick DM, Hinrichs A, Kwon J, Rice JP, Rohrbaugh J, Stock H, Wu W, Bauer LO, Chorlian DB, Crowe RR, Edenberg HJ, Foroud T, Hesselbrock V, Kuperman S, Nurnberger J Jr, O'Connor SJ, Schuckit MA, Stimus AT, Tischfield JA, Reich T, Begleiter H (2004) Linkage and linkage disequilibrium of evoked EEG oscillations with CHRM2 receptor gene polymorphisms: implications for human brain dynamics and cognition. Int J Psychophysiol 53:75–90

Jones KA, Porjesz B, Almasy L, Bierut L, Dick D, Goate A, Hinrichs A, Rice JP, Wang JC, Bauer LO, Crowe R, Foroud T, Hesselbrock V, Kuperman S, Nurnberger J Jr, O'Connor SJ, Rohrbaugh J, Schuckit MA, Tischfield J, Edenberg HJ, Begleiter H (2006) A cholinergic receptor gene (CHRM2) affects event-related oscillations. Behav Genet 36:627–639

Kammerer CM, Rainwater DL, Cox LA, Schneider JL, Mahaney MC, Rogers J, VandeBerg JL (2002) Locus controlling LDL cholesterol response to dietary cholesterol is on the baboon homologue of human chromosome 6. Arterioscler Thromb Vasc Biol 22:1720–1725

Kan Y, Dozy A (1978) Antenatal diagnosis of sickle-cell anaemia by DNA analysis of amniotic-fluid cells. Lancet 2:910–912

Kayser M, Nurnberg P, Berkovitch F, Nagy M, Roewer L (1995) Increased microsatellite variability in *Macaca mulatta* compared to humans due to a large scale deletion/insertion event during primate evolution. Electrophoresis 16:1607–1611

Kissebah AH, Sonnenberg GE, Myklebust J, Goldstein M, Broman K, James RG, Marks JA, Krakower GR, Jacob HJ, Weber J, Martin L, Blangero J, Comuzzie AG (2000) Quantitative trait loci on chromosome 3 and 17 influence phenotypes of the metabolic syndrome. Proc Natl Acad Sci USA 97:14478–14483

Knapp M, Seuchter SA, Baur MP (1994) Linkage analysis in nuclear families. 1: optimality criteria for affected sib-pair tests. Hum Hered 44:37–43

Knowlton RG, Cohen-Haguenauer O, Nguyen VC, Frézal J, Brown V, Barker D, Braman JC, Schumm JW, Tsui LC, Buchwald M, Donis-Keller H (1985) A polymorphic DNA marker linked to cystic fibrosis is located on chromosome 7. Nature 318:380

Kruglyak L, Lander ES (1995) Complete multipoint sib-pair analysis of qualitative and quantitative traits. Am J Hum Genet 57(2):439–454

Kruglyak L, Daly MJ, Reeve-Daly MP, Lander ES (1996) Parametric and nonparametric linkage analysis: a unified multipoint approach. Am J Hum Genet 58:1347–1363

Lander E, Kruglyak L (1995) Genetic dissection of complex traits: guidelines for interpreting and reporting linkage results. Nat Genet 11:241–247

Lappalainen J, Krupitsky E, Remizov M, Pchelina S, Taraskina A, Zvartau E, Somberg LK, Covault J, Kranzler HR, Krystal JH, Gelernter J (2005) Association between alcoholism and gamma-amino butyric acid alpha 2 receptor subtype in a Russian population. Alcohol Clin Exp Res 29:493–498

Lathrop GM, Lalouel JM (1984) Easy calculations of lod scores and genetic risks on small computers. Am J Hum Genet 36:460–465

Lathrop GM, Lalouel JM, Julier C, Ott J (1984) Strategies for multilocus linkage analysis in humans. Proc Natl Acad Sci USA 81:3443–3446

Lehman DM, Hunt KJ, Leach RJ, Hamlington J, Arya R, Abboud HE, Duggirala R, Blangero J, Goring HHH, Stern MP (2007) Haplotypes of transcription factor 7-like 2 (TCF7L2) gene and its upstream region are associated with type 2 diabetes and age of onset in Mexican Americans. Diabetes 56:389–393

Lejeune J, Lafourcade J, Berger R, Vialatte J, Bowswillwald M, Seringe P, Turpin R (1963) Trois cas de deletion partielle du bras court d'un chromosome 5. CR Acad Sci (Paris) 257:3098–3102

Lele KP, Penrose LS, Stallard HB (1963) Chromosome deletion in a case of retinoblastoma. Ann Hum Genet 27:171–174

Little PFR, Annison G, Darling S, Williamson R, Camba L, Modell B (1980) Model for antenatal diagnosis of β-thalassaemia and other monogenic disorders by molecular analysis of linked DNA polymorphisms. Nature 285:144–147

Luke A, Wu X, Zhu X, Kan D, Su Y, Cooper R (2003) Linkage for BMI at 3q27 region confirmed in an African-American population. Diabetes 52:1284–1287

Luo X, Kranzler HR, Zuo L, Wang S, Blumberg HP, Gelernter J (2005) CHRM2 gene predisposes to alcohol dependence, drug dependence and affective disorders: results from an extended case-control structured association study. Hum Mol Genet 14:2421–2434

Lyon MF, Searle AG (1989) In: Lyon MF, Searle AG (eds) Genetic variants and strains of the Laboratory Mouse, 2nd edn. Oxford University Press, Oxford

MacLean CJ, Bishop DT, Sherman SL, Diehl SR (1993) Distribution of lod scores under uncertain mode of inheritance. Am J Hum Genet 52(2):354–361

Mahaney MC, Rainwater DL, VandeBerg JL, Cox L, Rogers J, Blangero J, Hixson JE (1999) A quantitative trait locus for an HDL subfraction response to diet in pedigreed baboons: suggestive evidence for linkage to human chromosome 18q. Circulation 100(I):4–5

Martin LJ, Blangero J, Rogers J, Mahaney MC, Hixson JE, Carey KD, Comuzzie AG (2001) A quantitative trait locus influencing estrogen ratio in pedigreed baboons maps to a region homologous to human chromosome 19. Hum Biol 73:787–800

Maurer BA, Siwarski DF, Neefe JR (1979) Definition of two LD antigens in rhesus monkeys. Tissue Antigens 13:81–90

Mohr J (1951) A search for linkage between the Lutheran blood group and other hereditary characters. Acta Path Microbiol Scand 28:207–210

Mohr J (1954) A study of linkage in Man. Ejnar Munksgaard, Copenhagen

Morgan TH (1910) Sex limited inheritance in Drosophila. Science 32:120–122

Morin PA, Woodruff DS (1992) Paternity exclusion using multiple hypervariable microsatellite loci amplified from nuclear DNA of hair cells. In: Martin RD et al (eds) Paternity in primates: genetic tests and theories. Karger, Basel

Morton NE (1955) Sequential tests for the detection of linkage. Am J Hum Genet 7:277–318

Muller HJ (1916) The mechanism of crossing over. Am Nat 50:193–207

Murray JC, Buetow KH, Weber JL, Ludwigsen S, Scherpbier-Heddema T, Manion F, Quillen J, Sheffield VC, Sunden S, Duyk GM, Weissenbach GJ, Gyapay G, Dib C, Morrissette J, Lathrop GM, Vignal A, White R, Matsunami N, Gerken S, Melis R, Albertsen H, Plaetke R, Odelberg S, Ward D, Dausset J, Cohen D, Cann H (1994) A comprehensive human linkage map with centimorgan density. Cooperative Human Linkage Center (CHLC). Science 265:2049–2054

Nathans D, Smith H (1975) Restriction endonucleases in the analysis and restructuring of DNA molecules. Ann Rev Biochem 44:273–293

Neale MC, Neale BM, Sullivan PF (2002) Nonpaternity in linkage studies of extremely discordant sib pairs. Am J Hum Genet 70:526–529

Nowell PC, Hungerford DA (1960) A minute chromosome in human chronic granulocytic leukemia. Science 132:1497–1501

O'Connell JR, Weeks DE (1995) The VITESSE algorithm for rapid exact multilocus linkage analysis via genotype set-recording and fuzzy inheritance. Nat Genet 11:402–408

Ott J (1977) Linkage analysis with misclassification at one locus. Clin Genet 12:119–124

Ott J (1999) Analysis of human genetic linkage. Johns Hopkins University Press, Baltimore, p xxiii, 382

Penrose LS (1935) The detection of autosomal linkage in data which consists of pairs of brothers and sisters of unspecified parentage. Ann Eugen 6:133–138

Porjesz B, Almasy L, Edenberg HJ, Wang K, Chorlian DB, Foroud T, Goate A, Rice JP, O'Connor SJ, Rohrbaugh J, Kuperman S, Bauer LO, Crowe RR, Schuckit MA, Hesselbrock V, Michael Conneally P, Tischfield JA, Li T-K, Reich T, Begleiter H (2002) Linkage disequilibrium between the beta frequency of the human EEG and a GABAA receptor gene locus. Proc Natl Acad Sci USA 99:3729–3733

Rainwater DL, Kammerer CM, Mahaney MC, Rogers J, Cox LA, Schneider JL, VandeBerg JL (2003) Localization of genes that control LDL size fractions in baboons. Atherosclerosis 168:15–22

Reeders ST, Breuning MH, Davies KE, Nicholls RD, Jarman AP, Higgs DR, Pearson PC, Weatherall DJ (1985) A highly polymorphic DNA marker linked to adult polycystic kidney disease on chromosome 16. Nature 317:542

Renwick JH, Lawler SD (1955) Genetical linkage between the ABO and nail-patella loci. Ann Hum Genet 19:312–331

Reynisdottir I, Thorleifsson G, Benediktsson R, Sigurdsson G, Emilsson V, Einarsdottir AS, Hjorleifsdottir EE, Orlygsdottir GT, Bjornsdottir GT, Saemundsdottir J, Halldorsson S, Hrafnkelsdottir S, Sigurjonsdottir SB, Steinsdottir S, Martin M, Kochan JP, Rhee BK, Grant SFA, Frigge ML, Kong A, Gudnason V, Stefansson K, Gulcher JR (2003) Localization of a susceptibility gene for type 2 diabetes to chromosome 5q34-q35.2. Am J Hum Genet 73:323–335

Risch N (1990a) Linkage strategies for genetically complex traits: II. The power of affected relative pairs. Am J Hum Genet 46:229–241

Risch N (1990b) Linkage strategies for genetically complex traits: III. The effect of marker polymorphism on analysis of affected relative pairs. Am J Hum Genet 46:242–253

Risch N, Giuffra L (1992) Model misspecification and multipoint linkage analysis. Hum Hered 42:77–92

Risch N, Claus E, Giuffra L (1989) Linkage and mode of inheritance in complex traits. Prog Clin Biol Res 329:183–188

Rogers J, Kidd KK (1993) Nuclear DNA polymorphisms in a wild population of yellow baboons (*Papio hamadryas cynocephalus*) from Mikumi National Park, Tanzania. Am J Anthropol 90:477–486

Rogers J, Mahaney MC, Cox LA (2009) The development and status of the baboon genetic linkage map. In: VandeBerg JL et al (eds) The baboon in biomedical research, Springer, New York

Rogers J, Witte SM, Kammerer CM, Hixson JE, MacCluer JW (1995) Linkage mapping in Papio baboons: conservation of a synthetic group of six markers on human chromosome 1. Genomics 28:251–254

Rogers J, Garcia R, Shelledy W, Kaplan J, Arya A, Johnson Z, Bergstrom M, Novakowski L, Nair P, Vinson A, Newman D, Heckman G, Cameron J (2006) An initial genetic linkage map of the rhesus macaque (*Macaca mulatta*) genome using human microsatellite loci. Genomics 87:30–38

Rogers J, Mahaney MC, Witte SM, Nair S, Newman D, Wedel S, Rodriguez LA, Rice KS, Slifer SH, Perelygin A, Slifer M, Palladino-Negro P, Newman T, Chambers K, Joslyn G, Parry P, Morin PA (2000) A genetic linkage map of the baboon (*Papio hamadryas*) genome based on human microsatellite polymorphisms. Genomics 67:237–247

Rotimi CN, Comuzzie AG, Lowe WL, Luke A, Blangero J, Cooper RS (1999) The quantitative trait locus on chromosome 2 for serum leptin levels is confirmed in African-Americans. Diabetes 48:643–644

Rubin GM, Lewis EB (2000) A brief history of Drosophila's contributions to genome research. Science 287:2216–2218

Statistical Analysis for Genetic Epidemiology (S.A.G.E.) (2009) Statistical analysis for genetic epidemiology. Release 6.0.1: http://darwin.cwru.edu/. Accessed 9 Oct 2012

Saiki RK, Scharf S, Faloona F, Mullis KB, Horn GT, Erlich HA, Arnheim N (1985) Enzymatic amplification of beta-globin genomic sequences and restriction site analysis for diagnosis of sickle cell anemia. Science 230:1350–1354

Scott LJ, Mohlke KL, Bonnycastle LL, Willer CJ, Li Y, Duren WL, Erdos MR, Stringham HM, Chines PS, Jackson AU, Ludmila Prokunina-Olsson L, Ding C-J, Swift AJ, Narisu N, Hu T, Pruim R, Xiao R, Li X-Y, Conneely KN, Riebow NL, Sprau AG, Tong M, White PP, Hetrick KN, Barnhart MW, Bark CW, Goldstein JL, Watkins L, Xiang F, Saramies J, Buchanan TA, Watanabe RM, Valle TT, Kinnunen L, Abecasis GR, Pugh EW, Doheny KF, Bergman RN, Tuomilehto J, Collins FS, Boehnke M (2007) A genome-wide association study of type 2 diabetes in Finns detects multiple susceptibility variants. Science 316:1341–1345

Sham PC, Purcell S, Cherny SS, Abecasis GR (2002) Powerful regression-based quantitative-trait linkage analysis of general pedigrees. Am J Hum Genet 71:238–253

Sing CF, Haviland MB, Reilly SL (1996) Genetic architecture of common multifactorial diseases. Ciba Found Symp 197:211–229

Sladek R, Rocheleau G, Rung J, Dina C, Shen L, Serre D, Boutin P, Vincent D, Belisle A, Hadjadj S, Balkau B, Heude B, Charpentier G, Hudson TJ, Montpetit A, Pshezhetsky AV, Prentki M, Posner BI, Balding DJ, Meyre D, Polychronakos C, Froguel P (2007) A genome-wide association study identifies novel risk loci for type 2 diabetes. Nature 445:881–885

Snell GD (1941) Biology of the laboratory mouse, 1st edn. Blakiston Company, New York

Smith JR, Freije D, Carpten JD, Gronberg H, Xu J, Isaacs SD, Brownstein MJ, Bova GS, Guo H, Bujnovszky P, Nusskern DR, Damber J-E, Bergh A, Emanuelsson M, Kallioniemi OP, Walker-Daniels J, Bailey-Wilson JE, Beaty TH, Meyers DA, Walsh PC, Collins FS, Trent JM, Isaacs WB (1996) Major susceptibility locus for prostate cancer on chromosome 1 suggested by a genome-wide search. Science 274:1371–1374

Soyka M, Preuss UW, Hesselbrock V, Zill P, Koller G, Bondy B (2008) GABA-A2 receptor subunit gene (GABRA2) polymorphisms and risk for alcohol dependence. J Psychiatr Res 42:84–191

Sturtevant AH (1913) The linear arrangement of six sex-linked factors in Drosophila, as shown by their mode of association. J Exp Zool 14:43–59

Sun L, Wilder K, McPeek MS (2002) Enhanced pedigree error detection. Hum Hered 54:99–110

Tijio JH, Levan A (1956) The chromosome number of man. Hereditas 42:1–6

Tong Y, Lin Y, Zhang Y, Yang J, Zhang Y, Liu H, Zhang B (2009) Association between TCF7L2 gene polymorphisms and susceptibility to type 2 diabetes mellitus: a large Human Genome Epidemiology (HuGE) review and meta-analysis. BMC Med Genet 10:15. doi:10.1186/1471-2350-10-15

Wainwright BJ, Scambler P, Schmidtke J, Watson EA, Law H-Y, Farall M, Cooke HJ, Eiberg H, Williamson R (1985) Localization of cystic fibrosis locus to human chromosome 7cen-q22. Nature 318:382

Wang T, Elston RC (2005) Two-level Haseman-Elston regression for general pedigree data analysis. Genet Epidemiol 29:12–22

Wang JC, Hinrichs AL, Stock H, Budde J, Allen R, Bertelsen S, Kwon JM, Wu W, Dick DM, Rice J, Jones K, Nurnberger JI Jr, Tischfield J, Porjesz B, Edenberg HJ, Hesselbrock V, Crowe R, Schuckit M, Begleiter H, Reich T, Goate AM, Bierut LJ (2004) Evidence of common and specific genetic effects: association of the muscarinic acetylcholine receptor M2 (CHRM2) gene with alcohol dependence and major depressive syndrome. Hum Mol Genet 13:1903–1911

Weber JL, May PE (1989) Abundant class of human DNA polymorphisms which can be typed using the polymerase chain reaction. Am J Hum Genet 44:388–396

Weeks DE, Lange K (1992) A multilocus extension of the affected-pedigree-member method of linkage analysis. Am J Hum Genet 50:859–868

Weiss MC, Green H (1967) Human-mouse hybrid cell lines containing partial complements of human chromosomes and functioning human genes. Proc Natl Acad Sci USA 58:1104–1111

Weissenbach J (1993) A second generation linkage map of the human genome based on highly informative microsatellite loci. Gene 135:275–278

White R, Woodward S, Leppert M, O'Connell P, Nakmura Y, Hoff M, Herbst J, Lalouel J-M, Dean M, Vande Woude G (1985) A closely linked genetic marker for cystic fibrosis. Nature 318:382

Williams JT, Duggirala R, Blangero J (1997) Statistical properties of a variance-components method for quantitative trait linkage analysis in nuclear families and extended pedigrees. Genet Epidemiol 14:1065–1070

Wilson EB (1911) The sex chromosomes. Arch Mikrosk Anat Entwicklungsmech 77:249–271

Wu X, Cooper RS, Borecki I, Hanis C, Bray M, Lewis CE, Zhu X, Kan D, Luke A, Curb D (2002) A combined analysis of genome wide linkage scans for body mass index from the National Heart, Lung, and Blood Institute Family Blood Pressure Program. Am J Hum Genet 70:1247–1256

Yu A, Zhao C, Fan Y, Jang W, Mungall AJ, Deloukas P, Olsen A, Doggett NA, Ghebranious N, Broman KW, Weber JL (2001) Comparison of human genetic and sequence-based physical maps. Nature 409:951–953

4 Association Studies to Map Genes for Disease-Related Traits in Humans

Robert L. Hanson and Alka Malhotra

4.1 Introduction

Association studies have been widely used in human genetics as a way to map important disease-related traits. Genetic association generally refers to the tendency for particular alleles at a genetic marker to co-occur with particular trait values in a population. Historically, association studies were almost exclusively used for the investigation of regions in which there was reason to suspect that there were functional variants influencing the trait, with such suspicion coming from the known biology of the trait or from previous linkage studies. In recent years, however, technological developments have made it feasible to genotype hundreds of thousands of markers simultaneously across the entire genome. These genome-wide association studies have become powerful tools for mapping genes for susceptibility to human disease and for related traits.

R.L. Hanson (✉) · A. Malhotra
Diabetes Epidemiology and Clinical Research Section, National Institute of Diabetes and Digestive and Kidney Diseases, 1550 E. Indian School Road, Phoenix, AZ 85014, USA
e-mail: rhanson@phx.niddk.nih.gov

A. Malhotra
e-mail: alka@niddk.nih.gov

4.2 Heritability, Power, and Sample Size

4.2.1 Quantitative Traits

If there is a genetic variant with functional alleles that influence a quantitative trait, then one would expect to observe an association between genotypes at this variant and levels of the trait. The extent of the expected association depends on the frequencies of the functional alleles and the differences among genotypes in the level of the trait. The effects of these two parameters are described by the locus-specific heritability (h^2), which represents the proportion of trait variance explained by the association. For a diallelic polymorphism, under the assumption of Hardy-Weinberg equilibrium:

$$h^2 = \frac{f_H^2(\mu_{HH} - \mu)^2 + 2f_H(1 - f_H)(\mu_{HL} - \mu)^2 + (1 - f_H)^2(\mu_{LL} - \mu)^2}{\sigma^2} \quad (4.1)$$

where

- f_H frequency of the allele conferring high trait values,
- μ_{HH} trait mean for individuals homozygous for this allele,
- μ_{LL} mean for individuals homozygous for the allele conferring low trait values,
- μ_{HL} mean for heterozygotes,
- μ overall trait mean, and
- σ^2 total trait variance.

R. Duggirala et al. (eds.), *Genome Mapping and Genomics in Human and Non-Human Primates*, Genome Mapping and Genomics in Animals 5, DOI 10.1007/978-3-662-46306-2_4,

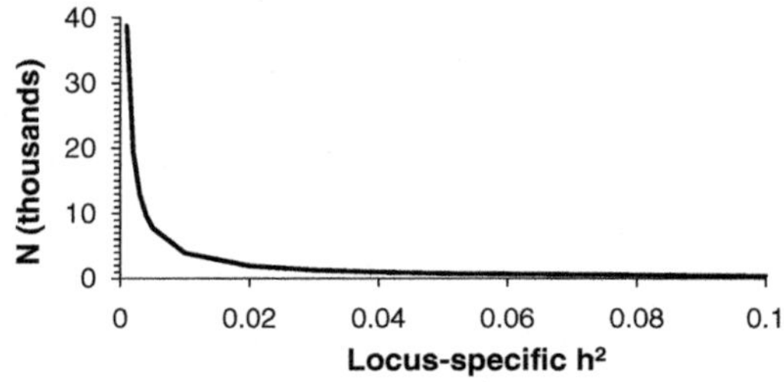

Fig. 4.1 Sample size required to map a quantitative trait by association according to the trait variance explained by the associated variant (locus-specific h^2) at $p < 7.2 \times 10^{-8}$ with 80 % power. Sample sizes are calculated on the basis of the correlation coefficient and range from 364 at $h^2 = 0.1$–38,762 at $h^2 = 0.001$

If representative individuals for a population are selected for the association study without regard for level of the trait, then the locus-specific h^2 will be the primary determinant of statistical power or sample size. Figure 4.1 shows the sample size required to detect a given locus-specific h^2 at genome-wide significance (taken here as $p < 7.2 \times 10^{-8}$, *v.i.*) with 80 % power. For variants with $h^2 > 0.05$ the sample size requirements are modest (<1,000), but substantial sample sizes are required for variants with $h^2 < 0.01$.

4.2.2 Dichotomous Traits

Most human diseases are dichotomous traits in which individuals are classified as affected or unaffected. Power and sample size calculations for dichotomous traits can be conducted in a fashion analogous to those for quantitative traits with the additional assumption that affection status is determined by being above or below a threshold on an underlying continuous liability scale. In population-based studies, the fact that the underlying quantitative trait is not observed can result in a substantial increase in the sample size required to map a dichotomous trait compared with that required for a quantitative trait. The magnitude of the increase depends on the prevalence of the disease. For example, if the disease prevalence is 0.01, a population of ~55,000 individuals is required to detect a locus conferring $h^2 = 0.01$ with genome-wide significance at 80 % power, compared with ~6,000 if the prevalence is 0.50 and ~4,000 if the quantitative trait can be directly analyzed.

Many disease mapping studies are conducted using a case-control design in which affected and unaffected individuals are selected in proportions that are different from those found in the general population, and this can greatly improve the efficiency of mapping studies for diseases with low prevalence. For example, if equal numbers of cases and controls are selected in such a study, ~1,800 individuals (900 of each) are required to detect a locus with $h^2 = 0.01$ for a disease with prevalence of 0.01, compared with ~55,000 if an unselected population is studied. For relatively rare diseases, the number of available cases can be a limiting factor, but power can be increased by including more than one control per case. For the example of disease prevalence of 0.01 and $h^2 = 0.01$, the total number of individuals required is 3,300 if cases and controls are matched 1:4 (650 cases and 2,650 controls). Case-control designs have also been used in the study of quantitative traits by selection of individuals for genotyping from the extreme values of the distribution (Hanson et al. 2006; Schork et al. 2000).

4.2.3 Allelic Odds Ratio: A Measure of Association

In case-control studies, it is customary to describe the association in terms of the allelic odds ratio (the increase in the odds of the disease per copy of the allele conferring high risk). If $f_{H\text{-case}}$ is the frequency of this high risk allele in cases and $f_{H\text{-cont}}$ the corresponding frequency in controls, the allelic odds ratio (OR_H) is approximately equal to $[f_{H\text{-case}}(1\text{-}f_{H\text{-cont}})]/[f_{H\text{-cont}}(1\text{-}f_{H\text{-case}})]$, though its exact value depends on the genotypic frequencies in cases and controls (Sasieni 1997). For a given sample size there is a strong dependence of power on frequency of the risk allele (f_H) and OR_H. At low values of minor allele frequency a larger odds ratio is required to achieve the same locus-specific h^2 as when the minor allele frequency is near 0.5. This is illustrated in Fig. 4.2.

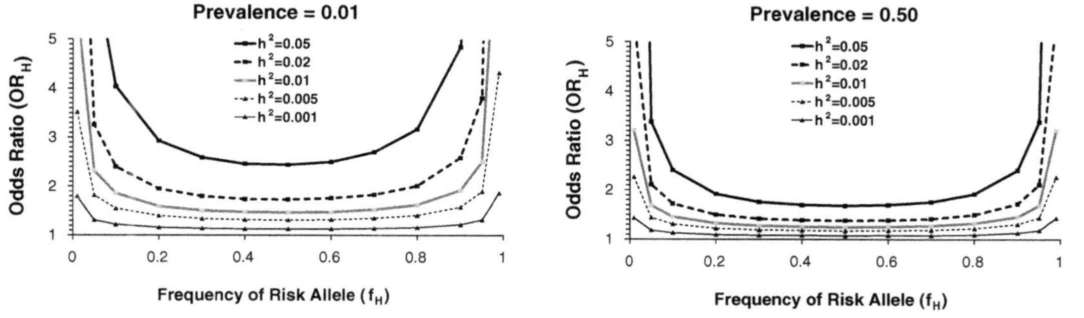

Fig. 4.2 Odds ratio per copy of the risk allele (OR_H) for a disease (dichotomous trait) according to the frequency of the risk allele (f_H) and the proportion of variance explained by the association in the underlying quantitative liability distribution (h^2). Results are shown for a disease with prevalence 0.01 and for one with prevalence 0.50. Note that, for the same h^2, OR_H is higher when the risk allele is rare or common than when it has frequency near 0.5. In addition, for the same h^2 and f_H, OR_H is higher when the disease is rare (or conversely when it is very common) compared to when it is near 0.50

4.3 Linkage Disequilibrium

4.3.1 Concordance with Functional Alleles

In the power and sample size calculations described in Sect. 4.2, the associations between marker alleles and trait values have been considered without regard to the functionality of the alleles. If the functional alleles (those that directly influence trait values due to their effects on molecular processes) can be genotyped, the parameters for the marker alleles reflect those of the underlying biologically important variants. In mapping studies, however, a selection of the known allelic variation is often assayed without any knowledge of whether the genotyped alleles are functional. In this situation, the power to detect an association with any given marker is dependent on the extent to which the marker alleles are concordant with functional alleles; allelic association that occurs between linked markers is termed "linkage disequilibrium." The degree of concordance between two loci is often described in terms of the parameter r^2. If a diallelic marker has alleles A and B, the frequencies of the haplotypes they comprise with the functional alleles, H and L, can be represented by f_{HA}, f_{HB}, f_{LA}, and f_{LB}. The degree of concordance is:

$$r^2 = \frac{(f_{HA}f_{LB} - f_{HB}f_{LA})^2}{f_H f_L f_A f_B} \tag{4.2}$$

(Hill and Weir 1994). r^2 is a measure of linkage disequilibrium that can range between 0 (no allelic association) and 1 (complete concordance), and the required sample size for a marker is inversely proportional to the r^2 with the functional alleles [assuming that trait association with the marker alleles is solely a function of linkage disequilibrium with the functional alleles] (Zondervan and Cardon 2004).

The human genome contains multiple small segments ("blocks") that are characterized by a high degree of linkage disequilibrium among the polymorphisms contained therein (Gabriel et al. 2002). The size of these blocks varies across human populations, in part because of historical recombination among loci and the multiple serial founder events from which these populations are thought to derive (Nordborg and Tavare 2002; Deshpande et al. 2009). Association studies to map disease-related traits are designed to exploit this linkage disequilibrium to identify genetic markers that are in close proximity to functional disease-causing variants (even though the functional polymorphism itself may not be directly genotyped). Since the resolution of association studies depends on historical recombination, rather than on the recent recombination that determines the resolution of family-based linkage studies, association studies typically identify smaller genomic regions of interest than do linkage studies. However, they also require a larger number of markers.

4.3.2 Linkage Disequilibrium in Human Populations

Since linkage disequilibrium patterns vary across human populations, the power of a given set of markers for association mapping will vary across populations as well. The International HapMap project has made an important contribution in identifying human linkage disequilibrium patterns and, thus, it has greatly facilitated the design of association studies (The International HapMap Consortium 2003). In this project, millions of single nucleotide polymorphisms (SNPs) have been genotyped in groups of individuals representing some of the major continental populations. The initial, and most extensive, genotyping has been done in Yoruba in Ibadan, Nigeria (YRI), Chinese in Beijing (CHB), Japanese in Tokyo (JPT), and individuals of European origin from the Centre d'Etude du Polymorphisme Humain collaboration (CEU). Other populations have been added more recently including: African ancestry in Southwest USA (ASW), Chinese in metropolitan Denver, Colorado (CHD), Gujarati Indians in Houston, Texas (GIH), Luhya in Webuye, Kenya (LWK), Mexican ancestry in Los Angeles, California (MEX), Maasai in Kinyawa, Kenya (MKK), and Tuscan in Italy (TSI). Examination of linkage disequilibrium patterns among a representative HapMap population allows for the selection of "tag" SNPs that capture the haplotypic variation within the population such that all SNPs are concordant with at least one of these tags with a specified r^2. Computer algorithms [such as the "tagger" program (de Bakker et al. 2005)] can be applied to select tag SNPs for genotyping.

The extent to which the HapMap populations are representative of other populations is variable, but analyses of data from the Human Genome Diversity Project suggest that common variation in most populations outside of Africa is well captured by one of the non-African HapMap populations [CEU, CHB or JPT] (Conrad et al. 2006). Although tags selected in the YRI population capture variation reasonably well in many African populations, there are African groups in whom variation is not well captured (Conrad et al. 2006). As the HapMap Project expands to include other populations, and as data from population-level sequencing studies are obtained from the 1000 Genomes Project (Altshuler et al. 2010), a more complete picture of linkage disequilibrium in human populations is emerging, and this will provide further utility in the design of association studies.

4.4 Genotyping

4.4.1 General Strategies

Prior to conducting an association mapping study, regions of interest must be defined and markers selected for genotyping. The genomic regions of interest may be confined to those that harbor genes that are strong candidates for containing variants that confer susceptibility to disease; such candidates may be defined on the basis of previous genetic or physiologic studies. Alternatively, investigators may choose to interrogate the entire genome to detect susceptibility variants without regard to whether they were predicted by existing biological knowledge. Investigators must also decide whether to attempt to assay all of the genetic variation in the regions of interest, by conducting sequencing studies, or to conduct targeted genotyping in an attempt to capture most of the relevant variation. Although large-scale sequencing experiments are rapidly becoming more feasible, they have generally been prohibitively costly, so most studies have attempted to type a subset of the variation. Most studies have also focused on common variants (minor allele frequency > 0.05), given the difficulties in assessing rare variants without sequencing studies.

4.4.2 Genome-Wide Arrays

A variety of arrays are available for conducting genotyping for genome-wide association studies. These arrays are able to simultaneously genotype large numbers of SNPs (80 K-5 M) across the entire human genome. Most of the commercially available arrays are associated with one or more computer algorithms that assign genotypes from the data generated by the experiment. Although these algorithms are generally accurate, a number of statistical quality control procedures need to be employed to ensure genotyping quality (Finner et al. 2010; Weale 2010). Examination of the sample-specific and SNP-specific call rates can help identify potentially problematic samples or SNPs; those with low rates of successful genotyping often produce erroneous values. Extreme deviations from Hardy-Weinberg equilibrium are often indicative of SNPs with systematic errors in genotype assignment. Individual samples which exhibit extreme deviations from the average rate of heterozygosity for the study may represent problems with DNA quality or unusual population structure. Inclusion of a number of samples typed in duplicate can help to identify SNPs where genotype assignment is uncertain due to a high degree of technical variation. Many of the genotyping algorithms have additional quality scores that can indicate degraded or contaminated samples, and additional methods have been developed to identify plates of samples that may contain systematic genotyping errors (Pluzhnikov et al. 2010).

By examination of the linkage disequilibrium patterns among SNPs on these arrays and those identified in the HapMap Project, one can determine the extent to which these arrays capture the known variation in the HapMap populations. These analyses suggest that for the most widely used dense arrays, which contain 300 K-1 M SNPs, 65–90 % of common SNPs that are not on the array have $r^2 > 0.80$ with a SNP on the array in non-African populations (CEU, CHB, JPT), while in Africans (YRI) the coverage is lower at 40–70 % (Li et al. 2008). Sequencing studies have indicated that 75–90 % of common SNPs that are not represented among the HapMap SNPs are highly concordant with a HapMap SNP and thus will generally be captured by these arrays; however, some common SNPs are not well captured (Bhangale et al. 2008; Takeuchi et al. 2008). Most rare variants are not well captured by these arrays and, since they are also often not present in HapMap databases, cannot be tagged using the HapMap populations (The International HapMap Consortium 2005; Xu et al. 2007). The development of large-scale sequencing technology should allow rare variants to be more easily included in association mapping studies in the future.

4.5 Data Analysis

4.5.1 Methods for Association Testing

As association mapping studies are often conducted in "unrelated" individuals, conventional statistical methods that assume independence of observations can often be applied, including analysis of variance for continuous traits and contingency table methods for dichotomous traits. To control for important covariates, methods such as linear or logistic regression can be employed. If family members are included in the study, then methods that account for the dependence among observations (such as the linear mixed model) are often used. For diallelic markers, an additive model is often used in which the odds of disease or the level of a quantitative trait is modeled as a function of the number of copies of a given allele (i.e., genotypes are coded 0, 1, or 2). More general models that test differences among all 3 genotypes can also be used, as can those that assume dominance for one of the alleles, but this comes at the cost of increasing the number of degrees of freedom or the number of statistical tests. Although the additive model captures much of the information in many scenarios, there are situations when associations can be missed if other models are not used (Slager and Schaid 2001), so investigators need to carefully consider the balance between missing some associations and increasing the multiple testing burden. A variety of software packages, such as PLINK and SNP-TEST, have been developed to conduct these analyses rapidly over a large number of markers and to extract the relevant summary statistics (Purcell et al. 2007; Marchini and Howie 2010).

4.5.2 Genotype Imputation

When dense sets of markers are genotyped, as in genome-wide association studies, it is possible to impute the genotypes for untyped markers using the linkage disequilibrium in a reference panel of individuals typed at all markers (Browning and Browning 2007; Marchini et al. 2007; Li et al. 2010). This essentially uses the genotypes at typed markers and the haplotype frequencies in the reference sample (usually one of the Hap Map populations) to assign the probability of a given genotype at the untyped marker. Analyses that include imputed markers can result in a modest increase in the power of association mapping studies (Hao et al. 2009; Li et al. 2010). Use of imputed genotypes is particularly important in meta-analyses that combine results over studies that may have used different genotyping arrays and that, thus, have different SNPs directly genotyped.

Imputation assumes that the linkage disequilibrium pattern in the reference population is representative of that in the study population, but fairly accurate imputation is possible in many populations by using combinations of HapMap populations as the reference (Huang et al. 2009a). However, even modest inaccuracy in imputation can reduce power and increase type I error in some situations, so care is necessary in applying these methods (Huang et al. 2009b; Almeida et al. 2011). It is often useful to apply the imputation techniques after masking some of the typed markers to quantify the accuracy of the imputation. When imputed genotypes are analyzed, the analyses need to account for uncertainty in the assignment of genotypes. Weighted likelihood-based methods, where the weights depend on the probability of the genotype, are probably optimal for these analyses, but may be time-consuming. A simple alternative is to use the posterior expectation of the genotype as the predictor variable in a conventional regression model. (For example, if p_2 and p_1 represent the probability of carrying 2 copies or 1 copy of a given allele assigned by the imputation procedure, the linear term in the additive model is taken as $2p_2 + p_1$). This approach provides an accurate approximation to the full likelihood calculation in many situations (Guan and Stephens 2008; Zheng et al. 2011).

4.5.3 Presentation of Genome-Wide Association Results

The results of a genome-wide association study are often presented in a "Manhattan" plot, in which $-\log_{10}(p\text{-value})$ for each SNP is plotted by location across the genome. An example of such a plot is shown in Fig. 4.3a [reflecting a 100 K genome-wide association study for type 2 diabetes in American Indians (Hanson et al. 2007)]. Although no association achieves genome-wide significance, there are a number of regions with evidence suggestive of association. The quantile–quantile plot is another way in which p-values are often presented for genome-wide association studies. In this plot, the cumulative distribution of the p-values in descending order is plotted in relation to a uniform distribution (the expected under the global null hypothesis of no association with any marker). The results for this example show little deviation from the expected values overall (Fig. 4.3b). However, there is some deviation at the levels associated with greater significance (Fig. 4.3c), which is expected if there are some truly positive associations.

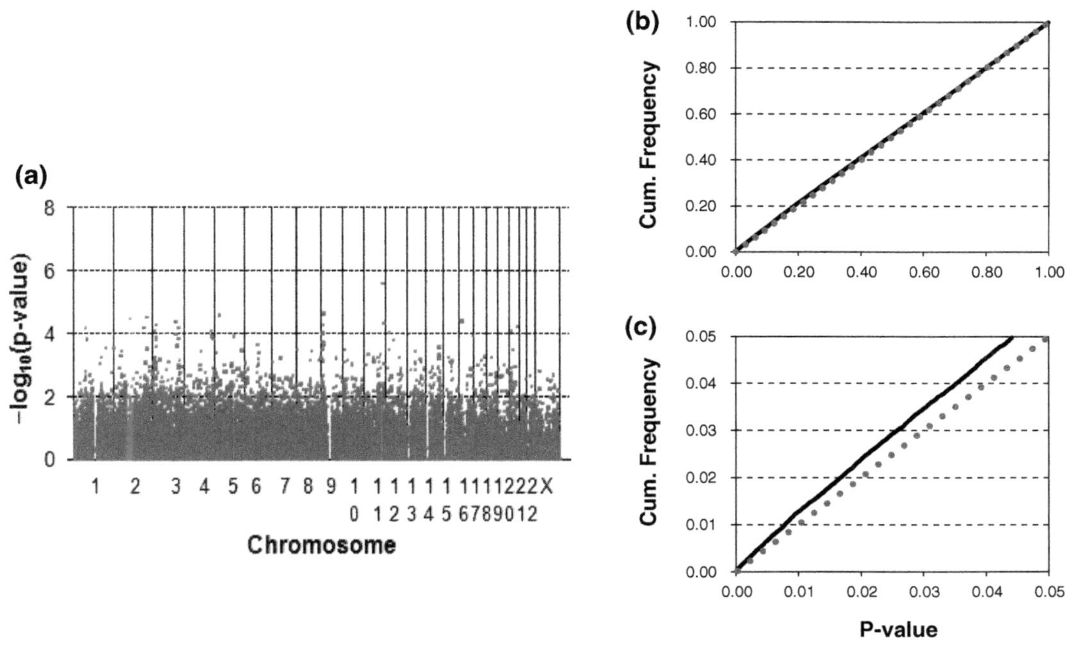

Fig. 4.3 **a** "Manhattan" plot showing significance of association in relation to genomic position; significance is plotted as $-\log_{10}(p\text{-value})$ so that higher levels indicate greater significance. **b** Quantile–quantile plot of the cumulative distribution of *p*-values across all markers; the *solid line* represents the observed distribution and the *dotted line* is that expected under the global null hypothesis. Overall there is little departure from the expected distribution, and this indicates little inflation of the test statistic. **c** The same quantile–quantile plot restricted to the region $p < 0.05$. There is a modest increase in significance beyond that expected, as expected if some markers were truly associated. All data are from Hanson et al. (2007)

4.6 Population Stratification

4.6.1 Confounding in Association Studies

Population stratification refers to a process that leads to a population being composed of different genetic subpopulations. This could result from ethnic admixture or undetected relationships among individuals; all human populations have some degree of stratification. This can be an important confounder in association mapping studies. If a population is composed of subpopulations that differ in risk of disease, then association may be observed with any allele that differs in frequency among subpopulations regardless of whether the marker is located in proximity to a polymorphism conferring disease susceptibility. An example of such confounding is that of the $Gm^{3;5,13,14}$ haplotype of immunoglobulin G, which is associated with a low prevalence of type 2 diabetes in American Indians [$p < 0.0001$] (Knowler et al. 1988). However, $Gm^{3;5,13,14}$ is at a much higher frequency in Europeans than in full-heritage American Indians, and Europeans have a lower prevalence of type 2 diabetes. There is no association with diabetes once the degree of European heritage is taken into account ($p = 0.30$), a finding which indicates that the overall association is confounded by European admixture.

4.6.2 Methods to Account for Population Stratification

A number of techniques have been developed to account for population stratification in association studies. "Genomic control" is a technique by which one estimates the extent of inflation in statistical significance in the study sample (Devlin and Roeder 1999; Devlin et al. 2004); this estimate is made either from the distribution of the chi-square (χ^2) test statistic for association in all markers or from that in a randomly selected set of markers. The inflation factor (λ) is estimated as the ratio of the mean or median value of the χ^2 statistic to its expected value under the global null hypothesis. This factor is then used to calculate the *p*-value for each marker corrected for the inflation, which is presumably caused by population stratification (the corrected χ^2 is taken as the uncorrected χ^2 divided by λ). This is a simple technique that is often useful, but it assumes that overall inflation is due to population stratification, whereas in polygenic traits it may be due in part to linkage disequilibrium with functional variants (Yang et al. 2011a). It also applies the same correction to all markers so it does not change their ranking. In theory, however, some forms of population stratification may not influence all markers equally, but will primarily affect those that differ in frequency among subpopulations.

Another approach is to use genetic marker data to generate covariates that reflect membership in subpopulations. Given known allele frequency differences between ancestral human populations, estimates of individual admixture can be derived (Hoggart et al. 2003; Tang et al. 2005; Alexander et al. 2009). Panels of markers with large differences in allele frequencies between major human continental populations are available for this purpose (Halder et al. 2008; Kosoy et al. 2009).

In genome-wide association studies, covariates that reflect subpopulation membership can be derived from the large number of markers available without a priori specification of the particular subpopulations. Principal components (linear combinations derived from all markers that account for a large part of the total variance among all individuals) can be generated. Controlling for the first few principal components is often an effective means of reducing the effects of population stratification (Price et al. 2006). Alternatively, clustering methods can be used to assign individuals to discrete subpopulations, and covariates for these can be incorporated (Pritchard et al. 2000). Methods that use marker data to estimate substructure have the advantage over genomic control that they do not assume that the effects of stratification apply equally to all markers. Nonetheless, there may still be markers that are confounded by aspects of population stratification that are not detected by these methods.

4.6.3 Family-Based Association Tests

Family-based association methods that are robust to population stratification can also be used, and a number of such tests have been developed (Spielman et al. 1993; Lake et al. 2000; Martin et al. 2000). These methods test whether a specific allele(s) and the trait of interest are co-transmitted within families. Family-based tests can be considered tests of linkage as well as association. Although these tests are robust to population stratification, they require collection of family data and are only informative for individuals in whom at least one parent is heterozygous for the marker. For a given number of individuals, therefore, they are generally less powerful than general association tests that do not require family data.

4.7 Statistical Significance

4.7.1 Multiple Comparisons

The issue of the appropriate level of statistical significance for genetic association studies has been somewhat controversial. A large number of statistical tests are often conducted, particularly in genome-wide studies. This results in a multiple comparison problem in that, under the global null hypothesis that none of the variants are truly associated with the trait, many variants with nominally significant associations (e.g., $p < 0.05$) will still be observed. Furthermore, most genetic variation is likely to be unassociated with any given trait, so, in the absence of any functional biologic information, the "prior" probability of a true association is low. Therefore, most nominal associations are likely to be false, even at levels of statistical significance conventionally taken as stringent. For these reasons, most statisticians recommend very stringent thresholds for statistical significance in association mapping studies. There are two basic approaches that have been used to establish the specific threshold: correction for the number of tests actually performed in any given experiment and correction for the number of potential tests across the entire genome.

4.7.2 Experiment-Wide Significance

In calculation of experiment-wide statistical significance, one might consider a Bonferroni correction in which the desired global p-value is simply divided by the number of markers to obtain the marker-specific threshold (e.g., if 300 K markers are typed $p < 0.05/300{,}000 = 1.7 \times 10^{-7}$ is considered experiment-wide significance at $p < 0.05$). However, this approach is probably overly stringent, because it does not consider linkage disequilibrium among the markers. The false discovery rate method (Benjamini and Hochberg 1995), in which the correction is applied in a stepwise fashion considering the overall distribution of p-values, is often less stringent but still assumes independence among markers.

Permutation methods, in which the observed nominal p-value is compared with the null distribution calculated by repeatedly permuting the trait at random among study participants, can effectively account for the correlation among markers. Permutation methods are computationally burdensome and, since they depend on the assumption of exchangeability among the individuals over whom the values are permuted, can be very difficult to implement in family studies or other study designs with complex dependencies among individuals. Alternative methods attempt to use the linkage disequilibrium among markers to estimate the number of effectively independent statistical tests (Nyholt 2004; Duggal et al. 2008), over which the Bonferroni method can be applied. These methods, while less computationally intensive than permutation, can still account for the correlation among markers.

4.7.3 Genome-Wide Significance

A potential disadvantage of all methods for calculating experiment-wide significance is that the thresholds are dependent on the number of markers typed, whereas the interpretation of the result for any given marker is intuitively not dependent on how many other markers have been tested. This limitation does not apply if one attempts to correct for the effective number of independent potential tests across the whole genome (if all variation was sampled). Several investigators have attempted to estimate this number empirically with extrapolation to an infinitely dense map [although the ascertainment of common SNPs may influence this estimate] (Dudbridge and Gusnanto 2008; Hoggart et al. 2008; Wellcome Trust Case Control Consortium 2007). Based on these estimates, the resulting thresholds for genome-wide significance range from 5×10^{-7} to 8×10^{-9} for non-African populations (with slightly lower values for YRI). Most investigators have, thus, used a threshold of in the vicinity of Dudbridge and Gusnanto's (2008) estimate of 7.2×10^{-8} for genome-wide significance.

4.8 Follow-Up Studies

4.8.1 Replication of Results

Once an association mapping study has been completed a number of steps are required to ensure the validity of the results and to understand their biological implications. Given technical and stochastic variation in the methods, replication of the findings in individuals from similar populations is important, regardless of whether genome-wide significance was obtained in the initial mapping study. Since variants with the strongest effects among multiple markers are typically selected, the effect estimates from the initial mapping study are likely biased and, thus, weaker effects will generally be seen in the replication study. In designing replication studies the effect of this "winner's curse" phenomenon ideally should be taken into account (Zhong and Prentice 2010). Replication studies may be performed by genotyping additional individuals or by conducting meta-analyses that compare results with previously performed association studies.

Meta-analyses that combine results across multiple mapping studies may help to identify consistent associations, some of which may achieve genome-wide significance only in the meta-analysis. For example, meta-analysis of the initial genome-wide association studies of type 2 diabetes helped to distinguish the consistent associations among them and to identify many susceptibility variants that were not immediately obvious in any individual study (Zeggini et al. 2008).

4.8.2 Fine-Mapping and Functional Studies

The identification of a variant that is reproducibly associated with a trait does not imply that this variant is itself functional. Analysis of other known variants in the region of interest and identification of additional variants through sequencing is needed to help identify the functional allele(s). Although association studies typically have high resolution, the pattern of linkage disequilibrium can be chaotic, so the fine-mapping and sequencing studies may still need to extend over several Mb, particularly if relatively rare variants underlie the signal (Dickson et al. 2010). In some cases, the relevant function may be clear, e.g., if a missense variant is identified in a gene with a known biological role in the disease, but in many cases the functional relevance of an association is not clear. In these cases, evidence for functionality may need to come from experimental systems, such as in vitro assays, and "knock-out" or transgenic animal models.

Prioritizing variants for functional studies can often benefit from statistical analyses. For example, analysis of the association of pairs of variants in linkage disequilibrium, each conditional on the effect of the other, can help to determine if the association with one of the variants explains that seen with the other or if each contributes independent information. Variants whose effects remain after conditioning on the effects of others are stronger candidates for functional studies. The sample sizes required for such studies, however, can often be prohibitively large when the linkage disequilibrium is strong, particularly when the amount of variance explained by the association is small (see Table 4.1), as is the case for many disease-associated traits in humans (for which h^2 = 0.001–0.01 is often typical). These studies could be facilitated if proximal traits can be identified in the relevant pathway that are more closely related to the genetic variation and that, thus, can be expected to have larger effects (e.g., h^2 = 0.1–0.3). Studies of gene expression levels in relevant human tissues are potentially useful in this respect. Obviously, however, the effects of variants in complete concordance ($r^2 \approx 1$) cannot be distinguished statistically.

4.8.3 Population and Biological Contexts

Bioinformatic analyses that integrate the genetic findings into known biochemical pathways may be useful for understanding the biological implications of the results of mapping studies. Many mapping studies are done in selected samples that are not representative of human populations, and population-based epidemiologic studies are necessary to quantify the population-level effects of the detected variants. The results of association mapping studies will depend on the underlying biology of the trait and how amenable the genetic factors are to the technology used.

Table 4.1 Sample size requirements to detect a functional allele at $p < 0.01$ with 80 % power conditional on a marker in linkage disequilibrium with the functional locus

	Locus-specific h^2					
r^2	0.001	0.01	0.05	0.10	0.20	0.30
0.50	23,338	2,316	447	214	97	58
0.60	29,172	2,894	558	266	120	72
0.70	38,894	3,858	743	354	159	94
0.80	58,340	5,785	1,113	529	237	140
0.90	116,676	11,566	2,223	1,055	470	276
0.95	233,349	12,127	4,441	2,105	937	548

r^2 represents the degree of linkage disequilibrium between functional and marker alleles; locus-specific h^2 is that associated with the functional alleles. Sample size was calculated by modification of the formula of Milton (1986) as:

$$n = 3 + \frac{(z_\alpha + z_\beta)^2(1 - h^2)}{(h^2 - r^2h^2)}$$

where z_α represents the normal deviate associated with the desired p-value and z_β represents the normal deviate associated with the desired power. If both functional and marker alleles are typed, the analysis would be conducted to determine if the functional allele is associated with a quantitative trait conditional on the association with the marker allele. The identities of functional and marker alleles are likely unknown, but the analysis would be conducted for each conditional on the other. One would expect the functional alleles to be associated conditional on the marker, while the marker would not be associated conditional on the functional alleles

The extent to which the findings of mapping studies can explain the overall heritability of the trait varies. For example, 3 robustly replicated loci account for 44 % of the variance in fetal hemoglobin levels in adults, which is 50 % of the overall heritability (Menzel and Thein 2009). On the other hand, while 32 variants with replicated associations with body mass index, a measure of obesity, have been identified, the effects of these variants are quite weak. Taken together the 32 variants explain only 1.5 % of the variation in body mass index or ~3 % of the overall heritability (Speliotes et al. 2010). Detection of variants with such small h^2 at genome-wide significance required genotyping over 200,000 individuals, and it is likely that many other variants exist that could not be robustly detected. For genome-wide association studies, methods are available to estimate the extent to which the markers are in linkage disequilibrium with variants that explain this "missing" heritability and to estimate the number of additional variants contributing to the trait that were not detected (So et al. 2010; Yang et al. 2011b). Methods to identify panels of large numbers of markers that are associated with the trait have also been used (Purcell et al. 2009). These methods, which use genomic information in aggregate, are not strictly mapping studies.

4.9 Conclusion

Genome-wide association studies have been successful for a variety of human diseases and related traits. Such studies have identified robustly reproducible associations for obesity, metabolic diseases, cardiovascular disease, gastrointestinal diseases, hematologic traits and human cancers, among many others (Easton et al. 2007; Eeles et al. 2008; Menzel and Thein 2009; Franke et al. 2010; Musunuru and Kathiresan 2010; Speliotes et al. 2010; Voight et al. 2010). An online catalogue of published genome-wide association studies is available at http://www.genome.gov/gwastudies/. In many cases, the biologic implications of the findings of these studies are not yet clear and additional studies are necessary to determine this.

As the molecular genetic methods move increasingly toward sequencing, association mapping studies will be able to make use of the entire range of genetic variation. Since the initial genome-wide association studies were conducted with arrays that mostly assayed common variants, the variants reproducibly associated with diseases have largely been common ones. As whole genome and whole exome (those targeting all the exons in the genome) sequencing studies become more widely used, it will be possible to determine the role of relatively rare, as well as common, variants in susceptibility to human disease (Cirulli and Goldstein 2010; Gibson 2012). With sequencing studies, however, issues of data quality, analytic methods, statistical significance and replication are likely to become increasingly complex. Association methods that simultaneously consider multiple variants within a gene or region may be particularly useful in the context where many rare variants influence the trait (Liu and Leal 2010). Methods that are not classical association techniques, such as those that employ measures of genomic similarity, may also be useful for sequence-based mapping studies (Bansal et al. 2010). However, in many situations association methods are likely to best capture the underlying biology and, thus, they will remain important tools for genetic mapping of diseases and related traits in humans and other primate species.

Acknowledgments Supported by the intramural research program of the National Institute of Diabetes and Digestive and Kidney Diseases.

References

Alexander DH, Novembre J, Lange K (2009) Fast model-based estimation of ancestry in unrelated individuals. Genome Res 19:1655–1664

Almeida MAA, Oliveira PSL, Pereira TV, Krieger JE, Alexandre C (2011) An empirical evaluation of imputation accuracy for association statistics reveals increased type-I error rates in genome-wide associations. BMC Genet 12:10

Altshuler DL, Durbin RM, Abecasis GR et al (2010) A map of human genome variation from population-scale sequencing. Nature 467:1061–1073

Bansal V, Libiger O, Ali Torkamani A, Schork NJ (2010) Statistical analysis strategies for association studies involving rare variants. Nat Rev Genet 11:773–785

Benjamini Y, Hochberg Y (1995) Controlling the false discovery rate: a practical and powerful approach to multiple testing. J Roy Stat Soc B 57:289–300

Bhangale TR, Rieder MJ, Nickerson DA (2008) Estimating coverage and power for genetic association studies using near-complete variation data. Nat Genet 40:841–843

Browning SR, Browning BL (2007) Rapid and accurate haplotype phasing and missing-data inference for whole-genome association studies by use of localized haplotype clustering. Am J Hum Genet 81:1084–1097

Cirulli ET, Goldstein DB (2010) Uncovering the roles of rare variants in common disease through whole-genome sequencing. Nat Rev Genet 11:415–425

Conrad DF, Jakobsson M, Coop G, Wen X, Wall JD, Rosenberg NA, Pritchard JK (2006) A worldwide survey of haplotype variation and linkage disequilibrium in the human genome. Nat Genet 38:1251–1260

de Bakker PIW, Yelensky R, Pe'er I, Gabriel SB, Daly MJ, Altshuler D (2005) Efficiency and power in genetic association studies. Nat Genet 37:1217–1223

Deshpande O, Batzoglou S, Feldman MW, Cavalli-Sforza LL (2009) A serial founder effect model for human settlement out of Africa. Proc Biol Sci 276:291–300

Devlin B, Bacanu SA, Roeder K (2004) Genomic control to the extreme. Nat Genet 36:1129–1130

Devlin B, Roeder K (1999) Genomic control for association studies. Biometrics 55:997–1004

Dickson SP, Wang K, Krantz I, Hakonarson H, Goldstein DB (2010) Rare variants create synthetic genome-wide associations. PLoS Biol 8:e1000294

Dudbridge F, Gusnanto A (2008) Estimation of significance thresholds for genomewide association scans. Genet Epidemiol 32:227–234

Duggal P, Gillanders EM, Holmes TN, Bailey-Wilson JE (2008) Establishing an adjusted p-value threshold to control the family-wide type 1 error in genome wide association studies. BMC Genomics 9:516

Easton DF, Pooley KA, Dunning AM et al (2007) Genome-wide association study identifies novel breast cancer susceptibility loci. Nature 447:1087–1093

Eeles RA, Kote-Jarai Z, Giles GG et al (2008) Multiple newly identified loci associated with prostate cancer susceptibility. Nat Genet 40:316–321

Finner H, Strassburger K, Heid IM, Herder C, Rathmann W, Giani G, Dickhaus T, Lichtner P, Meitinger T, Wichmann H-E, Illig T, Gieger C (2010) How to link call rate and p-values for Hardy-Weinberg equilibrium as measures of genome-wide SNP data quality. Stat Med 29:2347–2358

Franke A, McGovern DBP, Barrett JC et al (2010) Meta-analysis increases to 71 the tally of confirmed crohn's disease susceptibility loci. Nat Genet 42:1118–1125

Gabriel SB, Schaffner SF, Nguyen H, Moore JM, Roy J, Blumenstiel B, Higgins J, DeFelice M, Lochner A, Faggart M, Liu-Cordero SN, Rotimi C, Adeyemo A, Cooper R, Ward R, Lander ES, Daly MJ, Altshuler D (2002) The structure of haplotype blocks in the human genome. Science 296:2225–2229

Gibson G (2012) Rare and common variants: twenty arguments. Nat Rev Genet 13:135–145

Guan Y, Stephens M (2008) Practical issues in imputation-based association mapping. PLoS Genet 4: e1000279

Halder I, Shriver M, Thomas M, Fernandez JR, Frudakis T (2008) A panel of ancestry informative markers for estimating individual biogeographical ancestry and admixture from four continents: utility and applications. Hum Mutat 29:648–658

Hanson RL, Looker HC, Ma L, Muller YL, Baier LJ, Knowler WC (2006) Design and analysis of genetic association studies to finely map a locus identified by linkage analysis: sample size and power calculations. Ann Hum Genet 70:332–349

Hanson RL, Bogardus C, Duggan D, Kobes S, Knowlton M, Infante AM, Marovich L, Benitez D, Baier LJ, Knowler WC (2007) A search for variants associated with young-onset type 2 diabetes in American Indians in a 100 K genotyping array. Diabetes 56:3045–3052

Hao K, Chudin E, McElwee J, Schadt EE (2009) Accuracy of genome-wide imputation of untyped markers and impacts on statistical power for association studies. BMC Genet 10:27. doi:10.1186/1471-2156-10-27

Hill WG, Weir BS (1994) Maximum-likelihood estimation of gene location by linkage disequilibrium. Am J Hum Genet 54:705–714

Hoggart CJ, Parra EJ, Shriver MD, Bonilla C, Kittles RA, Clayton DG, McKeigue PM (2003) Control of confounding of genetic associations in stratified populations. Am J Hum Genet 72:1492–1504

Hoggart CJ, Clark TG, Iorio MD, Whittaker JC, Balding DJ (2008) Genome-wide significance for dense SNP and resequencing data. Genet Epidemiol 32:179–185

Huang L, Li Y, Singleton AB, Hardy JA, Abecasis G, Rosenberg NA, Scheet P (2009a) Genotype-imputation accuracy across worldwide human populations. Am J Hum Genet 84:235–250

Huang L, Wang C, Rosenberg NA (2009b) The relationship between imputation error and statistical power in genetic association studies in diverse populations. Am J Hum Genet 85:692–698

Knowler WC, William RC, Petitt DJ, Steinberg AG (1988) $Gm^{3;5,13,14}$ and type 2 diabetes mellitus: an association in American Indians with genetic admixture. Am J Hum Genet 43:520–526

Kosoy R, Nassir R, Tian C, White PA, Butler LM, Silva G, Kittles R, Alarcon-Riquelme ME, Gregersen PK, Belmont JW, De La Vega FM, Seldin MF (2009) Ancestry informative marker sets for determining continental origin and admixture proportions in common populations in America. Hum Mutat 30:69–78

Lake SL, Blacker D, Laird NM (2000) Family-based tests of association in the presence of linkage. Am J Hum Genet 67:1515–1525

Li M, Li C, Guan W (2008) Evaluation of coverage variation of SNP chips for genome-wide association studies. Eur J Hum Genet 16(5):635–643

Li Y, Willer CJ, Ding J, Scheet P, Abecasis GR (2010) MaCH: using sequence and genotype data to estimate haplotypes and unobserved genotypes. Genet Epidemiol 34:816–834

Liu DJ, Leal SM (2010) Replication strategies for rare variant complex trait association studies via next-generation sequencing. Am J Hum Genet 87:790–801

Marchini J, Howie B (2010) Genotype imputation for genome-wide association studies. Nat Rev Genet 11:499–511

Marchini J, Howie B, Myers S, McVean G, Donnelly P (2007) A new multipoint method for genome-wide association studies by imputation of genotypes. Nat Genet 39:906–913

Martin ER, Monks SA, Warren LL, Kaplan NL (2000) A test for linkage and association in general pedigrees: the pedigree disequilibrium test. Am J Hum Genet 67:146–154

Menzel S, Thein SL (2009) Genetic architecture of hemoglobin F control. Curr Opin Hematol 16:179–186

Milton S (1986) A sample size formula for multiple regression studies. Public Opin Quart 50:112–118

Musunuru K, Kathiresan S (2010) Genetics of coronary artery disease. Annu Rev Genomics Hum Genet 11:91–108

Nordborg M, Tavare S (2002) Linkage disequilibrium: what history has to tell us. Trends Genet 18:83–90

Nyholt DR (2004) A simple correction for multiple testing for single-nucleotide polymorphisms in linkage disequilibrium with each other. Am J Hum Genet 74:765–769

Pluzhnikov A, Below JE, Konkashbaev A, Tikhomirov A, Kistner-Griffin E, Roe CA, Nicolae DL, Cox NJ (2010) Spoiling the whole bunch: quality control aimed at preserving the integrity of high-throughput genotyping. Am J Hum Genet 87:123–128. doi:10.1016/j.ajhg.2010.06.005

Price AL, Patterson NJ, Plenge RM, Weinblatt ME, Shadick NA, Reich D (2006) Principal components analysis corrects for stratification in genome-wide association studies. Nat Genet 38:904–909

Pritchard JK, Stephens M, Donnelly P (2000) Inference of population structure using multilocus genotype data. Genetics 155:945–959

Purcell SM, Neale B, Todd-Brown K, Thomas L, Ferreira MAR, Bender D, Maller J (2007) PLINK: a tool set for whole-genome association and population-based linkage analyses. Am J Hum Genet 81:559–575

Purcell SM, Wray NR, Stone JL et al (2009) Common polygenic variation contributes to risk of schizophrenia and bipolar disorder. Nature 460:748–752

Sasieni PD (1997) From genotypes to genes: doubling the sample size. Biometrics 53:1253–1261

Schork NJ, Nath SK, Fallin D, Chakravarti A (2000) Linkage disequilibrium analysis of biallelic DNA markers, human quantitative trait loci, and threshold-defined case and control subjects. Am J Hum Genet 67:1208–1218

Slager SL, Schaid DJ (2001) Case-control studies of genetic markers: power and sample size approximations for Armitage's test for trend. Hum Hered 52:149–153

So HC, Yip BH, Sham PC (2010) Estimating the total number of susceptibility variants underlying complex diseases from genome-wide association studies. PLoS ONE 5:e13898

Speliotes EK, Willer CJ, Berndt SI et al (2010) Association analyses of 249,796 individuals reveal 18 new loci associated with body mass index. Nat Genet 42:937–948

Spielman RS, McGinnis RE, Ewens WJ (1993) Transmission test for linkage disequilibrium: the insulin gene region and insulin-dependent diabetes mellitus (IDDM). Am J Hum Genet 52:506–516

Takeuchi F, Serizawa M, Kato N (2008) HapMap coverage for SNPs in the Japanese population. J Hum Genet 53:96–99

Tang H, Peng J, Wang P, Risch NJ (2005) Estimation of individual admixture: analytical and study design considerations. Genet Epidemiol 28:289–301

The International HapMap Consortium (2003) The international HapMap project. Nature 426:789–796

The International HapMap Consortium (2005) A haplotype map of the human genome. Nature 437:1299–1320

Voight BF, Scott LJ, Steinthorsdottir V et al (2010) Twelve type 2 diabetes susceptibility loci identified through large-scale association analysis. Nat Genet 42:579–589

Weale ME (2010) Quality control for genome-wide association studies. Methods Mol Biol 628:341–372

Wellcome Trust Case Control Consortium (2007) Genome-wide association study of 14,000 cases of seven common diseases and 3,000 shared controls. Nature 447:661–667

Xu Z, Kaplan NL, Taylor JA (2007) Tag SNP selection for candidate gene association studies using HapMap and gene resequencing data. Eur J Hum Genet 15:1063–1070

Yang J, Weedon MN, Purcell S, Lettre G, Estrada K, Willer CJ, Smith AV, Ingelsson E, O'Connell JR, Mangino M, Mägi R, Madden PA, Heath AC, Nyholt DR, Martin NG, Montgomery GW, Frayling TM, Hirschhorn JN, McCarthy MI, Goddard ME, Visscher PM, GIANT Consortium (2011a) Genomic inflation factors under polygenic inheritance. Eur J Hum Genet 19:807–812. doi:10.1038/ejhg.2011.39

Yang J, Lee SH, Goddard ME, Visscher PM (2011b) GCTA: a tool for genome-wide complex trait analysis. Am J Hum Genet 88:76–82

Zeggini E, Scott LJ, Saxena R et al (2008) Meta-analysis of genome-wide association data and large-scale replication identifies additional susceptibility loci for type 2 diabetes. Nat Genet 40:638–645

Zheng J, Li Y, Abecasis GR, Scheet P (2011) A comparison of approaches to account for uncertainty in analysis of imputed genotypes. Genet Epidemiol 35:102–110

Zhong H, Prentice RL (2010) Correcting "winner's curse" in odds ratios from genomewide association findings for major complex human diseases. Genet Epidemiol 34:78–91

Zondervan KT, Cardon LR (2004) The complex interplay among factors that influence allelic association. Nat Rev Genet 5:89–100

Gene Expression Studies and Complex Diseases

5

Harald H.H. Göring

5.1 Introduction

The human genome contains a large number of genes, each of which may be viewed as a specific genomic segment that encodes information for one or several defined functions. Parts of the DNA sequence of a gene are transcribed into RNA that is then translated into protein. The vast majority of genes encode one or several proteins, and the intermediate RNA product is therefore referred to as messenger RNA, or mRNA. However, there are a number of genes producing non-coding RNA molecules, such as ribosomal RNA (rRNA), transfer RNA (tRNA), or micro RNA, among others, where the RNA molecules have a variety of functions by themselves. All forms of RNA molecules are the subject of gene expression studies but different technologies may be required to investigate different types of RNA.

Figure 5.1 provides a basic overview of the central information flow in biology and the various analytical techniques and areas of genetic and epidemiological investigation related to it. The main focus of human genetic epidemiological research is to identify the genes, and their variants, which influence our individual characteristics, with the most emphasis (and money) directed toward diseases and other clinically relevant traits. The statistical methods for correlating genotypes and phenotypes are referred to as linkage analysis and association analysis. For this type of analysis, genotype data must be generated. Over the last several years, aided by astounding progress in genomic and other laboratory technologies, a variety of additional approaches have gained popularity to investigate the genetic etiology of human diseases and their pathology. These approaches are complementary to genotype-based linkage and association analysis (and to each other), providing additional information that can be used to understand the biology of human conditions. These approaches include correlation analysis between a trait of interest and gene expression levels. This analysis requires quantification of gene expression levels rather than genotyping. Similar techniques include proteomic (Kooij et al. 2014; Van Eyk 2011) and metabolomic profiling (Kettunen et al. 2012; Sreekumar et al. 2009; Tukiainen et al. 2012) (or methylomic profiling (Mill and Heijmans 2013), which involves assessment of DNA methylation status; not shown in Fig. 5.1). In each case, the goal is to correlate a trait of interest to measured levels of transcripts, proteins, or metabolites, in order to identify any processes that are connected, in some manner, to the trait of interest.

H.H.H. Göring (✉)
Department of Genetics, Texas Biomedical Research Institute, PO Box 760549, San Antonio TX 78245-0549, USA
e-mail: hgoring@txbiomedgenetics.org

H.H.H. Göring
Department of Genetics, Texas Biomedical Research Institute, 7620 NW Loop 410, San Antonio TX 78227, USA

R. Duggirala et al. (eds.), *Genome Mapping and Genomics in Human and Non-Human Primates*, Genome Mapping and Genomics in Animals 5, DOI 10.1007/978-3-662-46306-2_5,

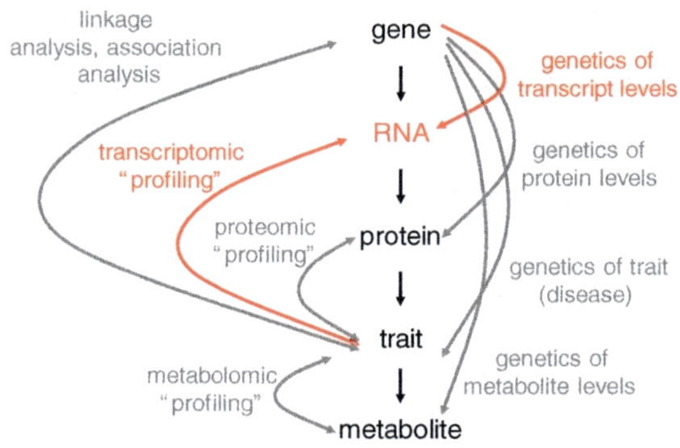

Fig. 5.1 Central information flow in biology and analytical techniques and investigations in genetic epidemiology. This chapter focuses on the investigations involving gene expression data, highlighted in *red*

It is now possible to simultaneously quantify the abundance of essentially all transcripts in a tissue sample (or even a single cell) using modern genomic technologies. The characterized transcriptome can then be used for two main purposes: First, the abundance of individual transcripts (or sets of transcripts) can be correlated with a trait of interest in order to identify those that are significantly correlated with the trait. The genes encoding these transcripts may possibly be involved in the etiology of the trait, and/or it may be that the trait in turn has an impact on the expression levels of these genes. I will refer to this type of investigation as transcriptional correlation analysis, transcriptional profiling, transcriptomic profiling, or gene expression profiling. Second, each transcript may be viewed as a trait whose genetic regulation can be investigated by statistical genetic technologies, in order to identify genomic variants that influence the transcriptional activity of the gene being examined. This second type of investigation is sometimes referred to as "genetical genomics" (de Koning and Haley 2005; Jansen and Nap 2001), a terminology which I view as confusing and which I will not use here. Both of these investigations involving gene expression data—transcriptional profiling and genetic regulation of gene expression—are the topics of this chapter.

Another way to describe the central analyses involving gene expression data in genetic epidemiological research is shown in Fig. 5.2. There are three central information sources available to us—trait phenotypes, gene expression levels, and genotypes—and these permit three types of correlations to be analyzed. The traditional association analysis (or linkage analysis) investigates the relationship between trait phenotypes and genotypes at polymorphic variants (shown on the left side of Fig. 5.2). Assuming that the genetic variants influencing a trait of interest exert their effect via modulation of gene expression, gene expression data may be viewed as an intermediate trait between genotype and clinical outcome of interest, and the overall correlation between genotype and trait phenotype may be viewed as

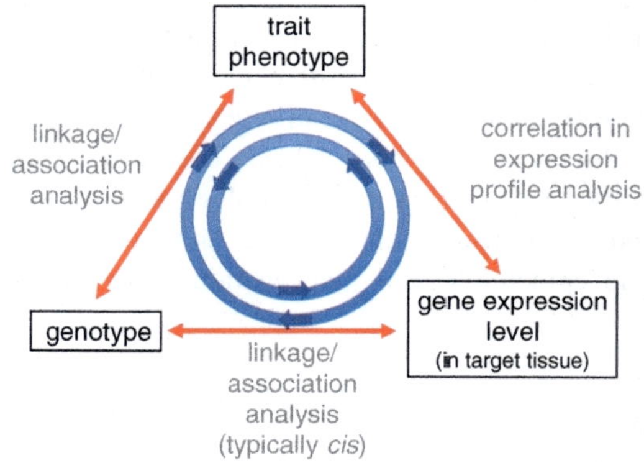

Fig. 5.2 Overview of analyses relating trait phenotypes, gene expression levels, and genotype data. Note that the circle of analyses is ideally conducted on a single sample, which requires that all three types of data are available. Alternatively, multiple different samples may be used if necessary. The analyses can be conducted in both directions

an amalgamation of the correlation between genotype and gene expression level and between gene expression level and the trait phenotype. Again, these are the two types of investigations involving gene expression data commonly undertaken, here referred to as transcriptional profile analysis and analysis of the genetic regulation of gene expression, respectively, as mentioned previously.

Perhaps the main motivation behind transcriptional investigations at the present time is the observation that most of the genetic variants that are significantly associated with complex traits (as typically identified in genome-wide association studies, or GWAS) do not alter the amino acid sequence of proteins or have any other obvious functional effect (Hindorff et al. 2009; Visscher et al. 2012; Welter et al. 2014). In many cases, the associated variants are located outside of genomic regions known to be part of a gene. This leads to the speculation that the variants underlying complex traits are often regulatory in nature. This is in contrast to the genetic defects underlying many Mendelian disorders, many of which directly impact the protein sequence, thereby altering or abolishing protein function. While the activity of genes is regulated at many different levels, including at the stage of transcription, translation, and post-translationally by modification of proteins, transcriptional regulation is a critical component, and the abundance of transcripts can be assessed more readily and more comprehensively than the amounts of proteins and their modifications, due to the pairing potential of the building blocks of DNA and RNA, which is what makes amplification via PCR possible. It is important to note that transcriptional regulation itself is a highly complex process, with multiple potential stages of regulation, including in the location, timing, and speed of transcription, in the decay of transcripts, in the usage of alternative promoters, transcriptional stops, and different splice sites (which lead to the existence of potentially many different transcripts per gene), in other RNA editing processes, and at other steps which are not yet understood. When gene expression level is measured using some molecular technique in the laboratory, we often do not know whether it is a higher transcription rate or a lower rate of decay (or both) that is behind the observed high abundance of a given transcript. Fortunately, this does not complicate the statistical analyses per se, but this uncertainty must be taken into account when interpreting the results. Note also that terms such as "gene expression level" or "transcript level" are often not well-defined in manuscripts, with sloppy wordage widespread. In most cases, what is measured is the abundance of RNA detected by some specific probe (as in microarray studies) or the number of RNA sequencing (RNAseq) reads belonging to a gene or parts of a gene (such as an exon). Quantifying specific transcripts is actually very challenging even now and is not routinely undertaken in most large-scale transcriptomic studies.

This book chapter focuses on the overall concepts related to gene expression studies. Details on the laboratory and analytical methods of these studies are beyond the scope here. This field of investigation is developing rapidly and gaining in popularity, and I will only focus on some key aspects of these types of studies. The references cited here are only a small selection of the rapidly expanding literature; the reader is encouraged to find additional, and potentially newer or better, references on his/her own.

5.2 Correlation Analysis Between Trait Phenotypes and Expression Levels

As mentioned, gene expression data permits us to correlate trait phenotypes with gene expression levels. Those genes whose expression level is significantly correlated (after appropriate multiple testing correction) with the trait of interest are presumably somehow related to that trait, and the set of genes may shed light on the biological pathways related to trait etiology or physiology. The large literature on complex trait gene mapping studies, in particular GWAS studies (Hindorff et al. 2009; Visscher et al. 2012; Welter et al. 2014), clearly shows that individual common variants associated with the risk of a complex disease can be localized. However, it should not be forgotten that this is a challenging undertaking even for common variants, typically requiring sample sizes that were unimaginable only a few years ago. The reason is that these variants individually only modestly influence the risk (of some disease) for a person, and at a population level account for only a small fraction of trait heritability. Common variants with a large influence on penetrance are very rare for complex traits. One example is ApoE4 and Alzheimer's disease (Corder et al. 1993; Strittmatter et al. 1993). It seems likely that for many complex diseases low frequency variants of high penetrance exist (Terwilliger and Göring 2009), but these variants are difficult to localize because of their rarity. Conceptually, transcriptional profile studies could be more powerful. One intuitive reason may be that many functional variants within a gene could be assessed simultaneously using this approach, as long as these variants influence the expression level that can be measured. In effect, the expression level is a read-out that combines the effects of all regulatory variants impacting transcription, and it might therefore be easier to identify genes related to a trait of interest. At least that is the argument that is often made (It may also be the case that the multiple testing burden is somewhat reduced compared to genome-wide analysis of sequence variants, but this may no longer hold as our molecular techniques for gene expression characterization become ever better, leading to discovery of many alternative transcripts per gene).

Despite the general promise and potential, correlation studies between a clinical trait and transcript data have substantial drawbacks. For illustration, let us contrast these studies with a linkage or association study that is based on genotype data. When significant evidence of linkage and/or association has been found, it is clear that a causal factor in the etiology of the trait being studied has been localized (assuming that the finding is not a false positive, the approximate risk of which can be gleaned from the obtained significance level). We may not know which gene(s) and which variant(s) in the mapped candidate region are causal, but it is certain that the identified genomic region harbors one or several variants that influence the trait being investigated. The only real concern related to study design is that the study subjects are matched for ethnicity, which really means that the genome-wide allele frequencies are very similar in cases and controls. If this concern is avoided by design from the outset, such as by ascertaining cases and controls from the same ethnically homogeneous population, or even by a variety of statistical control techniques after ascertainment, and if there is no difference in genotyping approach and quality between both groups, then a causal inference is warranted. [As an aside: One more caveat is that the relationship between genotype and trait risk may not be direct. One example is the FTO locus that was first identified as a risk locus for type II diabetes in a case-control study of that disease (Zeggini et al. 2007). It ultimately turned out that it is really an obesity locus (Frayling et al. 2007) that was mapped in the diabetes study because the cases and controls had different average body mass index levels because of the correlation between obesity and diabetes in the population.] Note that in such genetic studies it is not absolutely crucial (though it may help power and be therefore warranted) that cases and controls are matched for other characteristics, such as sex, age, socioeconomic status, or smoking habits. The key reason behind all these characteristics of linkage and association studies is that genotypes are constant throughout life. Ignoring many exceptions here for simplicity, genotypes are the same in all cells of an individual and are independent of environmental factors to which a person is exposed.

The situation is very different in studies relating a trait of interest to transcript abundance. The expression level of a given gene is not the same in all cells of a body, and the expression level is often influenced by factors of the external environment as well as of the internal environment (i.e., the body and its conditions). The ramifications of this are profound. First, if a significant correlation between disease and gene expression has been observed, the cause–effect relationship is not clear. To stay with diabetes as our example, it is possible that significant transcriptional correlates of the disease are involved in the etiology of the trait (as the loci identified in genetic association studies definitely are). It is also possible that the identified genes are themselves influenced in their expression by the disease. And both relationships could exist at the same time. The former would be useful to learn about trait etiology, while the latter would inform about pathophysiology. Upfront, it is not clear whether the identified transcripts "act upstream or downstream" of the studied disease. A prospective study design or other timeline techniques may help to clarify what is cause and what is effect. Perturbation studies, in which gene expression of a sample is taken before and after some type of manipulation, such as exposure to a chemical, may also help in addressing the ordering of the observed significant correlations.

Second, it is possible that confounder variables could explain the observed, statistically significant correlations between trait and transcripts. For example, it is possible that, say, the diabetic study participants take medications that the control individuals do not; or that the diabetics on average eat a different diet than the controls; or that they exercise less than the controls; and so on. All of these differences between cases and controls, individually or jointly, could potentially explain the observed differences in transcript levels, in which case the observed significant correlations would be artifacts caused by confounder variables. It is very difficult to exclude this possibility in transcriptional profile studies. It is advisable to match cases and controls for as many possible confounder variables as possible, or to measure known confounders and subsequently account for their effects analytically. However, the identity of these confounders is often not known or they cannot be measured accurately. A possible solution is to use the transcriptional profile data itself as a way to identify potential confounders. One example of this is to use the expression levels of "indicator genes" to infer the cell type composition of a tissue sample (such as blood) (Gaujoux and Seoighe 2013). A general approach could be implemented based on principal components analysis, and the top principal components (which tag the key sources of covariation in the expression data, many of which may be related to potential confounder variables) could be subsequently regressed out. The difficulty here is that this approach risks removing the very relationship between trait and gene expression that one seeks to identify; in essence, one may throw out the baby with the bathwater.

Third, tissue specificity of gene expression comes into play. In many cases, the appropriate target tissue is not accessible, or it may not be known. In those situations, investigators conduct their study using another source tissue, in the hope that it will serve as a suitable surrogate tissue. However, this is not a generalizable characteristic of different tissues, because it can vary from gene to gene (and potentially from genetic variant to genetic variant) whether two tissues are suitable proxies for one another. This is discussed in detail below.

We have grappled with these types of complications in a study of schizophrenia, where we contrasted the expression profiles from lymphoblastoid cell lines (LCLs) from cases with schizophrenia and controls without the disease (Sanders et al. 2013). While it is perhaps unlikely that differences in expression levels of cell lines are caused by the disease status (or related differences in environment, attributable to disease-related medication or, say, smoking)—after all, these cell lines are far removed from study subjects and their exposures—it is difficult to exclude the possibility that some aspects of the LCLs could vary between cases and controls, independent of disease. For example, could it be that the LCLs of cases and controls were generated in slightly different manners, which may be the cause of observed transcriptional differences? To guard against this, we measured various cell lines characteristics as part of the study and included these variables as covariates. However, the fact remains that using different sets of covariates leads to somewhat different findings, and one cannot know whether all relevant confounders are accounted for.

All these complications that arise in transcriptional correlation studies compared with genotypic correlation studies can be viewed as the difference between genetic epidemiology and epidemiology more generally. It is the former that is unusual. The nature of genetic inheritance endows genetic epidemiological studies with many advantages that are not shared in most areas of epidemiological research. When correlating transcript abundances with clinical phenotypes, we are no longer in the realm of genetic epidemiology, and thus face many systematic challenges that can be difficult to overcome.

5.3 Genetic Regulation of Gene Expression

Instead of correlating gene expression to a trait of interest, as discussed in the previous section, gene expression levels can also be subjected to statistical genetic dissection. The quantitative expression level of a gene, an exon, or a specific transcript may be viewed as a quantitative trait that is under the influence of genetic and environmental influence like any other trait. Linkage and association analyses can therefore be conducted on transcript abundance values in order to localize the genomic regions and variants that influence the amount of a transcript being present in a given sample.

Studies investigating the genetic regulatory machinery influencing gene expression levels have gained popularity for two main reasons: First, they permit us to study the basic biology underlying a key regulatory step in how our genes' activities are controlled. Second, knowledge about which genetic variants are significantly associated with the expression of a particular gene provides clues about the identity of likely functional variants and their regulatory potential. This sort of functional information can be used, potentially along with many other pieces of information, to prioritize which of the variants that were previously identified in a GWAS of a complex disease are most likely to be functional. This information is relevant in particular because of the hypothesis that most of the functional variants underlying complex traits are subtle regulatory variants. Ultimately, laboratory assays will generally be required to prove that a given variant causally influences a trait of interest, but by accumulating different sources of information on each variant, including whether or not (and in which direction and to which degree) it is associated with the expression of a particular gene, we can hone in on the true functional variants with a greater degree of precision, thereby reducing and speeding up the more time-consuming and much more expensive functional assays to be conducted in the laboratory. A genetic variant found to be significantly associated with some expression level is typically referred to as an expression quantitative trait locus (eQTL) or an expression quantitative trait nucleotide (eQTN) in the case of single nucleotide polymorphisms (SNPs). The term regulatory SNP, or rSNP, is also sometimes used (Guo et al. 2014).

Genetic studies of gene regulation have proven to be highly successful in many regards. This is perhaps not surprising because the expression level of a given gene is, after all, a very direct representation of gene action, and the impact of a regulatory genetic variant on gene expression thus could be pronounced. The relationship would certainly appear to be much closer than one would expect to exist between a gene's activity and a complex trait, where any one variant influences disease risk typically by only a very small amount. Thus, one might expect a priori that studying the genetic regulation of gene expression would be a fruitful undertaking. A variety of studies, conducted in families, in twins, and more recently in unrelated individuals, have shown that the vast majority of gene expression levels are significantly heritable (Göring et al. 2007; Nica et al. 2011; Price et al. 2011; Grundberg et al. 2012). This is clear evidence for the existence of genetic regulatory variants and their influence on gene expression levels in the aggregate. Note, however, that the estimated heritabilities for many expression traits are quite modest, similar to the estimates obtained for many complex diseases. This suggests that either there is substantial measurement error in quantifying gene expression levels, and/or that these traits are subject to myriad influences, including by environmental factors (both of the external environment acting upon a person as well as the internal environment, within the body, to which a given cell is exposed). Therefore, gene expression levels are best viewed as being fairly complex traits.

We conducted one of the largest genetic investigations of genome-wide gene expression at the time, measuring gene expression by Illumina microarrays in white blood mononuclear cells in 1,240 randomly ascertained Mexican American family members from around San Antonio, Texas, USA (Göring et al. 2007). A brief review of this study is provided here as an example of a genetic investigation of gene expression. After full processing, 20,413 (43 %) probes out of a total of 47,289 on the microarray detected significant expression at a false discovery rate (FDR) of 0.05. Among the autosomal probes with significant expression, the quantitative expression levels of the vast majority of probes (85 %) were significantly heritable at FDR 0.05. The median heritability estimate was 23 %, with higher heritabilities among RefSeq probes (which are much better designed and annotated, on average). These estimates support the view that gene expression traits are substantially controlled by genetic factors. We subsequently conducted linkage analysis in order to localize major loci influencing the expression traits. As described in the sections below, we broke the genome into two components—the gene locus targeted by a given probe itself and the remainder of the genome. At FDR 0.05, a large number of probes (1,345), though representing only a fairly small proportion of probes (7 %), showed significant evidence of linkage to the structural gene locus. This is clear evidence that genetic variants in and around a given gene, such as in the promoter region, have substantial influence on that gene's expression levels. For some genes, the heritability attributable to the structural gene locus explained virtually all the estimated overall heritability, suggesting that these genes' expression is largely monogenically controlled. The mean effect size of the structural locus was 5 % on average (with a median of 2 %). The structural locus thus appears to account for a substantial proportion of heritability (based on this particular study, the estimated proportion of genetic variance explained by the structural gene locus is in the range of 10–25 %). We largely failed to identify significant eQTLs elsewhere in the genome, far away from the structural gene, suggesting that these distant regulatory genetic factors, while clearly important in the aggregate, have individually very small effect sizes, making them difficult to detect. Later studies, using better molecular technologies, have largely supported our findings, refining estimates and identifying many more significant eQTLs, as described in the following paragraphs.

5.3.1 *Cis* eQTLs

Given the substantial heritabilities of most gene expression levels, the logical next step is to conduct linkage and in particular association analyses in order to localize specific variants that are significantly associated with expression traits. In contrast to studies of most clinically relevant traits, when investigating the genetic regulation of a particular gene there is an obvious genomic candidate region, namely the gene itself and its chromosomal vicinity (see Fig. 5.3). The reason for paying special attention to this small fraction of the genome is that genetic variants near a gene, and in particular in its promoter region, are quite likely to influence that gene's expression level, e.g., by interfering with the binding of proteins required for transcription. Many of these variants presumably act in *cis*. The difference between *cis* and *trans* is shown in Fig. 5.4. By *cis* (from Latin, meaning "on this side") we mean that a given gene expression regulatory variant influences the expression level only on the physical molecule—i.e., the chromosome—on which it resides, but not on the homologous sister chromosome. The reasons are likely structural, i.e., proteins and other factors bind to a particular chromosomal region to initiate, maintain, and regulate gene expression of that chromosomal molecule, and alleles on that entity thus influence the expression of the gene on only that chromosomal copy. Most

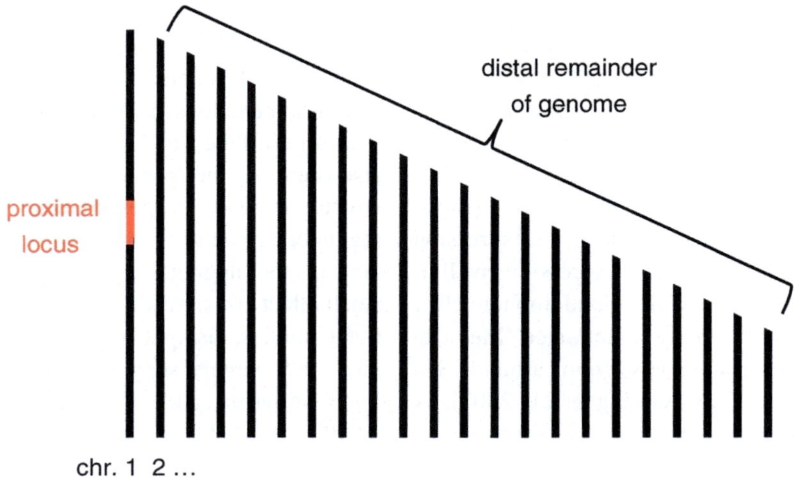

Fig. 5.3 Partitioning of the genome into the proximal locus and the distal remainder of the genome (relative to a gene). When investigating the genetic regulation of gene expression of a particular gene (shown here in *red*, located in chromosome 1), the genome-wide search may be conducted in two parts. The gene itself and its chromosomal surrounding area is a good candidate region to harbor *cis* eQTLs, while the rest of the genome may harbor *trans* eQTLs. Note the enormous differences in search area and associated multiple testing burden

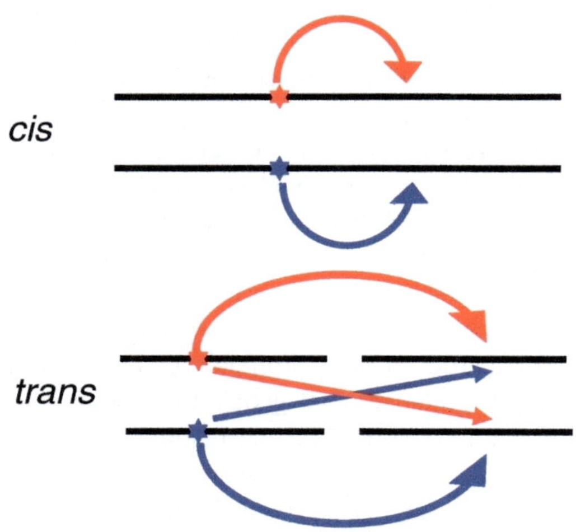

Fig. 5.4 Illustration of *cis* and *trans* effects. An allele acting in *cis* influences only the molecular molecule (chromosome) on which it resides. In contract, *trans* acting factors influence both sister chromosomes

variants located elsewhere in the genome are thought to act in *trans* (from Latin, meaning "on the other side"). These variants influence both chromosomal copies equally. For example, some variant may alter the structure of a transcription factor, which in turn alters the expression of both copies of a given gene—regardless on which sister chromosome a copy of the gene is located. In general, *cis*-acting variants are equated with those close to a given gene, sometimes referred to as proximal variants. Similarly, the term *trans*-acting variant is used for those variants located far away on the same chromosome or on a different chromosome, sometimes called distal or distant variants. While the assumption about how a variant acts based on where it is located relative to a gene will often be correct, it is nonetheless generally only an assumption until confirmed by other means (Gilad et al. 2008). Some variants close to a gene may turn out to be *trans* eQTLs. And some *cis* variants may be located far from a gene but on the same chromosome. (Could it be that *cis* variants may even be located on another chromosome, depending on how chromosomes are packed in three dimensions within the nucleus?) This caveat should be kept in mind when reading the literature, in which these terms are often used interchangeably.

Given the existence of a well-justified candidate region of interest around any gene, one may break a genome-wide search for eQTLs into two parts, one confined to a gene and its surrounding genomic region to search for *cis* variants, and one covering the remainder of the genome to localize *trans* variants (Göring et al. 2007). eQTL studies have proven to be highly successful in the search for *cis* variants (Cheung et al. 2005; Dixon et al. 2007; Göring et al. 2007; Stranger et al. 2007; Emilsson et al. 2008; Montgomery et al. 2010; Pickrell et al. 2010; Grundberg et al. 2012). With gradual improvements in measuring gene expression, and with increased sample sizes, the proportion of genes estimated to be *cis* regulated is creeping upwards, from a small number of *cis*-regulated genes observed at first to perhaps ultimately the majority of genes investigated. It appears likely that every gene is under *cis* regulation to at least some degree, with only power limiting our ability to detect significant proximal associations for all genes. *Cis* eQTLs are therefore frequent, and in fact likely universal for all genes in all tissues. Their impact on quantitative gene expression has frequently been shown to be substantial. In some cases, *cis*-regulatory variants appear to account for much of the estimated heritability of a gene, indicating that the expression level of such a gene is essentially a monogenic trait (though not necessarily influenced by only a single variant). For other genes, *cis*-regulatory effects account for only some proportion of overall genetic variation, suggesting a more complex mode of inheritance with substantial aggregate importance of *trans*-acting variants. *Cis* effects are fairly easy to detect for two reasons: Their commonly strong effect sizes, and because of the limited multiple testing correction that is required when searching for them, because only a very small proportion of the genome, namely the gene and its vicinity, needs to be screened. By now, there are catalogs listing many putative *cis* eQTLs for many genes in many tissues, and these databases are freely available http://www.ncbi.nlm.nih.gov/projects/gap/eqtl/index.cgi (Yang et al. 2010; Xia et al. 2012).

5.3.2 *Trans* eQTLs

In contrast to the local search for *cis* variants, mapping of *trans* variants requires searching systematically through the entire genome (excluding the small proximal region around a gene). *Trans* eQTL studies have proven to be quite difficult (Cheung et al. 2005; Dixon et al. 2007; Göring et al. 2007; Stranger et al. 2007; Emilsson et al. 2008; Montgomery et al. 2010; Pickrell et al. 2010; Grundberg et al. 2012), perhaps more difficult than some scientists had initially assumed. Part of the explanation is the enormous multiple testing burden incurred when searching the entire genome, especially when one does so on thousands of gene expression traits. Beyond that reason, the difficulty of finding *trans* eQTLs indicates that these regulatory variants individually only influence the expression level of a given gene modestly. In other words, the effect sizes of individual *trans* eQTLs are small. Thus, large sample sizes will be required to localize these variants robustly and comprehensively. At the present time, many studies report a small number of potential *trans* eQTLs, but the evidence is generally fairly weak and large-scale replication studies are still limited (Grundberg et al. 2012). There are some observations suggesting the existence of "master regulators", i.e., *trans* eQTLs that influence the expression of many genes. This makes intuitive sense if one, e.g., thinks of variants within a transcription factor that is involved in the expression of a whole range of genes. At the present time, much of the supporting data for master regulators is fairly modest, and future work is required to characterize master regulatory eQTLs better.

Figure 5.5 shows an example of a genome-wide joint linkage and association study for a particular gene (*PPA2*), in this case conducted on peripheral blood mononuclear cells from randomly ascertained participants belonging to multigenerational families (Göring et al. 2007). Note that this plot looks very different from a so-called Manhattan plot from a normal GWAS on a complex disease. Here, there is an enormous peak on chromosome 4, which is centered on the exact location where the studied gene is located in the human genome. This signal almost certainly points to *cis* variants in and near the gene, and the magnitude of the signal highlights the substantial effect size of the variants in the proximal gene region. In contrast, the remaining

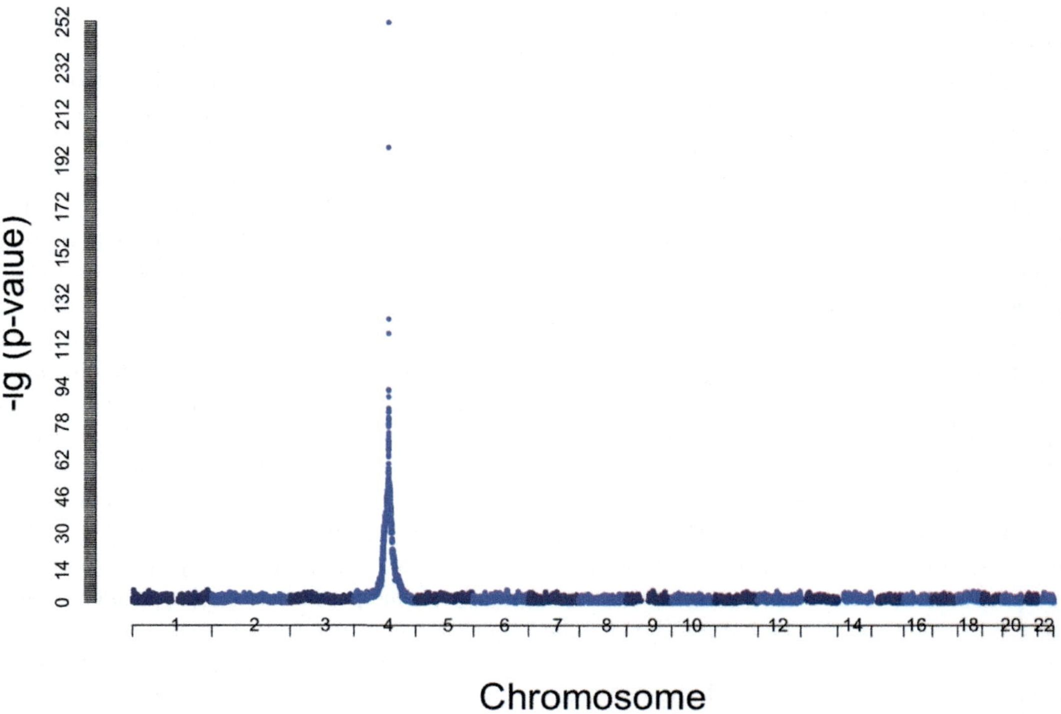

Fig. 5.5 An example of a genome-wide search for eQTLs. The particular example is the *PPA2* gene (inorganic pyrophosphatase 2 precursor), whose expression level in peripheral blood mononuclear cells was assessed with probe GI_31881619-A on a microarray. The large linkage/association peak on chromosome 4 is located at the position of the gene and demonstrates the strength of the effects of *cis* eQTLs on gene expression. In contrast, the plot does not contain outstanding peaks elsewhere in the genome, highlighting the small effect sizes of *trans* eQTLs and the associated difficulty in localizing them

genome yields a very flat pattern, without any outstanding peaks. This illustrates the small effect sizes of *trans* eQTLs and the associated difficulty in localizing them.

5.4 Integrative Genomic Studies

In the previous two sections, I have separately discussed studies correlating gene expression profiles to a trait of interest and studies investigating the genetic regulation of gene expression, respectively. These are two of the three branches of investigation shown in Fig. 5.2, with trait-genotype correlation analysis being the third, and most commonly performed investigation. Ideally, we would like to bring as many sources of information to bear to dissect the etiology of a trait and to identify the functional genetic variants. I have tried to illustrate this conceptually in Fig. 5.6. Thus, we would like to integrate the results obtained from different data sources, to comprehensively assess the evidence for genetic correlation of specific variants with a clinical trait of interest, and the likelihood that the associated variants are functional and of relevance. In the case of gene expression studies, the focus is on the three types of analyses shown in Fig. 5.2, but this is not meant as a suggestion that other sources of information, such as from proteins, metabolites, gene methylation, sequence conservation, predictions of deleteriousness of variants based on structural protein changes, etc., are unimportant or should not be used.

Such integration of the central three types of analyses shown in Fig. 5.2 is not easy, at least if

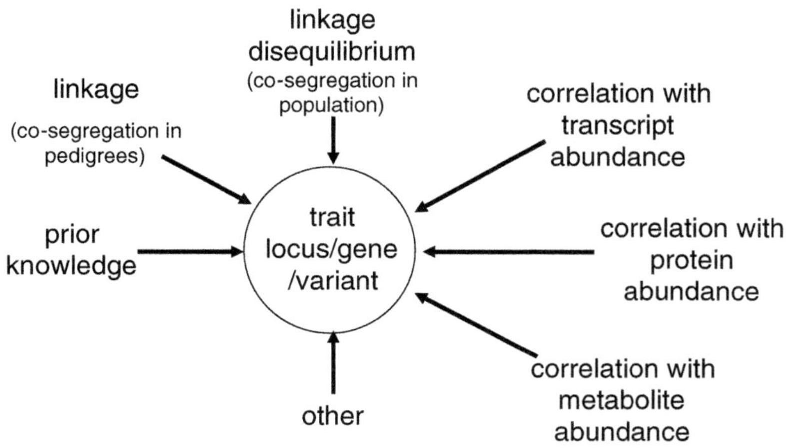

Fig. 5.6 Many different information sources can be used to identify the loci, genes, and genetic variants influencing a complex trait. Integration of these disparate data sources can be difficult, and this chapter mainly focused on the use of genotype data and transcriptional profile data

the goal is to use a comprehensive analytical approach (such as a Bayesian approach in which the posterior probabilities yielded by one type of data and analysis serve as the prior probabilities for the next data source and analytical step). In most cases, studies use more of an ad hoc approach. If data on trait phenotypes, genotypes, and gene expression levels are all available in the same dataset, one may first perform regular linkage and association analysis, and then subsequently examine whether the significant (or suggestive) trait-associated variants are also significantly associated with the expression level of nearby genes. This could suggest a possible way in which a positional variant may influence the trait of interest, and may increase the interest and attention devoted to this variant. Lastly, one may then investigate whether the gene(s) regulated by the eQTLs show evidence for correlation with the trait of interest, thereby closing the circle. Note that the three analyses comprising the circle of relationships must yield results that are consistent with one another (at least if undertaken on the same dataset). Either all three correlations are positive, or one correlation is positive while the two others are negative. If different datasets are used for association analysis of the trait, for eQTL discovery, and/or for transcriptional profiling of the trait, then one may still seek to combine the results from these separate investigations, but power may be reduced (e.g., because a given variant may have a large effect size in one dataset but not in another). In this case, it is possible that the results of the three analyses are no longer consistent with one another. However, truly existing, important relationships between a trait, genotype, and gene expression level should still yield consistent results (except in unusual situations).

It is also possible to perform the analyses in the opposite direction. One may start with a transcriptional profile study, then identify *cis* eQTLs influencing trait-correlated transcripts, and lastly examine whether the identified candidate variants truly show evidence of association with the clinical trait. When the analyses are conducted in this orientation, then the eQTL and trait association steps can be used to determine whether the observed transcriptional signatures reflect genes involved in the etiology of the trait, or whether the causal connection goes in the other direction, with the trait phenotype influencing the expression levels of these genes (Schadt et al. 2005). In general, there is ample opportunity for smart approaches to be developed to integrate different information sources and to use these data to order the relationships between the different variables being examined.

5.5 Tissue Specificity

Tissue specificity is an important consideration in gene expression studies. This is a very important topic, but it is also one that is complicated and for which there are no clear-cut answers that generalize to all traits, genes, and genetic variants. This issue mainly arises because the tissue that is the primary source of a disease is often not available for study (In fact, in many cases the true source tissue is not even certain). Often the reason is that the tissue is inaccessible to a certain degree, and ethical considerations preclude invasive procedures required to obtain it. A further, related complication is that most tissues comprise many different cell types, which vary from one another in their gene expression, and the relevant cell type is often not known or cannot be readily isolated (or only in a manner which possibly impacts those cells and their behavior greatly).

When talking about tissue specificity it is important to realize that it is not important whether the absolute expression level of a gene is the same in different tissues—as long as the gene is expressed sufficiently highly so that the expression level can be accurately measured. When assessing whether one (accessible) tissue can serve as a surrogate for another (inaccessible) tissue, what matters is whether the inter-individual variation in the expression of a gene is maintained between different study subjects. In eQTL studies, the critical question is whether the same genetic variants influence gene expression, and whether the direction and effect sizes of these variants are similar. This has been empirically examined in a number of studies (Ding et al. 2010; Greenawalt et al. 2011; Nica et al. 2011; Grundberg et al. 2012). In transcriptional profiling studies relating gene expression to some disease, the central concern is whether the relationship between the trait and gene expression level is the same in both tissues. My own opinion is that the suitability of a surrogate tissue will depend on the trait being studied, the particular gene(s) involved, and their underlying eQTL(s). It seems unlikely that any given tissue, or for that matter any given cell type, can universally serve as a proxy for another tissue or cell type. The best that we can hope for is to estimate the similarity of gene expression and its genetic regulatory machinery between as many different tissues and cell types as possible, in order to identify the best overall match, with the greatest overlap in gene expression patterns and eQTLs (Göring 2012).

An important project designed to assess tissue specificity of eQTLs is the US American National Institutes of Health sponsored Genotype-Tissue Expression Roadmap project now underway (GTEx Consortium 2013). The goal is to obtain tissue samples of >1,000 fatal accident victims, gathering as many tissues as quickly as possible after death. These tissues will then be expression profiled, and large-scale eQTL studies will be undertaken in each tissue separately and multiple tissues jointly. Ultimately, this will lead to a catalog of eQTLs, and their estimated effect sizes, in a large number of human tissues. This will permit us to identify best tissue matches, potentially even on a per-gene or per-genetic variant basis. An alternative approach, which is less centralized in scientific direction but which may ultimately be even more informative, is based on attempts to recreate different cell types (ultimately all cell types?) from induced pluripotent stem cells (iPSCs). This topic is well beyond this chapter, but this technology ultimately holds the promise that many (or even all?) cell types become accessible for gene expression study in each study participant or clinical patient (Robinton and Daley 2012).

At the present time, our knowledge of the tissue specificity of eQTLs is fairly limited. It appears to be the case that strong *cis* eQTLs, in particular those close to the transcriptional start site, are fairly universal between tissues, and that those further away are increasingly tissue-specific (Dimas et al. 2009; Grundberg et al. 2012). There are some indications that *trans* eQTLs may often be tissue specific (Grundberg et al. 2012). An important caveat to keep in mind when interpreting these results is that the real effect size of a true eQTL is correlated with our ability to detect it in the first place, as well as with our certainty that the finding is real. Thus, the weaker any *cis* eQTL variants are as we move away from the transcriptional start site, the less certain we are to detect them. It is thus not surprising that less consistency is observed for those more subtle variants. This caveat is even more important in the case of *trans* eQTLs, whose effect sizes are generally smaller and where our power of localization is further weakened owing to the enormous multiple testing burden. While it seems quite plausible that more distant *cis* eQTLs and *trans* eQTLs are more tissue-specific in their influence on gene expression than strong *cis* eQTLs close to the transcriptional start site, I do not find the supporting data wholly convincing at the present time.

Note that it is fully rational and also reasonable, at least in my opinion, to use proxy tissues in many scientific examinations at this point in time. While negative results may be difficult to interpret and may even be entirely uninformative, positive correlations observed between a clinical trait and a gene's expression, or the realization that a candidate variant may be an eQTLs, provide potentially interesting clues that can then be pursued in more detail in the laboratory and/or in a more appropriate tissue that is only available in few samples.

5.6 Microarray Versus RNAseq

Microarrays containing a large number of probes and RNA sequencing (RNAseq) are the two approaches that are now being used to characterize gene expression on a whole genome basis. Older methods, such as quantitative PCR, continue to be used as well, but they are limited to specific genes rather than assess the entire genome at once. Microarray and RNAseq technologies both have advantages and disadvantages. For a review, see (Majewski and Pastinen 2011). Some of the pros and cons of both approaches are the following: The drawbacks of microarrays are that they are limited to known transcripts; they assess only the expression level of a short stretch of RNA and generally cannot distinguish between alternative transcripts; they are susceptible to polymorphisms in the sequence targeted by a probe; they require many copies of RNA molecule for robust expression detection and quantification; they are somewhat susceptible to batch effects. On the plus side, however, microarray studies are fairly cheap, fast, and require limited annotation work by the investigators.

In contrast, RNAseq is (at least conceptually) able to identify all transcripts, including alternative transcripts of a gene; RNAseq is much better suited for the study of RNA editing; since there is no probe per se, RNAseq is less susceptible to polymorphisms in specific transcript regions (though the presence of polymorphisms may interfere with alignment); and RNAseq is much more sensitive in the detection of low frequency (even single copy?) RNA molecules. Downsides of the technology include its substantial cost, and the substantial annotation work that is required. Also, RNAseq is sensitive to sequencing problems and artifacts, and it is not clear whether low copy transcripts have biological function (even if they are reproducible).

Previously, most studies used microarrays, but increasingly the field is transitioning to RNAseq as the preferred choice of technology. As the cost of sequencing comes down more, and as the length of sequencing reads and sequence accuracy improves further, the benefits of RNAseq compared to microarrays will become more pronounced.

5.7 Allele-Specific Expression Analysis

Most eQTL studies conducted to date have searched for association between the genotype of a genetic variant and the overall expression level of a gene, exon, or particular transcript. In many cases, transcripts themselves include polymorphic sites, and the allele present in a given transcript molecule tells which of the two sister chromosome produced the transcript. It is thus possible to estimate the expression level separately for each chromosome. Each "allele's" expression level can then be genetically analyzed separately. It is particularly useful to analyze the proportion of transcripts of a given gene derived from one of the two sister chromosomes, to search for eQTLs, which is referred to as allele-specific expression (ASE) analysis (Almlöf et al. 2012; Pastinen 2010). The reason why allele-specific expression analysis is so useful is that there is a built-in internal control for many factors, including most environmental influences on gene expression and also *trans* eQTLs. In general, these factors will equally influence the expression of both copies of a gene, assuming there is no interaction between the factors and proximal eQTLs. (And the very meaning of the term *trans* refers to the fact that *trans* eQTLs impact the expression of both chromosomes). By taking the relative proportions of transcripts derived from one chromosome compared to its sister chromosome, one automatically controls for these chromosome-non-specific factors. For this reason, ASE studies are extremely powerful for the detection of *cis* eQTLs. It seems likely that many more ASE studies will be conducted in the future, and there is opportunity to refine the analytical methods and to combine ASE analyses with conventional eQTL studies.

5.8 Exposome Studies and Intervention Studies

This chapter has focused on the utility of transcriptional profile studies to investigate the etiology and pathophysiology of complex diseases and to study the genetic regulation of gene expression. More generally, gene expression profiles are one kind of "deep cellular phenotype", providing a highly detailed characterization of the state of a given cell type or tissue type from a study subject at the time of sample collection. Therefore, gene expression profiling is a very general tool that can be used to address many different research questions.

One area of great interest is to use transcriptional profiles for investigating the influence of environmental factors. The totality of the environmental factors to which we are exposed is sometimes referred to as the exposome (Wild 2005). Therefore, one can attempt to correlate gene expression data to measured environmental exposures to search for significant transcriptional correlates of the exposure. This can provide information how the exposure influences cellular biology. For example, one may contrast smokers to nonsmokers. Significant differences in transcriptional profiles may conceptually include genes that influence the probability that someone is a smoker (these genes would therefore be involved in the etiology of the smoking trait). Or the differences may reflect the consequences of smoke inhalation on cellular processes, which may be useful for understanding how smoking influences our body and what the pathophysiological consequences may be. We have conducted a transcriptomic study of smoking and found substantial differences between smokers and nonsmokers in the transcriptional profiles from PBMCs (Charlesworth et al. 2010). Our interpretation was that these differences largely reflect the consequences of smoking behavior rather than modulate the probability of smoking. An example of a transcriptomic study to investigate environmental pollutants is described in a recent manuscript (De Coster et al. 2013).

Intervention studies involving transcriptional profiling are a useful way to investigate the physiological consequences of an exposure. For example, one may measure gene expression in a relevant tissue in patients with a particular disease before and after administering a relevant drug. Such an investigation may provide information about the means of drug action. In addition, it may be possible to screen the patient pool for those individuals for whom the drug is likely to be effective and for those people in whom the drug may not work. Perturbation studies can also be conducted in cell lines and fresh tissue samples, in which case gene expression levels are measured before and after the perturbation has been administered to the samples. As an example from my own research (unpublished), we are currently conducting a study in which we expose lymphoblastoid cell lines derived from schizophrenics and controls to the neurotransmitter dopamine and measure the gene expression levels before and after exposure via RNAseq, with the goal of understanding the relationship of dopamine to the disease. This type of study provides a very high level of experimental control and permits the administration of highly topical perturbations. It seems likely that expression profiling will become a more common component of such studies, in order to assess the impact of an intervention on cellular activity and processes.

5.9 Concluding Remarks

Transcriptional profile data is now generated as part of many different types of studies. This chapter has mainly focused on using gene expression data to identify genes connected to the trait of inference (either etiologically or physiologically) and for examining the genetic regulation of gene expression in detail. Comprehensive genome-wide assessment of gene expression has only been possible for less than a decade or so, and transcriptional profile studies have quickly become part of the standard repertoire of investigative tools available to researchers and clinicians. Given the enormous range of studies involving gene expression profiling, it is difficult to give a comprehensive overview. I have purposefully not focused on details of the methodology (both on the laboratory side and on the analytical side). Instead, I have sought to highlight some of the basic concepts and how transcriptional profiling studies fit into the wider context of human genetic epidemiological investigations of complex diseases. This area of research has proven to be very fruitful, and it is clear that transcriptional profiling studies will become more common in the future.

References

Almlöf JC, Lundmark P, Lundmark A, Ge B, Maouche S, Göring HH, Liljedahl U, Enström C, Brocheton J, Proust C, Godefroy T, Sambrook JG, Jolley J, Crisp-Hihn A, Foad N, Lloyd-Jones H, Stephens J, Gwilliam R, Rice CM, Hengstenberg C, Samani NJ, Erdmann J, Schunkert H, Pastinen T, Deloukas P, Goodall AH, Ouwehand WH, Cambien F, Syvänen AC (2012) Powerful identification of cis-regulatory SNPs in human primary monocytes using allele-specific gene expression. PLoS ONE 7:e52260

Charlesworth JC, Curran JE, Johnson MP, Göring HH, Dyer TD, Diego VP, Kent JW Jr, Mahaney MC, Almasy L, MacCluer JW, Moses EK, Blangero J (2010) Transcriptomic epidemiology of smoking: the effect of smoking on gene expression in lymphocytes. BMC Med Genomics 3:29

Cheung VG, Spielman RS, Ewens KG, Weber TM, Morley M, Burdick JT (2005) Mapping determinants of human gene expression by regional and genome-wide association. Nature 437:1365–1369

Corder EH, Saunders AM, Strittmatter WJ, Schmechel DE, Gaskell PC, Small GW, Roses AD, Haines JL, Pericak-Vance MA (1993) Gene dose of apolipoprotein E type 4 allele and the risk of Alzheimer's disease in late onset families. Science 261:921–923

De Coster S, van Leeuwen DM, Jennen DG, Koppen G, Den Hond E, Nelen V, Schoeters G, Baeyens W, van Delft JH, Kleinjans JC, van Larebeke N (2013) Gender-specific transcriptomic response to environmental exposure in Flemish adults. Environ Mol Mutagen 54:574–588

de Koning DJ, Haley CS (2005) Genetical genomics in humans and model organisms. Trends Genet 21:377–381

Dimas AS, Deutsch S, Stranger BE, Montgomery SB, Borel C, Attar-Cohen H, Ingle C, Beazley C, Gutierrez Arcelus M, Sekowska M, Gagnebin M, Nisbett J, Deloukas P, Dermitzakis ET, Antonarakis SE (2009) Common regulatory variation impacts gene expression in a cell type-dependent manner. Science 325:1246–1250

Ding J, Gudjonsson JE, Liang L, Stuart PE, Li Y, Chen W, Weichenthal M, Ellinghaus E, Franke A, Cookson W, Nair RP, Elder JT, Abecasis GR (2010) Gene expression in skin and lymphoblastoid cells: refined statistical method reveals extensive overlap in cis-eQTL signals. Am J Hum Genet 87:779–789

Dixon AL, Liang L, Moffatt MF, Chen W, Heath S, Wong KC, Taylor J, Burnett E, Gut I, Farrall M, Lathrop GM, Abecasis GR, Cookson WO (2007) A genome-wide association study of global gene expression. Nat Genet 39:1202–1207

Emilsson V, Thorleifsson G, Zhang B, Leonardson AS, Zink F, Zhu J, Carlson S, Helgason A, Walters GB, Gunnarsdottir S, Mouy M, Steinthorsdottir V, Eiriksdottir GH, Bjornsdottir G, Reynisdottir I, Gudbjartsson D, Helgadottir A, Jonasdottir A, Styrkarsdottir U, Gretarsdottir S, Magnusson KP, Stefansson H, Fossdal R, Kristjansson K, Gislason HG, Stefansson T, Leifsson BG, Thorsteinsdottir U, Lamb JR, Gulcher JR, Reitman ML, Kong A, Schadt EE, Stefansson K (2008) Genetics of gene expression and its effect on disease. Nature 452:423–428

Frayling TM, Timpson NJ, Weedon MN, Zeggini E, Freathy RM, Lindgren CM, Perry JR, Elliott KS, Lango H, Rayner NW, Shields B, Harries LW, Barrett

JC, Ellard S, Groves CJ, Knight B, Patch AM, Ness AR, Ebrahim S, Lawlor DA, Ring SM, Ben-Shlomo Y, Jarvelin MR, Sovio U, Bennett AJ, Melzer D, Ferrucci L, Loos RJ, Barroso I, Wareham NJ, Karpe F, Owen KR, Cardon LR, Walker M, Hitman GA, Palmer CN, Doney AS, Morris AD, Smith GD, Hattersley AT, McCarthy MI (2007) A common variant in the FTO gene is associated with body mass index and predisposes to childhood and adult obesity. Science 316:889–894

Gaujoux R, Seoighe C (2013) CellMix: a comprehensive toolbox for gene expression deconvolution. Bioinformatics 29:2211–2212

Gilad Y, Rifkin SA, Pritchard JK (2008) Revealing the architecture of gene regulation: the promise of eQTL studies. Trends Genet 24:408–415

Göring HH (2012) Tissue specificity of genetic regulation of gene expression. Nat Genet 44:1077–1078

Göring HH, Curran JE, Johnson MP, Dyer TD, Charlesworth J, Cole SA, Jowett JB, Abraham LJ, Rainwater DL, Comuzzie AG, Mahaney MC, Almasy L, MacCluer JW, Kissebah AH, Collier GR, Moses EK, Blangero J (2007) Discovery of expression QTLs using large-scale transcriptional profiling in human lymphocytes. Nat Genet 39:1208–1216

Greenawalt DM, Dobrin R, Chudin E, Hatoum IJ, Suver C, Beaulaurier J, Zhang B, Castro V, Zhu J, Sieberts SK, Wang S, Molony C, Heymsfield SB, Kemp DM, Reitman ML, Lum PY, Schadt EE, Kaplan LM (2011) A survey of the genetics of stomach, liver, and adipose gene expression from a morbidly obese cohort. Genome Res 21:1008–1016

Grundberg E, Small KS, Hedman ÅK, Nica AC, Buil A, Keildson S, Bell JT, Yang TP, Meduri E, Barrett A, Nisbett J, Sekowska M, Wilk A, Shin SY, Glass D, Travers M, Min JL, Ring S, Ho K, Thorleifsson G, Kong A, Thorsteindottir U, Ainali C, Dimas AS, Hassanali N, Ingle C, Knowles D, Krestyaninova M, Lowe CE, Di Meglio P, Montgomery SB, Parts L, Potter S, Surdulescu G, Tsaprouni L, Tsoka S, Bataille V, Durbin R, Nestle FO, O'Rahilly S, Soranzo N, Lindgren CM, Zondervan KT, Ahmadi KR, Schadt EE, Stefansson K, Smith GD, McCarthy MI, Deloukas P, Dermitzakis ET, Spector TD (2012) Multiple tissue human expression resource (MuTHER) consortium. mapping cis- and trans-regulatory effects across multiple tissues in twins. Nat Genet 44:1084–1089

GTEx Consortium (2013) The genotype-tissue expression (GTEx) project. Nat Genet 45:580–585

Guo L, Du Y, Chang S, Zhang K, Wang J (2014) rSNPBase: a database for curated regulatory SNPs. Nucleic Acids Res 42(Database issue):D1033–1039

Hindorff LA, Sethupathy P, Junkins HA, Ramos EM, Mehta JP, Collins FS, Manolio TA (2009) Potential etiologic and functional implications of genome-wide association loci for human diseases and traits. Proc Natl Acad Sci U S A 106:9362–9367

Jansen RC, Nap JP (2001) Genetical genomics: the added value from segregation. Trends Genet 17:388–391

Kettunen J, Tukiainen T, Sarin AP, Ortega-Alonso A, Tikkanen E, Lyytikäinen LP, Kangas AJ, Soininen P, Würtz P, Silander K, Dick DM, Rose RJ, Savolainen MJ, Viikari J, Kähönen M, Lehtimäki T, Pietiläinen KH, Inouye M, McCarthy MI, Jula A, Eriksson J, Raitakari OT, Salomaa V, Kaprio J, Järvelin MR, Peltonen L, Perola M, Freimer NB, Ala-Korpela M, Palotie A, Ripatti S (2012) Genome-wide association study identifies multiple loci influencing human serum metabolite levels. Nat Genet 44:269–276

Kooij V, Venkatraman V, Tra J, Kirk J, Rowell J, Blice-Baum A, Cammarato A, Van Eyk J (2014) Sizing up models of heart failure: proteomics from flies to humans. Proteomics Clin Appl doi: 10.1002/prca.201300123 [Epub ahead of print]

Majewski J, Pastinen T (2011) The study of eQTL variations by RNA-seq: from SNPs to phenotypes. Trends Genet 27:72–79

Mill J, Heijmans BT (2013) From promises to practical strategies in epigenetic epidemiology. Nat Rev Genet 14:585–594

Montgomery SB, Sammeth M, Gutierrez-Arcelus M, Lach RP, Ingle C, Nisbett J, Guigo R, Dermitzakis ET (2010) Transcriptome genetics using second generation sequencing in a Caucasian population. Nature 464:773–777

Nica AC, Parts L, Glass D, Nisbet J, Barrett A, Sekowska M, Travers M, Potter S, Grundberg E, Small K, Hedman AK, Bataille V, Tzenova Bell J, Surdulescu G, Dimas AS, Ingle C, Nestle FO, di Meglio P, Min JL, Wilk A, Hammond CJ, Hassanali N, Yang TP, Montgomery SB, O'Rahilly S, Lindgren CM, Zondervan KT, Soranzo N, Barroso I, Durbin R, Ahmadi K, Deloukas P, McCarthy MI, Dermitzakis ET, Spector TD, MuTHER Consortium (2011) The architecture of gene regulatory variation across multiple human tissues: the MuTHER study. PLoS Genet 7:e1002003

Pastinen T (2010) Genome-wide allele-specific analysis: insights into regulatory variation. Nat Rev Genet 11:533–538

Pickrell JK, Marioni JC, Pai AA, Degner JF, Engelhardt BE, Nkadori E, Veyrieras JB, Stephens M, Gilad Y, Pritchard JK (2010) Understanding mechanisms underlying human gene expression variation with RNA sequencing. Nature 464:768–772

Price AL, Helgason A, Thorleifsson G, McCarroll SA, Kong A, Stefansson K (2011) Single-tissue and cross-tissue of gene expression via identity-by-descent in related or unrelated individuals. PLoS Genet 7: e1001317

Robinton DA, Daley GQ (2012) The promise of induced pluripotent stem cells in research and therapy. Nature 481:295–305

Sanders AR, Göring HH, Duan J, Drigalenko EI, Moy W, Freda J, He D, Shi JMGS, Gejman PV (2013) Transcriptome study of differential expression in schizophrenia. Hum Mol Genet 22:5001–5014

Schadt EE, Lamb J, Yang X, Zhu J, Edwards S, Guhathakurta D, Sieberts SK, Monks S, Reitman M,

Zhang C, Lum PY, Leonardson A, Thieringer R, Metzger JM, Yang L, Castle J, Zhu H, Kash SF, Drake TA, Sachs A, Lusis AJ (2005) An integrative genomics approach to infer causal associations between gene expression and disease. Nat Genet 37:710–717

Sreekumar A, Poisson LM, Rajendiran TM, Khan AP, Cao Q, Yu J, Laxman B, Mehra R, Lonigro RJ, Li Y, Nyati MK, Ahsan A, Kalyana-Sundaram S, Han B, Cao X, Byun J, Omenn GS, Ghosh D, Pennathur S, Alexander DC, Berger A, Shuster JR, Wei JT, Varambally S, Beecher C, Chinnaiyan AM (2009) Metabolomic profiles delineate potential role for sarcosine in prostate cancer progression. Nature 457:910–914

Stranger BE, Nica AC, Forrest MS, Dimas A, Bird CP, Beazley C, Ingle CE, Dunning M, Flicek P, Koller D, Montgomery S, Tavaré S, Deloukas P, Dermitzakis ET (2007) Population genomics of human gene expression. Nat Genet 39:1217–1224

Strittmatter WJ, Saunders AM, Schmechel D, Pericak-Vance M, Enghild J, Salvesen GS, Roses AD (1993) Apolipoprotein E: high-avidity binding to beta-amyloid and increased frequency of type 4 allele in late-onset familial Alzheimer disease. Proc Natl Acad Sci USA 90:1977–1981

Terwilliger JD, Göring HH (2009) Update to Terwilliger and Göring's "Gene mapping in the 20th and 21st centuries" (2000): gene mapping when rare variants are common and common variants are rare. Hum Biol 81:729–733

Tukiainen T, Kettunen J, Kangas AJ, Lyytikäinen LP, Soininen P, Sarin AP, Tikkanen E, O'Reilly PF, Savolainen MJ, Kaski K, Pouta A, Jula A, Lehtimäki T, Kähönen M, Viikari J, Taskinen MR, Jauhiainen M, Eriksson JG, Raitakari O, Salomaa V, Järvelin MR, Perola M, Palotie A, Ala-Korpela M, Ripatti S (2012) Detailed metabolic and genetic characterization reveals new associations for 30 known lipid loci. Hum Mol Genet 21:1444–1455

Van Eyk JE (2011) Overview: the maturing of proteomics in cardiovascular research. Circ Res 108:490–498

Visscher PM, Brown MA, McCarthy MI, Yang J (2012) Five years of GWAS discovery. Am J Hum Genet 90:7–24

Welter D, MacArthur J, Morales J, Burdett T, Hall P, Junkins H, Klemm A, Flicek P, Manolio T, Hindorff L, Parkinson H (2014) The NHGRI GWAS catalog, a curated resource of SNP-trait associations. Nucleic Acids Res 42(Database issue):D1001–1006

Wild CP (2005) Complementing the genome with an "exposome": the outstanding challenge of environmental exposure measurement in molecular epidemiology. Cancer Epidemiol Biomark Prev 14:1847–1850

Xia K, Shabalin AA, Huang S, Madar V, Zhou YH, Wang W, Zou F, Sun W, Sullivan PF, Wright FA (2012) seeQTL: a searchable database for human eQTLs. Bioinformatics 28:451–452

Yang TP, Beazley C, Montgomery SB, Dimas AS, Gutierrez-Arcelus M, Stranger BE, Deloukas P, Dermitzakis ET (2010) Genevar: a database and Java application for the analysis and visualization of SNP-gene associations in eQTL studies. Bioinformatics 26:2474–2476

Zeggini E, Weedon MN, Lindgren CM, Frayling TM, Elliott KS, Lango H, Timpson NJ, Perry JR, Rayner NW, Freathy RM, Barrett JC, Shields B, Morris AP, Ellard S, Groves CJ, Harries LW, Marchini JL, Owen KR, Knight B, Cardon LR, Walker M, Hitman GA, Morris AD, Doney AS, Wellcome Trust Case Control Consortium (WTCCC), McCarthy MI, Hattersley AT (2007) Replication of genome-wide association signals in UK samples reveals risk loci for type 2 diabetes. Science 316:1336–1341

6 Copy Number Variations and Chronic Diseases

August N. Blackburn and Donna M. Lehman

6.1 An Introduction to Copy Number Variation

Genetic variation ranges from single nucleotide changes to large chromosome level events. The most well characterized form of variation is single nucleotide variants with a minor allele frequency (MAF) > 1 %, referred to as single nucleotide polymorphisms (SNP). SNPs have been a workhorse of human genetics due to ease and reproducibility of genotyping. In contrast, structural variation encompasses variants with a broad range of sizes and complexity and has historically lagged behind the progress achieved using SNPs due to difficulty of genotyping. Structural variation is generally defined by 5 groups of variants: deletions, insertions, duplications, translocations, and inversions. In this chapter we focus on a subset of structural variation comprised of deletions and duplications, colloquially termed copy number variation (CNV).

A.N. Blackburn
Department of Cellular and Structural Biology, University of Texas Health Science Center, 7703 Floyd Curl Drive, San Antonio, TX 78229, USA
e-mail: BlackburnA@uthscsa.edu

D.M. Lehman (✉)
Departments of Medicine and Cellular and Structural Biology, University of Texas Health Science Center, 7703 Floyd Curl Drive, San Antonio, TX 78229, USA
e-mail: lehman@uthscsa.edu

Deletions are simply the absence of sequence when compared to a reference genome. When compared to a reference, the sequences flanking a deletion of reference sequence are continuous and juxtaposed in direct orientation. Deletions are unambiguous, having a single genomic location, although in some cases exact breakpoint locations are ambiguous if there is high sequence similarity, such as repetitive DNA, at the sights flanking the deletion. Duplication refers to sequences which share high sequence homology (>90 %) that are found in greater than two copies in the genome. Duplications are often broken into tandem and dispersed categories. Dispersed duplications are often further broken down into inter-chromosomal and intra-chromosomal categories. Intra-chromosomal duplications are further distinguished as being either in direct or inverted orientation with respect to each other.

Originally copy number variation referred only to variation larger than 1 Kb. However this size limitation has largely been disregarded due to its arbitrary nature when compared to the observed spectrum of variant sizes. Very small duplications and deletions (<50 bp) are often referred to as INDELs. Although they are part of a continuous spectrum of sizes of CNVs, these smaller variants are often distinguished from larger variants because they are almost exclusively identified using sequencing. Generally, the term copy number variation is reserved for submicroscopic events, therefore excluding monosomy and trisomy, but does include variants on

R. Duggirala et al. (eds.), *Genome Mapping and Genomics in Human and Non-Human Primates*, Genome Mapping and Genomics in Animals 5, DOI 10.1007/978-3-662-46306-2_6,

the scale of 1–5 Mb even though these are assessable by microscopic techniques such as fluorescent in situ hybridization (FISH).

6.2 The Brief History of Copy Number Variation in the Human Genome

The account of the role of CNV in human disease is appropriately centered on the publication of the human genome. Access to a reference sequence enabled new technologies which subsequently powered the discoveries described below. Examples of CNV and human disease were present prior to the availability of the reference sequence. However, the extent of CNV was unknown, and global investigation of CNVs contribution to human disease was not feasible. Thus, the publication and availability of the reference human genome represents a major inflection point in the rate which new discoveries regarding CNVs were made.

6.2.1 Prior to an Available Reference Human Genome

In Tajima's description of the use of the D statistic to test the Neutral Theory of molecular evolution, he postulates that insertions and deletions may be common in the genome of *Drosophila melanogaster* (Tajima 1989). If this observation is extended to reflect our understanding of genomes as a group at the time, it can be viewed as an early prediction that CNV, albeit small CNVs, would be common in the human genome. Twenty years later we are reaching a consensus understanding of the catalog of human genetic variation at the population level, and discovering that CNV is yet even more common than had been anticipated. The functional effects of these variants at all levels of biology, from cellular processes to disease susceptibility, are still being determined. However, even prior to an available reference sequence we could draw from known examples that CNV plays an important role in human disease.

Prior to the publication of the human genome, examples of CNVs in human disease etiology were known for single gene disorders. The role of deletions at the alpha gene locus in α-thalassemias was known since the 1970s (Ottolenghi et al. 1974; Higgs et al. 1979). In the early 1990s we learned that reciprocal deletions and duplications at 17p11.2 are involved in hereditary neuropathy with liability to pressure palsies (HNPP) and Charcot-Marie-Tooth neuropathy type 1A (CMT1A) respectively (Lupski et al. 1991; Chance et al. 1993). Examples such as these provided early evidence that CNV plays a role in human disease, yet there was no basic reference sequence for the human genome, and the extent of CNV between human genomes remained unknown.

6.2.2 Publication of the Human Genome

The publication of the first drafts of the human genome was a milestone event in human history (Lander et al. 2001; Venter et al. 2001). These first drafts were known to be incomplete because structural variation such as large segmental duplications, a course definition including both polymorphic and evolutionarily fixed duplication events generally defined as sequences with ≥90 % homology, complicated the process of sequencing and assembly. However, the next draft of the human reference sequence, published in 2004 by the International Human Genome Sequencing Consortium (IHGSC), drastically reduced the number of gaps in the human reference sequence (IHGSC 2004). A major discovery during the process of completing the human reference sequence was that an abundance of insertions and deletions, identified by observing length differences when using paired-end fosmid reads, appeared to represent polymorphisms (IHGSC 2004).

6.2.3 Early Discoveries of Genome-Wide Copy Number Variation

In 2002, Bailey et al. used computational methods and available data from sequencing efforts by IHGSC and Celera Genomics to identify segmental duplications in the human genome (Bailey et al. 2002). Interestingly, they observed patterns of both interchromosomal and intrachromosomal segmental duplications, and showed that unannotated segmental duplications created falsely identified SNPs in the public SNP database (dbSNP) (Bailey et al. 2002). Although this work showed that segmental duplications were present, what remained unclear is whether these segmental duplications were polymorphic or fixed in the human population. In the months prior to the 2004 publication by the IHGSC, two independent groups reported observing large-scale copy number variation which appeared to represent polymorphisms (Iafrate et al. 2004; IHGSC 2004; Sebat et al. 2004). Sebat et al. (2004) identified 221 copy number changes representing 76 copy number polymorphisms (CNP) in 20 individuals of geographically diverse ancestry using representational oligonucleotide microarray analysis (ROMA) (Sebat et al. 2004). Iafrate et al. (2004) identified 255 loci which appeared to be polymorphic in 55 unrelated individuals using array-based comparative genomic hybridization (array CGH). These regions were significantly enriched for overlapping segmental duplications and regions of gaps in the human genome sequence (Iafrate et al. 2004). These early studies, taken together with observations from the 2004 publication by the IHGSC provided evidence that structural variation, namely copy number variation (CNV), was common in the human genome.

Using 60 trios from the International HapMap Project (International HapMap 3 Consortium (IHM3C) et al. 2003) (30 European trios and 30 Yoruba trios) Conrad et al. (2006) identified 586 regions which had observed SNP genotypes consistent with genotyping artifacts caused by deletions. Concurrently, McCarroll and his collegues reported identifying 541 deletion variants in 269 HapMap individuals by identifying in SNP genotyping data the footprints of segregating deletions, such as Hardy-Weinberg disequilibrium, Mendelian inconsistency, and clusters of null genotypes (McCarroll et al. 2006). The SNPs that produce such errors are commonly disregarded in human genetic studies, so this study group performed additional molecular assays to confirm the presence of many of these deletion variants. Interestingly, they observed that common deletions were often in linkage disequilibrium with nearby SNPs, which was an early indication that CNVs could be tagged and indirectly assayed in genome wide association studies using SNPs (McCarroll et al. 2006).

In 2005, Tuzun et al. applied the paired-end mapping strategy developed for finishing the human genome sequence to a fosmid library from a second genome and discovered 139 insertions, 102 deletions, and 56 inversions when compared to the reference sequence (IHGSC 2004; Tuzun et al. 2005). Although most of the variants they identified were novel, they replicated the observation that CNVs were enriched for regions of segmental duplication. The approach used by Tuzun et al. (2005) was a significant advancement that was able to detect variants with a higher resolution than the CNVs discovered by Iafrate et al. (2004) and Sebat et al. (2004). Subsequently, Korbel et al. (2007) applied the paired-end sequencing approach to massively parallel shotgun sequencing on the 454 platform (Korbel et al. 2007). Using conservative thresholds for variant calling, they identified 1,297 Structural Variants (SV) events, the majority of which are CNVs, with an estimated breakpoint resolution of 644 base pairs (bp). The combination of paired-end sequencing and massively parallel sequencing technologies was a major contribution that set the foundation for future studies to move from identifying CNVs by comparing a handful of genomes to characterizing CNV at the population level.

With the exception of being enriched in regions of segmental duplication, poor consistency was observed between various methodological techniques for CNV identification, suggesting either high rates of Type I or Type II errors. Since many of these studies had performed conformational assays and estimated false discovery rates (FDR), it could be deduced that many variants remained undiscovered, likely due to methodological biases and conservative interpretation of results. Using a custom high-resolution array based comparative genomic hybridization (aCGH) approach to target previously identified CNV regions at ~ 1 Kb resolution, Perry et al. (2008) observed that all previous studies they investigated had overestimated the actual size of a substantial portion of the CNVs. Nonetheless, the methods developed for these early studies, and the results of the studies themselves, set the foundation for building a comprehensive understanding of structural variation at the population level, identifying mechanisms of formation, and discovering the role of CNVs in human disease.

6.3 High-Throughput Methods for Discovery and Genotyping of Copy Number Variation

Two general groups of technologies have primarily been used for CNV discovery and genotyping: array-based hybridization, and high-throughput sequencing. Array-based hybridization methods are affordable and are often superior for detecting very large deletions and duplications. However, array-based hybridization often misses smaller variants. High throughput sequencing is much more expensive than array-based hybridization, but is superior for detecting smaller variants, defining CNV boundaries, and for determining absolute copy number of high copy number duplications. Despite improvements in the methods which are currently available, the field is still in need of more comprehensive methods for CNV discovery and genotyping, and more accurate methods for CNV imputation in samples for which these technologies cannot be applied.

6.3.1 Array Based Methods

Microarrays are a group of technologies which rely on hybridization of prepared DNA samples to oligonucleotides designed to represent specific locations of the genome. Therefore, microarrays are a technology enabled by the availability of the reference sequence. There are two basic types of microarrays which are commonly used for CNV discovery and genotyping, aCGH and SNP genotyping microarrays (Alkan et al. 2011).

In aCGH two samples are fluorescently labeled and competitively hybridized to oligonucleotide arrays (Pinkel et al. 1998). Copy number variable regions are represented by imbalance in fluorescent intensity. As a QC measure the experiments are often repeated with swapped dyes. Since either of the two samples can carry copy number differences, well characterized reference genomes are preferred for comparison. Oligonucleotides are designed to uniquely identify specific locations along the genome. Signal intensities are normalized and converted to $\log_2$ ratio, a measurement representative of copy number. To identify CNVs, various algorithms can be applied that segment the genome into regions which appear to differ from the average, which is presumed to represent a copy number of 2. Deletions and duplications are detected as multiple consecutive probes which present similar decreases or increases in $\log_2$ ratio, respectively.

SNP-arrays also produce a measurement of signal intensity by comparing the hybridization intensities across samples (Peiffer et al. 2006). This measurement is known to have a lower signal to noise ratio than aCGH, but is still powerful enough to be useful. The relative intensity of each allele is informative for identifying copy number variation (Peiffer et al. 2006). If the SNP alleles are arbitrarily labeled A and B, the ratio of signal intensity of B to the sum of intensities of A and B, termed B-allele ratio, is informative of copy number. In the normal copy number state of 2, B-allele ratio will fall into three clusters: 0, 0.5, and 1, representing homozygous AA, heterozygous AB, and homozygous BB respectively. However, in the case of deletion the cluster at 0.5 will be lost indicating a loss of heterozygosity (LOH). Similarly, when there is copy number gain the cluster at 0.5 will split into two clusters of 0.33 and 0.66 representing the AAB and ABB genotypes respectively (Peiffer et al. 2006). Additional patterns of B-allele ratios are apparent for somatic copy number variation and other defined copy number states (Alkan et al. 2011). SNP arrays can also detect copy-neutral loss of heterozygosity, which is indicative of uniparental disomy or identity by descent (Alkan et al. 2011). Further, the application of SNP arrays to CNV calling benefits from the availability of SNP genotypes which can be used for phasing, tagging, and other imputation directed purposes which we will cover in Sect. 6.3.4.

Multiple statistical approaches have been implemented to identify CNVs from aCGH and SNP arrays, the most popular of which have been versions of circular binary segmentation and hidden Markov models (Olshen et al. 2004; Colella et al. 2007; Venkatraman and Olshen 2007; Wang et al. 2007; Coin et al. 2010). Comparative analyses of these algorithms indicate that using multiple algorithms to identify CNVs should be the preferred approach (Winchester et al. 2009; Dellinger et al. 2010; Pinto et al. 2011). Array-based approaches are known to be subject to variation in local DNA concentration that is correlated with GC content, which is often observed as "waviness" of $\log_2$ ratios for markers along the chromosome, and generally requires additional normalization procedures (Diskin et al. 2008). In a recent review, Pinto et al. (2011) showed that newer arrays tended to perform much better than legacy versions, that algorithms tended to perform best on the platforms they were designed for, and that current approaches generally underestimate CNV size, which is a shift from the observation of size overestimation reported by Perry et al. (2008). Overall, this suggests that many of the technical artifacts of CNV discovery have been addressed on newer chips and software pipelines.

CNV discovery differs from CNV genotyping in that in the discovery phase there is no a priori knowledge of the location of CNVs. Once copy number variable regions have been identified, common CNVs can often be genotyped more accurately by comparing marker intensities between samples within the region of interest. This approach has been implemented to perform association testing (Barnes et al. 2008; Welcome Trust Case Control Consortium (WTCCC) et al. 2010). Haplotype structure, determined from SNP genotypes, has also been used to improve CNV genotyping procedures (Coin et al. 2010). Taken together, these implementations indicate that, when available, added information available across samples improves CNV genotyping accuracy.

6.3.2 Sequencing Based Methods

Sequencing approaches to CNV discovery can be summarized into 4 approaches: split read, paired-end read, read depth, and de novo assembly approaches (Alkan et al. 2011). For the purpose of this chapter we will describe the benefits of de novo assembly in Sect. 6.3.3. Split-read approaches seek to identify variation that is captured within a single contiguous sequence read. By computationally "splitting" the alignment of a read to a reference sequence, split read approaches can find small deletions, insertions, and duplications. For split read approaches the upper bound on the size of sequence insertions and duplications that can be identified is the length of the read minus the sequence needed to map the read uniquely to a position in the genome, because the inserted or duplicated sequence must be contained within the length of the read. Given that most whole genome sequencing (WGS) approaches produce small reads, split read approaches are generally only able to detect very small insertions and duplications. In theory, split read approaches could detect very large deletions as long as there is a read that gaps the deletion breakpoints. However, in practice split read approaches are more effective for small deletions. In 2006, Mills et al. used a split read approach to identify 415,436 INDEL polymorphisms ranging in length between 1 and 9,989 base pairs using sequencing reads from 36 individuals (Mills et al. 2006). The overwhelming majority of these variants were 1–10 base pairs in length, and they observed little overlap with deletions identified by Conrad et al. (2006) and McCarroll et al. (2006), consistent with the observation of poor overlap between studies at the time.

For the paired-end read sequencing method, a DNA library is prepared such that the length of the DNA fragments to be sequenced fit into a tight distribution. Various approaches are available to produce libraries of different size distributions. Each DNA fragment is then sequenced from both ends. Each read, one from each end of the fragment, are then mapped back to the genome. The distance between read-pairs for regions not carrying large CNVs will fall into a tight distribution indicative of the distribution of sizes of the DNA fragments in the library prep. Deletions and duplications can be identified by abnormalities in the distance between read-pairs when they are mapped back to the reference. Deletions will create read-pairs that map further apart, and insertions will create read-pairs that map closer together than expected based on the distribution of the DNA library. As with a split-read approach there is an upper bound on the size of insertion/duplication that can be detected because the duplication has to be carried by the DNA molecule being sequenced. It is worth noting that paired-end sequencing can also detect inversion and novel sequence insertions by using one read as an anchor. As mentioned earlier, in the discovery of germ-line CNV, the paired-end read approach was first applied to fosmid libraries (Tuzun et al. 2005), and was later combined with WGS (Korbel et al. 2007).

In the read-depth approach, the coverage of the genome by sequencing reads is assumed to be uniformly distributed. Therefore regions with a loss or gain of genetic material are represented by loss or gain in the number of sequence reads. This approach was first applied to germ-line variants by Yoon et al. (2009). This approach is more effective and accurate with higher read-depth and is superior to split-read and paired-end read approaches for identifying large duplication events. Additionally, this method is superior to array based technologies for determining absolute copy number of high copy number duplications. However, similar to aCGH this approach requires correction for genomic "waviness" due to local GC content (Yoon et al. 2009). A comprehensive assessment of the platforms and computational strategies currently available using the read depth approach has recently been conducted (Magi et al. 2012).

6.3.3 Comprehensive Discovery and Genotyping

The methods described in Sects. 6.3.1 and 6.3.2 are not comprehensive. Identifying variants using array based hybridization is dependent on probe hybridization at the locus of the variant. Thus arrays with lower probe densities generally do not detect smaller variants. Arrays have poorer breakpoint definition than sequencing methods. All of the sequencing based approaches presented are powerful, but each is dependent on aligning reads to a reference sequence. There are situations where aligning reads to a reference sequence is not sufficient, such as the identification of unique sequence insertions, or variant calling in regions where short reads cannot be uniquely mapped. This suggests that an alternative approach may be necessary to genotype some CNVs.

De novo assembly followed by genome comparison is argued to be the most likely route to a comprehensive approach for variant discovery and genotyping because this approach is not dependent on aligning reads to a reference sequence (Alkan et al. 2011). This argument is very compelling, but current technologies produce short reads which limit the feasibility of this approach. However, this is a promising route to a truly comprehensive discovery and genotyping assuming technologies can be developed which produce extremely long reads, in the area of 100–200 Kb, with high accuracy.

In 2010, Pang et al. used and compared multiple approaches, including de novo assembly and comparison of genomes to identify CNVs in HuRef (Venter et al. 2001) DNA (Pang et al. 2010). Overall, de novo assembly was the most comprehensive method for CNV identification, but did miss known CNVs identified using other techniques. Each method used had its own distribution of sizes of variants in which it performed best, as expected based on the methods described above. CNVs between 1 and 10 Kb had the highest proportion of overlap of variants detected between technologies (Pang et al. 2010).

In the absence of a technology which produces long reads with high accuracy, there is growing momentum toward considering various forms of data in combined models. By combining read-depth with high resolution aCGH developed a method for correcting the reference copy number biases in aCGH alluded to in Sect. 6.3.1 (Ju et al. 2010). This approach was then applied to identify common Asian copy number variants with high accuracy (Park et al. 2010). The 1000 genomes project has also taken the approach of combining multiple lines of evidence for CNV identification (Mills et al. 2011). More specifically, Mills et al. combined results from multiple algorithms representing split read, read-pair, read-depth, assembly, and a combination read-pair/read-depth approach to identify CNVs in low coverage sequencing and trio sequencing data generated using three different sequencing platforms.

6.3.4 Imputation of CNVs

Comprehensive variant discovery through WGS is currently prohibitively expensive, which has motivated the development of methods to impute unobserved genotypes in samples using a framework of known genotypes (Howie et al. 2012). The feasibility of imputing di-allelic CNVs was demonstrated using data generated with SNP genotyping platforms and HapMap samples (International HapMap 3 Consortium (IHM3C) et al. 2010; Surakka et al. 2010). Not surprisingly, imputation performs more effectively with population-specific reference panels, especially for polymorphisms with lower Minor Allele Frequencies (MAFs) (IHM3C et al. 2010; Surakka et al. 2010). Since 2010, major strides have been made toward better computational approaches for imputation (Li et al. 2010; Howie et al. 2011, 2012), and toward more comprehensive reference panels (Mills et al. 2011; 1000 Genomes Project Consortium et al. 2012). In population samplings, imputation is currently limited to simple forms of CNV with higher MAFs. However, examples are beginning to indicate that complex regions of the genome containing CNV may be amenable to imputation as well (Boettger et al. 2012). Additionally, in pedigrees, where phase can be determined with high accuracy for entire chromosomes, imputation of complex regions should also be achievable.

6.4 Mechanisms of CNV Formation and Mutation Rates

Mechanisms of CNV formation can be broken into two broad categories: those which involve long homologous sequences, such as non-allelic homologous recombination (NAHR), and those which involve non-homologous repair (NHR), which often entail micro-homology at the breakpoint sites (Hastings et al. 2009).

NAHR between segmental duplications and Variable Number of Tandem Repeats (VNTR) shrinkage/expansion produce copy number variants with overlapping, but distinct, size distributions (Conrad et al. 2010). Using arrayCGH data with highly accurate breakpoint resolution, Conrad et al. (2010) investigated mechanisms of formation for CNVs genotyped in 450 individuals and determined that NAHR between segmental duplications contributed more frequency for larger variants than VNTR shrinkage/expansion, which had a greater relative contribution to formation of smaller CNVs. Interestingly, formation of duplications appeared more likely to be sequence dependent than formation of deletions, yet without knowledge of the exact sequence at the breakpoints the precise mechanisms of formation for those which could not be attributed to one of these two mechanisms remained unclear (Conrad et al. 2010).

WGS has provided information for those mechanisms of formation that requires knowledge of the exact sequence at CNV breakpoints. Mills et al. (2011) investigated sequencing data generated during the pilot phase of the 1000 genomes project observed that micro-homology/homology between 2 and 376 bases were present in the sequence flanking 70.8 and 89.6 % of deletions and insertion/duplications respectively.

Interestingly, for tandem duplications, duplication size was linearly correlated with the length of homologous sequence flanking the duplication. Mobile element insertions (MEI) were the predominant mechanisms of formation of insertion/duplications, while non-homologous repair (NHR) mechanisms such as micro-homology mediated break induced repair (MMBIR) were the predominant mechanism of deletion formation. NAHR was the second most predominant mechanism of formation for both insertion/duplications and deletions, making up a substantial portion of both. NAHR and NHR contribute to variants across the spectrum of CNV sizes, yet VNTR-mediated events were enriched for smaller events, which was consistent with the aCGH study by Conrad et al. (2010). Among MEI mediated duplications, there are enrichments of variants at 300 bp and 6 Kb, representing *Alu* and long interspersed elements (LINEs). It is important to note that very large duplications and deletions, greater than 100 Kb for deletions and 10 Kb for duplications are likely underrepresented in this study due to difficulty detecting CNVs beyond these limits using sequencing methods.

Among mechanisms of formation, NAHR between segmental duplications is of high clinical relevance. Through this mechanism, a large portion ($\sim$10 %) of the genome is predisposed to recurrent mutational events (Mefford and Eichler 2009). Recurrent copy number variants resulting from this mechanism are often large enough with sufficient shared genetic material that they can be presumed to exert similar effects on phenotypes of interest. Although NAHR is an important mechanism for recurrent mutation the effect of other mechanisms of formation should not be discounted. There are examples of Mendelian disorders which show that additional mechanisms, which normally mediate non-recurrent CNV mutations, produce similar phenotypic effects as the observed CNVs generated through NAHR between segmental duplications. For example, NAHR between segmental duplications is recognized to contribute to 99 % of CMT1A and HNPP cases, yet Zhang et al. (2010) identified 17 unique CNVs in this same region formed by additional mechanisms that produced phenotypic effects consistent with CMT1A and HNPP (Zhang et al. 2010).

6.5 Common CNVs and Disease

In recent years there has been much discussion over the role of common and rare variants in complex trait variation, with strong arguments being presented in support of both (Gibson 2011). The common disease common variant (CDCV) hypothesis has been tested through GWAS, the hallmark of which was published by the Wellcome Trust Case Control Consortium (WTCCC) in 2007. GWASs have identified over 1,000 SNPs which are associated with human disease-related phenotypes (Hindorff et al. 2009). However, these associations only explain a small portion of the additive heritability of the majority of traits investigated (Gibson 2011).

Given this observation, one may hypothesize that common CNVs, which we will refer to as copy number polymorphisms (CNPs), accounts for a portion of this "missing heritability". However, as discussed above CNPs are generally well tagged by SNPs, and therefore have already been indirectly interrogated through GWAS and are unlikely to explain the observed "missing heritability" (Hinds et al. 2006; McCarroll et al. 2006; Conrad et al. 2010). This observation was confirmed through direct interrogation of 3,432 CNPs in eight disease traits by the WTCCC, in which all significantly associated CNPs were tagged by SNPs already detected in GWAS (WTCCC et al. 2010). CNPs generally make more compelling candidates for functional alleles than SNPs because of their size and increased likelihood to overlap genes. Yet due to LD, proof of functionality requires additional information in the form of biological assays. In addition, associated CNPs are also subject to the possibility of synthetic association similar to those observed with GWAS using SNPs (Dickson et al. 2010).

Despite the observation that common CNVs do not appear to account for a large portion of the missing additive heritability of common complex disorders, there are common structural variants which are associated with complex disease phenotypes. A hallmark example is a study in which the authors hypothesized and confirmed that copy number variation at the gene CCL3L1 is associated with risk for HIV/AIDS susceptibility (Gonzalez et al. 2005). Further they showed that CCL3L1 copy number is highly population differentiated with higher CCL3L1 being more prevalent in Africans than non-Africans (Gonzalez et al. 2005). Among CNVs investigated by Conrad et al. (2010) this variant was the most highly population differentiated CNV overlapping gene exons.

A second hallmark example comes from age-related macular degeneration. In 2005, three groups independently identified a common SNP coding variant, Y402H, in complement factor H (CFH) which was strongly associated with risk for age-related macular degeneration (AMD) (Edwards et al. 2005; Haines et al. 2005; Klein et al. 2005). The population attributable risk of this variant to AMD was independently estimated to be 43 and 50 % (Edwards et al. 2005; Haines et al. 2005). This work strongly implicated the complement system in age-related macular degeneration. The following year, Hughes et al. was investigating SNPs in the complex region containing CFH and the related receptor genes CFHR1, CFHR2, CFHR3, CFHR4, and CFHR5. During their investigation they discovered a deletion of CFHR1 and CFHR3 on a common haplotype in Europeans that conferred reduced risk (odds ratio of 0.4) for age-related macular degeneration. Using multiple techniques they found that the deletion was 84,682 bases and flanked by two nearly identical 29 Kb segmental duplications implicating NAHR as the mechanism of formation.

A compelling, largely untested mechanism which fits into the CDCV hypothesis is the role of CNPs in mediating recurrent mutational events. As described above, segmental duplication predisposes ~10 % of the genome to recurrent mutation. As we just described, there are known examples in which complex disease risk CNVs were formed by NAHR between segmental duplications. Further as will be presented in Sect. 6.6, there are known examples in which complex disorders are caused by NAHR between segmental duplications which are polymorphic. It is therefore possible, depending on the size of the gap between the duplicated sequences, and the haplotype structure of the population, for these duplications to be carried on separate haplotype blocks. If both locations are polymorphic, this mechanism represents a form of potential epistatic interaction which has not yet been tested.

6.6 De Novo and Low MAF Variants

De novo and low MAF CNVs are known to play a role in multiple complex disorders. Large rare CNVs are enriched in patients with schizophrenia (Malhotra et al. 2011; Walsh et al. 2008; Xu et al. 2008), bipolar disorder (Malhotra et al. 2011), autism spectrum disorders (Sebat et al. 2007; Mefford and Eichler 2009; Pinto et al. 2010; Levy et al. 2011; Sanders et al. 2011), congenital heart disease (Soemedi et al. 2012), and developmental delay (Cooper et al. 2011). NAHR between segmental duplications is known to be a major driving force of de novo and low MAF CNVs across the size spectrum of CNVs, including those identified in these enrichments. Recurrent deletions mediated by NAHR between segmental duplications are hypothesized to predispose to disease (Sharp et al. 2006; Cooper et al. 2011). Micro-deletion syndromes can provide an example through which NAHR derived rare CNVs affect human disease. They are clinically heterogeneous phenotypes which are associated with specific recurrent deletions mediated by NAHR between segmental duplications. In general there is a correlation between the size of deletions and the severity of the phenotypes observed in individuals carrying these deletions due to increased likelihood for large deletions to overlap genes or create effects on gene expression, as discussed in Sect. 6.8. Since currently known micro-deletion syndromes are caused by very large deletions, it is likely that the micro-deletion syndromes represent the tail-end of a continuous distribution of phenotypic effects that are caused by deletions resulting from NAHR between segmental duplications.

A well-known micro-deletion syndrome is Koolen syndrome involving chromosome 17q21.31, for which carriers present with mental retardation, hypotonia, recognizable facial characteristics, and other heterogeneous phenotypes (Koolen et al. 2006, 2008; Sharp et al. 2006). Flanking segmental duplications predisposing to deletion fall on an inversion polymorphism shown to be under positive selection in Europeans (Stefansson et al. 2005), and has been shown to be a genetic determinate of meiotic recombination rates (Stefansson et al. 2005; Chowdhury et al. 2009; Fledel-Alon et al. 2011). Further investigation of this region has delineated at least 9 unique common haplotypes determined by inversion and duplication status of two unique sequences (Boettger et al. 2012; Steinberg et al. 2012). Segmental duplications containing the gene KANSL1 have independently derived and risen to high population frequencies in two unique instances suggesting positive selective pressure for increased copy number of KANSL1 (Boettger et al. 2012; Steinberg et al. 2012). The duplicated sequences are in direct orientation in only one of these two distinct segmental duplications events, and therefore only one of the two segmental duplication events predisposes to 17q21.31 micro-deletion syndrome. The haplotype carrying this segmental duplication is only observed at an appreciable frequency in Caucasian individuals (Boettger et al. 2012; Steinberg et al. 2012). This example and others, such as the complexity of the regions harboring CFH, CFHR1, and CFHR3 genes, suggest that similar complexity can be expected to underlie currently unidentified regions responsible for the heritable component of complex disease.

Taken together with the enrichment of large de novo and low MAF CNVs in complex disorders it is likely that additional micro-deletions will account for a portion of the apparent missing heritability of complex traits. As much as 5 % of schizophrenia and autism have been attributed to rare copy number variation at only a half dozen genomic locations (Gibson 2011). The strong known role of rare and de novo copy number variation is cited as support of rare variation in the etiology of common complex diseases (Gibson 2011).

As an example, micro-deletion at the 17q12 locus causes renal cyst and diabetes syndrome, also referred to as maturity onset diabetes of the young 5 (MODY5) (Nagamani et al. 2010). The effect of this deletion is sufficiently strong such that we were able to predict diabetes status in 2 of 3 related women carrying a ~1.44 Mb deletion in this region (Blackburn et al. 2013). Interestingly, the age of onset of the women with diabetes were 17 and 22.4 years respectively, representing the tail end of the distribution, while one woman was diabetes free at age 31, indicating incomplete penetrance (Blackburn et al. 2013). This observation fits the described scenario in which currently identified micro-deletion syndromes represent the tail-end of a continuous distribution of phenotypic effects, and supports the hypothesis that recurrent micro-deletions account for a portion of the observed phenotypic variation in complex disorders.

6.7 Population Studies of CNV

CNV has been investigated in multiple populations including Caucasians (Conrad et al. 2010; International HapMap 3 Consortium (IHM3C) et al. 2010; Mills et al. 2006, 2011), Asians (IHM3C et al. 2010; Ku et al. 2010; Park et al. 2010; Lou et al. 2011; Mills et al. 2011), Africans (IHM3C et al. 2010; Mills et al. 2011; Wineinger et al. 2011), and admixed populations such as Mexican Americans (IHM3C et al. 2010; Itsara et al. 2010; Mills et al. 2011; Blackburn et al. 2013). Following expected results according to population genetics theory, populations which have undergone bottlenecking carry the lowest number of polymorphisms, followed by admixed populations and populations which have not undergone recent bottlenecking such as Africans. Smaller CNVs are more frequent in individual genomes than larger CNVs, which may be attributable to selective forces, but may also be a byproduct of the mechanisms of formation. Deletions overlapping genes are enriched for lower minor allele frequencies (Conrad et al. 2010; Mills et al. 2011). Additionally, there is an inverse relationship between the size of deletions and their individual minor allele frequencies, which suggests that large deletions are under stronger purifying selection (Blackburn et al. 2013). Interestingly, CNPs in segmental duplication regions appear to be more population differentiated than CNPs in unique regions, and biallelic CNPs show greater population stratification than frequency matched SNPs (Campbell et al. 2011). Taken together these observations suggest that large deletions in regions of segmental duplication generally produce stronger effects and are under stronger selective pressure than SNPs, smaller deletions, and less complex regions of the genome. It is also observed that low MAF CNVs are more likely to be population specific (Mills et al. 2011) which is consistent with an enrichment of rare variants due to recent population expansion. These low MAF variants may contribute significantly to heritability of and ethnic differences in complex disorders. In summary, population genetic studies of CNVs provide evidence that suggest that large deletions and regions of segmental duplication may be especially deleterious and that these are good candidate regions to affect complex disease. Their low MAF may explain why their role has remained undiscovered to date.

6.8 CNV and Gene Expression

The role of gene expression in gene mapping is extensively covered in Chap. 5. Briefly, a mechanism through which disease variants can exert their effect is by affecting gene transcript abundance. Further, the expression quantitative trait loci (eQTL) with the strongest effect sizes have been observed to act primarily in *cis* (Göring et al. 2007). As a result, transcript abundances are of great interest as highly map able endophenotypes, and as a model of disease gene mapping. Given this, it is important to briefly address the role of CNVs in heritable gene expression.

Currently there are only a few comprehensive reports regarding investigating the role of copy number variation in heritable variation in gene expression. Schlattl et al. (2011), used Bacterial Artificial Chromosomes (BAC) arrays and 500 k SNP arrays to ascertain CNVs in 210 unrelated HapMap individuals, and attempted to identify a relative contribution of CNVs and SNPs to gene expression (Stranger et al. 2007). They determined that there was little overlap between eQTL associated with SNPs and those associated with CNVs (Stranger et al. 2007). However, a more comprehensive study by Schlattl et al. used CNV calls from the 1000 genomes project data, and equally high quality gene expression data and concluded that $\sim$48 % of CNV-associated eQTL genes are also identified using SNPs (Schlattl et al. 2011), an observation that is consistent with LD between common CNVs and SNPs. As with SNP associations from GWAS, it often remains unclear whether the CNVs identified are causally related to the expression phenotypes of interest because they could simply be tagging truly causal variants through LD. Schlattl et al. (2011) showed that significant CNV-gene pairs in which the CNV and gene overlap were enriched for positive correlations, strongly suggesting causality. Further, Gamazon et al. (2011) found that SNPs tagging CNVs are significantly enriched for *cis* eQTLs, and are overrepresented in the National Human Genome Research Institute (NHGRI) catalog of GWAS SNPs. Taken together; this evidence suggests that CNVs overlapping genes make very compelling candidate variants in eQTL, QTL, and GWAS regions.

The authors of this book chapter, and others, have reported that larger CNVs appear at lower frequencies, which suggest purifying selection (Blackburn et al. 2013). Similarly, it is observed that larger CNVs are more likely to influence the expression of nearby genes, which provides a mechanism through which larger CNVs could be under stronger purifying selection (Schlattl et al. 2011). Currently there remain many aspects of the relationship between CNV, heritable gene expression, and complex disease that remain undetermined. Presumably there is a plethora of unidentified CNV-associated eQTL to be discovered. Further, we don't know the contribution of dispersed duplications to heritable gene expression at their insertion sites since, for some duplications, the insertion sites currently remain unknown. We also do not know if CNVs which affect the expression of a gene are more likely to affect the expression of a second non-overlapping gene, and if so, what the predominant mechanisms for this effect are. The relative contribution of common and rare variants on gene expression in most human tissues is also unknown at this time.

6.9 Somatic CNVs, Aging, and Cancer

Somatic mosaicism of copy number variation is an understudied aspect of the heritable component of human disease. Initially, these two areas seemed divorced from each other, as somatic events were thought to be stochastic. However, elucidations of the mechanisms which determine copy number mutational events suggest these may be related to each other. As we discussed, some regions of the genome are predisposed to recurrent mutational events. One can now image a scenario in which CNVs predispose a region of the genome harboring a tumor suppressor or oncogene to deletion or duplication through mechanisms outlined in Sect. 6.4 above, the end result being predisposition to the specific recurrent somatic mutational events observed in cancer. Multiple lines of evidence already strongly suggest that mutations in genes regulating DNA repair predispose to cancer phenotypes. However, little work has been done to identify whether the somatic events observed in cancer are themselves heritable, and if so what the genetic determinates of this heritable component are. The high heritability estimates of some cancers and the observed recurrent causal mutation events might suggest that the recurrent mutational events themselves are heritable, although to our knowledge this hypothesis has not been directly tested.

In 2008, two studies reported results indicating somatic mosaicism of copy number variation. Bruder et al. (2008) reported observing discordant CNVs between monozygotic twins, a clear indication of somatic mosaicism. Piotrowski et al. (2008) reported observing copy number differences between otherwise healthy differentiated tissues. However, three sets of twins studied using WGS did not appear to harbor discordant copy number variation (Baranzini et al. 2010). These observations seemed to be at odds. However, improved methods for detection of somatic mosaicism from SNP array and array-CGH data have been developed (Gonzalez et al. 2011), which is beginning to lead to a more refined understanding of somatic structural changes (Forsberg et al. 2012). The primary observations thus far are that somatic structural changes increase with age and that there appears to be self-removal of these aberrant cells in blood (Forsberg et al. 2012). Both of these observations may potentially explain the apparent discrepancies observed in the previous studies. Interestingly, these observations are consistent with late age of onset somatic diseases such as cancer.

6.10 Final Remarks

Technological advances following the publication of the human genome have allowed us to begin to investigate copy number variation in human populations in a genome-wide fashion. Early studies investigating copy number variation showed poor overlap of identified variants between studies, but provided important methods which are now commonly used in the field. Comprehensive methods for CNV discovery and genotyping are a necessity for thorough investigation, and these methods are still in development. Genome wide association studies of copy number variation have provided examples of variants that fit the CDCV hypothesis; however the observed associations are not sufficient to account for the estimated additive heritability of complex disorders. Rare and de novo CNVs have been strongly associated with multiple complex disorders, and have provided evidence for recurrent mutation as a mechanism of disease. Although little work has been done to elucidate the relationship between CNV and heritable gene expression, early investigations from this area of research indicate that CNVs which overlap genes make especially enticing functional variant candidates in complex disease loci. The role of heritable predisposition to somatic mosaicism of CNV in complex disease is a wholly unstudied research area which is empirically promising based on observations from studies in cancer and the mechanisms of formation of CNVs. Initial studies indicate that somatic mosaicism of CNV is likely ripe with undiscovered disease mechanisms. In summary, the field of investigation of the role of CNV in common complex disorders is immature, yet early work indicating that CNV is a major source of genetic and heritable phenotypic variation between individuals suggests that those willing to investigate these more complicated regions of the genome in complex diseases should be prepared for interesting discoveries.

References

1000 Genomes Project Consortium, Abecasis GR, Altshuler D, Auton A, Brooks LD, Durbin RM, Gibbs RA, Hurles ME, McVean GA (2010) A map of human genome variation from population-scale sequencing. Nature 467:1061–1073. doi: 10.1038/nature09534

1000 Genomes Project Consortium, Abecasis GR, Auton A, Brooks LD, DePristo MA, Durbin RM, Handsaker RE, Kang HM, Marth GT, McVean GA (2012) An integrated map of genetic variation from 1,092 human genomes. Nature 491:56–65. doi:10.1038/nature11632

Alkan C, Coe BP, Eichler EE (2011) Genome structural variation discovery and genotyping. Nat Rev Genet 12:363–376. doi:10.1038/nrg2958

Bailey JA, Gu Z, Clark RA, Reinert K, Samonte RV, Schwartz S, Adams MD, Myers EW, Li PW, Eichler EE (2002) Recent segmental duplications in the human genome. Science 297:1003–1007

Baranzini SE, Mudge J, van Velkinburgh JC, Khankhanian P, Khrebtukova I, Miller NA, Zhang L, Farmer AD, Bell CJ, Kim RW, May GD, Woodward JE, Caillier SJ, McElroy JP, Gomez R, Pando MJ, Clendenen LE, Ganusova EE, Schilkey FD, Ramaraj T, Khan OA, Huntley JJ, Luo S, Kwok PY, Wu TD, Schroth GP, Oksenberg JR, Hauser SL, Kingsmore SF (2010) Genome, epigenome and RNA sequences of monozygotic twins discordant for multiple sclerosis. Nature 464(7293):1351–1356. doi:10.1038/nature08990

Barnes C, Plagnol V, Fitzgerald T, Redon R, Marchini J, Clayton D, Hurles ME (2008) A robust statistical method for case-control association testing with copy number variation. Nat Genet 40:1245–1252. doi:10.1038/ng.206

Blackburn A, Göring HH, Dean A, Carless MA, Dyer T, Kumar S, Fowler S, Curran JE, Almasy L, Mahaney M, Comuzzie A, Duggirala R, Blangero J, Lehman DM (2013) Utilizing extended pedigree information for discovery and confirmation of copy number variable regions among Mexican Americans. Eur J Hum Genet 21:404–409. doi:10.1038/ejhg.2012.188

Boettger LM, Handsaker RE, Zody MC, McCarroll SA (2012) Structural haplotypes and recent evolution of the human 17q21.31 region. Nat Genet 44:881–885. doi:10.1038/ng.2334

Bruder CE, Piotrowski A, Gijsbers AA, Andersson R, Erickson S, Diaz de Ståhl T, Menzel U, Sandgren J, von Tell D, Poplawski A, Crowley M, Crasto C, Partridge EC, Tiwari H, Allison DB, Komorowski J, van Ommen GJ, Boomsma DI, Pedersen NL, den Dunnen JT, Wirdefeldt K, Dumanski JP (2008) Phenotypically concordant and discordant monozygotic twins display different DNA copy-number-variation profiles. Am J Hum Genet 82:763–771. doi:10.1016/j.ajhg.2007.12.011

Campbell CD, Sampas N, Tsalenko A, Sudmant PH, Kidd JM, Malig M, Vu TH, Vives L, Tsang P, Bruhn L, Eichler EE (2011) Population-genetic properties of differentiated human copy-number polymorphisms. Am J Hum Genet 88:317–332. doi:10.1016/j.ajhg.2011.02.004

Chance PF, Alderson MK, Leppig KA, Lensch MW, Matsunami N, Smith B, Swanson PD, Odelberg SJ, Disteche CM, Bird TD (1993) DNA deletion associated with hereditary neuropathy with liability to pressure palsies. Cell 72:143–151

Chowdhury R, Bois PR, Feingold E, Sherman SL, Cheung VG (2009) Genetic analysis of variation in human meiotic recombination. PLoS Genet 5(9): e1000648. doi:10.1371/journal.pgen.1000648

Coin LJ, Asher JE, Walters RG, Moustafa JS, de Smith AJ, Sladek R, Balding DJ, Froguel P, Blakemore AI (2010) cnvHap: an integrative population and haplotype-based multiplatform model of SNPs and CNVs. Nat Methods 7:541–546. doi:10.1038/nmeth.1466

Colella S, Yau C, Taylor JM, Mirza G, Butler H, Clouston P, Bassett AS, Seller A, Holmes CC, Ragoussis J (2007) QuantiSNP: an objective Bayes hidden-Markov model to detect and accurately map copy number variation using SNP genotyping data. Nucleic Acids Res 35:2013–2025

Conrad DF, Andrews TD, Carter NP, Hurles ME, Pritchard JK (2006) A high-resolution survey of deletion polymorphism in the human genome. Nat Genet 38:75–81

Conrad DF, Pinto D, Redon R, Feuk L, Gokcumen O, Zhang Y, Aerts J, Andrews TD, Barnes C, Campbell P, Fitzgerald T, Hu M, Ihm CH, Kristiansson K, Macarthur DG, Macdonald JR, Onyiah I, Pang AW, Robson S, Stirrups K, Valsesia A, Walter K, Wei J; Wellcome Trust Case Control Consortium, Tyler-Smith C, Carter NP, Lee C, Scherer SW, Hurles ME (2010) Origins and functional impact of copy number variation in the human genome. Nature 464:704–712. doi: 10.1038/nature08516

Cooper GM, Coe BP, Girirajan S, Rosenfeld JA, Vu TH, Baker C, Williams C, Stalker H, Hamid R, Hannig V, Abdel-Hamid H, Bader P, McCracken E, Niyazov D, Leppig K, Thiese H, Hummel M, Alexander N, Gorski J, Kussmann J, Shashi V, Johnson K, Rehder C, Ballif BC, Shaffer LG, Eichler EE (2011) A copy number variation morbidity map of developmental delay. Nat Genet 43:838–846. doi:10.1038/ng.909

Dellinger AE, Saw SM, Goh LK, Seielstad M, Young TL, Li YJ (2010) Comparative analyses of seven algorithms for copy number variant identification from single nucleotide polymorphism arrays. Nucleic Acids Res 38:e105. doi:10.1093/nar/gkq040

Dickson SP, Wang K, Krantz I, Hakonarson H, Goldstein DB (2010) Rare variants create synthetic genome-wide associations. PLoS Biol 8:e1000294. doi:10.1371/journal.pbio.1000294

Diskin SJ, Li M, Hou C, Yang S, Glessner J, Hakonarson H, Bucan M, Maris JM, Wang K (2008) Adjustment of genomic waves in signal intensities from whole-genome SNP genotyping platforms. Nucleic Acids Res 36:e126. doi:10.1093/nar/gkn556

Edwards AO, Ritter R 3rd, Abel KJ, Manning A, Panhuysen C, Farrer LA (2005) Complement factor H polymorphism and age-related macular degeneration. Science 308:421–424

Fledel-Alon A, Leffler EM, Guan Y, Stephens M, Coop G, Przeworski M (2011) Variation in human recombination rates and its genetic determinants. PLoS ONE 6:e20321. doi:10.1371/journal.pone.0020321

Forsberg LA, Rasi C, Razzaghian HR, Pakalapati G, Waite L, Thilbeault KS, Ronowicz A, Wineinger NE, Tiwari HK, Boomsma D, Westerman MP, Harris JR, Lyle R, Essand M, Eriksson F, Assimes TL, Iribarren C, Strachan E, O'Hanlon TP, Rider LG, Miller FW, Giedraitis V, Lannfelt L, Ingelsson M, Piotrowski A, Pedersen NL, Absher D, Dumanski JP (2012) Age-related somatic structural changes in the nuclear genome of human blood cells. Am J Hum Genet 90:217–228. doi:10.1016/j.ajhg.2011.12.009

Gamazon ER, Nicolae DL, Cox NJ (2011) A study of CNVs as trait-associated polymorphisms and as expression quantitative trait loci. PLoS Genet 7: e1001292. doi:10.1371/journal.pgen.1001292

Gibson G (2011) Rare and common variants: twenty arguments. Nat Rev Genet 13:135–145. doi:10.1038/nrg3118

Gonzalez E, Kulkarni H, Bolivar H, Mangano A, Sanchez R, Catano G, Nibbs RJ, Freedman BI, Quinones MP, Bamshad MJ, Murthy KK, Rovin BH, Bradley W, Clark RA, Anderson SA, O'connell RJ, Agan BK, Ahuja SS, Bologna R, Sen L, Dolan MJ, Ahuja SK (2005) The influence of CCL3L1 gene-containing segmental duplications on HIV-1/AIDS susceptibility. Science 307:1434–1440

González JR, Rodríguez-Santiago B, Cáceres A, Pique-Regi R, Rothman N, Chanock SJ, Armengol L, Pérez-Jurado LA (2011) A fast and accurate method to detect allelic genomic imbalances underlying mosaic rearrangements using SNP array data. BMC Bioinf 12:166. doi:10.1186/1471-2105-12-166

Göring HH, Curran JE, Johnson MP, Dyer TD, Charlesworth J, Cole SA, Jowett JB, Abraham LJ, Rainwater DL, Comuzzie AG, Mahaney MC, Almasy L, MacCluer JW, Kissebah AH, Collier GR, Moses EK, Blangero J (2007) Discovery of expression QTLs using large-scale transcriptional profiling in human lymphocytes. Nat Genet 39:1208–1216

Haines JL, Hauser MA, Schmidt S, Scott WK, Olson LM, Gallins P, Spencer KL, Kwan SY, Noureddine M, Gilbert JR, Schnetz-Boutaud N, Agarwal A, Postel EA, Pericak-Vance MA (2005) Complement factor H variant increases the risk of age-related macular degeneration. Science 308:419–421

Hastings PJ, Ira G, Lupski JR (2009) A microhomology-mediated break-induced replication model for the origin of human copy number variation. PLoS Genet 5:e1000327. doi:10.1371/journal.pgen.1000327

Higgs DR, Pressley L, Old JM, Hunt DM, Clegg JB, Weatherall DJ, Serjeant GR (1979) Negro alpha-thalassaemia is caused by deletion of a single alpha-globin gene. Lancet 2:272–276

Hindorff LA, Sethupathy P, Junkins HA, Ramos EM, Mehta JP, Collins FS, Manolio TA (2009) Potential etiologic and functional implications of genome-wide association loci for human diseases and traits. Proc Natl Acad Sci USA 106:9362–9367. doi:10.1073/pnas.0903103106

Hinds DA, Kloek AP, Jen M, Chen X, Frazer KA (2006) Common deletions and SNPs are in linkage disequilibrium in the human genome. Nat Genet 38:82–85

Howie B, Fuchsberger C, Stephens M, Marchini J, Abecasis GR (2012) Fast and accurate genotype imputation in genome-wide association studies through pre-phasing. Nat Genet 44:955–959. doi:10.1038/ng.2354

Howie B, Marchini J, Stephens M (2011) Genotype imputation with thousands of genomes. G3 (Bethesda) 1:457–470. doi:10.1534/g3.111.001198

Iafrate AJ, Feuk L, Rivera MN, Listewnik ML, Donahoe PK, Qi Y, Scherer SW, Lee C (2004) Detection of large-scale variation in the human genome. Nat Genet 36:949–951

International HapMap 3 Consortium, Altshuler DM, Gibbs RA et al (2010) Integrating common and rare genetic variation in diverse human populations. Nature 467:52–58. doi: 10.1038/nature09298

International Human Genome Sequencing Consortium (2004) Finishing the euchromatic sequence of the human genome. Nature 431:931–945

Itsara A, Wu H, Smith JD, Nickerson DA, Romieu I, London SJ, Eichler EE (2010) De novo rates and selection of large copy number variation. Genome Res 20:1469–1481. doi:10.1101/gr.107680.110

Ju YS, Hong D, Kim S, Park SS, Kim S, Lee S, Park H, Kim JI, Seo JS (2010) Reference-unbiased copy number variant analysis using CGH microarrays. Nucleic Acids Res 38:e190. doi:10.1093/nar/gkq730

Klein RJ, Zeiss C, Chew EY, Tsai JY, Sackler RS, Haynes C, Henning AK, SanGiovanni JP, Mane SM, Mayne ST, Bracken MB, Ferris FL, Ott J, Barnstable C, Hoh J (2005) Complement factor H polymorphism in age-related macular degeneration. Science 308:385–389

Koolen DA, Sharp AJ, Hurst JA et al (2008) Clinical and molecular delineation of the 17q21.31 microdeletion syndrome. J Med Genet 45:710–720. doi:10.1136/jmg.2008.058701

Koolen DA, Vissers LE, Pfundt R, de Leeuw N, Knight SJ, Regan R, Kooy RF, Reyniers E, Romano C, Fichera M, Schinzel A, Baumer A, Anderlid BM, Schoumans J, Knoers NV, van Kessel AG, Sistermans EA, Veltman JA, Brunner HG, de Vries BB (2006) A new chromosome 17q21.31 microdeletion syndrome associated with a common inversion polymorphism. Nat Genet 38:999–1001

Korbel JO, Urban AE, Affourtit JP, Godwin B, Grubert F, Simons JF, Kim PM, Palejev D, Carriero NJ, Du L, Taillon BE, Chen Z, Tanzer A, Saunders AC, Chi J, Yang F, Carter NP, Hurles ME, Weissman SM, Harkins TT, Gerstein MB, Egholm M, Snyder M (2007) Paired-end mapping reveals extensive structural variation in the human genome. Science 318:420–426

Ku CS, Pawitan Y, Sim X, Ong RT, Seielstad M, Lee EJ, Teo YY, Chia KS, Salim A (2010) Genomic copy number variations in three Southeast Asian populations. Hum Mutat 31:851–857. doi:10.1002/humu.21287

Lander ES, Linton LM, Birren B et al (2001) International human genome sequencing consortium initial sequencing and analysis of the human genome. Nature 409:860–921

Levy D, Ronemus M, Yamrom B, Lee YH, Leotta A, Kendall J, Marks S, Lakshmi B, Pai D, Ye K, Buja A, Krieger A, Yoon S, Troge J, Rodgers L, Iossifov I, Wigler M (2011) Rare de novo and transmitted copy-number variation in autistic spectrum disorders. Neuron 70:886–897. doi:10.1016/j.neuron.2011.05.015

Li Y, Willer CJ, Ding J, Scheet P, Abecasis GR (2010) MaCH: using sequence and genotype data to estimate haplotypes and unobserved genotypes. Genet Epidemiol 34:816–834. doi:10.1002/gepi.20533

Lou H, Li S, Yang Y, Kang L, Zhang X, Jin W, Wu B, Jin L, Xu S (2011) A map of copy number variations in Chinese populations. PLoS ONE 6:e27341. doi:10.1371/journal.pone.0027341

Lupski JR, de Oca-Luna RM, Slaugenhaupt S, Pentao L, Guzzetta V, Trask BJ, Saucedo-Cardenas O, Barker DF, Killian JM, Garcia CA, Chakravarti A, Patel PI (1991) DNA duplication associated with Charcot-Marie-Tooth disease type 1A. Cell 66:219–232

Magi A, Tattini L, Pippucci T, Torricelli F, Benelli M (2012) Read count approach for DNA copy number variants detection. Bioinformatics 28:470–478. doi:10.1093/bioinformatics/btr707

Malhotra D, McCarthy S, Michaelson JJ, Vacic V, Burdick KE, Yoon S, Cichon S, Corvin A, Gary S, Gershon ES, Gill M, Karayiorgou M, Kelsoe JR, Krastoshevsky O, Krause V, Leibenluft E, Levy DL, Makarov V, Bhandari A, Malhotra AK, McMahon FJ, Nöthen MM, Potash JB, Rietschel M, Schulze TG, Sebat J (2011) High frequencies of de novo CNVs in bipolar disorder and schizophrenia. Neuron 72:951–963. doi:10.1016/j.neuron.2011.11.007

McCarroll SA, Hadnott TN, Perry GH, Sabeti PC, Zody MC, Barrett JC, Dallaire S, Gabriel SB, Lee C, Daly MJ, Altshuler DM; International HapMap Consortium (2006) Common deletion polymorphisms in the human genome. Nat Genet 38:86–92

Mefford HC, Eichler EE (2009) Duplication hotspots, rare genomic disorders, and common disease. Curr Opin Genet Dev 19:196–204. doi:10.1016/j.gde.2009.04.003

Mills RE, Luttig CT, Larkins CE, Beauchamp A, Tsui C, Pittard WS, Devine SE (2006) An initial map of insertion and deletion (INDEL) variation in the human genome. Genome Res 16:1182–1190

Mills RE, Walter K, Stewart C et al 1000 Genomes Project (2011) Mapping copy number variation by population-scale genome sequencing. Nature 470:59–65. doi: 10.1038/nature09708

Nagamani SC, Erez A, Shen J, Li C, Roeder E, Cox S, Karaviti L, Pearson M, Kang SH, Sahoo T, Lalani SR, Stankiewicz P, Sutton VR, Cheung SW (2010) Clinical spectrum associated with recurrent genomic rearrangements in chromosome 17q12. Eur J Hum Genet 18:278–284. doi:10.1038/ejhg.2009.174

Olshen AB, Venkatraman ES, Lucito R, Wigler M (2004) Circular binary segmentation for the analysis of array-based DNA copy number data. Biostatistics 5:557–572

Ottolenghi S, Lanyon WG, Paul J, Williamson R, Weatherall DJ, Clegg JB, Pritchard J, Pootrakul S, Boon WH (1974) The severe form of alpha thalassaemia is caused by a haemoglobin gene deletion. Nature 1974 251:389–392

Pang AW, MacDonald JR, Pinto D, Wei J, Rafiq MA, Conrad DF, Park H, Hurles ME, Lee C, Venter JC, Kirkness EF, Levy S, Feuk L, Scherer SW (2010) Towards a comprehensive structural variation map of an individual human genome. Genome Biol 11:R52. doi:10.1186/gb-2010-11-5-r52

Park H, Kim JI, Ju YS, Gokcumen O, Mills RE, Kim S, Lee S, Suh D, Hong D, Kang HP, Yoo YJ, Shin JY, Kim HJ, Yavartanoo M, Chang YW, Ha JS, Chong W, Hwang GR, Darvishi K, Kim H, Yang SJ, Yang KS, Kim H, Hurles ME, Scherer SW, Carter NP, Tyler-Smith C, Lee C, Seo JS (2010) Discovery of common Asian copy number variants using integrated high-resolution array

CGH and massively parallel DNA sequencing. Nat Genet 42:400–405. doi:10.1038/ng.555

Peiffer DA, Le JM, Steemers FJ, Chang W, Jenniges T, Garcia F, Haden K, Li J, Shaw CA, Belmont J, Cheung SW, Shen RM, Barker DL, Gunderson KL (2006) High-resolution genomic profiling of chromosomal aberrations using Infinium whole-genome genotyping. Genome Res 16:1136–1148

Perry GH, Ben-Dor A, Tsalenko A, Sampas N, Rodriguez-Revenga L, Tran CW, Scheffer A, Steinfeld I, Tsang P, Yamada NA, Park HS, Kim JI, Seo JS, Yakhini Z, Laderman S, Bruhn L, Lee C (2008) The fine-scale and complex architecture of human copy-number variation. Am J Hum Genet 82:685–695. doi:10.1016/j.ajhg.2007.12.010

Pinkel D, Segraves R, Sudar D, Clark S, Poole I, Kowbel D, Collins C, Kuo WL, Chen C, Zhai Y, Dairkee SH, Ljung BM, Gray JW, Albertson DG (1998) High resolution analysis of DNA copy number variation using comparative genomic hybridization to microarrays. Nat Genet 20:207–211

Pinto D, Darvishi K, Shi X, Rajan D, Rigler D, Fitzgerald T, Lionel AC, Thiruvahindrapuram B, Macdonald JR, Mills R, Prasad A, Noonan K, Gribble S, Prigmore E, Donahoe PK, Smith RS, Park JH, Hurles ME, Carter NP, Lee C, Scherer SW, Feuk L (2011) Comprehensive assessment of array-based platforms and calling algorithms for detection of copy number variants. Nat Biotechnol 29:512–520. doi:10.1038/nbt.1852

Pinto D, Pagnamenta AT, Klei L et al (2010) Functional impact of global rare copy number variation in autism spectrum disorders. Nature 466:368–372. doi:10.1038/nature09146

Piotrowski A, Bruder CE, Andersson R, Diaz de Ståhl T, Menzel U, Sandgren J, Poplawski A, von Tell D, Crasto C, Bogdan A, Bartoszewski R, Bebok Z, Krzyzanowski M, Jankowski Z, Partridge EC, Komorowski J, Dumanski JP (2008) Somatic mosaicism for copy number variation in differentiated human tissues. Hum Mutat 29:1118–1124. doi:10.1002/humu.20815

Sanders SJ, Ercan-Sencicek AG, Hus V et al (2011) Multiple recurrent de novo CNVs, including duplications of the 7q11.23 Williams syndrome region, are strongly associated with autism. Neuron 70:863–885. doi:10.1016/j.neuron.2011.05.002

Schlattl A, Anders S, Waszak SM, Huber W, Korbel JO (2011) Relating CNVs to transcriptome data at fine resolution: assessment of the effect of variant size, type, and overlap with functional regions. Genome Res 21:2004–2013. doi:10.1101/gr.122614.111

Sebat J, Lakshmi B, Malhotra D, Troge J, Lese-Martin C, Walsh T, Yamrom B, Yoon S, Krasnitz A, Kendall J, Leotta A, Pai D, Zhang R, Lee YH, Hicks J, Spence SJ, Lee AT, Puura K, Lehtimäki T, Ledbetter D, Gregersen PK, Bregman J, Sutcliffe JS, Jobanputra V, Chung W, Warburton D, King MC, Skuse D, Geschwind DH, Gilliam TC, Ye K, Wigler M (2007) Strong association of de novo copy number mutations with autism. Science 316:445–449

Sebat J, Lakshmi B, Troge J, Alexander J, Young J, Lundin P, Månér S, Massa H, Walker M, Chi M, Navin N, Lucito R, Healy J, Hicks J, Ye K, Reiner A, Gilliam TC, Trask B, Patterson N, Zetterberg A, Wigler M (2004) Large-scale copy number polymorphism in the human genome. Science 305:525–528

Sharp AJ, Hansen S, Selzer RR, Cheng Z, Regan R, Hurst JA, Stewart H, Price SM, Blair E, Hennekam RC, Fitzpatrick CA, Segraves R, Richmond TA, Guiver C, Albertson DG, Pinkel D, Eis PS, Schwartz S, Knight SJ, Eichler EE (2006) Discovery of previously unidentified genomic disorders from the duplication architecture of the human genome. Nat Genet 38:1038–1042

Soemedi R, Wilson IJ, Bentham J et al (2012) Contribution of global rare copy-number variants to the risk of sporadic congenital heart disease. Am J Hum Genet 91:489–501. doi:10.1016/j.ajhg.2012.08.003

Stefansson H, Helgason A, Thorleifsson G, Steinthorsdottir V, Masson G, Barnard J, Baker A, Jonasdottir A, Ingason A, Gudnadottir VG, Desnica N, Hicks A, Gylfason A, Gudbjartsson DF, Jonsdottir GM, Sainz J, Agnarsson K, Birgisdottir B, Ghosh S, Olafsdottir A, Cazier JB, Kristjansson K, Frigge ML, Thorgeirsson TE, Gulcher JR, Kong A, Stefansson K (2005) A common inversion under selection in Europeans. Nat Genet 37:129–137

Steinberg KM, Antonacci F, Sudmant PH, Kidd JM, Campbell CD, Vives L, Malig M, Scheinfeldt L, Beggs W, Ibrahim M, Lema G, Nyambo TB, Omar SA, Bodo JM, Froment A, Donnelly MP, Kidd KK, Tishkoff SA, Eichler EE (2012) Structural diversity and African origin of the 17q21.31 inversion polymorphism. Nat Genet 44:872–880. doi:10.1038/ng.2335

Stranger BE, Forrest MS, Dunning M, Ingle CE, Beazley C, Thorne N, Redon R, Bird CP, de Grassi A, Lee C, Tyler-Smith C, Carter N, Scherer SW, Tavaré S, Deloukas P, Hurles ME, Dermitzakis ET (2007) Relative impact of nucleotide and copy number variation on gene expression phenotypes. Science 315:848–853

Surakka I, Kristiansson K, Anttila V, Inouye M, Barnes C, Moutsianas L, Salomaa V, Daly M, Palotie A, Peltonen L, Ripatti S (2010) Founder population-specific HapMap panel increases power in GWA studies through improved imputation accuracy and CNV tagging. Genome Res 20:1344–1351. doi:10.1101/gr.106534.110

Tajima F (1989) Statistical method for testing the neutral mutation hypothesis by DNA polymorphism. Genetics 123:585–595

The International HapMap Consortium (2003) The International HapMap Project. Nature 426:789–796

Tuzun E, Sharp AJ, Bailey JA, Kaul R, Morrison VA, Pertz LM, Haugen E, Hayden H, Albertson D, Pinkel D, Olson MV, Eichler EE (2005) Fine-scale structural variation of the human genome. Nat Genet 37:727–732

Venkatraman ES, Olshen AB (2007) A faster circular binary segmentation algorithm for the analysis of array CGH data. Bioinformatics 23:657–663

Venter JC, Adams MD, Myers EW et al (2001) The sequence of the human genome. Science 291:1304–1351

Walsh T, McClellan JM, McCarthy SE et al (2008) Rare structural variants disrupt multiple genes in neurodevelopmental pathways in schizophrenia. Science 320:539–543. doi:10.1126/science.1155174

Wang K, Li M, Hadley D, Liu R, Glessner J, Grant SF, Hakonarson H, Bucan M (2007) PennCNV: an integrated hidden Markov model designed for high-resolution copy number variation detection in whole-genome SNP genotyping data. Genome Res 17:1665–1674

Wellcome Trust Case Control Consortium (2007) Genome-wide association study of 14,000 cases of seven common diseases and 3,000 shared controls. Nature 447:661–678

Wellcome Trust Case Control Consortium, Craddock N, Hurles ME, et al. (2010) Genome-wide association study of CNVs in 16,000 cases of eight common diseases and 3,000 shared controls. Nature 464:713–720. doi: 10.1038/nature08979

Winchester L, Yau C, Ragoussis J (2009) Comparing CNV detection methods for SNP arrays. Brief Funct Genomic Proteomics 8:353–366. doi:10.1093/bfgp/elp017

Wineinger NE, Pajewski NM, Kennedy RE, Wojczynski MK, Vaughan LK, Hunt SC, Gu CC, Rao DC, Lorier R, Broeckel U, Arnett DK, Tiwari HK (2011) Characterization of autosomal copy-number variation in African Americans: the HyperGEN study. Eur J Hum Genet 19:1271–1275. doi:10.1038/ejhg.2011.115

Xu B, Roos JL, Levy S, van Rensburg EJ, Gogos JA, Karayiorgou M (2008) Strong association of de novo copy number mutations with sporadic schizophrenia. Nat Genet 40:880–885. doi:10.1038/ng.162

Yoon S, Xuan Z, Makarov V, Ye K, Sebat J (2009) Sensitive and accurate detection of copy number variants using read depth of coverage. Genome Res 19:1586–1592. doi:10.1101/gr.092981.109

Zhang F, Seeman P, Liu P, Weterman MA, Gonzaga-Jauregui C, Towne CF, Batish SD, De Vriendt E, De Jonghe P, Rautenstrauss B, Krause KH, Khajavi M, Posadka J, Vandenberghe A, Palau F, Van Maldergem L, Baas F, Timmerman V, Lupski JR (2010) Mechanisms for nonrecurrent genomic rearrangements associated with CMT1A or HNPP: rare CNVs as a cause for missing heritability. Am J Hum Genet 86:892–903. doi:10.1016/j.ajhg.2010.05.001

7 Applications of Genomic Methods to Studies of Wild Primate Populations

Mary A. Kelaita

7.1 Introduction: Molecular Tools for Field Primatology

Nonhuman primates have been instrumental subjects of research intended to investigate the genetic basis of human health and diseases, given that many of their similarities with humans were inherited from a recently shared ancestor. Recent genomic technological advances have facilitated biomedical research conducted on captive nonhuman primates. Therefore, genomic efforts have focused on key nonhuman primate taxa considered to be the model species for biomedical research, such as macaques (Gibbs et al. 2007) and baboons (Rogers et al. 2009). For example, the use of the baboon as a model to determine how genetic variation influences a complex disease is discussed in this volume (e.g., Chap. 16; Comuzzie's chapter). In addition, special attention has also been given to apes owing to their status as our closest relatives (Stone and Verrelli 2006), which contributes to the understanding of what makes humans unique.

This chapter, however, is dedicated to studies of wild primate populations. Field studies of wild primates can be equally important, offering answers to ecological and evolutionary questions that cannot be addressed with data from captive populations. For example, the extent of genotypic and phenotypic variation both within and among populations can often only be observed in the wild. In addition, an evolutionary framework often requires the relevant ecological and environmental factors that shape primate lineage diversity (Tung et al. 2010). Wild population studies allow researchers to determine the effects of different environmental factors on an individual's phenotype. Some phenotypic variation may only be observed in the presence of specific genotype-by-environment interactions, and could suggest the need for the investigation of gene regulating mechanisms in that developmental pathway (Tung et al. 2011). It is also possible to test hypotheses about how some genotypes influence survival and reproduction (and therefore fitness) in wild populations given that they are under natural selective pressures (Bradley and Lawler 2011).

Apart from providing an ecological and evolutionary context, the diversity of wild primate populations suggests that phylogenetic comparisons within the primate order (including model and nonmodel species and their subpopulations) can shed light on when and how unique human adaptations evolved and what processes resulted in the observed current human-wide genomic variation. Phylogenetic relationships can be more accurately ascertained when samples are obtained with special considerations for the geographic distribution of and the extent of variation within wild populations (Luikart et al.

M.A. Kelaita (✉)
Department of Anthropology, University of Texas at San Antonio, One UTSA Circle, San Antonio, TX 78249, USA
e-mail: mary.kelaita@utsa.edu

R. Duggirala et al. (eds.), *Genome Mapping and Genomics in Human and Non-Human Primates*, Genome Mapping and Genomics in Animals 5, DOI 10.1007/978-3-662-46306-2_7,

2003; Thalmann et al. 2007). In these cases, genomic data can inform debates regarding the taxonomic placement of certain populations. Primate phylogenomics is a field that is already well underway in taking advantage of genomic tools (Moulin et al. 2008; Siepel 2009; Ting and Sterner 2013).

Bradley and Lawler (2011) present a comprehensive review on how field primatologists can take advantage of genomic tools to uncover genetic variation underlying primate adaptations, including candidate gene approaches, genome-wide association studies, and expression analyses. Additionally, Ting and Sterner (2013) recently reviewed the status of primate molecular phylogenetics after the introduction of genomic tools. This chapter will build on this existing body of knowledge by providing a brief background on genomic methods but will focus on other tools and applications of relevance to field primatologists that center on detecting variation in wild populations. Population genetics methods can inform studies of primate conservation (Chaves et al. 2011), hybridization (Kelaita and Cortés-Ortiz 2013), behavior and social organization (Di Fiore 2003), and demographic history (Lawler 2011). DNA can now be obtained through noninvansive sampling methods, which are preferred by many primatologists knowing the potential harm that could result during capture. Many primates cannot be habituated to human presence, leaving DNA as the only viable method for identifying individuals and measuring real and effective population sizes (Vigilant 2009). To that effect, molecular biology has already revolutionized the field of primatology, providing tools such as gel electrophoresis, restriction enzyme mapping, polymerase chain reaction (PCR), and finally DNA sequencing (Charlesworth 2010; Di Fiore 2003).

Thus far, the majority of the methods utilized by field primatologists have for the most part relied on inferences made from a few loci discovered through inefficient methods (Raveendran et al. 2006). This could result in inaccurate measures of variation, inability to discern relationships in parentage analysis, or unreliable estimates of divergence, given that various parts of the genome may have been under different evolutionary pressures. The most significant promise of the genomic revolution is the potential to acquire massive amounts of genetic data. Now, with the ability to study thousands to millions of genetic markers, field primatologists will be able to answer questions that they have been unable to with only a limited number of loci. Indeed, the decreasing costs of new technologies and the discovery of novel methods have generated a great deal of interest in determining how genomics can benefit wildlife biology and ecology studies (Thomas and Klaper 2004; Ryder 2005; Primmer 2009; Allendorf et al. 2010; Avise 2010; Ouborg et al. 2010; Steiner et al. 2013). While primates have been at the forefront of genomic sequencing efforts relative to other organisms (Tung et al. 2010), wild primate studies have been slow to incorporate many of the methods reviewed in the ecological genomics literature.

Neutral markers are used to generate estimates of parameters such as effective population size (Ne) and migration rate (m) (Allendorf et al. 2010) as well as nucleotide diversity and recombination rates (Steiner et al. 2013); therefore, the inclusion of a large number of markers from across the entire genome is necessary for accurate parameter estimation. For example, sequencing of the Sumatran and Bornean orangutan genomes revealed a much larger effective population size and greater genetic diversity in the Sumatran species and a divergence time that is more recent than those proposed by previous studies (Locke et al. 2011). Larger data sets enable researchers to test for outlier loci before estimating population parameters, thereby testing assumptions of neutrality (Luikart et al. 2003). Larger data sets also have the potential to uncover historical events such as population bottlenecks and expansions (Ryder 2005), especially given the "mosaic" nature of the genome, different regions of which may have undergone recombination and been subject to different selective pressures (Degnan and Rosenberg 2009). A greater number of markers would reveal linked loci and can improve haplotype inference in order to detect the extent and directionality of migration (Allendorf et al.

2010). With whole genome data, comparisons of the entire genome can be made across taxa which can shed light on the processes generating diversity in primate lineages (Hudson 2008).

7.2 Making the Transition from Genetics to Genomics

Primatologists who plan on taking advantage of the genomic revolution may find it difficult to make the transition, considering that overall few eukaryote species have received attention for genomic resource development (Hudson et al. 2008). When the first genome of a species is assembled and published, it serves as a reference map for assembling genomes of other individuals from the same species (Baker 2012). In addition, it can be scanned for the identification of polymorphic markers, as has been done for rhesus macaques (Raveendran et al. 2006). Many nonmodel primate species lack a fully assembled reference genome. Obtaining a fully sequenced genome in the absence of a reference genome requires a great investment in time and resources for *de novo* genome assembly. This is the case even despite recent advances in assembling genomes on a massively parallel scale (Wheeler 2008). Primatologists interested in using genomic tools currently have two options: either work with model organisms that already have significant genomic resources available or use the resources available from a closely related species for which a reference genome exists and apply them to a species of interest (Thomas and Klaper 2004).

Recently, after the sequencing of the first complete human genome, efforts have been in full force to sequence whole genomes of nonhuman primates, beginning with some species identified as sequencing targets for various reasons. Some were assigned the highest priority, owing to their taxonomic placement as index species in the primate phylogeny, their use in biomedical research (Marques-Bonet et al. 2009), or their conservation status (Ryder 2005). Currently, there are 32 ongoing primate genome projects (reviewed in Bradley and Lawler 2011, and listed on http://www.genome.gov/10002154). Field primatologists can begin to take advantage of published data by accessing a number of available online databases with built-in alignment search tools. Some researchers are conducting partial genome sequencing projects in an effort to provide more sequence data resources for nonmodel primate species for which no whole genome sequencing is currently planned. For example, Jameson et al. (2012) developed and annotated sequence reads from three platyrrhine species from genomic shotgun libraries of 3,000 individual sequences. These data can provide a resource for marker discovery in other related New World taxa.

Once a genome project is completed, the assembled and annotated genomes can be used as reference sequences in what is termed "massively parallel" or "next generation sequencing (NGS) technology", allowing for millions of simultaneous reads in each run. For some nonmodel species, an assembled genome of a closely related species can serve as a scaffold. These "genome-enabled" species studies can benefit from many of the currently available resources (Thompson et al. 2010), but must factor in genome assembly errors that result from low coverage and actual variation between the two species (Bradley and Lawler 2011). There remains a number of nonhuman primate species which have been ecologically well characterized but have not received much attention in sequencing projects (e.g., howler monkeys), possibly due to the perceived lack of their research's direct implications for understanding human health and evolution as well as their conservation status. Given the predicted reductions in costs and effort needed to assemble new genomes, this may change in the near future. Until then, primatologists can take steps toward making the transition from the genetic to the genomic era.

The first step for many primatologists is recognizing the different types of newly developed genomic technologies. This can be daunting given the accelerated rate at which new technologies are being introduced and utilized. The traditional Sanger technology provided sequence data of up to 2 kilobases through the detection of labeled nucleotides as they are incorporated during DNA synthesis (Zhang et al. 2011). Given the sequence length limitations, "shotgun" sequencing was introduced, so-called because DNA was sheared and inserted into cloning vectors, which were randomly fragmented and sequenced to produce short reads. Whole genomes were originally acquired in this manner, through the assembly of these reads into larger fragments, thereby generating sequence data for the entire genome of the individual. The challenge with assembling the first genome for any species is therefore the correct spatial mapping of reads in the absence of a fully mapped genome that can serve as a comparative reference. This is by no means a simple task; the assembly of a draft genome requires considerable bioinformatics know-how and computing resources. The task is further complicated by the presence of structural variation in the genome, including gene duplication (Davey et al. 2011).

NGS technology similarly accomplishes sequencing of the entire genome through the random fragmentation of DNA followed by their sequencing. The use of cloning is eliminated, and sequences are instead bound to adapters (Zhang et al. 2011). However, NGS technology actually comprises several types including Roche 454 pyrosequencing Illumina sequencing by synthesis, ABI SOLiD sequencing by ligation, and Helicos tSMS single-molecule sequencing, whose advantages and disadvantages have been compared (Hudson 2008; Ekblom and Galindo 2011). These technologies have a number of different applications which will be discussed below, but all come with their own set of challenges (Pool et al. 2010). When a reference genome is available, sequencing other individuals of the same species to uncover variation in the population is referred to as "resequencing" (Bentley 2008). This is most preferable given that complete genomic information for each individual is obtained, including coding and noncoding regions, allowing for inferences to be made about the evolutionary pressures that shaped genomes of extant species and uncovering sequence as well as structural variation. For some nonhuman primate species, *de novo* whole genome assembly remains impractical considering the amount of time, funding, expertise, and infrastructure necessary. An additional challenge is that the NGS instruments' data analysis software is usually designed to assemble and annotate human, rat, and mouse sequences. Working with other species requires further development of sequence assembly and annotation pipelines even when a fully assembled reference genome is available. Finally, analyzing a large number of individuals is essential for addressing population genetics questions, but obtaining whole genome sequences for each individual in a sample remains an unfeasible and costly endeavor.

A useful tool for nonmodel species research is expressed sequence tags (ESTs), which are short sequences produced by translating mRNA transcripts into complementary DNA, and represents only protein coding regions (Rudd 2003). ESTs are relatively inexpensive to produce and have been used extensively by molecular ecologists (Bouck and Vision 2007). Therefore, an alternative to genomics involves an analysis of the transcriptome, the mRNA obtained from different tissues at different life stages (Vera et al. 2008). Assembly of a species' transcriptome can be more feasible than that of the genome, given that it only involves mapping of coding sequences. This approach is often recommended for ecologists who plan to begin genomics projects for species that lack a reference genome (Cahais et al. 2012). Transcriptome characterization can be carried out on model organisms with available reference genomes or EST data, but can also involve *de novo* assembly (Cahais et al. 2012; Vera et al. 2008). In fact, Perry et al. (2012) developed a method for *de novo* transcriptome assembly and assembled thousands of sequences for 16 mammalian species, including 11 primate species. Interestingly, RNA comparisons revealed that endangered lemur populations exhibit considerable genetic variation, likely since factors that have impacted lemur populations occurred too recently to be reflected in observed genetic diversity measures. Such comparisons can now be made by accessing publicly available data. For example, Pipes et al.

(2012) developed a nonhuman primate reference transcriptome resource (http://nhprtr.org) presently hosting RNA sequence data for 13 primate species.

Random-primed cDNA libraries can be created and used to analyze nucleotide variation or they can provide information on whether and to what degree genes are expressed. In addition to the potential for massive, parallel investigations of gene expression, NGS can be used to produce the actual mRNA sequences for later assembly (Hudson 2008). Once a transcriptome is assembled, it can be used as a template for further resequencing or the development of markers and constructions of microarrays for expression profiling (Ekblom and Galindo 2011). The transcriptome, therefore, can be a viable method for generating genetic markers for wild population studies.

NGS technologies can be used to generate large amounts of sequence data even without assembling them into a full genome, and these data can be further interrogated for marker discovery. Also, given the difficulty in obtaining whole genome data for many individuals, there are a number of methods utilizing NGS technologies that sample some of the overall variation present in a population, sometimes referred to as genome complexity reduction (GCR) methods (Davey et al. 2011; Dou et al. 2012). For example, a number of known loci can be targeted through the selective capture of DNA prior to sequencing but high coverage sequencing of these regions provides intraspecific variation information that can be useful for population genetics analyses (Ekblom and Galindo 2011). Bi et al. (2012) performed an exon capture in chipmunks relying on a low-coverage draft genome of the ground squirrel that is 30 mya divergent from the chipmunk. They developed transcripts from different tissues and identified $\sim 12{,}000$ exons for capture from these transcripts. Unfortunately, this approach is limited to functional regions, although "exon-primed intron-crossing" (EPIC) markers were developed which can also span intron regions. EPIC markers have the unique property of being variable but also generally conserved across a broad range of species (Thompson et al. 2010). Finally, targeted sequencing of variable parts of the genome can be used as a barcoding approach as well (Ekblom and Galindo 2011), a method that can be of use for identifying plant and bacterial species from fecal samples.

Yet another GCR method ideal for population genetics analyses is called restriction site-associated DNA sequencing (RADSeq, Davey and Blaxter 2010). After genomic DNA is sheared with restriction enzymes, adapters with unique molecular identifiers for each individual are ligated to the fragments, allowing them to bind to the Ilumina flow cell. These fragments are then pooled, randomly sheared, and ligated to a second adapter with a divergent end that can only be amplified upon the amplification of the first adapter containing the molecular identifier. The resulting library is sequenced, generating sequence data of the adapters and the DNA flanking the restriction site, where polymorphisms can be found (Davey and Blaxter 2010). A similar method involves RNA sequencing (RNASeq) where cDNA libraries are used instead of genomic DNA (Wang et al. 2009).

Single nucleotide polymorphisms (SNPs) are especially suited for measuring genetic diversity, a large number of which can be discovered through resequencing (Hudson 2008). SNPs can be utilized as neutral markers for measuring genetic diversity but can also occur in coding or regulatory regions. SNPs can be employed in genome-wide association studies in pedigreed populations which are designed to discover statistically significant correlations between particular regions of the genome and the phenotype in question (Slate et al. 2009). The most feasible high-throughput method for SNP discovery is likely to be through transcriptome sequencing and resequencing (Hudson 2008) or through capture of sequences using EPIC markers, so that SNPs can be identified in a number of species related to the focal organism even without existing sequence data (Slate et al. 2009). Central to many population genetics analyses are measures of linkage disequilibrium (LD), which provides information about historical and demographic events, and can be determined from SNP data through the construction of linkage maps, which incidentally also aid in locating genes under selection (Thompson et al. 2011).

Recently, Bergey et al. (2013) applied the RADSeq technique to five primate species, including humans, representing major lineages within the primate order. They were able to detect a large number of SNPs that can be compared across closely related species at a relatively low cost. Therefore, the method can be adopted to search for SNPs that exhibit intraspecific variation, but also SNPs that can be used in phylogenetic analyses of relatively shallow trees. The RADSeq method requires high-quality DNA, preferably obtained from tissue or blood samples. However, there are promising methods for extracting DNA from fecal samples for genomic analyses (Perry et al. 2010), and together these studies show real promise for the ability of primatologists to work with large-scale genomic data when resources are scarce.

It is important to note that while using a subset of the genome through GCR methods for marker discovery is more feasible, whole genome sequences could still be more advantageous for demographic analyses given the presence of rare variants and could provide a more complete picture of allele frequencies (Pool et al. 2010).

7.3 Further Applications for Wild Primate Populations

7.3.1 Pedigree Reconstruction

To date, a large number of wild primate population studies lack pedigree information. Long-term studies of wild primate populations tracking several generations are rare. Knowing relatedness among individuals is important for identifying quantitative trait loci and measuring heritability (Pemberton 2008), as well as for measuring reproductive skew and for studying kin-directed behaviors (Di Fiore 2009). Many wild population studies have relied on microsatellite markers, which are highly variable, to infer relationships among individuals (Di Fiore 2009). However, the power to accurately determine pedigree relationships not only depends on how polymorphic a marker is but also the number of markers employed (Blouin 2003). SNP markers, while having lower power than microsatellites for resolving relationships, can be identified using high-throughput methods, providing ample numbers of markers for parentage analysis, and are less prone to genotyping errors (Hauser et al. 2011). For example, large numbers of SNPs have helped to determine relatedness among individuals in a zebra fish population (Santure et al. 2010). SNPs can potentially provide power for determining different categories of kinship beyond those of parent–offspring pairs or full sibs (Avise 2010). Microsatellites have so far remained the marker of choice for wild primate relatedness inference but with the availability of SNP discovery methods, primatologists can begin to construct accurate and specific relationships in natural populations.

7.3.2 Metagenomics

The field of metagenomics has allowed comparisons of microbial ecosystems across primate taxa, encompassing gastrointestinal and vaginal microbiomes. Microbial ecosystems reflect different species' phylogenetic history, dietary quality and availability, and even health outcomes in response to their respective environments (Amato et al. 2013). Gut microbes are thought to influence the evolution of their host, given their role in metabolizing certain nutritional components. Metagenomics studies have thus far provided evidence that microbial community composition is often not only species-specific but can also reflect habitat differences. Given that gut bacteria are largely parentally inherited, gut microbiota evolutionary history should coincide with that of their hosts (Ochman et al. 2010). Yildirim et al. (2010) utilized pyrosequencing technology of the small subunit rRNA (a region of the 16S rRNA gene) of different nonhuman primate species. They found greater similarity in microbial community composition within species than between species, and that gastrointenstinal microbiomes are highly associated with their host taxa. Overall, gut microbiota among great ape species was found to be phylogentically conserved (Ochman et al. 2010). However, a number of factors including ecological differences among the hosts' environments may shape gut microbial composition. The role of habitat differences (and further, dietary differences) was further confirmed by Amato and colleagues (2013), who assessed microbial community composition from howler monkey fecal samples by sequencing the same region of the rRNA gene. They found habitat specific microbial taxa composition, diversity, and richness, which is predicted by habitat type and shaped by the availability of plants in the diet.

7.3.3 Hybridization

Hybridization in primates has been garnering a great deal of attention recently as molecular tools have made it possible to detect more instances of gene flow across established taxonomically distinct primate taxa (Cortés et al. 2007). Debate regarding the importance of the role of hybridization in primate evolution continues (Zinner et al. 2011), and is receiving renewed interest given the finding that a number of genes have introgressed from Neanderthals into modern humans (Green et al. 2010). So far, researchers have been able to detect hybrid primate individuals using relatively few diagnostic microsatellite loci (Cortés-Ortiz et al. 2007; Tung et al. 2008; Kelaita and Cortés-Ortiz 2013). Yet, initial identification of these loci and subsequent testing is time consuming and cumbersome. Not only must loci successfully amplify and be highly variable, they must also possess fixed allelic differences between the parental species. SNPs, which instead can be identified with high-throughput methods, can also serve as diagnostic loci in hybridization studies (Finger et al. 2009; Hohenlohe et al. 2011).

Further, while few microsatellite loci can aid in the detection of hybrids, understanding the dynamics of gene flow and introgression across the hybrid zone is important for determining mechanisms of reproductive isolation and barriers to gene flow. Such an endeavor requires the use of a much greater number of loci (Allendorf et al. 2010). Teeter et al. (2009) discovered selection against hybrid genotypes and for some introgressed genotypes in a mouse hybrid zone using 41 SNPs. Whole genome data could potentially address the role of the number of loci and the size of their effects, dominance, epistasis, or chromosomal rearrangements in causing outbreeding depression. Hybridization has been shown to produce highly variable morphological characteristics in nonhuman primates (Ackermann et al. 2010; Kelaita and Cortés-Ortiz 2013) and it remains unclear what genetic interactions are the cause of this variability. Genomic approaches could also produce more accurate estimates of each hybrid's proportion of admixture (Allendorf et al. 2010; Steiner et al. 2013). With this information, morphology, behavior, and fitness can be compared across individuals of varying genomic background. Finally, genomic data promises to uncover past hybridization events that could have led to the formation of new species and the emergence of novel adaptations (Keller et al. 2012).

7.4 Concluding Remarks

It is likely that the number of genome-enabled nonhuman primate species will increase in the near future. This chapter has outlined a number of approaches that are feasible for wild population studies, some of which are relatively inexpensive and require little effort. These methods enable making evolutionary and functional inferences for a broader range of species, including nonmodel primate species that have generally received less attention in genomic resource development. However, field primatologists are likely to still face a number of obstacles to fully engaging in this type of research. A consistent concern in wild primate population studies is access to high-quality DNA, which is harder to obtain from noninvasive sampling methods. In addition, as Tung et al. (2010) recommend, considerable statistical and programming skill is required to undertake genome-scale analyses. Successful genomic endeavors often involve collaborations with researchers who have access to the infrastructure (both laboratory and computing) necessary or who possess expertise in these areas, but building on these skills as more resources become available is necessary given that technological discoveries are enabling investigators to conduct genomic studies with the budget and equipment of a small laboratory. Finally, while primatologists may be eager to acquire massive amounts of genetic data for a seemingly unlimited potential to answer important evolutionary and ecological questions, a well-designed project can help identify the minimum number of loci necessary for the analysis, the ideal sequencing technologies with the least amount of error produced, and the most time- and cost-efficient approaches for achieving one's goals.

References

Allendorf FW, Hohenlohe PA, Luikart G (2010) Genomics and the future of conservation genetics. Nat Rev Genet 11:697–709

Amato KR, Yeoman CJ, Kent A, Righini N, Carbonero F, Estrada A, Gaskins HR, Stumpf RM, Yildirim S, Torralba M, Gillis M, Wilson BA, Nelson KE, White BA, Leigh SR (2013) Habitat degradation impacts black howler monkey (*Alouatta pigra*) gastrointestinal microbiomes. Int Soc Microb Ecol 7:1344–1353

Avise J (2010) Perspective: conservation genetics enters the genomics era. Conserv Genet 11:665–669

Baker M (2012) *De novo* genome assembly: what every biologist should know. Nat Methods 9:333–337

Bentley DR (2008) Whole-genome re-sequencing. Curr Opin Genet Dev 16:545–552

Bergey CM, Pozzi L, Disotell TR, Burrell AS (2013) A new method for genome-wide marker development and genotyping holds great promise for molecular primatology. Int J Primatol 34:303–314

Bi K, Vanderpool D, Singhal S, Linderoth T, Moritz C, Good JM (2012) Transcriptome-based exon capture enables highly cost-effective comparative genomic data collection at moderate evolutionary scales. BMC Genom 13:403–417

Blouin MS (2003) DNA-based methods for pedigree reconstruction and kinship analysis in natural populations. Trends Ecol Evol 18:503–511

Bouck A, Vision T (2007) The molecular ecologist's guide to expressed sequence tags. Mol Ecol 16:907–924

Bradley BJ, Lawler RR (2011) Linking genotypes, phenotypes, and fitness in wild primate populations. Evol Anthropol 20:104–119

Cahais V, Gayral P, Tsagkogeorga G, Melo-Ferreira J, Ballenghien M, Weinert L, Chiari Y, Belkhir K, Ranwez V, Galtier N (2012) Reference-free transcriptome assembly in non-model animals from next-generation sequencing data. Mol Eco Res 12:834–845

Charlesworth B (2010) Molecular population genomics: a short history. Genet Res 92:397–411

Chaves PB, Alvarenga CS, Possamai CB, Dias LG, Boubli JP, Strier KB, Mendes SL, Fagundes V (2011) Genetic diversity and population history of a critically endangered primate, the Northern muriqui (*Brachyteles hypoxanthus*). PLoS ONE 6:e20722

Cortés-Ortiz L, Duda TF, Canales-Espinosa D, García-Orduña F, Rodríguez-Luna E, Bermingham E (2007) Hybridization in large-bodied New World primates. Genetics 176:2421–2425

Davey JW, Blaxter ML (2010) RADSeq: next generation population genetics. Brief Funct Genomics 9:416–423

Davey JW, Hohenlohe PA, Etter PD, Boone JQ, Catchen JM, Blaxter ML (2011) Genome-wide genetic marker discovery and genotyping using next-generation sequencing. Nat Rev Genet 12:499–510

Degnan JH, Rosenberg NA (2009) Gene tree discordance, phylogenetic inference and the multispecies coalescent. Trends Ecol Evol 24(6):332–340

Di Fiore A (2003) Molecular genetic approaches to the study of primate behavior, social organization, and reproduction. Yearb Phys Anthropol 46:62–99

Di Fiore A (2009) Genetic approaches to the study of dispersal and kinship in New World primates. In: Garber PA, Estrada A, Bicca-Marques JC, Heymann EW, Strier KB (eds) South American primates, comparative perspectives in the study of behavior, ecology, and conservation, series: developments in primatology: progress and prospects. Springer, Heidelberg, pp 211–250

Dou J, Zhao X, Fu X, Jiao W, Wang N, Zhang L, Hu X, Wang S, Bao Z (2012) Reference-free SNP calling: improved accuracy by preventing incorrect calls from repetitive genomic regions. Biol Direct 7:17–26

Ekblom R, Galindo J (2011) Applications of next generation sequencing in molecular ecology of non-model organisms. Heredity 107:1–15

Finger AJ, Stephens MR, Clipperton NW, May B (2009) Six diagnostic single nucleotide polymorphism markers for detecting introgression between cutthroat and rainbow trouts. Mol Ecol Resour 9:759–763

Green R, Krause J, Briggs A, Maricic T, Stenzel U, Kircher M, Patterson N, Li H, Zhai W, Fritz M (2010) A draft sequence of the Neandertal genome. Science 328:710–722

Gibbs RA, Rogers J, Katze MG, Bumgarner R, Weinstock GM, Mardis ER, Remington KA, Strausberg RL, Venter JC (2007) Evolutionary and biomedical insights from the rhesus macaque genome. Science 316:222–234

Hauser L, Baird M, Hildborn R, Seeb LW, Seeb JS (2011) An empirical comparison of SNPs and microsatellites for parentage and kinship assignment in a wild sockeye salmon (*Oncorhynchus nerka*) population. Mol Ecol Resour 11(S1):150–161

Hohenlohe PA, Amish SJ, Catchen JM, Allendorf FW, Luikart G (2011) Next-generation RAD sequencing identifies thousands of SNPs for assessing hybridization between rainbow and westslope cutthroat trout. Mol Ecol Resour 11(S1):117–122

Hudson ME (2008) Sequencing breakthroughs for genomic ecology and evolutionary biology. Mol Ecol Resour 8:3–17

Jameson NM, Xu K, Yi SV, Wildman DE (2012) Development and annotation of shotgun sequence libraries from New World monkeys. Mol Ecol Resour 12:950–955

Kelaita MA, Cortés-Ortiz L (2013) Morphological Variation of Genetically Confirmed *Alouatta pigra* x *A. palliata* hybrids from a natural hybrid zone in Tabasco Mexico. Am J Phys Anth 150:223–234

Keller I, Wagner CE, Greuter L, Mwaiko S, Selz OM, Sivasundar A, Wittwer S, Seehausen O (2012) Population genomic signatures of divergent adaptation, gene flow and hybrid speciation in the rapid radiation of Lake Victoria cichlid fishes. Mol Ecol 22:2848–2863

Lawler RR (2011) Demographic concepts and research pertaining to the study of wild primate populations. Yearb Phys Anthropol 54:63–85

Locke DP, Hillier LW, Warren WC, Worley KC, Nazareth LV et al (2011) Comparative and demographic analysis of orangutan genomes. Nature 469:529–533

Luikart G, England PR, Tallmon D, Jordan S, Taberlet P (2003) The power and promise of population genomics: from genotyping to genome typing. Nat Rev 4:981–994

Marques-Bonet T, Ryder OA, Eichler EE (2009) Sequencing primate genomes: what have we learned? Annu Rev Genomics Hum Genet 10:355–386

Moulin S, Gerbault-Seureau M, Dutrillaux B, Richard FA (2008) Phylogenomics of African guenon. Chrom Res 16:783–799

Ochman H, Worobey M, Kuo C, Ndjango JN, Peeters M, Hahn BH, Hugenholtz P (2010) Evolutionary relationships of wild hominids recapitulated by gut microbial communities. PLOS Biol 8:e1000546

Ouborg NJ, Pertoldi C, Loeschcke V, Bijlsma R, Hedrick PW (2010) Conservation genetics in transition to conservation genomics. Trends Genet 26:177–187

Pemberton JM (2008) Wild pedigrees: the way forward. Proc R Soc B 275:613–621

Perry GH, Marioni J, Pall M, Gilad Y (2010) Genomic-scale capture and sequencing of endogenous DNA from feces. Mol Ecol 19:5328–5331

Perry GH, Melsted P, Marioni J, Wang Y, Bainer R, Pickrell JK, Michelini K, Zehr S, Yoder AD, Stephens M, Pritchard JK, Gilad Y (2012) Comparative RNA sequencing reveals substantial genetic variation in endangered primates. Genome Res 22:602–610

Pipes L, Li S, Bozinoski M, Palermo R, Peng X, Blood P, Kelly S, Weiss JM, Thierry-Mieg J, Thierry-Mieg D, Zumbo P, Chen R, Schroth GP, Mason CE, Katze MG (2012) The non-human primate reference transcriptome resource (NHPRTR) for comparative functional genomics. Nucleic Acids Res 41: D906–D914

Pool JE, Hellmann I, Jensen JD, Nielsen R (2010) Population genetic inference from genomic sequence variation. Genome Res 20:291–300

Primmer CR (2009) From conservation genetics to conservation genomics. Ann N Y Acad Sci 1162:357–368

Raveendran M, Harris RA, Milosavljevic A, Johnson Z, Shelledy W, Cameron J, Rogers J (2006) Designing new microsatellite markers for linkage and population genetic analyses in rhesus macaques and other non-human primates. Genomics 88:706–710

Rogers J, Mahaney MC, Cox LA (2009) The development and status of the baboon genetic linkage map. In: Barrett L (ed) The baboon in biomedical research. Developments.in primatology: progress and prospects. Springer, New York

Rudd S (2003) Expressed sequence tags: alternative or complement to whole genome sequences? Trends Plant Sci 8:321–329

Ryder OA (2005) Conservation genomics: applying whole genome studies to species conservation efforts. Gytogenet Genome Res 108:6–15

Santure AW, Stapley J, Ball AD, Birkhead TR, Burke T, Slate J (2010) On the use of large marker panels to estimate inbreeding and relatedness: empirical and simulation studies of a pedigreed zebra finch population typed at 771 SNPs. Mol Ecol 19:1439–1451

Siepel A (2009) Phylogenomics of primates and their ancestral populations. Genome Res 19:1929–1941

Slate J, Gratten J, Beraldi D, Stapley J, Hale M, Pemberton J (2009) Gene mapping in the wild with SNPs: guidelines and future directions. Genetica 136:97–107

Steiner CC, Putnam AS, Hoeck PEA, Ryder OA (2013) Conservation genomics of threatened animal species. Annu Rev Anim Biosci 1:261–281

Stone A, Verrelli B (2006) Focusing on comparative ape population genetics in the postgenomic age. Curr Opin Genet Dev 16:586–591

Thalmann O, Fischer A, Lankester F, Pääbo S, Vigilant L (2007) The complex evolutionary history of gorillas: Insights from genomic data. Mol Biol Evol 24:146–158

Thomas MA, Klaper R (2004) Genomics for the ecological toolbox. Trends Ecol Evol 19:441–445

Thompson RC, Wang IJ, Johnson JR (2010) Genome-enabled development of DNA markers for ecology, evolution and conservation. Mol Ecol 19:2184–2195

Ting N, Sterner KN (2012) Primate molecular phylogenetics in a genomic era. Mol Phylogenet Evol 66:565–568

Tung J, Charpentier MJ, Garfield DA, Altmann J, Alberts SC (2008) Genetic evidence reveals temporal change in hybridization patterns in a wild baboon population. Mol Ecol 17:1998–2011

Tung J, Alberts S, Wray G (2010) Evolutionary genetics in wild primates: combining genetic approaches with field studies of natural populations. Trends Genet 26:353–362

Tung J, Akinyi MY, Mutura S, Altmann J, Wray GA, Alberts SC (2011) Allele-specific gene expression in a wild nonhuman primate population. Mol Ecol 20:725–739

Vera JC, Wheat CW, Fescemyer HW, Frilander MJ, Crawford DL, Hanski I, Marden JH (2008) Rapid transcriptome characterization for a nonmodel organism using 454 pyrosequencing. Mol Ecol 17: 1636–1647

Vigilant L, Guschanski K (2009) Using genetics to understand the dynamics of wild primate populations. Primates 50:105–120

Wang Z, Gerstein M, Snyder M (2009) RNA-Seq: a revolutionary tool for transcriptomics. Nat Rev Genet 10:57–63

Wheeler DA, Wheeler DA, Srinivasan M, Egholm M, Shen Y, Chen L, McGuire A, He W, Chen YJ, Makhijani V, Roth GT, Gomes X, Tartaro K, Niazi F, Turcotte CL, Irzyk GP, Lupski JR, Chinault C, Song XZ, Liu Y, Yuan Y, Nazareth L, Qin X, Muzny DM, Margulies M, Weinstock GM, Gibbs RA, Rothberg JM (2008) The complete genome of an individual by massively parallel DNA sequencing. Nature 452:872–876

Yildirim S, Yeoman CJ, Sipos M, Torralba M, Wilson BA, Goldberg TL, Stumpf RM, Leigh SR, White BA, Nelson KE (2010) Characterization of the fecal microbiome from non-human wild primates reveals species specific microbial communities. PLoS ONE 5: e13963

Zhang J, Chiodini R, Badr A, Zhang G (2011) The impact of next-generation sequencing on genomics. J Gene Genomics 38:95–109

Zinner D, Arnold ML, Roos C (2011) The strange blood: natural hybridization in primates. Evol Anthropol 20:96–103

8 Comparative Genomics: Tools for Study of Complex Diseases

Laura A. Cox

8.1 Introduction

Comparative genomics is a research approach that uses bioinformatics tools to integrate data from multiple genomes for numerous types of information from each genome and is used to address questions ranging from the role played by gene x environment interactions in human health and disease to evolutionary relationships of species using phylogenetic analysis methods. The identification of genomic similarities and/or genomic differences can be used to build models or "systems" to develop a better understanding of specific biological systems. Completion of each successive mammalian genome sequence, each sequence assembly, each whole genome expression study, and each whole genome polymorphism study has exponentially increased the information available and the bioinformatics tools available that support comparative genomic analyses. The databases from which data are drawn for comparative analyses include low-resolution physical maps, high-resolution physical maps, statistical genomic maps, genomic assembly maps, restriction enzyme maps, recombination maps, gene expression profile data, noncoding RNA annotation, alternative gene structure annotation, conserved coding and noncoding elements, insertion/deletion elements, orthologous chromosomal regions, polymorphisms, and structural variants (Fig. 8.1). Different types of tools have been developed that use assembled genomes as frameworks for mapping disease-related quantitative trait loci and mapping genetic associations with human diseases and model organisms, for predicting expressed genes, noncoding RNAs, protein-DNA binding sites, RNA-DNA binding sites, and for aligning genomic regions from multiple species for phylogenetic analysis such as phylogenetic shadowing. Furthermore, these publicly available databases link to other websites that provide detailed information on each data point (e.g., gene, protein, single nucleotide polymorphism, etc.). Consequently, initial investigation of a genetic element (chromosome, gene, promoter, etc.,) using basic bioinformatics tools will typically reveal extensive information about the system of interest before a single laboratory experiment has begun.

In this chapter, I will discuss comparative genomic tools used for the study of gene x environment interactions underlying complex diseases by comparison of a model organism genome with the human genome. In studies of pedigreed baboons, my colleagues and I are using comparative genomic tools combined with classical genetic tools to identify genes that influence variation in complex disease. I will present examples of the use of these new tools for the identification of concordant quantitative

L.A. Cox (✉)
Department of Genetics, Texas Biomedical Research Institute, 7620 NW Loop 410, San Antonio, TX 78227, USA
e-mail: lcox@txbiomed.org

R. Duggirala et al. (eds.), *Genome Mapping and Genomics in Human and Non-Human Primates*, Genome Mapping and Genomics in Animals 5, DOI 10.1007/978-3-662-46306-2_8,

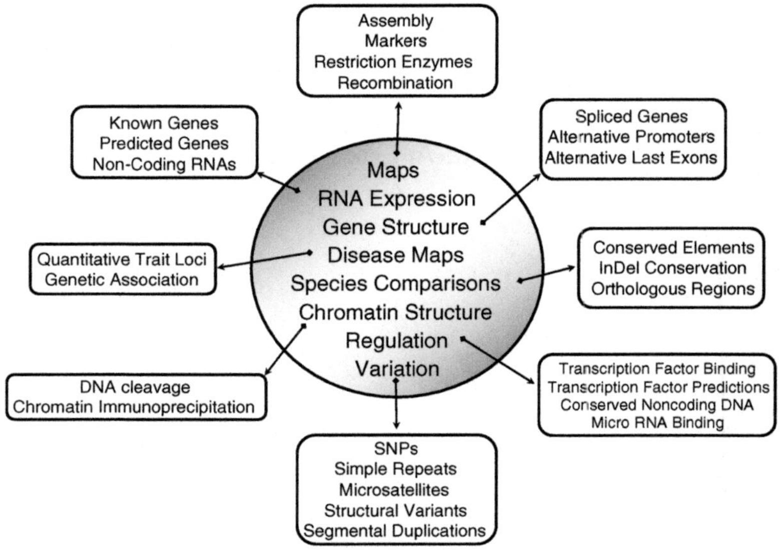

Fig. 8.1 Overview of comparative genomic resources. Each central element is found in public databases and each connected element links out to one or more specialty databases

trait loci, regulatory elements, conserved gene domains, gene expression profiles, and gene networks relevant to cardiovascular disease. In addition, I will present how these tools can be used for identification of specific polymorphic nucleotides that influence variation in a cardiovascular disease-related quantitative trait in a nonhuman primate. I will also show how these results can be used for identification of polymorphisms that are likely to influence human quantitative trait variation and complex disease. The marked reduction in sequencing costs will soon provide additional data on numerous individuals in multiple species that will again significantly expand the information gained using comparative genomics tools. This information will provide greater power to predict the genes and the polymorphisms in these genes that influence quantitative traits, which will decrease discovery time for the identification of these genes, and their functional polymorphisms.

Our laboratory is using the baboon as a model to determine how genetic variation influences atherogenesis. Central to these studies is the identification of genes underlying variation in cholesterol metabolism. The commonly used methods for positionally cloning novel genes are labor- and time-intensive. In addition, these methods are complicated when localizing and identifying genes regulating multigenic traits. In order to identify novel genes encoding QTLs, we have developed an efficient strategy to identify candidate genes. This strategy uses information from the baboon linkage map, the human genome sequence, including annotated and predicted genes, the pedigreed baboon colony at the Southwest National Primate Research Center (SNPRC), the quantitative measures for atherosclerosis-related traits, and the human genome database in concert with gene expression array methods. Using this approach we identified the gene and variants within the gene that influence variation in a size fraction of high-density lipoprotein cholesterol (HDL_1-C) (Cox et al. 2007).

To identify novel cardiovascular related genes, that is, genes not previously known to contribute to atherogenesis or dyslipidemia, we initially used classical genetic methods to identify chromosomal regions containing loci that influence the trait of interest. The foundation resource for these studies is a baboon genetic linkage map that we constructed using 284 random microsatellite markers from the human linkage map (Cox et al. 2007). In addition to the linkage map, scientists in the Department of Genetics at the Texas Biomedical Research Institute have collected quantitative trait data on more than 150 lipid and lipoprotein quantitative traits in the same 951 pedigreed baboons that were used to construct the linkage map. Genome scans were performed for each quantitative trait to identify quantitative trait loci (QTL) influencing each atherosclerosis-related trait (e.g., Cox et al. 2002; Kammerer et al. 2001, 2003; Rainwater et al. 2003; Vinson et al. 2007; Voruganti et al. 2007). After QTL identification, QTL regions of interest were fine mapped to reduce the chromosomal region of interest (e.g., Cox et al. 2007). After identifying and refining the QTL region of interest, we used a modified genomic expression profiling method integrated with bioinformatics analyses to prioritize candidate genes in the QTL region of interest. The evaluation of candidate genes in the QTL region of interest is all-inclusive, with analysis of both annotated and predicted genes. Prioritized candidate genes were then analyzed in detail by identification and genotyping of polymorphisms that may regulate variation in the quantitative trait. Functional polymorphisms were identified by statistical functional analyses and validated by molecular functional analyses (Cox et al. 2007). Furthermore, we used transcriptome profiling data analyzed using bioinformatics tools to identify genetic pathways and networks underlying each phenotype. In this chapter, I will describe the methods used for each of these steps and provide examples from our work studying genes underlying variation in atherosclerosis-related traits.

8.2 QTL Identification and Fine Mapping

As with many model organisms, no physical map for baboon exists and the genome sequence map is currently in draft form with numerous gaps. In the absence of a well-annotated baboon reference genome, we use comparative genomic methods to: (1) decrease the QTL region of interest by fine mapping (Sect. 8.2.3) and (2) determine known and predicted genes in the reduced region of interest for each QTL (Sect. 8.3.1). When we began our QTL gene identification projects, the rhesus genome had not yet been sequenced; therefore, we performed the comparative genomic examples presented here using the human genome map as the reference genome. These methods, however, can be used for any species with a nonsequenced genome (target) against a species with a sequenced genome (reference). To identify gene(s) encoding QTLs, the closer the two species are related evolutionarily the more informative the comparison.

8.2.1 QTL Identification

As mentioned above, data were collected from the SNPRC pedigreed baboons for quantitative traits related to atherosclerosis. Genome scans were performed for these quantitative traits using the baboon linkage map and a number of QTLs were identified. In this chapter, we use results from our work on two QTLs identified from the genome scans as examples, one QTL influencing HDL_1-C (a size fraction of HDL-C) (Cheng et al. 1988) on baboon chromosome 18 (Mahaney et al. 1998) and one QTL influencing the hypertension trait sodium lithium counter transport (SLC) activity on baboon chromosome 5 (Kammerer et al. 2001). For the HDL_1-C QTL on baboon chromosome 18, the two-point linkage analysis showed a peak LOD score of 7.32 at marker D18S72. We defined the region of interest for the QTL, which is the chromosomal region most likely to include the gene(s) influencing HDL_1-C variation, as the two LOD support interval, i.e., the region included in the area under the QTL curve from the peak out to the two LOD drop in the curve (Cox et al. 2002). For the SLC QTL on baboon chromosome 5, we obtained evidence for an SLC QTL with a peak LOD score of 9.3 located near marker D4S1645 (human chromosome 4 is the orthologue of baboon chromosome 5). This QTL accounts for approximately two thirds of the total additive genetic variation in SLC activity in baboons.

8.2.2 Sequence Alignment for Fine Mapping Chromosomal Regions

DNA sequence alignment of the target and reference genomes is necessary for the identification of repetitive elements that can be used to fine map the region of interest and reduce the number of

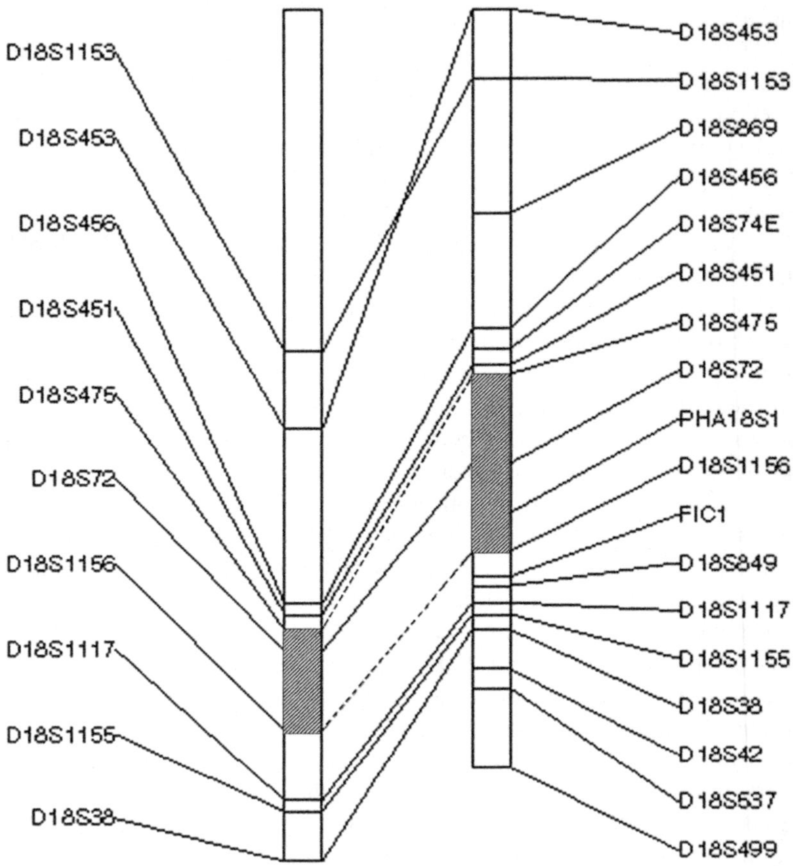

Fig. 8.2 Alignment of baboon chromosome 18 (*left*) HDL QTL region of interest (*hashed lines*) with human chromosome 18 (*right*) using genotyped microsatellite markers for baboon (modified from http://baboon.txbiomedgenetics.org/Bab_Results/GraphicMaps/chrom18.php)

candidate genes that must be analyzed. By genotyping microsatellite markers and repetitive elements common to both the target and reference genomes it is possible to align the target and reference linkage maps (e.g., Fig 8.2). Because the reference genome has both a linkage map and whole genome sequence, the alignment of the reference genome syntenic block with the target genome provides a physical map for the target species' QTL region of interest. The underlying assumption is that for conserved syntenic regions, repetitive elements, encoded genes, noncoding RNAs, regulatory elements, etc., are conserved between target and reference genomes. Multiple species' genome sequences can be aligned (Vista Genome Browser; http://pipeline.lbl.gov/cgi-bin/gateway2) (Frazer et al. 2004; Shah et al. 2004) for the region of interest to test the extent of element conservation between reference and target syntenic regions. Based on our work using human, rhesus, and baboon microsatellite markers in the baboon genome and the human genome, we know that repetitive elements common to two species may be polymorphic in one species but not the other. Therefore, sequence alignment will provide a list of repetitive elements that are good candidates for microsatellite markers based on repeat length; however, variation in a repetitive element length must be tested empirically (e.g., Cox 2002; Cox et al. 2007).

8.2.3 Fine Mapping a QTL Region of Interest

To fine map a QTL region of interest, we must identify microsatellite markers that are amplifiable and polymorphic in our target (baboon) species. When we first began fine mapping baboon QTLs, we screened human microsatellite markers in baboon that were included in the human genome linkage maps (Cox et al. 2006a; Rogers et al. 2000). Although this strategy was successful identifying new markers for the baboon linkage map, it was extremely inefficient with less than a 25 % success rate for marker identification. In addition, some of these markers did not yield clean PCR products making genotyping difficult and some markers were not very polymorphic. Therefore, we devised a comparative genomics approach to identify and test putative baboon microsatellite markers. First we defined the genomic sequence included in our region of interest by identifying the physical map location of microsatellite markers flanking the region of interest using the reference genome. We entered the microsatellite identifiers into the University of California Santa Cruz (UCSC) Genome Bioinformatics browser (http://genome.ucsc.edu/); (Kent et al. 2002) query box and retrieved the genomic locations delimiting the QTL region of interest. We then scanned human genomic DNA sequence in the region of interest at 1 million basepair (Mbp) blocks in 5 Mbp intervals for repetitive elements of 12 or more di, tri, or tetra repeats using the UCSC Genome Bioinformatics, Table Browser function (http://genome.ucsc.edu/cgi-bin/hgTables) (Karolchik et al. 2004) to list all microsatellite and simple repetitive elements in the region of interest including 300 bp of flanking sequence 5′ and 3′ of each repeat. After excluding 1 Mbp regions that already contained microsatellite markers in the baboon linkage map, putative markers were prioritized by proximity to annotated genes, providing another link to the reference genome map. Because we know there are sequence differences between human and baboon but we don't know what nucleotides differ, we designed two pairs of PCR primers for amplification of each repetitive element (Oligo v6.89, Molecular Biology Insights, Inc.). Parameters for primer design included PCR product length from 150 to 300 bp, PCR primer length of 24 nucleotides (nt), GC content greater than 55 %, and a T_m of 55–68 °C Also, the stability (ΔG) of primer-template duplexes must be less than 10 °C difference between the T_m of each primer and no primer/dimer pair formation is allowed. We used the BLAT alignment tool (http://genome.ucsc.edu/cgi-bin/hgBlat; Kent 2002) with the human genome to ensure primer specificity (Cox et al. 2009).

With the recent availability of baboon genomic sequence in the NCBI Trace Archive (http://www.ncbi.nlm.nih.gov/Traces/trace.cgi?), we have added an additional step to this procedure. After repetitive element identification, we use the BLAST tool (http://www.ncbi.nlm.nih.gov/BLAST/Blast.cgi?PAGE=Nucleotides&PROGRAM=blastn&BLAST_SPEC=TraceArchive&BLAST_PROGRAMS=megaBlast&PAGE_TYPE=BlastSearch; Altschul et al. 1990) with the predicted PCR product sequence from the human genome against the baboon Trace Archive to determine the repetitive element repeat number in baboon and to identify baboon flanking sequence for primer design. Since many species now have genomic sequence data available in the NCBI Trace Archive, but the genomes have not yet been assembled, this tool is also useful as a second reference sequence when the first reference sequence is not as evolutionarily close to the target as the second "Trace Archive" reference.

To optimize the chances of identifying polymorphic baboon microsatellite markers for the pedigreed baboon colony, we used a panel of 12 baboons that represent a large portion of the genetic diversity in the pedigreed colony. Genomic DNA was amplified by PCR for each target region using a fluorescently labeled forward primer and unlabeled reverse primer. PCR products were size-fractionated in an automated sequencer Applied Biosystems, Inc. (ABI) and genotyped using Genotyper software. Heritability was tested for each polymorphic marker by genotyping 2–3 baboon nuclear families (i.e., sire, dam, 2–3 offspring). If multiple polymorphic, heritable markers were identified for a chromosomal interval, the most polymorphic marker was selected for genotyping. Selected microsatellite markers were genotyped for the phenotyped, pedigreed baboons. The new markers were then included in the linkage map and the genome scan for the quantitative trait repeated.

8.3 Characterizing the Refined Region of Interest

8.3.1 Sequence Alignment

After refining the QTL region of interest, the chromosomes must be aligned and the genomic sequence in the region of interest must be retrieved for the reference sequence. Although microsatellite markers from the linkage map were used to align the chromosomes before fine mapping the region of interest, the same analysis must be performed including the new markers. It is possible that small chromosomal rearrangements not apparent with the original alignment are apparent with the new markers. As described in Sect. 8.2.2, microsatellite markers common to both target and reference genomes were used to align the target and reference genomic regions. If repetitive elements were used for genotyping the target genome, these contain physical map "addresses" in the reference genome that can also be used to tie the target genome to the framework of the reference genome. To do so, the microsatellite sequence including flanking region sequence was entered into the reference genome BLAT search tool in the UCSC Genome Bioinformatics browser (http://genome.ucsc.edu/cgi-bin/hgBlat; Kent 2002). Output from this search will include the sequence alignment and physical map location in the reference genome. Once the reference genome region of interest has been defined, it is possible to identify all known and predicted genes in the interval as well as noncoding RNAs and regulatory elements. In addition, it is possible to determine if other scientists have mapped QTLs (Rapp 2000) or genetic disease associations (Becker et al. 2004) with that chromosomal interval.

8.3.2 Sequence Alignment for Rearranged Chromosomal Regions

It is not unusual to find chromosomal rearrangements such as inversions when comparing target and reference linkage maps. In addition, it is not unusual to find QTL regions of interest that include areas of rearrangement between target and reference chromosomes. A central element in the identification of genes encoding QTLs is to include all possible candidate genes in the initial screening process. Therefore, chromosomal rearrangements and the portions of chromosome that overlap with QTL regions of interest must be defined as clearly as possible for inclusion and exclusion of candidate genes. That is, all possible genes that can be excluded should be excluded in order to reduce the number of genes that must be interrogated; however, because one does not want to exclude a candidate gene that lays in the region of interest the region must be defined as clearly as possible.

With this in mind, we often see rearranged regions with inadequate linkage map data (i.e., number of microsatellite markers in the linkage map) to precisely identify the chromosomal regions of interest in the reference chromosome. An example of this is shown in Fig. 8.3a, where the QTL region of interest includes baboon chromosome 5 from D4S414 to D4S2365. This QTL region of interest spans a chromosomal inversion when comparing baboon against human. Using the mapped microsatellite markers, the region of the orthologous human chromosome (chromosome 4) for the region from D4S414 to D4S1645 is clear; all the markers included between these markers are the same for baboon and human with the order from p to q reversed. Whereas, the DNA that should be included in the region from D4S1645 to D4S2365 is not as clear. D4S2365 borders the baboon region of interest and D4S413 is outside the region of interest for baboon and this is consistent in human. So, the conserved chromosomal region should be p-ter to D4S2365; however, D4S414 is outside the region of interest in baboon but flanks the region likely to include QTL region of interest DNA. In this case, the investigator has 2 choices: (1) fine map additional markers between D4S2365 and D4S414 or (2) include all genes and expressed genes as candidates for the D4S2365–D4S414 region knowing that some genes are likely to be outside the region of interest. Due to the required time and resources for candidate gene interrogation

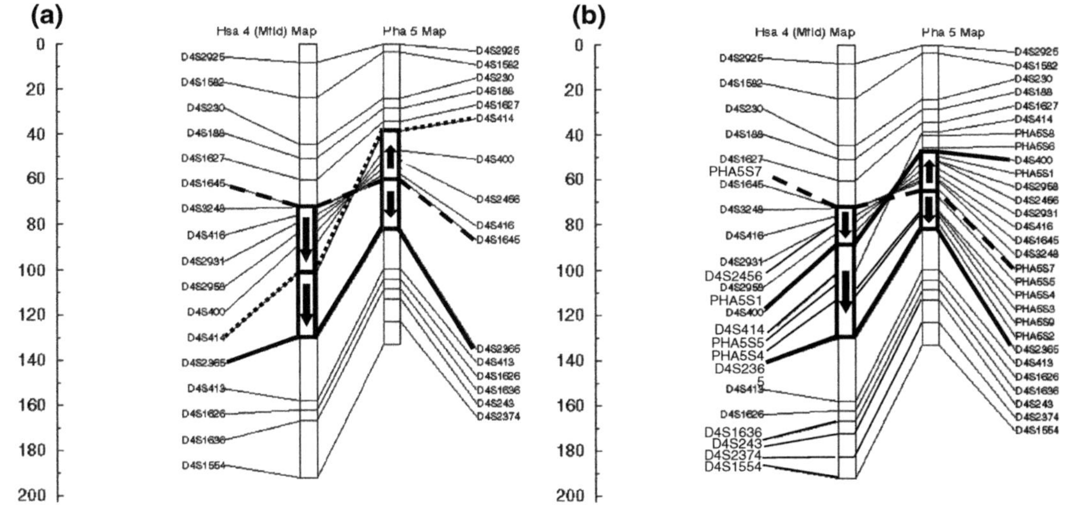

Fig. 8.3 Identification of target DNA in region of interest with rearranged reference chromosome. Human chromosome 4 (Hsa 4) is on the *left*, baboon chromosome 5 (Pha), the ortholog to Hsa 4 is on the *right*. *Lines* between the chromosomes show marker order. *Boxes* show chromosome segment conservation. *Arrows* indicate relative directions. **a** shows chromosome comparison by mapped microsatellite markers with region of interest in *bold boxes* and chromosome direction indicated by *arrows*. **b** shows reduced region of interest with inclusion of additional microsatellite markers in the linkage map

and prioritization, the first option is usually worth the time invested.

In the example of a QTL mapping to baboon chromosome 5, we chose to fine map the region of interest and more clearly define the region of rearrangement as described in Sect. 8.2.3. Figure 8.3b shows the QTL region of interest after including additional markers in the linkage map. Markers that were identified specifically from baboon genomic sequence as described in Sect. 8.2.3 are indicated by the "PHA" identifier; all of these sequences can be assigned locations relative to the mapped human microsatellite markers using the human genome sequence for the human orthologous chromosome using the UCSC Genome Browser BLAT alignment tool. Addition of the new markers more narrowly defines the chromosomal breakpoint between the human and baboon orthologous chromosomes, shown by the dashed line PHA5S7 and narrows the QTL region of interest more than 11 Mbp by moving the p-ter border of the QTL from D4S414 to D4S400 and based on the March 2006 human genome assembly (http://genome.ucsc.edu/staff.html) reduces the number of candidate annotated genes by 78 and predicted genes by 38.

8.3.2.1 Identifying Known and Predicted Genes in Region of Interest

The UCSC Genome Browser (Kent et al. 2002) tool is used to identify the physical map location of genes and predicted genes within a QTL region of interest. To do so, the microsatellite identifiers for the two markers delimiting the borders of the QTL region of interest are entered into the UCSC Genome Browser query box and the genomic locations retrieved. These two locations define physical map location of the QTL region of interest on the reference genome. When both of these locations are entered into the query box together, the UCSC Genome Browser window will display the entire genomic region of interest.

To list all genes in the defined region, select the "Tables" link in the top bar in Fig. 8.4 to load the Table Browser tool (Fig. 8.5). This new window defaults to selecting a positional table

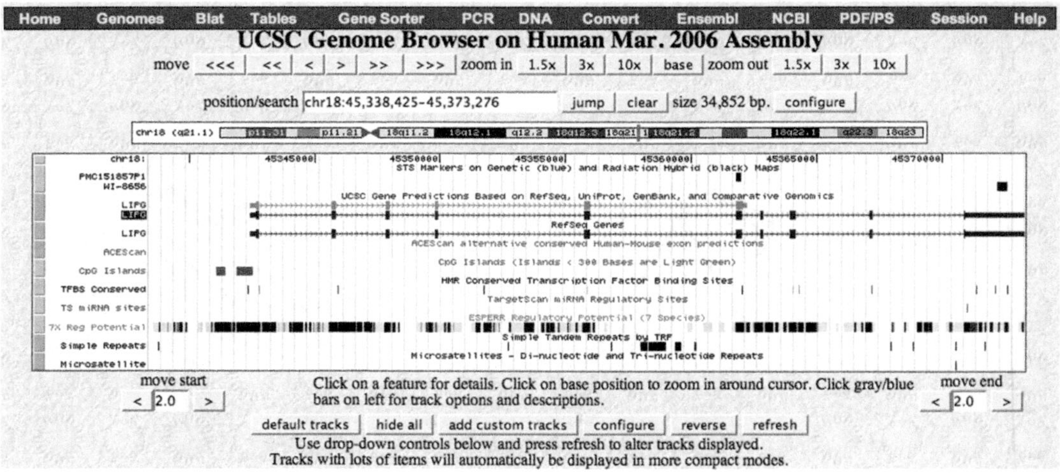

Fig. 8.4 UCSC Genome browser showing annotated gene and predicted gene tracks for the HDL_1-C QTL region of interest. From *top* to *bottom*, genome browser function links, the genome assembly version, navigational tools, chromosome position numerically, and graphically on the chromosome diagram. The browser window shows the base number, the track name, and the contents of the track for annotated genes (UCSC based on RefSeq, UniProt, and GenBank) and predicted gene (N scan)

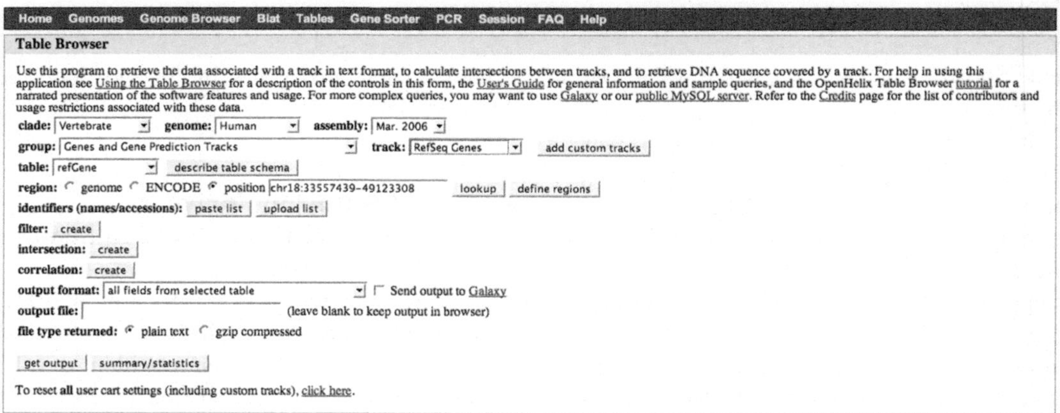

Fig. 8.5 UCSC Table browser function. This tool can be used to provide tabular data on any tracks included in the UCSC genome browser and can be used for defined chromosomal regions or for the entire genome. In this figure, the table browser has been selected to generate lists of genes and predicted genes for the QTL chromosomal region of interest

for the region of interest viewed in the Genome Browser (Fig. 8.4). The Table Browser tool can be used to provide tabular data on any tracks included in the UCSC genome browser and can be used for defined chromosomal regions as positional data, can be used for non-positional data, or can be used to retrieve data for the entire genome. In this example, the Table Browser is used to download in tabular form all annotated genes for the region of interest. In addition, using the gene predicted tracks combined with Expressed Sequence Tag (EST) and spliced EST tracks, all high confidence predicted genes can be listed. The gene array tracks, indicating expressed genes detected by whole genome expression profiling, may also be included to gain additional information regarding expression levels and tissue type expression of genes in the region of interest. The output data for the RefSeq gene track and the Gene Scan Prediction track includes GenBank ID number, exon start site and exon stop site for each exon in the annotated or predicted gene and gene sequence. The UCSC Table Browser webpage includes links to descriptions of all table functions and links to tutorials for use of the Table Browser features including merging, filtering, and intersecting data from multiple tracks for output of data in tabular form (http://genome.ucsc.edu/cgi-bin/hgTables).

8.4 Prioritizing Candidate Genes Using Expression Profiling

Central to our strategy for the identification of genes encoding QTLs is based on the premise that the gene regulating the QTL must be encoded within the region of the QTL signal and the gene must be expressed in the relevant tissue. Consequently, we developed a Chromosomal Region Expression Array (CREA) strategy that allows us to evaluate all DNA sequences in the region of interest that may encode the gene influencing the QTL We do not limit our approach to the analysis of known genes; the CREA is inclusive for all genes, ESTs and predicted genes within the QTL region of interest. To interrogate the arrays, we use heterologous RNA from the tissue most likely to be relevant to the quantitative trait. In addition, we collect tissues from sibling baboons discordant for the quantitative trait in order to minimize genetic variation due to genetic background and to maximize genetic differences for the gene(s) encoding the QTL. Using this approach, we can significantly reduce the number of candidate genes in the QTL region of interest (Cox et al. 2002).

Since developing the custom arrays we have moved to using current sequencing methods to analyze gene expression in QTL intervals. The advantage of RNA Seq methods (Illumina GA IIx platform) is that we are able to identify and quantify all transcripts expressed in a sample. In addition, quantification of each transcript is not dependent on knowing the precise gene or non-coding RNA sequence beforehand, as is the case when designing primers for array-based methods. For candidate gene prioritization, only those genes in the QTL interval are used from the RNA Seq data. However, with the use of network analysis, it is possible to strengthen candidate gene priority by inclusion or exclusion of the candidate genes in networks constructed from the entire transcriptome.

8.4.1 Discordant Sibs CREA Analysis

To identify baboons for the positional cloning of the gene encoding the HDL_1-C QTL study, we performed phenotypic and genotypic analysis of the pedigreed baboon population and identified baboon sib-pairs discordant for HDL_1-C serum concentrations. The sib-pairs differed by at least one standard deviation for HDL_1-C values. In addition, members of each selected sib-pair did not share IBD (identical-by-descent) alleles, or for some markers shared only one IBD allele, in the chromosomal region of interest. For details of sib-pair HDL_1-C phenotype data see Cox et al. (2002). Because the QTL peak LOD score is greater for the high cholesterol high fat diet than the chow diet, we predicted that the gene influencing HDL_1-C would be differentially expressed between the two diets. Therefore, we collected liver biopsies from baboons before and after a 7-week, high cholesterol, high fat (HCHF) diet challenge. RNA was extracted from the liver biopsies and used to measure expression of all known and predicted genes in the QTL region of interest. In addition, gene expression was compared between the chow and the high cholesterol, high fat diets (Cox et al. 2002).

8.4.2 Designing a CREA

The CREA approach can be achieved by either constructing a custom array or by analyzing RNA Seq data for the chromosomal region of interest. The CREA method is less expensive but may miss a gene or noncoding RNA due to probe mismatches or may miss novel genes or noncoding RNAs not predicted in the reference genome. Both of these methods are consistent with a conservative approach to positionally cloning the gene encoding a QTL where one evaluates all genes, noncoding RNAs and predicted genes in a QTL region of interest. The analysis relies on an annotated reference genome such as the human genome.

For the CREA method, after defining the physical map locations of the markers delimiting the QTL region of interest, we use the UCSC Table Browser to identify all genes and predicted genes in the QTL region of interest and use the Table Function to download all exon sequences for each of these genes and predicted genes. The exon sequence of each gene is then used to design 65-mer oligonucleotides specific for each gene. The oligonucleotides are then arrayed and used for chromosome region specific expression profiling. To design gene specific primers for a list of genes for which the cDNA sequence has not yet been determined, we use a comparative genomics approach. First we align the human cDNA sequence with the rat and or mouse cDNA sequence (USCS Genome Browser, NCBI-Search Nucleotide, GeneLynx and Rat Genome Database). We assume that nucleotides conserved between human and rodent will be conserved between human and baboon. We then import both sequences into Sequencer (Gene Codes, Inc.), align the cDNAs, and design oligonucleotides for the gene based on conserved coding regions using Oligo Primer Analysis Software (Molecular Biology Insights, Inc). Oligonucleotide design constraints include: (1) oligonucleotide $\geq$ 65 nucleotides long; (2) less than 8 mismatches between species; (3) 45–55 % GC content; (4) no tetranucleotide repeats; (5) no significant hairpin loops (less than 7 bonds in a hairpin); and (6) optimal probe with highest T_m and the highest negative ΔG value for GC clamp. After oligonucleotide design, sequence specificity is confirmed by performing an NCBI-BLAST search and uniqueness of the oligonucleotide is confirmed allowing less than 90 % maximum identity with nontarget sequences. After gene orientation is confirmed, oligonucleotides are synthesized and nylon-based arrays printed with oligonucleotides spotted in triplicate (Northcott et al. 2012).

Some investigators use a modified CREA approach where they perform whole genome expression profiling using a commercial gene array and analyze genes in the QTL region of interest. For species that do not have a commercial array available, this presents a problem for QTL candidate gene prioritization. If the investigator uses an array from a different species, such as a human gene array for baboon gene expression, there are likely to be sequence differences between human and baboon for some genes resulting in some array probes that do not cross react with baboon sequence. In these instances, the lack of signal for a specific gene may be because the gene is not expressed but it may also be because the gene probe does not cross react. Another limitation of some commercial arrays is that they include only annotated genes and not predicted genes. We know from previous experience that some "predicted genes" in one assembly of the human genome can become annotated genes in later assemblies. Therefore, if a commercial array is used, an investigator should supplement those data with a custom array that includes all predicted genes in the QTL interval and includes all genes that did not give a signal using the commercial array.

The Next Gen sequencing platforms provide a means to sequence and determine abundance of all transcripts (cDNAs) expressed in a tissue. Using the RNA Seq method, genomic DNA in the QTL region of interest is used to map all expressed transcripts. Genome annotations are used to annotate known and predicted genes and noncoding RNAs. In addition, because transcript abundance is measured using this method it is possible to identify differentially expressed transcripts in response to a challenge or that differ among groups with variation in the phenotype of interest. Using this method, we have identified novel baboon transcripts that were not known or predicted in human (Cox et al. unpublished data).

8.4.3 Prioritizing Genes in Region of Interest

Regardless of the method used in Sect. 8.4.2, to quantify expression of genes in the QTL region of interest, genes are prioritized based on expression profiles, proximity to the peak LOD score, biological relevance to the trait of interest, and association with cardiovascular disease QTLs from other studies. A positional table is generated using the UCSC table browser that includes annotated genes, expressed genes, and QTLs. The QTL track includes human, mouse, and rat QTL data annotated as a component of the rat genome database project (Rapp 2000). The table is then filtered to retain all CREA expressed genes. Mean values for both groups (e.g., low and high HDL_1-C from chow and HCHF diets) are added to the table for each CREA expressed gene. In addition, GeneCards (http://www.genecards.org/; Rebhan and Prilusky 1997) and Online Mendelian Inheritance in Man (OMIM) (http://www.ncbi.nlm.nih.gov/omim; OMIM 2008) databases are accessed for known function(s) of each annotated gene.

Genes are then ranked first by consistency of each gene expression profiles with the QTL signal. In this example, the QTL signal was observed for the HCHF diet but not the chow diet. Because these baboons were selected based on their contribution to the QTL signal and because these baboons are discordant for HDL_1-C, we predicted that the gene influencing HDL_1-C would be differentially expressed between low and high responders on the HCHF diet, but not the chow diet. Therefore, in this case the highest priority genes were differentially expressed between low and high responders on the HCHF diet and showed either no differences between low and high responders on the chow diet or no differences in expression for the low responders comparing chow and HCHF diets. Genes included in this group were further prioritized based on biological relevance to the genes' known function with the quantitative trait and proximity to the peak LOD score. Predicted genes cannot be prioritized based on known function and are therefore prioritized by expression profiles and location relevant to related QTLs mapped to the QTL region of interest. Using this approach for the chromosome 18 QTL influencing HDL_1-C, we began with 354 genes and predicted genes in the region of interest and reduced the number of candidates to 3 genes (Cox et al. 2005).

8.5 Functional Polymorphism Identification

After prioritization of candidate genes, functional polymorphism(s) in the gene, that is, the polymorphisms that influence variation in the quantitative trait must be identified. To date, there are no good prediction tools for the identification of functional polymorphisms. In our baboon HDL_1-C QTL candidate gene study of endothelial lipase (*LIPG*), we evaluated the orthologous human gene for conserved noncoding sequences (Vista Genome Browser, http://pipeline.lbl.gov/cgi-bin/gateway2). These analyses showed conservation from mouse to human for two regions in the 5′ flanking region of *LIPG* One region was immediately upstream to the 5′ untranslated region and the other region was located −2,446 bp from the transcription start site. No polymorphisms were identified in the conserved region proximal to the 5′ untranslated region and none of the polymorphisms located in the upstream conserved region influenced *LIPG* expression of HDL_1-C variation. Furthermore, our study of *LIPG* revealed two functional single nucleotide polymorphisms (SNPs) and one deletion-insertion polymorphism (DIP). SiteSeer (Boardman et al. 2003) was used to determine predicted transcription factor binding to the *LIPG* promoter binding for the functional DIP and SNPs in the 5′ flanking region. One SNP was located in a predicted transcription factor binding site and the insertion for the DIP included a predicted transcription binding site; however, the second SNP was not located in any predicted or annotated regulatory element (Cox et al. 2007). Therefore, traditional methods must still be used to identify polymorphisms in each candidate gene, all polymorphisms must be genotyped in the population from which the QTL was detected, and quantitative trait nucleotide analyses must be performed on each polymorphism to identify functional polymorphisms. In cases where candidate genes are predicted to be differentially expressed and the variation in gene expression influences variation in the quantitative trait of interest, polymorphisms in potential regulatory regions as well as the coding regions must be identified (Curran et al. 2005). In addition, resequencing is most likely to reveal informative polymorphisms if animals representative of variation in the quantitative trait of interest are resequenced for polymorphism identification.

To limit the number of polymorphisms that must be genotyped, we used the panel of discordant baboons for resequencing. Because these baboons differ by at least one standard deviation for the quantitative trait of interest and each selected sib-pair in the panel does not share identical-by-descent (IBD) alleles in the chromosomal region of interest, then polymorphisms that may influence variation in the gene encoding the QTL will be present in this group of animals.

8.5.1 Sequencing Candidate Genes

In our baboon HDL_1-C QTL example, the baboon candidate genes in the QTL region of interest had not yet been sequenced. Therefore, we used gene and genome sequence information from the human reference genome to isolate and sequence the baboon gene. To sequence each candidate gene for which no gene sequence exists, we first isolated Bacterial artificial Chromosome (BAC) clones containing the gene from a baboon BAC library (BACPAC Resources; BACPACorders@chori.org). We used the human DNA sequence to design primers for amplification of a fragment from each candidate gene using Oligo software (Molecular Biology Insights, Inc). The gene fragment was amplified using these primers and the fragment was then used as a probe to isolate a baboon BAC clone containing the gene. The baboon gene of interest was then sequenced from the BAC clone using sequencing primers based on the reference sequence gene data. To download the human gene sequence, we used the Genome Browser "get DNA" feature (http://genome.ucsc.edu/cgi-bin/hgc?hgsid=107910572&o=33557438&g=getDna&i=mixed&c=chr18&l=33557438&r=49123308&db=hg18&hgsid=107910572; Kent et al. 2002). To do so, we first entered the gene name or GenBank ID number into the "position/search" box in the Browser window (Fig. 8.4). The Browser displays a link for that gene. After activating the link, the Browser displays the gene from the transcription start site to the end of the 3′UTR and shows intron–exon structure of the gene. The genomic location was indicated in the "position/search" box. The lower number in the "position/search" box could be changed to 4,000 bp less than the number displayed and the "jump" button used to display the gene including 4,000 bp of promoter (Fig. 8.4).

Clicking on the "DNA" link along the top of the page loaded a new page asking for display preferences; the gene position was auto filled into the "position" box (Fig. 8.6). If there was a preference for DNA display such as lower case for noncoding and upper case for coding sequences, this can be selected using the "extended case/color options" feature. Selecting "get DNA" will prompt the browser to display the DNA sequence for the gene or region of interest (Fig. 8.7). The DNA sequence was copied from the display and pasted into the Oligo software program for design of sequencing primers. In addition, the DNA sequence could be pasted into Sequencher software (GeneCodes, Inc., Ann Arbor, MI) for alignment and distribution analysis of sequencing primers. Each exon for the reference gene was acquired in the same manner and included in the Sequencher alignment as landmarks of coding regions in the candidate gene. For the top priority candidate genes, we sequenced the introns, exons, untranslated regions, and ~4,000 bp of the promoter. We chose to sequence beyond the traditional 1,000 bp of promoter sequence because strong enhancer elements have frequently been found in gene promoters between −4,000 and −1,000 bp from the transcription start site.

If RNA Seq methods are used to sequence and quantify gene expression in a relevant panel of animals, then resequencing will only need to be done for the introns and promoter regions. In

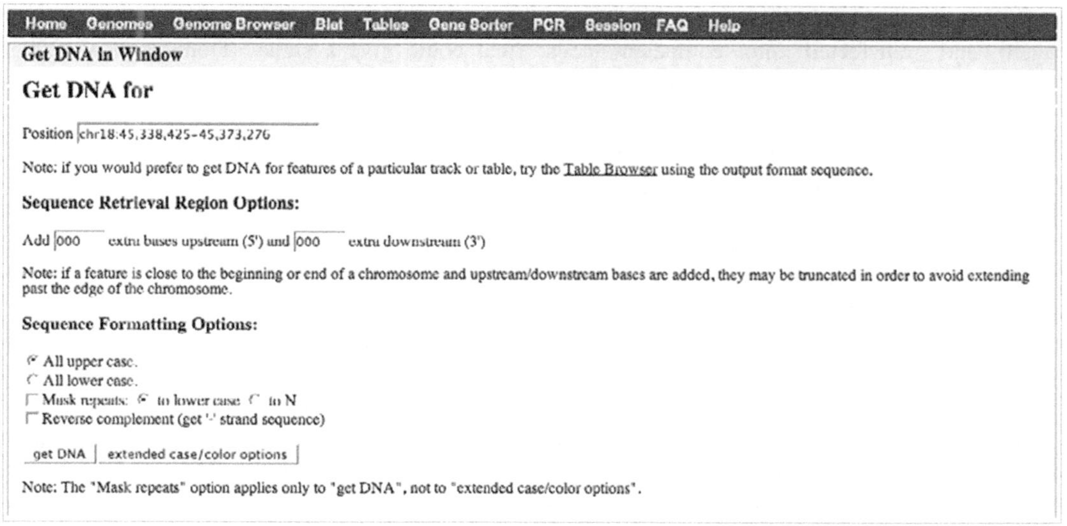

Fig. 8.6 The "Get DNA" feature in the UCSC genome browser can be used to download sequences from any track. This tool allows the user to define the chromosomal interval from which the sequence will be retrieved and annotation of the retrieved sequence

```
>hg18_dna range=chr18:45338425-45373276 5'pad=0 3'pad=0 strand=+ repeatMasking=none
TATTTATTTGGGTAGAGATGAGGTCTCCTTATGTTGCCCTGGCTGGTCTC
AAATGCCTAGCCTCAAGCCATCCTTCCACTTTGGCCTCCCAAAGTGCCAG
GATTACAGGCGTGAGCCACCACACCCAGCCACTTAATTTTAATTTCATGT
GTTTCTTTTTACCTTTATAATAGGACCACTAGGAAACATAAAATTATACA
TGTGCCGCTATGGACTGAATTGGGACCCCCTCAAATTCCTATGTTGAAGC
CCTAACTCCCTATGATGTATTTGGAGATGGGGCCTGTGGGAGATCATTTG
GTTTAGATGAGGTTATGAGGTGAGGCACCATGATGGGATTAGAGTTGTTA
TTGGAAGAGACATCAGCGTGTTCTTTCTCTCTCTCTCTCTCTCTCTCTCT
CTCTCTCCTCTCTCTCTCTCTGCCATGTGAGGACACAGTGAGAAGGCAGC
CATCTATAAGCCAGAAAGAGGGCCCTCACCAGAAACCGACTATGCTGGTA
CCCTGATCTTGGACTTCTAGTCTCCAGAACTATGAGAAAATAAATTTCTG
TTTAAAGCCACTAAGTCTATGGTATACTGTTCTGGCAGCCCAAACTGACC
AAGACATGCGGTTTGTATTATATATTTCTGTTGGACAGCATTGGTCCAGA
TATCTGGGAACTCCTACATACCAGCCAGCCTTCTGGCACTTGTAACTTTC
TGTATTGTGTCTGTGAGAGCACAGGCATGGTCTCCAGGCCAGTGTTTCCT
CCCTAGCAGCTGCTCAATAAGCTTCTGGCAATTAAGTCATTTTCTTGGTT
GTAAGAATATAAACAGTCTCTGGATAATGTATGTAAAAGAGGACCTCATT
AAGAGAATATTGGGTAACAACTGGAATCTGCAGGGCAGAAGAACCAGGCT
```

Fig. 8.7 Output from genomic DNA sequence retrieval from the "get DNA" feature in the UCSC Genome Browser. The *top line* describes the chosen parameters. Sequence can be copied and pasted into any sequence analysis program or word processing file

addition, the exon sequences from the RNA Seq can be used to design primers for intron sequencing.

8.5.2 Resequencing Candidate Genes

Sequence data from Sect. 8.5.1 is used to design sequencing primers for resequencing the gene. Sequence polymorphisms are identified by sequencing the candidate gene in a panel of animals discordant for the quantitative trait of interest. To ensure that polymorphisms are identified, resequencing is performed on single alleles; all genomic DNA fragments that will be sequenced from the panel of discordant animals are subcloned and 8 clones for each animal in the panel are sequenced. Briefly, genomic DNA (50 ng) is amplified using species specific gene primers, PCR buffer, and Taq DNA Polymerase.

PCR products are subcloned into pTOPO (Invitrogen) and transfected into competent cells (Invitrogen). Plasmid DNA is purified (Qiagen) and sequenced (Applied Biosystems, Inc. (ABI)). Sequencing products are purified using Exonuclease I (USB) and Shrimp Alkaline Phosphatase (USB) and size fractionated. Sequence data are imported into Sequencher, Gene Codes, Inc. (GCI) for alignment and identification of polymorphisms. Nucleotides and insertion/deletions are considered polymorphic if they are validated by their presence in either (1) two or more baboons in the sib-pair panel and data are consistent using primers from both directions, or (2) one baboon and the data were consistent for sequence data from multiple clones, i.e., 4 clones with one variant and 4 clones for a second variant.

8.6 Cross Species Use of Whole Genome Expression Arrays

We use whole genome expression profiling (from gene arrays or RNA Seq) to provide additional data for QTL candidate genes. Ontological pathway (http://www.geneontology.org/) (Ashburner et al. 2000) and KEGG Pathway (www.genome.jp/kegg/) (Kanehisa et al. 2004) analysis of whole genome expression data provide detailed data on individual genes in the context of that gene's role in described biological/biochemical pathways and may reveal insights into molecular mechanisms by which a gene influences a QTL. Cross species use of whole genome expression arrays provides a list of genes that provide quality signal for the RNA samples of interest. These experiments provide extensive information about expression of many genes regardless of the species specificity of the array. One caveat of the cross species use of gene arrays is that the lack of signal for a gene could be due to either low gene expression or lack of cross species hybridization for that gene. From the perspective of simply studying expressed genes, this is not a problem. However, if the investigator wishes to perform pathway analyses for the dataset then the issue of "no-signal" genes becomes an issue. Z-score calculations defining significant gene categories and pathways are based on the total number of genes on the array that could give a signal (Doniger et al. 2003). Thus, to accurately calculate z-scores, the array of baboon genes for which expression was detected on the human gene chip must be defined. Therefore, for our baboon gene expression studies, we evaluated both human Affymetrix (Affymetrix U133A 2.0) and human Illumina (Illumina Human WG-6 v2) gene arrays for whole genome expression profiling of baboon RNA samples.

To evaluate each human whole genome expression array, we used baboon RNA from 12 baboons for 13 different tissues including liver, kidney, lymphocytes, fat, placenta, 0.5 gestation (G) and 0.9G fetal liver, 0.5G and 0.9G fetal frontal cortex, 0.5G and 0.9G fetal kidney, and 0.5G and 0.9G fetal adrenal. Whole genome expression profiling was performed for each sample and the samples were quality filtered based on 0.5 for Affymetrix (for details see Cox et al. 2006b) and 0.95 for Illumina gene arrays. Because Affymetrix and Illumina use different types of probes and different measures to assess signal quality, the quality filter setting differs for these two platforms. The lists of quality genes from each tissue were merged for each array to generate a list of genes providing a quality signal for baboon RNA on that array platform. Using this method, 16,186 of the 22,227 genes on the Affymetrix GeneChip and 17,231 of 25,538 annotated genes and 4,916 of 20,658 predicted genes on the Illumina BeadChip were detected with quality signal. The merged list for each array platform is the virtual "custom" baboon array for that platform. After determining the genes included in each custom array, the list is uploaded into Genesifter (VizX Labs, Seattle, WA) as the "custom" baboon array and used to perform pathway analyses on the whole genome expression profiling datasets.

8.7 Conclusion

Comparative genomic methods provide a wealth of data for many genetic questions before the first laboratory experiment begins. A basic knowledge of the central data repositories, available databases, and basic analytical tools will help determine what is known about a system, what can be inferred using data from multiple species, and generate specific hypotheses and questions to address the hypotheses. The UCSC Genome Browser has become a central repository for annotated genome data. In addition, the Genome Browser links out to more detailed information for all included data types. The database is continually updated and new tools are continually developed and added to the Genome Browser. With that said, information in the UCSC Genome Browser depends on data that are provided by other investigators. For example, baboon genome sequence is routinely downloaded to the NCBI trace archive (http://www.ncbi.nlm.nih.gov/Traces/trace.cgi?) from the Genome Sequencing Center (Baylor Genome Sequencing Center for baboon) as the data are generated. However, these data will not be downloaded to the UCSC Genome Browser until the baboon genome has been assembled. Consequently, species-specific genome sequence data may be found prior to release to the UCSC Genome Browser. This is the case for bottlenose dolphin, kangaroo rat, and echinoid genome sequences to name just a few. A list of ongoing genome sequencing projects can be found at the NCBI Entrez Genome Project website (http://www.ncbi.nlm.nih.gov/genomes/leuks.cgi). In addition, early genome sequence data may be found at different websites (different databases) than cDNA databases with the two datasets generated by laboratories independent from each other. Scientists involved in sequencing a species' genome may not be part of the community of scientists who routinely use that species as a model system. For this reason, often scientists who use a particular model organism may not be aware that the genome sequencing for that organism is underway. The search for sequences specific to your species of interest, even if they are unassembled and unannotated, will add confidence to your comparative genomic analyses and are worth the time spent searching to see if they exist.

References

Altschul SF, Gish W, Miller W, Myers EW, Lipman DJ (1990) Basic local alignment search tool. J Mol Biol 215:403–410

Ashburner M, Ball CA, Blake JA, Botstein D, Butler H, Cherry JM, Davis AP, Dolinski K, Dwight SS, Eppig JT, Harris MA, Hill DP, Issel-Tarver L, Kasarskis A, Lewis S, Matese JC, Richardson JE, Ringwald M, Rubin GM, Sherlock G (2000) Gene ontology: tool for the unification of biology. The Gene Ontology Consortium. Nat Genet 25:25–29

Becker KG, Barnes KC, Bright TJ, Wang SA (2004) The genetic association database. Nat Genet 36:431–432

Boardman PE, Oliver SG, Hubbard SJ (2003) Siteseer: visualisation and analysis of transcription factor binding sites in nucleotide sequences. Nucleic Acids Res 31:3572–3575

Cheng ML, Kammerer CM, Lowe WF, Dyke B, VandeBerg JL (1988) Method for quantitating cholesterol in subfractions of serum lipoproteins separated by gradient gel electrophoresis. Biochem Genet 26:657–681

Cox LA (2002) The FIC1 gene: structure and polymorphisms in baboon. J Med Prim 31:1–12

Cox LA, Birnbaum S, VandeBerg JL (2002) Identification of candidate genes regulating HDL cholesterol using a chromosomal region expression array. Genome Res 12:1693–1702

Cox L, Birnbaum S, Mahaney M, VandeBerg J (2005) Characterization of candidate genes regulating HDL-C using expression profiling. In: Proceedings of the XIII International Congress on Genes, Gene Families, and Isozymes. Medimond, Bologna Italy, pp. 177–180

Cox LA, Mahaney MC, VandeBerg JL, Rogers J (2006a) A second-generation genetic linkage map of the baboon (*Papio hamadryas*) genome. Genomics 88:274–281

Cox LA, Nijland MJ, Gilbert JS, Schlabritz-Loutsevitch NE, Hubbard GB, McDonald TJ, Shade RE, Nathanielsz PW (2006b) Effect of 30 per cent maternal nutrient restriction from 0.16 to 0.5 gestation on fetal baboon kidney gene expression. J Physiol 572:67–85

Cox LA, Birnbaum S, Mahaney MC, Rainwater DL, Williams JT, VandeBerg JL (2007) Identification of promoter variants in baboon endothelial lipase that regulate HDL-cholesterol levels. Circulation 116:1185–1195

Cox LA, Glenn J, Ascher S, Birnbaum S, VandeBerg JL (2009) Integration of genetic and genomic methods for identification of genes and gene variants encoding QTLs in the nonhuman primate. Methods 49:63–69

Curran JE, Jowett JB, Elliott KS, Gao Y, Gluschenko K, Wang J, Abel Azim DM, Cai G, Mahaney MC, Comuzzie AG, Dyer TD, Walder KR, Zimmet P, MacCluer JW, Collier GR, Kissebah AH, Blangero J (2005) Genetic variation in selenoprotein S influences inflammatory response. Nat Genet 37:1234–1241

Doniger SW, Salomonis N, Dahlquist KD, Vranizan K, Lawlor SC, Conklin BR (2003) MAPPFinder: using gene ontology and genMAPP to create a global gene-expression profile from microarray data. Genome Biol 4(1):R7

Frazer KA, Pachter L, Poliakov A, Rubin EM, Dubchak I (2004) VISTA: computational tools for comparative genomics. Nucleic Acids Res 32:W273–W279

Kammerer CM, Cox L, Mahaney MC, Rogers J, Shade R (2001) Sodium lithium counter transport activity is linked to chromosome 5 in baboons. Hypertension 37:398–402

Kammerer CM, Rainwater DL, Schneider JL, Cox LA, Mahaney MC, Rogers J, VandeBerg JL (2003) Two loci affect angiotensin I-converting enzyme activity in baboons. Hypertension 41:854–859

Kanehisa M, Goto S, Kawashima S, Okuno Y, Hattori M (2004) The KEGG resource for deciphering the genome. Nucleic Acids Res 32 (Database Issue): D277–D280. DOI: 10.1093/nar/gkh063

Karolchik D, Hinrichs AS, Furey TS, Roskin K, Sugnet CW, Haussler D, Kent WJ (2004) The UCSC table browser data retrieval tool. Nucl Acids Res 32:D493–D496

Kent WJ (2002) BLAT-The BLAST-like alignment tool. Genome Res 12:656–664

Kent WJ, Sugnet CW, Furey TS, Roskin KM, Pringle TH, Zahler AM, Haussler D (2002) The human genome browser at UCSC. Genome Res 12:994–1006

Mahaney MC, Rainwater DL, Rogers J, Cox LA, Blangero J, Almasy L, VandeBerg JL, Hixson JE (1998) A genome search in pedigreed baboons detects a locus mapping to human chromosome 18q that influences variation in serum levels of HDL and its subfractions. Circulation 98:I5 (Abstract)

Northcott CA, Glenn JP, Shade RE, Kammerer CM, Hinojosa-Laborde C, Fink GD, Haywood JR, Cox LA (2012) A custom rat and baboon hypertension gene array to compare experimental models. Exp Biol Med 237:99–110

OMIM (2008) Online Mendelian Inheritance in Man. Johns Hopkins University, Baltimore, MD Retrieved MIM Number: {606945}, http://www.ncbi.nlm.nih.gov/omim/. Accessed June 12, 2007

Rainwater DL, Kammerer CM, Mahaney MC, Rogers J, Cox LA, Schneider JL, VandeBerg JL (2003) Localization of genes that control LDL size fractions in baboons. Atherosclerosis 168:15–22

Rapp JP (2000) Genetic analysis of inherited hypertension in the rat. Physiol Rev 80:135–172

Rebhan M, Prilusky J (1997) Rapid access to biomedical knowledge with GeneCards and HotMolecBase: implications for the electrophoretic analysis of large sets of gene products. Electrophoresis 18:2774–2780

Rogers J, Mahaney MC, Witte SM, Nair S, Newman D, Wedel S, Rodriguez LA, Rice KS, Slifer SH, Perelygin A, Slifer M, Palladino-Negro P, Newman T, Chambers K, Joslyn G, Parry P, Morin PA (2000) A genetic linkage map of the baboon (*Papio hamadryas*) genome based on human microsatellite polymorphisms. Genomics 67:237–247

Shah N, Couronne O, Pennacchio LA, Brudno M, Batzoglou S, Bethel EW, Rubin EM, Hamann B, Dubchak I (2004) Phylo-VISTA: interactive visualization of multiple DNA sequence alignments. Bioinformatics 20:636–643

Vinson A, Mahaney MC, Cox LA, Rogers J, VandeBerg JL, Rainwater DL (2007) A pleiotropic QTL on 2p influences serum Lp-PLA(2) activity and LDL cholesterol concentration in a baboon model for the genetics of atherosclerosis risk factors. Atherosclerosis 196:667–673

Voruganti VS, Tejero ME, Proffitt JM, Cole SA, Freeland-Graves JH, Comuzzie AG (2007) Genome-wide scan of plasma cholecystokinin in baboons shows linkage to human chromosome 17. Obesity 15:2043–2050

9 Genetic Structure and Its Implications for Genetic Epidemiology: Aleutian Island Populations

Michael H. Crawford

9.1 Introduction

The concept of genetic population structure has been defined a number of different ways (Crawford 1998):

1. The relationship between elements (genes, genotypes, phenotypes, and individuals) that comprise populations (Workman and Jorde 1980).
2. Corrections of ideal population (Hardy-Weinberg-Castle equilibrium) with properties such as panmixis, infinite size, and equal genetic contributions of genotypes (Cavalli-Sforza and Bodmer 1971).
3. All demographic and genetic attributes or parameters of a population (Schull and MacCluer 1968).
4. Population subdivision by geography, religion, language, and ethnicity (Jorde 1980). Roughly, population genetic structure can be characterized as the distribution of genes within populations and/or among subpopulations.

Although the majority of the observed variation in humans is found within populations, rather than between geographical entities, the genes are not distributed randomly and there is "structure" in the arrangement and distribution of the genes within populations. Based on blood groups and protein variation, Lewontin (1972) through an analysis of variance determined that most of the variation ($\sim$85 %) occurs within populations, while a small amount of variation is between larger entities—such as continental populations. These results, based on standard genetic markers, were further substantiated by Barbujani et al. (1997) using AMOVA on short tandem repeats (STRs). However, the use of single nucleotide polymorphisms (SNPs), particularly those located on the Y chromosome, for AMOVA analysis indicates that there is higher variance when comparing continental populations. These results reflect the smaller effective population sizes (Ne) using nonrecombining Y-chromosome markers (NRY; with three of the four sex chromosomes in a breeding pair being X) and the genetic information contained in SNP distributions.

Documented social and geographic factors that impact on the distribution of genes include: (1) geographical distances being correlated to genetic distances (both standard markers and mtDNA sequences) in Siberia and Aleutian Islands (Crawford 2007); (2) language and geography in Siberian and native American populations (Crawford et al. 1997); (3) religion, geography, and economics in small fishing villages (outports) of Newfoundland and agricultural communities of Tiszahat, Hungary (Koertvelyessy et al. 1993; Martin et al. 2000); (4) the establishment of new political boundaries in the Tiszahat region of Hungary after World War II, which separated the traditional Hungarian villages from their relatives

M.H. Crawford (✉)
Laboratory of Biological Anthropology, Department of Anthropology, University of Kansas, 1415 Jayhawk Blvd., Lawrence, KS 66045, USA
e-mail: crawford@ku.edu

R. Duggirala et al. (eds.), *Genome Mapping and Genomics in Human and Non-Human Primates*, Genome Mapping and Genomics in Animals 5, DOI 10.1007/978-3-662-46306-2_9,

across the former Soviet border, thus modifying the migration patterns and altering the predicted genetic structure (Crawford et al. 1999); (5) the genetic repercussions of the subdivision of an Altai population of Central Asia into patrilineal clans as revealed by discriminant function analyses of frequencies of variable number tandem repeat (VNTR) markers (Crawford et al. 2002); and (6) gene flow as a result of unique historical events in northern Mexico and the Caribbean (Crawford 1976, 1978, 1984). Beuten et al. (2011) demonstrated using 64 ancestry informative markers (AIMs) that those significant differences in substructure were present in two cohorts of Mexican American subjects from the San Antonio area of Texas. Substantial differences in admixture proportions were observed between 706 participants of the San Antonio Family Diabetes Study (SAFDS) and 586 male samples from San Antonio Center for Biomarkers of Risk of Prostate Cancer (SABOR), although the participants from these studies are from the same geographic region.

The genetic structure of populations, sculpted by the actions of the forces of evolution, determines the presence of specific genotypes, which interact with the environment to produce complex phenotypes such as chronic diseases like atherosclerosis, diabetes, hypertension, gall bladder disease, alcoholism, and osteoporosis. This interaction of environment, genotype, mutation, genetic structure, and complex phenotype is a basic model utilized in genetic epidemiology and diagrammatically represented in Fig. 9.1.

The complex interactions among geography, unique historical events, environmental and demographic factors, and the genes are difficult to document in large, continuously distributed human populations because of the underlying genetic structure. However, island populations offer geographically discrete aggregates, isolated by physical barriers that limit population migration and result in genetic differentiation. Complex populations are often stratified by ethnic or racial groups that differ genetically from surrounding subpopulations. Turakulov and Esteal (2003), based on SNP distributions across the entire genome and different populations, posed the question: How many SNPs are required to detect population structure? They concluded that for a U.S. population more than 65 random SNPs are necessary to detect distinct geographically separated populations. A total of 100 SNPs raises the probability of correct assignment to over 90 %. Using autosomal STRs, Jorde et al. (1997) demonstrated that at least 50 STRs are necessary to statistically detect genetic structure based on the major U.S. ethnic groups.

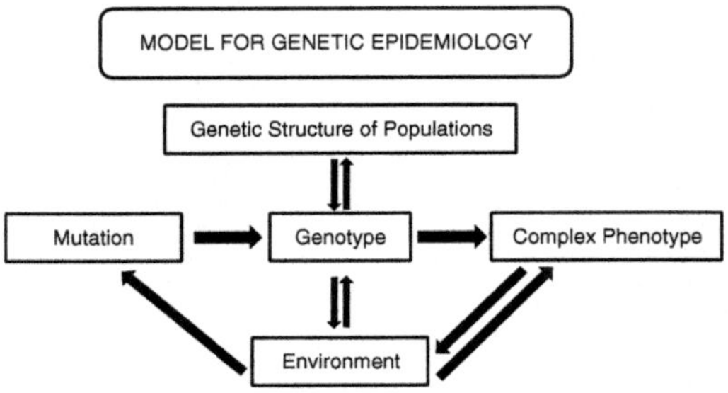

Fig. 9.1 Schematic representation of the relationship among the genetic structure of human populations, environment, and the complex phenotype

Nonrecombining Y-chromosome markers have been used to detect population structure in the United States. Hammer et al. (2006) using a set of 61 Y-chromosome SNPs on 2,517 individuals from 38 U.S. regional and ethnic samples found considerable interethnic admixture in these regional samples. They noted that "continental origin rather than current location in the U.S. determines major patterns of Y-chromosome variation for most ethnic groups" (Hammer et al. 2006). They also observed that despite intermarriage among the ethnic groups, which would "erode" population structure, only a small proportion of all participants were derived from two or more "racial" groups. Thus, despite admixture, population structure persists in U.S. regional groups.

9.1.1 Population Structure and Its Relevance to Mapping Genes for Complex Traits

In recent years, the genome-wide association study (GWAS) method has become a popular design with increased successes in localizing susceptibility genes/variants for common, complex diseases such as diabetes (Manolio et al. 2009; Prokopenko et al. 2008). This approach is based on the common variant/common disease hypothesis, and generally uses data from large samples of unrelated cases and controls. Although most of the GWAS's have involved populations of European ancestry, there have been efforts to examine additional populations, including recently admixed populations such as African Americans and Hispanics and isolated or founder populations such as the Old Order Amish, given attention to the potential issues such as allele frequency and linkage disequilibrium (LD) differences among diverse populations (Cooper et al. 2008; Pasaniuc et al. 2011; Rosenberg et al. 2010; Shen et al. 2010). Because hidden population structure can potentially lead to spurious findings in genetic association studies, it is necessary to adjust for population structure to avoid bias in association findings. The advent of high-throughput genotyping technologies has made it possible to adjust for the effects of population stratification using information from thousands of unselected SNPs across the genome or selected sets of AIMs (Tian et al. 2008; Kosoy et al. 2009). In consideration of the genome as a mosaic of chromosomal segments originating from different ancestral populations, a number of approaches including genomic control (GC) method or use of the principal components (PCs) of the observed SNP variation as covariates in an association analysis have been used to account for population substructure by paying attention to the issues of "global ancestry" and "local ancestry" (Baran et al. 2012; Hao et al. 2010; Pääbo 2003; Qin et al. 2010).

The purpose of this chapter is to provide the genetic substructure of the Aleutian Island populations which are molded by historical founder effects and one-way gene flow from the European males to the Aleutian females. This information should aid in any future genetic investigations of complex diseases in the Aleutian Island populations, given the current burden of diseases such as hypertension and diabetes in the Alaskan populations (please see Chap. 11 by MacCluer et al. in this volume regarding genetics of complex diseases in Eskimos) including the Aleuts (Torrey et al. 1979; Schraer et al. 1988; Naylor et al. 2003; Amparo et al. 2011).

9.1.2 Population Structure of the Aleutian Islands

9.1.2.1 Geography and Subsistence Background

There are more than 200 Aleutian Islands stretching westward from Alaska almost 2,000 km toward Siberia (Fig. 9.2, a map of the Aleutian Archipelago). The estimated size of the pre-Contact Aleutian population is between 15,000 and 20,000 inhabitants. The Aleut form of subsistence depended almost entirely on marine resources, including sea mammals, fish, and invertebrates. They fished, hunted in the sea, and collected sea urchins and mussels on reefs off the coasts of the islands. The Aleuts built swift seagoing kayaks (baidarkas) for hunting seals and other sea mammals. They lived in subterranean complexes to protect themselves from the harsh island environments, high winds, cold, and high humidity.

9.1.2.2 Archeological Background

The earliest evidence for human habitation in the Aleutian Islands is in the eastern Fox Islands at Anagula and Hog Island, located off the larger island Unalaska, dating approximately 9,000–8,000 years before present. Archeological sites in the Central Islands occur significantly later, beginning with those on the Andreanof Islands at approximately 5,000 years ago. The western islands were not settled until approximately 3,500 years before present (Rat Islands) with evidence for the occupation of Attu 2,210 years ago and Shemya 3,255 years ago. Apparently, the earliest Aleuts crossed the Bering land bridge more than 9,000 years ago and colonized (through kin-structured expansion) the islands from the Alaska Peninsula in a westward

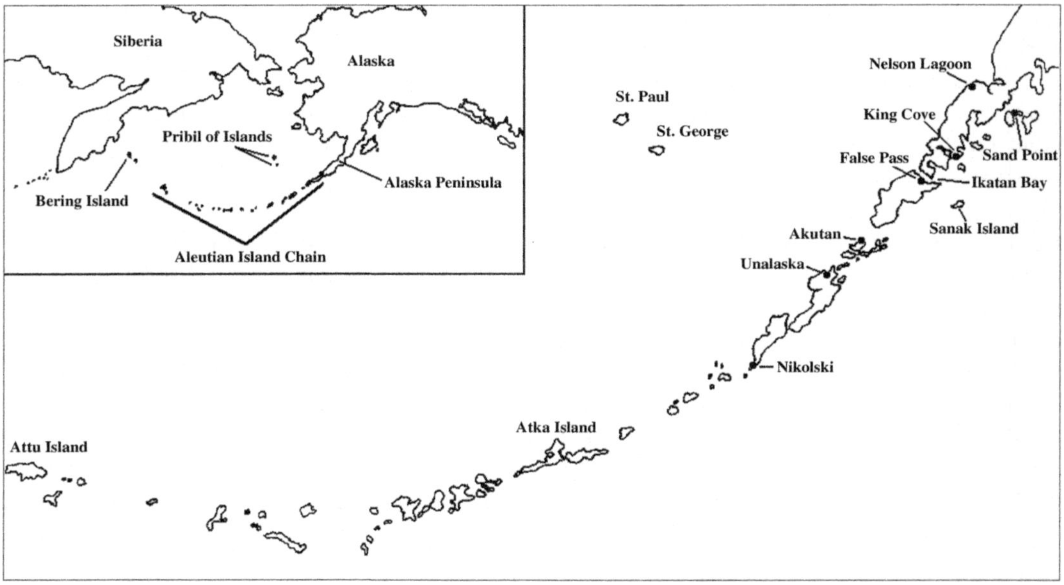

Fig. 9.2 Map of the Aleutian Archipelago

direction, reaching the Far Islands only 3,500 years ago (West et al. 2007). The early Aleut populations apparently failed to reach the most westerly islands of the Archipelago, the Commander Islands (Bering and Medni), since there is no archeological evidence of Aleut settlements on these island off the coast of Kamchatka. The initial expansion of the Aleuts into the Archipelago was followed by regional dialectic differentiation, unique archeological assemblages, and distinctive cultural innovations in the Eastern, Central, and Western Islands. Alternative hypotheses have been generated by Russian archeologists claiming that some island-hopping Paleo-Aleuts originated from Kamchatka and island-hopped to the Western Aleutian Islands. However, neither the archeological nor molecular data support these suppositions (Rubicz et al. 2007).

9.1.2.3 Russian Contact

Russian contact with Aleutian Native populations dates back to the eighteenth century, with voyages of exploration by Vitus Bering and Aleksei Cherikov. Disease epidemics (smallpox, tuberculosis, measles, and influenza) and warfare reduced the Aleutian Native population from an estimated 15,000–20,000 persons to 2,000 persons. In addition, Aleut males and some families from the surrounding islands were forcibly relocated by the Russians to previously uninhabited islands containing breeding grounds for seal populations. These relocations to the Commander Islands (consisting of Medni and Bering Islands) in 1825–1828 and to the Pribilof Islands (St. George and St. Paul) in 1825–1830 established settlements involved in harvesting seal furs and provisioning expeditions into the Americas (Rubicz et al. 2010).

9.1.2.4 Disruption During World War II

The original population structure of the Aleutian Islands was further disrupted by: (1) the purchase of Alaska by the United States from Russia and the political separation of the Aleuts of the Commander Islands from their kin distributed along the remainder of the Archipelago; (2) the occupation of Attu during World War II by the Japanese army, which rounded up the inhabitants and transported them to a camp in Japan (few who survived the detention in Japan returned to the Aleutian Islands); and (3) ongoing war in the Aleutian Archipelago, which also forced the relocation of Aleuts from the western and central islands to camps in mainland Alaska. Most of the Aleuts returned to their home islands with the cessation of wartime activities in the islands.

9.2 Methodology

9.2.1 Sampling

A total of 11 island populations (Akutan, Atka, Bering—Commander Islands, False Pass, King Cove, Nelson Lagoon, Nikolski on Umnak, Sand Point, St. George, St. Paul—Pribilof Islands, and Unalaska) were sampled from 1999 to 2007 (Fig. 9.2). In addition, Aleut volunteers residing in Anchorage were sampled and their DNA was assigned on the basis of their villages of birth. Samples were collected from indigenous populations of Kamchatka (Koryaks, Evens, and Itel'men) for comparative purposes to test a Russian hypothesis concerning origins of Aleuts from Kamchatka.

9.2.2 DNA Collection and Analysis

Buccal swabs, sputum samples, and blood specimens were collected from participants and DNA was extracted using standard phenol chloroform extraction methods (Chomczynski and Sacchi 1987) and Chelex-based extraction method in the field (Walsh et al. 1991). Type restriction fragment length polymorphism (RFLP) cut site analyses in the coding region were conducted for mitochondrial DNA haplotype assignment. The A haplogroup was defined by the presence of *+HaeIII 663*, haplogroup B by the presence of *9 bp deletion*; C was characterized by the absence of *HincII 13259* and the presence of *AluI 13262,* and D was defined by the absence of *AluI 5176* (Rubicz 2001). MtDNA samples were sequenced using Sanger dideoxy cycle protocols on a Beckman CEQ 8,000 autoanalyzer for the hypervariable region HVS-I.

Y-chromosome haplogroups and haplotypes were constructed using single nucleotide polymorphisms (SNPs) and short tandem repeats (STRs). A total of 10 of the following SNPs were amplified and identified: Q (P36); Q3 (M3); C (RPS4Y); I (M170); I1a (M253); J (12f2); N (M231); R1a (SRY1038); R1b (M269); and E3 (P2). The following STRs were genotyped: DYS 19; DYS 3891; DYS 38911; DYS 390; DYS 391; DYS 392; DYS 393; DYS 385a,b; DYS 438; and DYS 439 (for a description of the methodology, see Zlojutro 2008).

9.2.3 Analytical Methods

Genetic discontinuity was detected using spatial analysis of molecular variance (SAMOVA) for defining population groupings that are geographically homogeneous and maximally differentiated from each other (Dupanloup et al. 2002). In addition, Delaunay triangulation methods of Monmonier (1973) and the BARRIER computer program ver. 2.2 (Manni and Guérard 2004; Manni et al. 2004) were utilized to construct a geographic network of sampling locations. Voronoi tessalation was used to derive Delaunay triangulation. Based on this triangulation connectivity network, Monmonier's algorithm was used to identify genetic boundaries, i.e., those geographic zones that have the greatest differences between populations. SAMOVA analysis was performed on D_A distances based on HVS-I with 2–7 population groups selected a priori. The highest U_{CT} value (greatest genetic variance between the K number of groups) was used to determine K number of groups.

9.3 Results

Based on RFLP analyses and hypervariable segment (HVS-I) sequences of Aleut samples from the western and central islands, Aleut mtDNA haplogroups belong to two of the five New World founding haplogroups, 71.5 % D and 28.5 % A (Rubicz 2001, 2007; Rubicz et al. 2003; Zlojutro et al. 2006). The eastern Aleutian Islands and the Alaska Peninsula Aleut populations contain a higher incidence of A haplogroups; plus, there is greater admixture with females of European ancestry (Zlojutro 2008).

The results from the Y-chromosome analyses differ significantly from the observed mtDNA patterns. The majority of the Y-chromosome haplogroups identified in the Aleut samples represent European lineages (85 %), whereas mtDNA haplogroups A and D are two of the four major Native American matrilineages believed to have been associated with the original peopling of the New World. In Fig. 9.3, the Y-chromosome haplogroup frequencies for the sampled Aleut communities are presented. Overall, haplogroups R1a (25.5 %) and R1b (21.9 %) have the highest frequencies. In Europe, haplogroup R1a is predominantly found in Eastern European and Russian populations, while haplogroup R1b is most common in Western Europe and the British Isles. Haplogroup I1a has the third highest frequency (13.9 %) in the Aleuts and is common in Scandinavian populations (Karlsson et al. 2006). Of the remaining haplogroups identified in the Aleuts, only Q* and Q3 are believed to be Native American lineages. Thus, 85 % of the total male sample has non-Aleut Y-chromosomes, of European origin (primarily Russian, Scandinavian, and England).

Figure 9.4 illustrates the east–west distribution of mtDNA haplogroups in the Aleutian Archipelago and the surrounding circumpolar region. On the western edge of this continuum, haplogroup D reaches fixation on Bering Island, a population aggregated by the Russians in the 1825–1828 period, while the higher frequencies of the A haplogroup are seen in the eastern region (Rubicz et al. 2010). The higher incidence of A haplogroup (observed in both Alaskan Yupik and the Athapaskans of the Alaskan mainland) apparently reflects gene flow from groups across the Alaskan boundary into the eastern Aleut populations. Archeological evidence supports this interpretation because of the

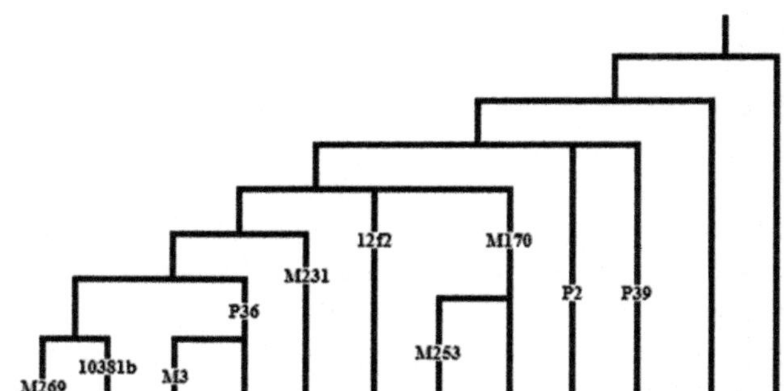

	R1b	R1a	Q3	Q*	N	J	I1a	I*	E3	C3b
Nelson Lagoon	6	1				1	3	3		
Sand Point	5	5	1			1	6	2	1	
King Cove	5	9	2	1			3			
False Pass		1	3			1	3		1	
Akutan	4			4	2			1		
Unalaska	4	5	2	1			1	3		
St. George	2	7	1							
St. Paul	3	3	2		5	1	2			
Bering Island	1	4	1	3	3	3	1	1	3	
TOTAL	30	35	12	9	10	7	19	10	5	0
Percentage	21.9	25.5	8.8	6.6	7.3	5.1	13.9	7.3	3.6	0.0

Fig. 9.3 Y-chromosome haplotypes based on SNPs for the Aleutian Islands (Zlojutro et al. 2008)

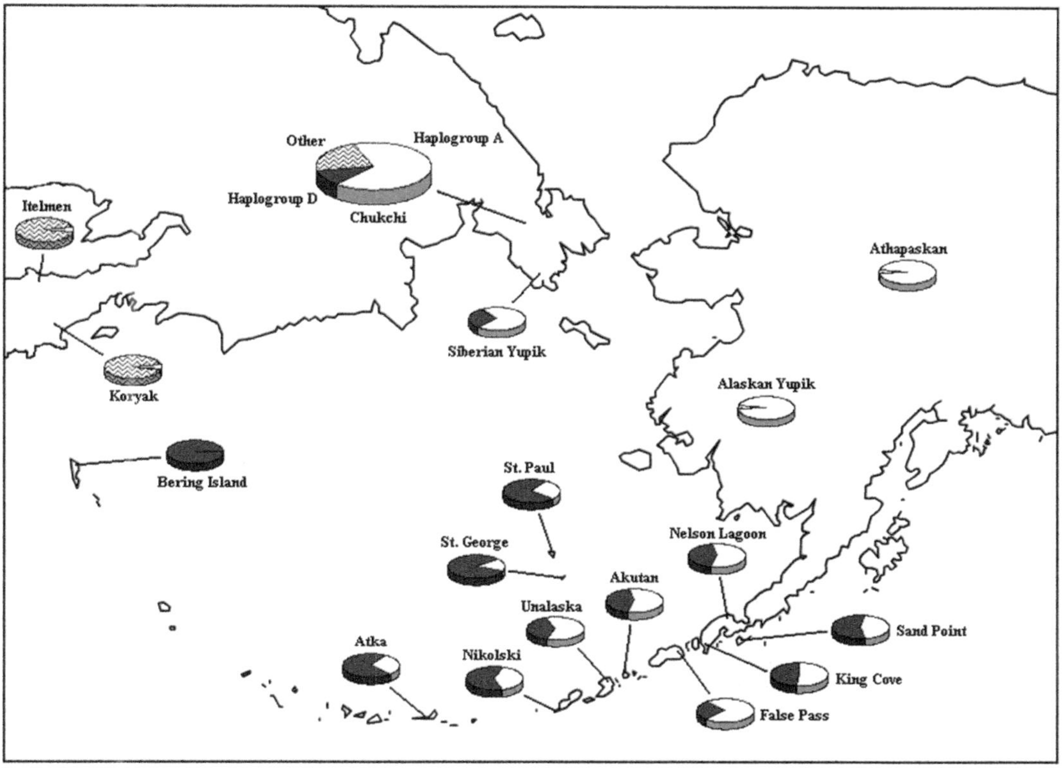

Fig. 9.4 MtDNA haplogroup east–west gradient in the Aleutian Archipelago represented by *pie charts*

similarities of trade objects, suggesting cultural exchanges (Dumond 2001).

An exceptionally high correlation was observed between genetic and geographic distance matrices in the Aleutian Islands. Mantel tests of matrices based on intermatch mtDNA D sequences and geographic distances (as the crow flies) among 11 Aleutian Islands yielded a correlation $r = 0.72$; $p < 0.000$. The relationship between genetics and geography is highly significant, indicating the preservation of the genetic structure along the maternal lineages. By contrast, there is no significant correlation between Y-chromosome haplotype based distances and geographical distances in kilometers between the islands. Mantel test for intermatch mtDNA D versus Y STR–based distances (Nei's D_a) = 0.26 ($p = 0.164$) n.s. It is surprising that the maternal pre-Contact population structure has been maintained, despite depopulation, relocation of populations, admixture with Russians in the eastern islands and with English and Scandinavians in the western islands (Crawford et al. 2010).

Figure 9.5 displays a plot of autocorrelation indices on the ordinate and geographic distances in kilometers on the abscissa. The plot indicates that a highly significant $p < 0.000$ relationship exists between geographic distance and mtDNA sequences. However, the negative correlations in the autocorrelation results do not support an isolation-by-distance model. As the geographic distances become of greater magnitude, the autocorrelations become negative. The most parsimonious explanation is that the demic expansion of the Aleuts was kin-selected, followed by founder effect, genetic drift, and subpopulation differentiation (Crawford et al. 2010).

The highest U_{CT} (variance among groups relative to total variance in the sample) was obtained when K was set to four groups (0.326; $p = 0.000$). The SAMOVA analysis reveals that the Kamchatkan populations (Itelmen and

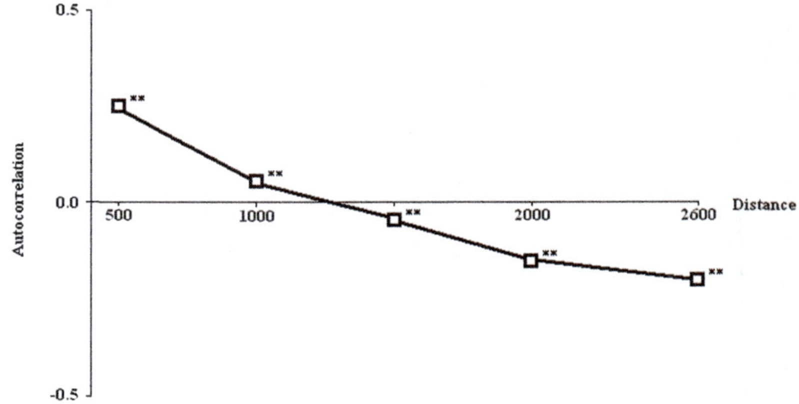

Fig. 9.5 Spatial patterns of mtDNA sequence diversity in the Aleutian Islands, following the methods of Bertorelle and Barbujani (1995)

Koryaks) and Alaskan Yupik show genetic discontinuity from the other circumpolar populations (Fig. 9.6). The aggregate populations, Bering, St. George and St. Paul, cluster with Atka—an island that forcibly contributed substantially to the founding of these aggregates. Surprisingly, in the SAMOVA analysis, the Athapaskans, Chukchi, and Siberian Yupik are included in the fourth group and do not warrant the separation observed in the Delaunay triangulation method.

The genetic barriers revealed by the Delaunay triangulation method (Fig. 9.7; Manni and Guérard 2004) are: (1) Alaskan Yupik from Chukchi, Siberian Yupik, and Aleut populations; (2) between Kamchatkan populations and the remainder of the circumpolar groups; (3) Aleuts from Alaskan Eskimo populations; and (4) eastern Aleuts from central and western island populations. These barriers clearly reflect the evolutionary history of the circumpolar populations on both sides of the Bering Strait. The Athapaskans cluster with the populations of the Chukchi Peninsula and a barrier exists between them and the Alaskan Yupik and the Aleut populations. The barrier between the easternmost Aleuts and the western groups reflects the temporal pause (until climatic change) by ancestral Aleutian populations in their westward expansion (West et al. 2007). The genetic discontinuity

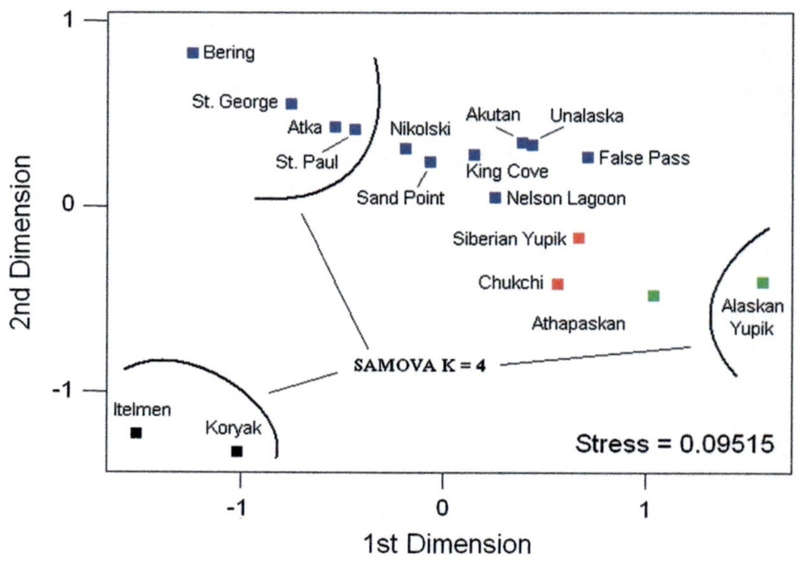

Fig. 9.6 Multidimensional scaling plot of mismatch–intermatch distances based on mtDNA sequences (HVS-I). SAMOVA groupings (K = 4) are indicated in the MSD plot by *arcs* separating population aggregates. Stress is a measure of goodness-of-fit between the distances in the projected MDS to the function of the original distances

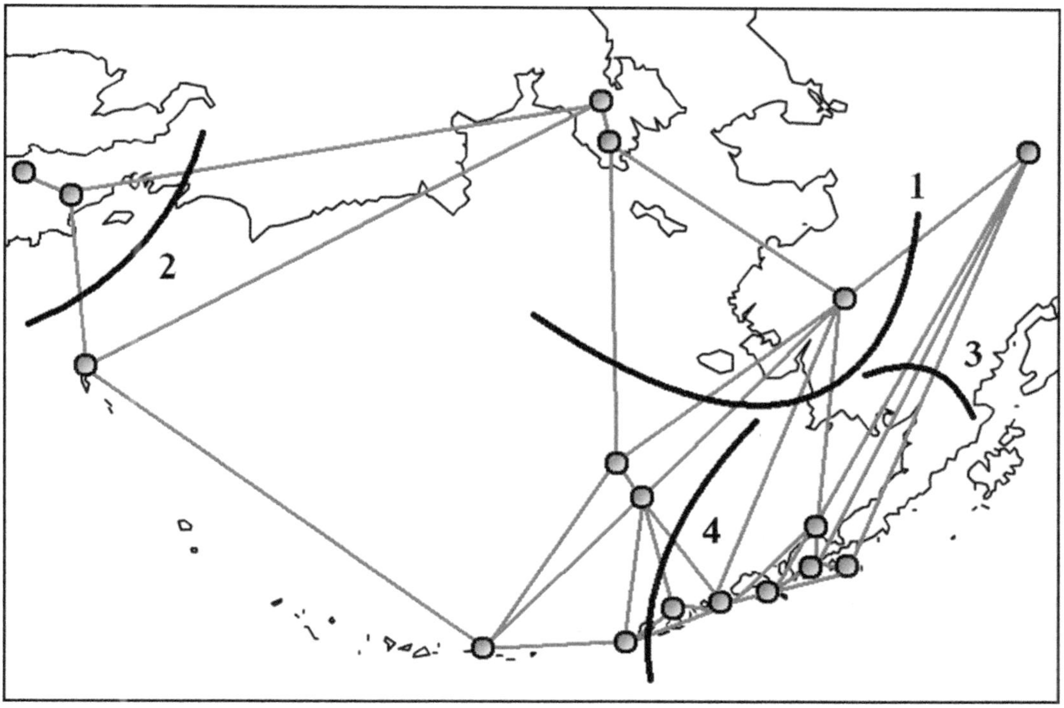

Fig. 9.7 Delaunay triangulation method for genetic barriers based on mtDNA sequences (HVS-I) among Aleutian Island residents and other circumpolar populations

reflects the high frequency of haplogroup D in Atka and the aggregated Aleut settlements. This study documents regional genetic microdifferentiation, following the expansion from the eastern and central regions to the western islands, resulting in the founder effect and the action of other stochastic processes (Crawford et al. 2010).

9.4 Conclusions

The genetic structure of the indigenous populations of the Aleutian Islands is preserved in the maternal lineage—characterized by mitochondrial DNA sequences. In the western islands, women with known Aleut ancestry display only the A and D mitochondrial haplogroups. European mtDNA haplogroups were not detected in the western and central islands, but a low frequency was observed in the eastern islands from non-Aleut women marrying into the communities. In contrast, the Y-chromosome markers reveal a predominantly east European haplogroup (85 %) and only 15 % of Aleut males exhibited Native American Q or Q3 haplogroups. Thus, the gene flow from Russian colonialists was primarily in one direction, Russian males marrying Aleut females. Therefore, to accurately detect admixture and population structure of the Aleutian Islands, both NRY and mtDNA markers are required to detect the asymmetry of the gene flow. Autosomal, recombining STRs provide an intermediate picture of the relationships among the population subdivisions.

How can we be certain that the genetic structure revealed by mtDNA sequences is indeed the result of the original settlement of the Aleutian Islands and subsequent genetic microdifferentiation? Mantel tests indicate that an intimate and statistically significant relationship ($r = 0.7$ $p < 0.000$) still persists between the geography of the archipelago (as measured in kilometers as the crow flies between the islands) and the genetic distances (measured as intermatch distances using mtDNA sequences). On the other hand, there is no indication of a statistically significant relationship between genetic distances measured by NRY markers and geographical distances. History plays a prominent role in explaining the distribution of genes throughout the Aleutian Archipelago. Russian Y chromosomes and surnames occur in the western and central islands, while Scandinavian and English Y chromosomes are distributed throughout the eastern islands. The western European incursion into the eastern Aleutian Islands and the Alaska Peninsula reflect the purchase of Alaska in the nineteenth century from Russia by the United States, followed by settlement of west European fishermen on specific eastern islands. Thus, it is essential in disease association studies that an adequate number of randomly selected SNPs, distributed throughout the genome, be used to measure and reflect the actual genetic structure of the population and its subdivisions. In addition, nonrecombining portions of the genome are useful to elucidate the history and chronology of the observed population structure.

Acknowledgments I would like to thank the Aleut people who generously supported this study and provided hospitality and goodwill throughout the decade of research. The Aleut Corporation and the Aleutian/Pribilof Island Association and the tribal councils were instrumental in the success of this research. I thank Alice Petrovelli (tribal elder) and Liza Mack (graduate student in Anthropology at Idaho State University) for their assistance in the field. I am indebted to my former research assistants, Rohina Rubicz and Mark Zlojutro, who assisted me in the field and conducted most of the laboratory analyses. This research was supported by grants from the National Science Foundation: OPP-990590 and OPP-0327676.

References

Amparo P, Farr SL, Dietz PM (2011) Chronic disease risk factors among American Indian/Alaska Native women of reproductive age. Prev Chronic Dis 8:A118

Baran Y, Pasaniuc B, Sankararaman S, Torgerson DG, Gignoux C, Eng C, Rodriguez-Cintron W, Chapela R, Ford JG, Avila PC, Rodriguez-Santana J, Burchard EG, Halperin E (2012) Fast and accurate inference of local ancestry in Latino populations. Bioinformatics 28:1359–1367

Barbujani G, Magagni A, Minch E, Cavalli-Sforza LL (1997) An apportionment of human DNA diversity. Proc Natl Acad Sci USA 94:4516–4519

Bertorrelle G, Barbujani G (1995) Analysis of DNA diversity by spatial autocorrelation. Genetics 140:811–819

Beuten J, Halder I, Fowler SP, Goring HHH, Duggirala R, Arya R, Thompson IM, Leach RJ, Lehman DM (2011) Wide disparity in genetic admixture among Mexican Americans from San Antonio, Texas. Ann Hum Genet 75:529–538

Cavalli-Sforza LL, Bodmer WF (1971) The genetics of human populations. WH Freeman, San Francisco

Chomczynski P, Sacchi N (1987) Single-step method of RNA isolation by acid Guanidinium Thiocynate-Penol-Chloroform extraction. Anal Biochem 162:156–159

Cooper RS, Tayo B, Zhu W (2008) Genome-wide association studies: implications for multiethnic samples. Hum Mol Genet 15: 17(R2):R151–R155

Crawford MH (1976) The Tlaxcaltecans: prehistory, demography, morphology and genetics. No. 7, Department of Anthropology, University of Kansas, Lawrence

Crawford MH (1978) Population dynamics in Tlaxcala, Mexico: the effects of gene flow, selection, and geography on the distribution of gene frequencies. In: Otten C, Hamed B, Meier R (eds) Genetic studies on human populations. Mouton Press, The Hague, pp 215–225

Crawford MH (1984) Current developments in anthropological genetics Vol. III In: Crawford MH (ed) Black Caribs: a case study of biocultural adaptation. Plenum Press, New York

Crawford MH (1998) Population structure of Native Americans. In: Origins of Native Americans: evidence from anthropological genetics. Cambridge University Press, Cambridge, UK, Ch 5, pp 149–193

Crawford MH (2007) Genetic structure of circumpolar populations: a synthesis. Am J Hum Biol 19:203–217

Crawford MH, Koertvelyessy T, Pap M, Szilagyi K, Duggirala R (1999) The effects of a new political border on the migration patterns and predicted kinship (Phi) in a subdivided Hungarian agricultural population: Tiszahat. Homo 50:201–210

Crawford MH, McComb J, Mitchell RJ (2002) Genetic structure of pastoral populations of Siberia: the Evenki of Central Siberia and the Kizhi of Gorno Altai. In: Human biology of the pastoral populations. Cambridge University Press, Cambridge, UK, Ch 2, pp 10–49

Crawford MH, Rubicz R, Zlojutro M (2010) Origins of the Aleuts and the genetic structure of populations of the archipelago: molecular and archaeological perspective. Hum Biol 82:695–718

Crawford MH, Williams J, Duggirala R (1997) Genetic structure of Siberian indigenous populations. Am J Phys Anthro 104:177–192

Dumond D (2001) Toward a (yet) newer view of the (pre) history of the Aleutians. In: Dumond D (ed) Archaeology in the Aleut zone of Alaska, some recent research. University of Oregon Press, Eugene, OR, pp 289–309

Dupanloup I, Schneider S, Excoffier L (2002) A simulated annealing approach to define the genetic structure of populations. Mol Ecol 11:2571–2581

Hammer MF, Chamberlain VF, Kearney VF, Stover D, Zhang G, Karafet T, Walsh B, Redd AJ (2006) Population structure of Y chromosome SNP haplogroups in the United States and forensic implications for constructing Y chromosome STR databases. Forensic Sci Int 164:45–55

Hao K, Chudin E, Greenawalt D, Schadt EE (2010) Magnitude of stratification in human populations and impacts on genome wide association studies. PLoS ONE 5:e8695. doi:10.1371/journal.pone.0008695

Jorde LB (1980) The genetic structure of subdivided human populations. In: Mielke JH Crawford MH (eds) Current developments in anthropological genetics vol. 1 Theory and Methods. Plenum Press, New York pp 135–208

Jorde LB, Rogers AR, Bamshad M, Watkins WS, Krakowiak P, Sung S, Kere J, Harpending HC (1997) Microsatellite diversity and the demographic history of modern humans. Proc Natl Acad Sci USA 94:3100–3103

Karlsson AO, Wallerstrom T, Gotherstrom T, Holmlund G (2006) Y-chromosome diversity in Sweden—a long-time perspective. Eur J Hum Genet 14:963–970

Koertvelyessy T, Crawford MH, Pap M, Szilagyi K (1993) The influence of religious affiliation on surname repetition (RP) in marriages of Marokpapi, Hungary. Antropologischen Anzieger 51:309–316

Kosoy R, Nassir R, Tian C, White PA, Butler LM, Silva G, Kittles R, Alarcon-Riquelme ME, Gregersen PK, Belmont JW, De la Vega FM, Seldin MF (2009) Ancestry informative marker sets for determining continental origin and admixture proportions in common populations in America. Hum Mutat 30:69–78

Lewontin RC (1972) The apportionment of human diversity. Evol Biol 61:381–398

Manni F, Guérard E (2004) Barrier. In: Manual of the user version 2.2. Population Genetics Team, Musée de l'homme, Paris, France

Manni F, Guérard E, Heyer E (2004) Geographic patterns of (genetic, morphologic, linguistic) variation: how barriers can be detected by 'Monmonier's Algorithm'. Hum Biol 76:173–190

Manolio TA, Collins FS, Cox NJ, Goldstein DB, Hindorff LA, Hunter DJ, McCarthy MI, Ramos EM, Cardon LR, Chakravarti A, Cho JH, Guttmacher AE, Kong A, Kruglyak L, Mardis E, Rotimi CN, Slatkin M, Valle D, Whittemore AS, Boehnke M, Clark AG, Eichler EE, Gibson G, Haines JL, Mackay TF, McCarroll SA, Visscher PM (2009) Finding the missing heritability of complex diseases. Nature 461:747–753

Martin LJ, Crawford MH, Koertvelyessy T, Keeping D, Collins M, Huntsman R (2000) The population structure of ten Newfoundland out ports. Hum Biol 72:997–1016

Monmonier M (1973) Maximum-difference barriers: an alternative numerical regionalization method. Geogr Anal 3:245–261

Naylor JL, Schraer CD, Mayer AM, Lanier AP, Treat CA, Murphy NJ (2003) Diabetes among Alaska Natives: a review. Int J Circumpolar Health 62:4

Pääbo S (2003) The mosaic that is our genome. Nature 421:409–412

Pasaniuc B, Zaitlen N, Lettre G, Chen GK, Tandon A, Kao WH, Ruczinski I, Fornage M, Siscovick DS, Zhu W, Larkin E, Lange LA, Cupples LA, Yang Q, Akylbekova EL, Musani SK, Divers J, Mychaleckyj J, Li M, Papanicolaou GJ, Millikan RC, Ambrosone CB, John EM, Bernstein L, Zheng W, Hu JJ, Ziegler RG, Nyante SJ, Bandera EV, Ingles SA, Press MF, Chanock SJ, Deming SL, Rodriguez-Gil JL, Palmer CD, Buxbaum S, Ekunwe L, Hirschhorn JN, Henderson BE, Myers S, Haiman CA, Reich D, Patterson N, Wilson JG, Price AL (2011) Enhanced statistical tests for GWAS in admixed populations: assessment using African Americans from CARe and a Breast Cancer Consortium. PLoS Genet 7:e1001371

Prokopenko I, McCarthy MI, Lindgren CM (2008) Type 2 diabetes: new genes, new understanding. Trends Genet 24:613–621

Qin H, Morris N, Kang SJ, Li M, Tayo B, Lyon H, Hirschhorn J, Cooper RS, Zhu W (2010) Interrogating local population structure for fine mapping in genome-wide association studies. Bioinformatics 26:2961–2968

Rosenberg NA, Huang L, Jewett EM, Szpiech ZA, Jankovic I, Boehnke M (2010) Genome-wide association studies in diverse populations. Nat Rev Genet 11:356–366

Rubicz RC (2001) Origins of the Aleuts: molecular perspectives. Master's Thesis. University of Kansas, Lawrence, Kansas

Rubicz RC (2007) Evolutionary consequences of recently founded Aleut communities in the Commander and Pribilof Islands, Ph.D. Dissertation. University of Kansas, Lawrence, Kansas

Rubicz R, Melton P, Crawford MH (2007) Molecular markers in anthropological genetic studies. In: Anthropological Genetics, Cambridge University Press, Cambridge, UK. Ch 6 pp 141–186

Rubicz R, Schurr TG, Babb PL, Crawford MH (2003) Mitochondrial DNA variation and origins of the Aleuts. Hum Biol 75:809–835

Rubicz R, Zlojutro M, Sun G, Spitsyn V, Deka R, Young K, Crawford MH (2010) Genetic architecture of a small, recently aggregated Aleut population: Bering Island. Hum Biol 82:719–736

Schraer CD, Lanier AP, Boyko EJ, Gohdes D, Murphy NJ (1988) Prevalence of diabetes mellitus in Alaskan Eskimos, Indians, and Aleuts. Diabetes Care 11:693–700

Schull WJ, MacCluer JW (1968) Human genetics: structure of populations. Annu Rev Genet 2:279–304

Shen H, Damcott CM, Rampersaud E, Pollin TI, Horenstein RB, McArdle PF, Peyser PA, Bielak LF, Post WS, Chang YP, Ryan KA, Miller M, Rumberger JA, Sheedy PF 2nd, Shelton J, O'Connell JR, Shuldiner AR, Mitchell BD (2010) Familial defective apolipoprotein B-100 and increased low-density lipoprotein cholesterol and coronary artery calcification in the old order Amish. Arch Intern Med 170:1850–1855

Tian C, Gregersen PK, Seldin MF (2008) Accounting for ancestry: population substructure and genome-wide association studies. Hum Mol Genet 17(R2):R143–R150

Torrey EF, Reiff FM, Noble GR (1979) Hypertension among Aleuts. Am J Epidemiol 110:7–14

Turakulov R, Esteal S (2003) Number of SNPs loci needed to detect population structure. Hum Hered 55:37–45

Walsh PS, Metzger DA, Higuchi R (1991) Chelex^{R} 100 as a medium for simple extraction of DNA for PCR-based typing from forensic material. BioTechnique 10:506–513

West D, Savinitsky A, Crawford MH (2007) Aleutian Islands: Archaeology, molecular genetics and ecology. Trans Roy Soc Edinb: Earth Environ Sci 98:1–11

Workman PL, Jorde LB (1980) The genetic structure of the Aland Islands. In: Eriksson AW, Forsius H, Nevalinna HR, Workman PL (eds) Population structure and genetic disease. Academic Press, New York, pp 487–508

Zlojutro M (2008) Mitochondrial DNA and Y-chromosome variation of Aleut populations: genetic implications of founder effect, asymmetrical gene flow and the original peopling of the Aleutian Archipelago, Ph.D. dissertation, University of Kansas, Lawrence, Kansas

Zlojutro M, Rubicz R, Devor EJ, Spitsyn VA, Makarov SV, Wilson K, Crawford MH (2006) Genetic structure of the Aleuts and Circumpolar populations based on mitochondrial DNA sequences: a synthesis. Amer J Phys Anthrop 129:446–464

Mapping Genes in Isolated Populations: Lessons from the Old Order Amish 10

Braxton D. Mitchell, Alejandro A. Schäffer, Toni I. Pollin, Elizabeth A. Streeten, Richard B. Horenstein, Nanette I. Steinle, Laura Yerges-Armstrong, Alan R. Shuldiner, and Jeffrey R. O'Connell

B.D. Mitchell (✉) · T.I. Pollin · E.A. Streeten · R.B. Horenstein · N.I. Steinle · L. Yerges-Armstrong · A.R. Shuldiner · J.R. O'Connell
Department of Medicine, University of Maryland School of Medicine, 22 S Greene Street, Baltimore, MD 21201, USA
e-mail: bmitchel@medicine.umaryland.edu

T.I. Pollin
e-mail: tpollin@medicine.umaryland.edu

E.A. Streeten
e-mail: estreete@medicine.umaryland.edu

R.B. Horenstein
e-mail: rhorenst@medicine.umaryland.edu

N.I. Steinle
e-mail: nsteinle@medicine.umaryland.edu

L. Yerges-Armstrong
e-mail: lyerges@medicine.umaryland.edu

A.R. Shuldiner
e-mail: ashuldin@medicine.umaryland.edu

J.R. O'Connell
e-mail: joconnel@medicine.umaryland.edu

A.A. Schäffer
National Center for Biotechnology Information, National Library of Medicine, National Institutes of Health, DHHS, 8600 Rockville Pike, Bethesda, MD 20894, USA
e-mail: aschaffe@helix.nih.gov

E.A. Streeten · A.R. Shuldiner
Geriatric Research and Education Clinical Center, Veterans Administration Medical Center, Baltimore, MD 21201, USA

N.I. Steinle
Diabetes and Endocrinology Section, Veterans Administration Medical Center, Baltimore, MD 21201, USA

10.1 Introduction

The application of high throughput "–omics" technologies (e.g., genomics, transcriptomics, and proteomics) to human and medical genetics in recent years has led to numerous gene discoveries for a variety of complex diseases and traits. Many of the studies utilizing these technologies have used samples obtained from large population-based (or in some cases family-based) studies to identify DNA sequence variants and gene expression and protein profiles associated with the trait of interest. Insights about the genetic underpinnings of trait variation and disease susceptibility have come from studies of different populations. The goal of this review is to describe how unique insights can be provided through studies carried out in isolated populations, that is, populations that are relatively genetically homogeneous because they descend from a relatively small number of ancestors (founders) and thus the individuals in the population are genetically similar. To illustrate this point, we provide in this chapter: (1) a brief description of the OOA community in Lancaster County, PA, an isolated population whom our group at the University of Maryland School of Medicine has been studying since 1993 with the help of scientists at the NIH; (2) a description of several genetic discoveries we have made in this population by virtue of the fact that it is an isolated population; and (3) the relevance of these discoveries to our understanding of health and biology.

R. Duggirala et al. (eds.), *Genome Mapping and Genomics in Human and Non-Human Primates*, Genome Mapping and Genomics in Animals 5, DOI 10.1007/978-3-662-46306-2_10,

10.2 Old Order Amish of Lancaster County, PA: History and Background

Amish are Anabaptists, meaning that they believe in adult baptism by choice. The Anabaptist movement was started in the 1520s in Germany and the cantons of Switzerland by Jakob Hutter (after whom the Anabaptist Hutterites are named), Conrad Grebel, Felix Manz and Blaurock (Gingerich and Kreider 2002). The founders of Anabaptism were contemporaries in both time and place of Martin Luther (1483–1546) and John Calvin (1509–1564), who are prominent among the founders of the more mainstream Protestant (Reformation) movement. Among the central tenets of the Anabaptists were pacifism, separation of church and state, and adult baptism. These were radical views (interesting to contrast with the modern perception that Anabaptists are traditionalists) for war-torn sixteenth-century Europe and caused the Anabaptists to be persecuted and socially isolated. The social and religious isolation had the side effect of creating a genetic bottleneck, which is one of the reasons that genetic studies in Anabaptist societies have proven so fruitful centuries later.

The Amish community takes its name from the founder Jakob Ammann (ca. 1644–1730). Ammann was a leader in the Swiss Anabaptist (often called Swiss Brethren) church who felt that adherence to church rules was being enforced too loosely. Ammann instituted a practice of strict "shunning" by which his followers were to reject socially those who did not strictly practice the church rules. Ammann also introduced a practice of plain dress, which is followed rigidly by the OOA to this day.

Because of persecution, Anabaptists began to escape from Europe by boat to the British colonies (later to become the United States) beginning in the early eighteenth century (Gingerich and Kreider 2002). In 1737, there was a large group of unambiguously Amish immigrants to what is now Lancaster County, Pennsylvania; some earlier Anabaptist immigrants may have also been Amish (Gingerich and Kreider 2002). In the late seventeenth century, Eastern Pennsylvania had been opened for settlement by the efforts of William Penn (1644–1718), for whom Pennsylvania was named. This area may have been particularly attractive to the Anabaptist immigrants because Lancaster has some of the most fertile land in the eastern United States and because Penn was a (pacifist) Quaker, suggesting correctly that the Anabaptists' religious views would be more tolerated in their new land. Hundreds of Amish immigrants settled in this area during an approximately 100-year period beginning in the early eighteenth century (Gingerich and Kreider 2002). There was another burst of Amish immigration from Europe after the war of 1812 approximately coinciding with a Pennsylvania Amish migration westward to newer states, especially Ohio and Indiana (Gingerich and Kreider 2002). These states were newly attractive because of policies by the United States government that increasingly excluded the Native American residents and encouraged new settlement of western areas of the rapidly growing country.

There are no longer any Amish in Europe. Since their relocation to the United States, there have been several divisions within the Amish. A defining characteristic of the different Amish groups and their offshoots remains their belief in adult baptism and a humble lifestyle. For the purpose of genetic studies of complex traits, this aspect of the Amish culture may be useful because confounding environmental factors such as diet and socioeconomic status are relatively homogeneous in the Amish. Furthermore, rates of alcohol drinking and tobacco usage (two environmental risk factors for many complex diseases) are much lower among the Amish than in the United States as a whole (Ferketich et al. 2008). Several excellent histories and descriptions of the Amish have been written, including one by Kraybill (2001).

Amish culture is unique among most Western cultures in that its core beliefs in church, community and social cohesiveness, and selflessness permeate daily life. Amish society has been traditionally agrarian, but with dwindling availability of farmland and large family sizes, many have taken on other occupations, including carpentry and shop-keeping. Compared to non-Amish, Amish are very physically active; they maintain a traditional lifestyle, still utilizing the horse and buggy as their main mode of transportation, and do not use electricity in their homes.

Another aspect of Anabaptist culture that makes these groups especially suited to genetic studies is that, like the Mormons, they are fascinated with knowing and recording their genealogies. Hundreds, if not thousands, of Anabaptist genealogy books have been published, and there are libraries in Pennsylvania and Indiana focused on housing these books. The Swiss Anabaptist Genealogical Association (SAGA) actively maintains a website of genealogies and holds annual meetings. To make these resources more directly useful in genetic studies, we have systematically (removing many errors and duplicates) combined several of these genealogy sources into the Anabaptist Genealogy Database (AGDB) (Agarwala et al. 2003). The present AGDB version 5 includes over 530,000 distinct individuals, including all the approximately 106,000 individuals in the 2009 edition of the Descendants and History of Christian Fisher (Beiler 2009), which is generally considered the most comprehensive genealogy book for the Lancaster County OOA. We have also developed a software package, PedHunter (Agarwala et al. 1998; Lee et al. 2010), that makes it possible to automatically construct pedigrees in various ways and answer a variety of queries that medical genetics researchers find useful. PedHunter is formally separate from AGDB, so that PedHunter can be and has been used in genetic studies of other groups with available genealogic records. Unlike the SAGA databases, the development, maintenance and usage of AGDB is considered human subjects research and has been covered since 1997 by an IRB-approved protocol at the National Institutes of Health. Access to AGDB is granted to researchers at other institutions (e.g., University of Maryland, Medical School) who have their own IRB-approved human subjects' protocols for such studies.

Based on the publication of the 2002 Church Directory of the Lancaster County Amish (Gallagher and Beiler 2002), which enumerates households within the OOA Lancaster County church districts, we have estimated the population size of the Lancaster County Amish community at that time to be ~21,900 individuals, of whom ~7,500 are age 25 years or older (Tolea 2007). Due to the high birth rates within the OOA community, the population size had likely increased to 34,000–36,000 by 2010.

10.3 The Genetic Architecture of the Old Order Amish

In addition to being geographically localized, the OOA offer particular advantages for genetic studies because of their unique ancestral history. From a population genetics perspective, the three major forces shaping the genetic variation in the present Amish population are genetic drift, recombination, and mutation. Genetic drift increases or decreases the population frequency of any particular allele, recombination shuffles haplotypes across each chromosome, and mutation introduces new alleles. Each of our chromosomes is a mosaic of ancestral chromosomes, and the resolution of the mosaic depends on the number of ancestral generations insofar as this shapes the forces influencing genetic variation. One useful metric of that mosaic is the average kinship coefficient between founders and living descendants. For the OOA, we have developed simulation and analysis tools to investigate how genetic drift shapes the founder and allelic architecture of our study subjects in greater detail. Founder architecture refers to the distribution of founder alleles in the present-day population and thus represents identity-by-descent. Because we know the identities of the founders via the genealogy, we can measure founder architecture (i.e., identity-by-descent relationships throughout the population) by assigning to each founder two distinct alleles, randomly dropping them through the pedigree many times, and then computing frequencies for which particular founder alleles are observed and the numbers of different founder alleles present in the present-day population at a particular locus. In particular, these simulations provide insight as to how the genealogical relationships shape the frequency of rare population alleles.

Based on 3,480 adult Amish subjects who have participated in one or more of our Amish studies (representing nearly one-third of the total number of Amish adults projected to be in the community as of 2002) and their known ancestors dating back to the initial Amish founders of the Lancaster County settlement, we estimated the number of founders contributing to allelic variation at a single locus. These subjects can be connected into a single 10,124 subject pedigree with 364 founders, thus generating 728 founder alleles. Analysis of 1,000 replicates in which unique founder alleles are assigned and dropped down through the pedigree revealed that at a single locus there are on average 129.5 distinct founder alleles with a minimum of 114 and maximum of 148; this average can be regarded as the effective sample size of the 3,480 subjects. Thus, for any specified locus, on average 600 founder alleles (~728–129), or 82 % of the genetic variation, are lost. While this sounds high, recall that 50 % is lost from parent-to-child transmission. Figure 10.1 shows the frequency spectrum of the variation that survives in the top ten ranked founder alleles averaged over each replicate.

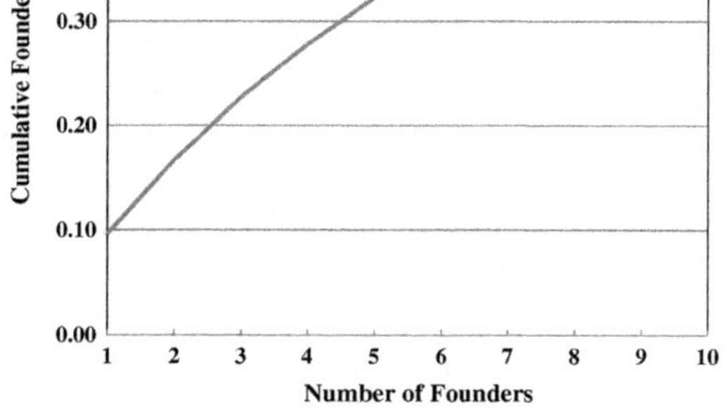

Fig. 10.1 Mean founder contribution to the present-day OOA gene pool

Because some founders have more descendants than others, the contribution of each to genetic variation in the current population is unequal. In fact, only 128 founders (78 females and 50 males) accounted for over 95 % of the mean relative founder contribution among living OOA descendants (Lee et al. 2010). Fifty percent of the total genetic variation is accounted for by only 10 distinct founders; however, the actual number of founders that account for that variation varies across replicates. Across the 1,000 replicates, 19 distinct individuals assumed the role of highest contributing founder, with a maximum contribution of 12 % and an average contribution of 9.6 %. If that founder allele having a non-negligible frequency in the Amish genealogy represents a variant that is rare in the general population, then these results highlight how genetic drift can shape the rare and low-frequency allele spectrum present in the isolated population of known genealogy. It is also anticipated that the number of rare-disease alleles will be smaller in the founder population compared to large outbred populations, but any rare allele that survives into the living descendants of the founders of the isolated population will be present in many carriers, unless it is very deleterious to fitness, in the heterozygous state. For example, the founder allele *APOB* R3500Q, described below, has a 6 % frequency in our population, versus <1 % in the general population.

Another important result of the simulation is that the remaining 50 % of the variation not shown in Fig. 10.1 is accounted for on average by 120 founder alleles (129.5 − 10) in decreasing frequency, down to a handful of copies of an allele. For example, we have found in our sample of 3,480 Amish subjects only 7 copies of a mutation (rs121918387, allele frequency 0.1 %) in the *APOB* gene leading to a truncated species (apoB67) that has been previously reported in an Amish community from the Midwest (Welty et al. 1991). Thus, founder rare-allele frequency in the current population can range from >10 % down to <0.1 %, which impacts the statistical power to detect them. Finally, another important result of the simulation shows that most low-frequency alleles in the current population (<3 %) will be founder effects; that is, of the few copies that might have been present in different founders all but one is lost or only one reaches high frequency. Thus, low-frequency population alleles on single nucleotide polymorphism (SNP) chips that are not monomorphic in the Amish tend to track founder chromosome segments and as a consequence any rare causal variant on that segment. Precisely because rare alleles in the OOA generally represent founder alleles, there tends to be long-range linkage disequilibrium (LD) among low-frequency alleles in the OOA, although the same is not true for common alleles that have entered the population in multiple founders (Van Hout et al. 2009). The long-range LD between low-frequency alleles in the Amish has facilitated detection of both major genes for triglycerides (*APOC3* R19X) and low density lipoprotein (LDL) levels (*APOB* R3500Q) in the Amish, where the most associated SNPs were >300 kb from the gene, respectively. Thus, in essence one is detecting linkage with the nearest genotyped, polymorphic SNP, i.e., a within-Amish association, not a population association.

10.4 Approaches for Mapping Genes in the Old Order Amish

From the above discussion, it should be evident that isolated populations such as the Amish can be invaluable resources for identifying variants that may be rare in the general population, but whose frequencies are increased due to such forces as genetic drift. While the frequencies of numerous rare variants may be increased in the Amish, many others will be lost. In contrast, most variants that are common in European Caucasian populations are also likely to be common in the Amish (Van-Hout et al. 2009).

One might expect that among the rare variants over-represented in the OOA will be some having high penetrance and large effect sizes, provided that the associated phenotype is not deleterious to childbearing directly or indirectly. Within the OOA and other isolated populations, identifying carriers of these types of variants may be important for clinical or risk prediction purposes. But equally important, however, such rare variant discovery may provide exciting insights about biology extending far beyond the OOA population alone, for example, by implicating the involvement of a gene in a novel biological pathway.

A variety of approaches are amenable for mapping rare strong-effect variants in isolated populations such as the OOA. Candidate gene approaches can be fruitful for investigating extreme phenotypes since the genetic variants involved may be exonic mutations that disrupt gene function and the causative mutations may be more easily identifiable because of their predictive effects on gene function. Conventional model-based genetic linkage analysis of large pedigrees is also an attractive approach and is facilitated in the OOA by the genealogy resources described in Sect. 10.2. Other pedigree-based approaches are also well-suited for gene mapping in isolated populations. For example, because of the relatively small number of founders, husbands and wives typically share common ancestors, making it possible to utilize homozygosity by descent mapping (Lander and Botstein 1987) to map genes for very rare recessively inherited traits. This approach, which involves identifying stretches of homozygosity within affected individuals, has been used to find the causative genes for several monogenic, recessive disorders in the OOA [reviewed in Strauss and Puffenberger (2009)], and we have used an extension of this approach for complex traits to map genes for blood pressure variation in the OOA (McArdle et al. 2007).

One of the known pitfalls of homozygosity mapping is allelic heterogeneity, meaning that there are multiple deleterious variants of the same gene in the population and hence affected individuals may be compound heterozygotes for two mutations (Miano et al. 2000). Indeed, allelic heterogeneity has been documented for at least two disorders prevalent in the OOA: phenylalanine hydroxylase deficiency (Wang et al. 2007) and Cohen syndrome (Taban et al. 2007); in the case of Cohen syndrome, there are affected OOA individuals who are compound heterozygotes for two mutations of the causative gene, *VPS13B* (Taban et al. 2007). If one wants to avoid the allelic heterogeneity pitfall, one can use PedHunter (Agarwala et al. 1998; Lee et al. 2010) to automatically and systematically extract from AGDB (Agarwala et al. 2003) pedigrees that connect all the obligate carriers and affected individuals. Then, one can compute conventional model-based LOD scores to find evidence of linkage, as was done for example, for the disorder Amish microcephaly (Rosenberg et al. 2002).

For complex traits, some statistical geneticists have argued that model-free (a.k.a. nonparametric) test statistics such as NPL scores (Whittemore and Halpern 1994; Kruglyak et al. 1996) are preferable to model-based LOD scores. However, this statistical preference creates a computational problem because the software packages that compute NPL scores, e.g., (Kruglyak et al. 1996) use the Lander-Green method of pedigree analysis, which can handle many markers, but only pedigrees of very limited size. OOA pedigrees are not of limited size. For this purpose, the size is measured as the number of "bits," which is twice the number of non-founders minus the number of founders. One solution, which has been used in some OOA studies of complex traits, e.g., (Cummings et al. 2012), is to partition the large AGDB-derived pedigrees using the software PedCut (Liu et al. 2008). PedCut uses the criterion of bits explicitly, so that the output pedigrees fit just within the bit constraint of whichever linkage analysis software package will be used to compute the model-free linkage test statistics.

Some case studies we describe below show by example that variants of large effect but low frequency can also be mapped using genome-wide association (GWA) analysis provided that the frequencies of the variants are high enough to be in linkage disequilibrium with SNPs represented on the genotyping chip used in the GWAS. Through genetic drift and/or other mechanisms variants entering the OOA population through only a single or small number of founders have, in fact, increased in frequency sufficiently such that they have been mapped to disease traits using the GWAS approach. But regardless of how they have been mapped, the identification of highly penetrant variants can provide important insights into biology. In the section below, we briefly describe four highly penetrant variants that have been mapped to cardiovascular or bone-related traits in the OOA and describe some lessons learned from these exciting discoveries.

10.5 Highly Penetrant Variants Present in the Old Order Amish and Some Lessons Learned

10.5.1 *ABCG8* Gly574Arg and Cardiovascular Risk

Sitosterolemia is an autosomal recessive disorder characterized by excess accumulation of dietary plant sterols in the circulation. The disease was initially described in 1974 in two Amish-Mennonite sisters (Bhattacharyya and Connor 1974) and was later identified in other members of the Amish community including a 13-year-old boy who died of coronary artery disease (Kwiterovich et al. 1981). As of 2001, only 45 sitosterolemia cases have been reported worldwide in the literature (Lee et al. 2001b). The disease is caused by a defect in the transport of plant sterols from cells in the intestinal mucosa and liver into the gut lumen for excretion from the body. The molecular defect arises from biallelic mutations in either of two genomically adjacent transporter genes (*ABCG5* or *ABCG8*) that mediate this process. At least 18 different mutations causing sitosterolemia have been identified in these two genes (Berge et al. 2000; Lee et al. 2001a). Sitosterolemia leads to excessively high concentrations of sitosterol in the blood, mild to significant elevation in cholesterol levels, and development of early atherosclerotic heart disease.

The cause of the increased cardiovascular risk in sitosterolemia is thought to be related to the direct effects of sitosterol or its metabolites on vessel walls or on lipoprotein cholesterol transport, but exact mechanisms have yet to be fully elucidated. This has prompted a debate about whether modest elevation of sitosterol levels is associated with increased cardiovascular risk at the population level. Associations have been reported between elevated phytosterol levels and cardiovascular disease in some (Assmann et al. 2006; Glueck et al. 1991; Miettinen et al. 1998), but not other (Silbernagel et al. 2009; Wilund et al. 2004) studies. While individuals heterozygous for one of the sitosterol-causing mutations do not have the sitosterol disease, they do have markedly elevated plasma levels of sitosterol. Taking advantage of the relatively high frequency of the *ABCG8* sitosterol-causing mutation in the Amish (a glycine to arginine substitution at position 574, Gly574Arg) afforded us the opportunity to assess the relation of elevated sitosterol levels to cardiovascular risk in a homogeneous population with high variation in plasma sitosterol levels.

We initially identified 15 G574R carriers from our Amish biobank (from which we calculated an allele frequency of 0.8 % in the OOA) and then recruited 99 additional carriers who were close relatives of the initial 15. Compared to age-and sex-matched non-carriers, carriers had 35 % higher plasma levels of plant sterols, including sitosterol, but no difference in body mass index or cholesterol and triglyceride levels. Moreover, carriers had slightly lower values of carotid wall thickness, corresponding to *lower*, not higher, levels of subclinical atherosclerosis (Horenstein et al. 2013). This sitosterol example demonstrates not only the enrichment of a variant in an isolated population that in its homozygous state can have clinically significant consequences, but also how study of the heterozygous state can provide important insights into epidemiological issues—in this case whether modestly elevated plant sterol levels increase cardiovascular risk.

10.5.2 *APOB* R3500Q: Isolated High LDL and Subclinical Atherosclerosis

Elevated low density lipoprotein cholesterol (LDL-C) level is a major cardiovascular disease (CVD) risk factor. To identify genes influencing variation in plasma LDL-C levels, we carried out a genome-wide association study on LDL-C in 841 relatively healthy Amish adults. We identified a cluster of SNPs highly associated ($p < 10^{-68}$) with variation in LDL-C levels on chromosome 2 in the region of the *APOB* gene, a strong positional candidate gene because of its role in lipid metabolism (Shen et al. 2010). Sequencing of this gene revealed the presence of a nonsynonymous mutation (R3500Q), a previously known mutation that is responsible for familial defective apolipoprotein B-100 (Soria et al. 1989). This mutation interferes with the folding of the apoB protein, thus impairing its ability to bind with the LDL particle and impeding LDL-C clearance (Borén et al. 2001). While the frequency of this mutation is <0.5 % in European Caucasians (Austin et al. 2004), its frequency in the Lancaster county OOA is ~6 %, translating into an overall carrier frequency of ~12 %. All carriers of the R3500Q mutation shared a common haplotype surrounding the variant that extended for ~300 kbp, suggesting that the variant entered the population on a single founder (or perhaps multiple related founders), where it has been passed down through generations. In the Amish, each copy of the variant allele is associated with a 58 mg/dL increase in LDL-C level and overall the variant accounts for 26 % of the variation in LDL-C levels. Those Amish homozygous for the variant (of whom we have identified 5) have ~115 mg/dl higher LDL-C levels compared to those with no copies of this allele.

In the general population, elevated LDL-C rarely occurs in isolation; rather it occurs against the backdrop of other metabolic disturbances that may include obesity, hypertension, low HDL-C and high triglycerides, and inflammation. These concomitant conditions have made it difficult to quantify the effects of individual components of this metabolic cluster on cardiovascular risk. One of the unique features of the R3500Q mutation is that it causes high LDL-C levels in the absence of any of these metabolic abnormalities. Indeed, Amish carriers of the R3500Q variant did not differ from non-carriers in terms of body mass index, blood pressure, or any other lipid component that we measured, including LDL-C subclass particle patterns or appreciably with HDL-C levels. This provided us with the unique opportunity to evaluate the association of this LDL-C-elevating variant on development of subclinical atherosclerosis in the absence of other complicating metabolic factors. We therefore compared extent of coronary artery calcification, previously measured by electron beam computed tomography, between R3500Q carriers and non-carriers. R3500Q carriers were significantly more likely to have both detectable (odds ratio = 4.4) and extensive (odds ratio = 9.3) coronary artery calcification compared to non-carriers (Shen et al. 2010). These results highlight in a direct way the strong atherogenic role of elevated LDL-C levels.

The relatively high frequency of R3500Q carriers among the Amish also afforded us the opportunity to evaluate the effects of this mutation on other traits. In particular, multiple studies have reported associations between osteoporosis and cardiovascular disease, leading to the speculation that hyperlipidemia may predispose to both atherosclerotic heart disease and accelerated bone turnover. To evaluate this hypothesis, we compared bone mineral density (BMD) measured by dual-energy X-ray absorptiometry between subjects with and without the R3500Q mutation. We observed a 1.7–2.3 % lower BMD at the femoral neck, lumbar spine, and total body skeletal sites in carriers of the R3500Q variant than in noncarriers, implying a causal relationship between prolonged, high LDL-C and low BMD (Yerges-Armstrong et al. 2013). While a similar study of pleiotropy could be envisioned where one recruited participants with high LDL-C mutations from lipid clinics, this investigation was greatly facilitated by working with the OOA because of (1) the relatively high frequency of individuals with the same mutation and (2) the ability to screen efficiently for a mutation at a population level as opposed to only ascertaining participants with severe enough clinical manifestations to be referred for specialized care.

10.5.3 *APOC3* R19X: Cardioprotection from a Loss of Function Mutation

The features of the Amish discussed here have enabled the discovery and characterization of a unique mutation in the *APOC3* gene, R19X, which in turn is providing access to biological insights previously unavailable. The Heredity and Phenotype Intervention (HAPI) Heart Study was begun in 2002 to evaluate the role of genetic and non-genetic risk factors in the response to four short-term interventions affecting cardiovascular risk factors (Mitchell et al. 2008). These interventions included: a single high-fat meal, a cold pressor stress test, a dietary intervention altering salt intake, and short-term aspirin therapy.

As part of the HAPI Heart Study, we carried out a GWAS of triglyceride (TG) response to the single high-fat meal and found a single SNP to be associated after Bonferroni correction with both fasting ($p = 4 \times 10^{-14}$) and postprandial ($p = 3 \times 10^{-10}$) TG levels (Pollin et al. 2008). The SNP itself was located in an intron of a gene called *DSCAML1* but was also 800 kbp away from a cluster of genes playing a key role in lipid metabolism, the *APOA1/C3/A4/A5* region. These genes were considered to be viable positional candidate genes in the Amish because the relatedness of the subjects through recent founders was expected to lead to longer regions of allele sharing than would be expected in a population sample. The minor allele (frequency = 0.028) of the SNP was associated with considerably lower TG levels and postprandial response, mimicking the effect previously shown of knocking out the *APOC3* gene in the mouse (Maeda et al. 1994). Sequencing the coding region of the *APOC3* gene led to the finding that the GWAS-associated SNP was tagging a founder mutation (*APOC3* R19X) that resulted in an insertion of a premature stop codon within the signal peptide region of the gene, effectively serving as a human *APOC3* knockout allele. So far, only heterozygotes have been observed, but in these individuals, in addition to hypotriglyceridemia, we observed an overall favorable lipid profile, including increased HDL cholesterol (not seen in the mouse due to species differences in lipoprotein biology) as well as reduced prevalence of subclinical cardiovascular disease as measured by reduced coronary calcification in comparison to Amish individuals without the mutation.

Prior to the discovery of this mutation, only a few coding mutations in *APOC3* had been reported, in only a handful of individuals each (Karathanasis et al. 1983, 1987; Liu et al. 2000; Norum et al. 1982; Ordovas et al. 1989; von Eckardstein et al. 1991). Some evidence that these mutations could produce cardioprotective lipid profiles was observed, but the numbers were too small to be conclusive, and assessment of association with an actual disease phenotype was not possible. Notably, the R19X mutation could be traced back in the OOA to a single couple and the resulting pedigree could be used to identify and recruit living individuals who had a high probability of being carriers. Because of the limited gene pool in the Amish and consequent founder effect, we were thus able to study an adequate sample size of mutation carriers to obtain conclusive evidence of favorable effects of lowering apoC-III in humans. This finding has enhanced interest in developing pharmaceutical agents that directly lower apoC-III production (Visser et al. 2012) as therapy for dyslipidemia in those not lucky enough to carry an apoC-III lowering mutation, making the finding of this otherwise rare mutation of great interest to the general population.

10.5.4 *COL1A2* G610C: Characterization of Variable Disease Penetrance

Osteogenesis imperfecta (OI) is a heritable form of bone disease characterized by high fracture risk. Additional characteristics that are variably present include short stature, dentinogenesis imperfect (tooth malformation resulting from defects in dentin), blue sclerae and hearing loss. The disease is typically associated with mutations in one of the genes encoding the α chains of the type I procollagen molecules, *COL1A1* or *COL1A2*. Over 1,000 mutations in these genes have been identified in OI patients to date, and there is marked variability in the clinical expression of the disease ranging from mild to severe.

As part of our Amish complex disease research program, we measured bone mineral density in a large number of subjects to screen for osteoporosis. From this screen, one subject was identified with particularly low BMD and a history of multiple fragility fractures and short stature suggestive of OI. We screened this subject's DNA for mutations in *COL1A1* and *COL1A2* and identified a variant in *COL1A2* that alters the glycine-610 codon (GGT) to a cysteine (TGT) codon. This mutation has not been reported in patients with OI outside of our kindred. From other *COL1A2* OI mutations that have been characterized, it has been established that amino acid substitutions at invariant glycine residues typically result in clinically apparent phenotypes (Marini et al. 2007). Recruitment of family members of our index patient with OI led to the identification of 63 additional carriers of this mutation (for a total of 64). Using the PedHunter tool described above, ancestors of the 64 carriers were tracked through AGDB and their relationships established. This exercise revealed the pattern of inheritance to be consistent with a de novo mutation that occurred over 150 years ago (Daley et al. 2009).

OI is characterized by considerable phenotypic variability with the disease graded from type 1 (mild) to 4 (most extreme), based on clinical presentation, radiography, and mode of inheritance (van Dijk et al. 2011). The G610C mutation in the Amish represents the largest reported collection of OI patients with an identical collagen mutation. This has provided particular insights into the degree of phenotypic variability among a large group of subjects having the same OI-causing mutation and also having a relatively homogenous lifestyle. Compared to Amish without the OI mutation, those with OI had shorter stature and lower bone mineral density ($\sim$2 ½ standard deviation, SD, units lower BMD at the spine and $\sim$1 SD unit lower at the hip), although there was considerable overlap between those with the OI mutation and those without. Moreover, even among those with OI, there was substantial variation in BMD, with some subjects having near normal levels of spine BMD and others whose BMD was up to 5 SD units lower than unaffected subjects. Overall, 73 % of those with the OI mutation were judged to have moderate to severe disease (spine BMD <2 SD units below that of the general population), 23 % were judged to have mild disease (1–2 SD units lower than that of the general population), and 4 % had BMDs in the unaffected range (BMD > -1 SD below that of the general population).

Identification of the G610C mutation in this population has furthered our understanding of phenotypic variation within OI to a degree not previously appreciated. All carriers had the exact same mutation and this mutational homogeneity combined with their similar lifestyle underscores the involvement of multiple factors in determining phenotype. Further study of this collection of subjects offers opportunities for additional mechanistic insights, as for example, through efforts to identify genes that modify the effects of the G610C mutation.

10.6 Conclusion

The OOA of Lancaster County have been active participants in medical genetics studies since the 1960s, beginning with the studies of Dr. Victor McKusick, a pioneer of medical genetics (McKusick 1978). Dr. Charles Eugene “Gene” Jackson, another pioneer of medical genetics, did early studies in the Indiana Amish community, e.g., (Jackson et al. 1974). Amish genetic studies have led to many discoveries and the development of new research methods for genealogic studies over the past five decades. Insight has been gained into the role specific genes play in the pathophysiology of cardiovascular and bone disease and provided molecular targets for diagnosis and treatment. In more recent decades, the pioneering clinical efforts of Dr. Holmes Morton among the Pennsylvania Amish, and in the past decade of Dr. Heng Wang among the Ohio Amish and Dr. Amy Shapiro among the Indiana Amish have led to the establishment of Amish clinics. At these clinics, Amish patients can receive specialized medical attention, including genetic diagnoses and for a few disorders, individualized therapies. An overarching goal of medical genetic studies since McKusick’s early work has been to enable personalized medicine, informed by genetic and genomic understanding. It is fitting that since the Lancaster County OOA were participants in those early studies, they are also among the early patients in the US to receive personalized genomic medical care.

Because of their unique ancestral history, a modest number of founders account for all genetic variation present in the current OOA population. As a consequence, while many rare variants entering the population on only a single founder chromosome have become lost, others have increased in frequency through genetic drift and can be found at appreciable frequencies in the current population. Once a rare variant of interest is found, the genealogy makes it feasible to rapidly identify other likely carriers. The phenomenon of founder alleles of high frequency and strong effect has made the OOA a very rich population for genetic study. Some of the rare variant discoveries made to date have provided important biologic insights underscoring the overall importance to the broader scientific community of studying rare variants in the OOA.

Acknowledgments This research has been supported by NIH grants R01 DK54261, R01 AR46838, U01 HL072515, P30 DK072488, U01 HL084756, R01 HL088119, and R01 CA122844. Additional funding for the Amish sitosterol study was provided by an unrestricted research grant from the Investigator Initiated Studies Program of Merck Sharp & Dohme Corp. The research of A.A.S. on AGDB and PedHunter is supported by the Intramural Research Program of the National Institutes of Health, NLM.

References

Agarwala R, Biesecker LG, Hopkins KA, Francomano CA, Schäffer AA (1998) Software for constructing and verifying pedigrees within large genealogies and an application to the Old Order Amish of Lancaster County. Genome Res 8:211–221

Agarwala R, Biesecker LG, Schäffer AA (2003) Anabaptist genealogy database. Am J Med Genet Semin Med Genet 121C:32–37

Assmann G, Cullen P, Erbey J, Ramey DR, Kannenberg F, Schulte H (2006) Plasma sitosterol elevations are associated with an increased incidence of coronary events in men: results of a nested case-control analysis of the Prospective Cardiovascular Munster (PROCAM) study. Nutr Metab Cardiovasc Dis 16:13–21

Austin MA, Hutter CM, Zimmern RL, Humphries SE (2004) Genetic causes of monogenic heterozygous familial hypercholesterolemia: a HuGE prevalence review. Am J Epidemiol 160:407–420

Beiler K (2009) Descendants and history of Christian Fisher (1757–1838). In: Beiler K, Gordon H (eds) 4th edn. Grand Rapids

Berge KE, Tian H, Graf GA, Yu L, Grishin NV, Schultz J, Kwiterovich P, Shan B, Barnes R, Hobbs HH (2000) Accumulation of dietary cholesterol in sitosterolemia caused by mutations in adjacent ABC transporters. Science 290:1771–1775

Bhattacharyya AK, Connor WE (1974) Beta-sitosterolemia and xanthomatosis. A newly described lipid storage disease in two sisters. J Clin Invest 53:1033–1043

Borén J, Ekström U, Agren B, Nilsson-Ehle P, Innerarity TL (2001) The molecular mechanism for the genetic disorder familial defective apolipoprotein B100. J Biol Chem 276:9214–9218

Cummings AC, Jiang L, Velez-Edwards DR, McCauley JL, Laux R, McFarland LL, Fuzzell D, Knebusch C, Caywood L, Reinhart-Mercer L, Nations L, Gilbert JR, Konidari I, Tramontana M, Cuccaro ML, Scott WK, Pericak-Vance MA, Haines JL (2012) Genome-wide association and linkage study in the amish detects a novel candidate late-onset Alzheimer disease gene. Ann Hum Genet 76:342–351

Daley E, Streeten EA, Sorkin JD, Kuznetsova N, Shapses SA, Carleton SM, Shuldiner AR, Marini JC, Phillips CL, Goldstein SA, Leikin S, McBride DJ (2009) Variable bone fragility associated with an Amish COL1A2 variant and a knock-in mouse model. J Bone Miner Res 25:247–261

Ferketich AK, Katz ML, Kauffman RM, Paskett ED, Lemeshow S, Westman JA, Clinton SK, Bloomfield CD, Wewers ME (2008) Tobacco use among the Amish in Holmes County, Ohio. J Rural Health 24:84–90

Gallagher TE Jr, Beiler K (2002) Church directory of the Lancaster County Amish. Pequea Publishers, Gordonville

Gingerich HF, Kreider RW (2002) Amish and Amish Mennonite genealogies. Pequea Publishers, Gordonville

Glueck CJ, Speirs J, Tracy T, Streicher P, Illig E, Vandegrift J (1991) Relationships of serum plant sterols (phytosterols) and cholesterol in 595 hypercholesterolemic subjects, and familial aggregation of phytosterols, cholesterol, and premature coronary heart disease in hyperphytosterolemic probands and their first-degree relatives. Metabolism 40:842–848

Horenstein RB, Mitchell BD, Post WS, Leutjohann D, von Bergmann K, Ryan KA, Terrin M, Shuldiner AR, Steinle NI (2013) The ABCG8 G574R variant, serum plant sterol levels, and cardiovascular disease risk in the Old Order Amish. Arterioscl Thromb Vasc Biol 33:413–419

Jackson CE, Weiss L, Watson JH (1974) "Brittle" hair with short stature, intellectual impairment and decreased fertility: an autosomal recessive syndrome in an Amish kindred. Pediatrics 54:201–207

Karathanasis SK, Ferris E, Haddad IA (1987) DNA inversion within the apolipoproteins AI/CIII/AIV-encoding gene cluster of certain patients with premature atherosclerosis. Proc Natl Acad Sci USA 84:7198–7202

Karathanasis SK, Norum RA, Zannis VI, Breslow JL (1983) An inherited polymorphism in the human apolipoprotein A-I gene locus related to the development of atherosclerosis. Nature 301:718–720

Kraybill D (2001) The riddle of Amish culture. Johns Hopkins University Press, Baltimore

Kruglyak L, Daly MJ, Reeve-Daly MP, Lander ES (1996) Parametric and nonparametric linkage analysis: a unified multipoint approach. Am J Hum Genet 58:1347–1363

Kwiterovich PO Jr, Bachorik PS, Smith HH, McKusick VA, Connor WE, Teng B, Sniderman AD (1981) Hyperapobetalipoproteinaemia in two families with xanthomas and phytosterolaemia. Lancet 1:466–469

Lander ES, Botstein D (1987) Homozygosity mapping: a way to map human recessive traits with the DNA of inbred children. Science 236:1567–1570

Lee MH, Lu K, Hazard S, Yu H, Shulenin S, Hidaka H, Kojima H, Allikmets R, Sakuma N, Pegoraro R, Srivastava AK, Salen G, Dean M, Patel SB (2001a) Identification of a gene, ABCG5, important in the regulation of dietary cholesterol absorption. Nat Genet 27:79–83

Lee MH, Lu K, Patel SB (2001b) Genetic basis of sitosterolemia. Curr Opin Lipidol 12:141–149

Lee WJ, Pollin TI, O'Connell JR, Agarwala R, Schäffer AA (2010) PedHunter 2.0 and its usage to characterize the founder structure of the Old Order Amish of Lancaster County. BMC Med Genet 11:68

Liu F, Kirichenko A, Axenovich TI, van Duijn CM, Aulchenko YS (2008) An approach for cutting large and complex pedigrees for linkage analysis. Eur J Hum Genet 16:854–860

Liu H, Labeur C, Xu CF, Ferrell R, Lins L, Brasseur R, Rosseneu M, Weiss KM, Humphries SE, Talmud PJ (2000) Characterization of the lipid-binding properties and lipoprotein lipase inhibition of a novel apolipoprotein C-III variant Ala23Thr. J Lipid Res 41:1760–1771

Maeda N, Li H, Lee D, Oliver P, Quarfordt SH, Osada J (1994) Targeted disruption of the apolipoprotein C-III gene in mice results in hypotriglyceridemia and protection from postprandial hypertriglyceridemia. J Biol Chem 269:23610–23616

Marini JC, Forlino A, Cabral WA, Barnes AM, San Antonio JD, Milgrom S, Hyland JC, Körkkö J, Prockop DJ, De Paepe A, Coucke P, Symoens S, Glorieux FH, Roughley PJ, Lund AM, Kuurila-Svahn K, Hartikka H, Cohn DH, Krakow D, Mottes M, Schwarze U, Chen D, Yang K, Kuslich C, Troendle J, Dalgleish R, Byers PH (2007) Consortium for osteogenesis imperfecta mutations in the helical domain of type I collagen: regions rich in lethal mutations align with collagen binding sites for integrins and proteoglycans. Hum Mutat 28:209–221

McArdle PF, Dytch H, O'Connell JR, Shuldiner AR, Mitchell BD, Abney M (2007) Homozygosity by descent mapping of blood pressure in the Old Order Amish: evidence for sex specific genetic architecture. BMC Genet 8:66

McKusick VA (1978) Medical genetic studies of the Amish. Johns Hopkins University Press, Baltimore

Miano MG, Jacobson SG, Carothers A, Hanson I, Teague P, Lovell J, Cideciyan AV, Haider N, Stone EM, Sheffield VC, Wright AF (2000) Pitfalls in homozygosity mapping. Am J Hum Genet 67:1348–1351

Miettinen TA, Gylling H, Strandberg T, Sarna S (1998) Baseline serum cholestanol as predictor of recurrent coronary events in subgroup of Scandinavian simvastatin survival study. Finnish 4S investigators. BMJ 316:1127–1130

Mitchell BD, McArdle PF, Shen H, Rampersaud E, Pollin TI, Bielak LF, Jaquish C, Douglas JA, Roy-Gagnon MH, Sack P, Naglieri R, Hines S, Horenstein RB, Chang YP, Post W, Ryan KA, Brereton NH, Pakyz RE, Sorkin J, Damcott CM, O'Connell JR, Mangano C, Corretti M, Vogel R, Herzog W, Weir MR, Peyser PA, Shuldiner AR (2008) The genetic response to short-term interventions affecting cardiovascular function: rationale and design of the Heredity and Phenotype Intervention (HAPI) Heart Study. Am Heart J 155:823–828

Norum RA, Lakier JB, Goldstein S, Angel A, Goldberg RB, Block WD, Noffze DK, Dolphin PJ, Edelglass J, Bogorad DD, Alaupovic P (1982) Familial deficiency of apolipoproteins A-I and C-III and precocious coronary-artery disease. N Engl J Med 306:1513–1519

Ordovas JM, Cassidy DK, Civeira F, Bisgaier CL, Schaefer EJ (1989) Familial apolipoprotein A-I, C-III, and A-IV deficiency and premature atherosclerosis due to deletion of a gene complex on chromosome 11. J Biol Chem 264:16339–16342

Pollin TI, Damcott CM, Shen H, Ott SH, Shelton J, Horenstein RB, Post W, McLenithan JC, Bielak LF, Peyser PA, Mitchell BD, Miller M, O'Connell JR, Shuldiner AR (2008) A null mutation in human APOC3 confers a favorable plasma lipid profile and apparent cardioprotection. Science 322:1702–1705

Rosenberg MJ, Agarwala R, Bouffard G, Davis J, Fiermonte G, Hilliard MS, Koch T, Kalikin LM, Makalowska I, Morton DH, Petty EM, Weber JL, Palmieri F, Kelley RI, Schaffer AA, Biesecker LG (2002) Mutant deoxynucleotide carrier is associated with congenital microcephaly. Nat Genet 32:175–179

Shen H, Damcott CM, Rampersaud E, Pollin TI, Horenstein RB, McArdle PF, Peyser PA, Bielak LF, Post WS, Chang YP, Ryan KA, Miller M, Rumberger JA, Sheedy PF 2nd, Shelton J, O'Connell JR, Shuldiner AR, Mitchell BD (2010) Familial defective apolipoprotein B-100 and increased low-density lipoprotein cholesterol and coronary artery calcification in the old order amish. Arch Intern Med 170:1850–1855

Silbernagel G, Fauler G, Renner W, Landl EM, Hoffmann MM, Winkelmann BR, Boehm BO, Marz W (2009) The relationships of cholesterol metabolism and plasma plant sterols with the severity of coronary artery disease. J Lipid Res 50:334–341

Soria LF, Ludwig EH, Clarke HR, Vega GL, Grundy SM, McCarthy BJ (1989) Association between a specific apolipoprotein B mutation and familial defective apolipoprotein B-100. Proc Natl Acad Sci USA 86:587–591

Strauss KA, Puffenberger EG (2009) Genetics, medicine, and the Plain people. Annu Rev Genomics Hum Genet 10:513–536

Taban M, Memoracion-Peralta DS, Wang H, Al-Gazali LI, Traboulsi EI (2007) Cohen syndrome: report of nine cases and review of the literature, with emphasis on ophthalmic features. J AAPOS 11:431–437

Tolea M (2007) Patterns of hospital utilization in the Old Order Amish. PhD dissertation, University of Maryland, Baltimore

van Dijk FS, Cobben JM, Kariminejad A, Maugeri A, Nikkels PG, van Rijn RR, Pals G (2011) Osteogenesis

imperfecta: a review with clinical examples. Mol Syndromol 2:1–20
Van Hout CV, Levin AM, Rampersaud E, Shen H, O'Connell JR, Mitchell BD, Shuldiner AR, Douglas JA (2009) Extent and distribution of linkage disequilibrium in the Old Order Amish. Genet Epidemiol 34:146–150
Visser ME, Witztum JL, Stroes ES, Kastelein JJ (2012) Antisense oligonucleotides for the treatment of dyslipidaemia. Eur Heart J 33:1451–1458
von Eckardstein A, Holz H, Sandkamp M, Weng W, Funke H, Assmann G (1991) Apolipoprotein C-III (Lys58—Glu). Identification of an apolipoprotein C-III variant in a family with hyperalphalipoproteinemia. J Clin Invest 87:1724–1731
Wang H, Nye L, Puffenberger E, Morton H (2007) Phenylalanine hydroxylase deficiency exhibits mutation heterogeneity in two large old order Amish settlements. Am J Med Genet A 143A:1938–1940
Welty FK, Hubl ST, Pierotti VR, Young SG (1991) A truncated species of apolipoprotein B (B67) in a kindred with familial hypobetalipoproteinemia. J Clin Invest 87:1748–1754
Whittemore AS, Halpern J (1994) A class of tests for linkage using affected pedigree members. Biometrics 50:118–127
Wilund KR, Yu L, Xu F, Vega GL, Grundy SM, Cohen JC, Hobbs HH (2004) No association between plasma levels of plant sterols and atherosclerosis in mice and men. Arterioscler Thromb Vasc Biol 24:2326–2332
Yerges-Armstrong LM, Shen H, Streeten EA, Shuldiner AR, Mitchell BD (2013) Decreased bone mineral density in subjects carrying familial defective Apolipoprotein B-100. J Clin Endocrinol Metab 12:E1999–E2005

Genetics of Cardiovascular Disease in Minority Populations

11

Jean W. MacCluer, John Blangero, Anthony G. Comuzzie, Sven O.E. Ebbesson, Barbara V. Howard, and Shelley A. Cole

11.1 Introduction

Cardiovascular disease (CVD) is the leading cause of mortality in the United States and an important contributor to morbidity worldwide. Both genetic and environmental factors are generally recognized to influence susceptibility to CVD. Much research has been directed at identification of the genes involved in susceptibility and their interactions with environment. However, most of the genetic research on CVD and its risk factors has been done in populations of Northern European ancestry. In the United States and elsewhere, less attention has been paid to identifying genetic contributors to CVD in minority populations. Here we describe large-scale family studies of Mexican Americans, American Indians, and Alaskan Eskimos being conducted by geneticists at the Texas Biomedical Research Institute (Texas Biomed) in collaboration with investigators throughout the country. The San Antonio Family Heart Study (SAFHS) is examining the risk of CVD, diabetes, and obesity in >1,400 members of 42 large Mexican American families in San Antonio; the Strong Heart Family Study (SHFS) involves >3,800 individuals in 13 American Indian communities in Arizona, Oklahoma, and the Dakotas, members of 94 families; and the newest of the three studies, Genetics of Coronary Artery Disease in Alaska Natives (GOCADAN), includes >1,200 Alaskan Eskimos in the Norton Sound region of Alaska, who are primarily members of a single extended family. All three studies were approved by the institutional review boards of all collaborating institutions. In addition, the Strong Heart Family Study was approved by the Indian Health Service Institutional Review Board and by the 13 participating American Indian communities, and GOCADAN was approved by the Science Advisory Board of the Norton Sound Health Corporation. All participants in the three studies gave informed consent.

J.W. MacCluer (✉) · A.G. Comuzzie · S.A. Cole
Department of Genetics, Texas Biomedical Research Institute, 7620 NW Loop 410, PO Box 760549, San Antonio, TX 78227, USA
e-mail: jean@txbiomedgenetics.org

A.G. Comuzzie
e-mail: tony@txbiomedgenetics.org

S.A. Cole
e-mail: scole@txbiomedgenetics.org

S.O.E. Ebbesson
Norton Sound Health Corporation, PO Box 966, Nome, AK 99762, USA
e-mail: ffsoe@uaf.edu

B.V. Howard
MedStar Research Institute, 100 Irving Street NW, Washington, DC 20010, USA
e-mail: barbara.v.howard@medstar.net

J. Blangero
South Texas Diabetes and Obesity Institute, Regional Academic Health Center, University of Texas Health Science Center, 2102 Treasure Hills Blvd Harlingen, San Antonio, TX 78550, USA
e-mail: blangero@uthscsa.edu

R. Duggirala et al. (eds.), *Genome Mapping and Genomics in Human and Non-Human Primates*, Genome Mapping and Genomics in Animals 5, DOI 10.1007/978-3-662-46306-2_11,

11.2 Study Design and Collection of Family Data

The studies of Mexican Americans, American Indians, and Alaskan Eskimos all involve extended families that were recruited without regard to disease status. This strategy was possible because CVD and related disorders (diabetes and obesity) are common in these populations, thus assuring that the families recruited would exhibit substantial variability in CVD risk factors and would provide ample variation for genetic analysis. However, the recruitment strategy differed between populations because of their different environments, ranging from an urban setting (San Antonio Mexican Americans) to small towns, rural areas, and reservations (American Indians), to small, isolated villages (Alaskan Eskimos).

11.2.1 The San Antonio Family Heart Study

The San Antonio Family Heart Study (SAFHS) began in 1991 as a collaborative effort of geneticists and epidemiologists at Texas Biomed and at the University of Texas Health Science Center at San Antonio (UTHSCSA). We selected a predominantly Mexican American census tract on the west side of San Antonio and used the Polk Directory (Polk 1989) to obtain a complete listing of all addresses in the census tract. We created a computer file containing, for each address, the street name, street number, apartment number if applicable, name of head of household, and telephone number if available. We then randomized the list of addresses so that households could be contacted by field staff in random order. Recruitment letters were mailed to the heads of household and any other adult residents identified during the verification process. A few days later, a field worker went to the selected address and ascertained whether an eligible proband resided there. Each previously verified household on the randomly ordered address list was approached and enumerated. Enumeration consisted of a listing of all persons living in the household with their complete names, dates of birth, age, sex, and relationship to other household members.

Once an eligible proband agreed to participate in the study, the interviewer administered a home interview, which included pedigree information on all first, second, and third degree relatives known to the proband. Full names, addresses, and phone numbers of these relatives were recorded to facilitate future contact with them. After completion of the home interview, the proband was scheduled for a clinic examination and the interviewer began contacting other relatives in the pedigree. These other family members were administered a home (or in some cases, a telephone) interview and scheduled for a clinic examination. A graphic representation of each pedigree was reviewed with the proband at the clinic examination and revised if needed. This information also was reviewed with each family member subsequently contacted and was revised and/or expanded as needed. This procedure assured that all pedigree information was cross-checked and verified by multiple family members.

We recruited more than 1,400 members of large Mexican American families living in San Antonio, ascertained through a randomly chosen 40- to 60-year-old resident of the selected census tract. Probands were Mexican Americans with a spouse and at least six offspring and/or siblings who were at least 16 years old and living in San Antonio. Our pedigrees included the proband, his or her spouse, and all first, second, and third degree relatives of both, as well as spouses who had married into the family. Participants are members of 42 families, ranging in size from 3 to 102 examined individuals.

In a recall of more than 850 family members beginning in 1997, we repeated many of the clinic measures and also did ultrasound measurement of carotid artery wall thickness. We again obtained information on lifestyle variables. In a second recall begun in 2002, we examined 950 of the family members. Thus we obtained longitudinal data spanning a period of approximately 10 years.

11.2.2 The Strong Heart Family Study

The Strong Heart Study (SHS) was begun in 1989 as a traditional epidemiologic study, with the goal of estimating CVD mortality and morbidity rates and the prevalence of known or suspected CVD risk factors in American Indians. Approximately 1,500 American Indians aged 45–74 were recruited at each of the three field centers in Arizona (Pima/Maricopa and Tohono O'odham), North and South Dakota (Lakota and Oglala Sioux), and Oklahoma (Kiowa, Apache, Caddo, Comanche, Delaware, Fort Sill Apache, and Wichita). The participating communities are representative of the diverse environments in which nonurban American Indians live: The tribes in Arizona and the Dakotas live on reservations, and those in Oklahoma live among the general population. Lists of eligible persons were obtained from tribal rolls and verified by local residents. In Arizona and Oklahoma, every eligible person was invited personally or by letter to participate. In the Dakotas, a cluster sampling technique was used to obtain the 1,500 subjects needed. Participants were recruited from 13 communities: three in Arizona, three in the Dakotas, and seven in Oklahoma, where the majority of participants were in Kiowa. As in SAFHS, participants were recruited at random, without regard to disease status.

In 1996, we began a genetic study of the families of Strong Heart participants. When participants were recruited into the Strong Heart Study beginning in 1989, each filled out a family history form listing the names of their parents, offspring, and full and half siblings. The information on these family history forms provided the basis for our recruitment of families in the Strong Heart Family Study (SHFS). We matched names across the computerized records—for example, identifying a person who is an offspring in one record and a parent in another. In this way, we constructed large families, with each person named.

Because there were so many large families, we restricted our recruitment to those that had (a) a total of at least 5 sibs, of whom ≥3 were members of the original Strong Heart cohort, and (b) ≥12 age-eligible offspring of the cohort members. In two phases, we recruited ≥1,200 family members from each field center: approximately, 300 from each center beginning in 1995, and another ≥900 from each center beginning in 2000. Recruited family members were at least 15 years old. Across the three field centers, we recruited 94 extended families, ranging in size from 5 to 110 examined individuals. A reexamination of all of the 3,812 family members began in 2005. The information collected for each participant included birth date, sex, and tribal affiliation, as well as the lifestyle variables and CVD risk factors as described below.

11.2.3 GOCADAN

GOCADAN was initiated in the year 2000 by a group of Strong Heart Study investigators in collaboration with Dr. Sven Ebbesson of the University of Alaska—Fairbanks, to study the determinants of CVD, diabetes, and obesity in a group of primarily Inupiat Eskimos in the Norton Sound region of Alaska. Eskimos in this region live in isolated villages of 200–600 inhabitants and in the population center of Nome, with approximately 3,000 inhabitants. Each village includes residents who have relatives in other villages and/or in Nome. Because of the small size and isolation of the villages and the interrelatedness among villages, the approach used to identify and recruit participants differed from that used in the Mexican American and American Indian studies. Each village was visited by two investigators, who first described the study to the village leaders and at a community meeting. With the approval of the leaders and the community, every household in each village was visited. Every member of each household was enumerated and the names and addresses of parents, siblings, and offspring of each household member were recorded and later computerized. This approach was not used in Nome, where the only individuals recruited were relatives of participants who lived in the villages. Matching of names and birthdates on the records from different households enabled us to link household members from all of the villages into larger family units; most household members were ultimately linked into a single large pedigree. Between 2000 and 2004, we recruited a total of 1,214 participants at least 18 years of age from eight villages and Nome. Age eligibility was determined from U.S. census records. With one exception, at least 73 % of age-eligible residents were recruited in each village. The exception was a village in which recruitment was attempted during the summer hunting and fishing period when most residents were not available.

For all interested family members who were recruited into GOCADAN, informed consent was obtained, a questionnaire was administered, a physical exam was done in a clinic located in the village, and blood was drawn for assessment of CVD and associated risk factors and extraction of DNA for genetic studies (Ebbesson et al. 2006; Howard et al. 2005). A second round of examinations of each family member was completed in 2010.

11.3 Methods

11.3.1 Collection of Demographic, Lifestyle, and Phenotypic Data

The studies differed in the types of clinic sites used for participant examinations. In the SAFHS, in which participants were examined three times over a period of 10 years, clinic exams were done in a mobile clinic situated on the campus of UTHSCSA; in a neighborhood clinic near the census tract where most participants lived; and at the General Clinical Research Center (GCRC) in the Audie Murphy VA Hospital near UTHSCSA. Transportation to clinic exams was provided for participants who needed it. In the SHFS, temporary clinics (sometimes a mobile clinic) were set up at each field site; three in Arizona, two in Oklahoma, and two in the Dakotas. The GOCADAN study required that a temporary clinic be set up in each village for the period during which clinical exams were being done in the village. Ingenuity was required in identifying clinic sites in villages that may have had no more than 200 residents.

In each study a monetary incentive was provided. In all three studies, each family member was invited to the clinic, where he or she was given a physical examination and a questionnaire was administered. Information obtained in all of the studies is shown in Table 11.1. Informed consent was obtained either before the clinic visit or at its beginning. From the personal interview, information was obtained on demographic characteristics, pedigree relationships, acculturation and socioeconomic status, medical history, reproductive history, and environmental risk factors: diet, smoking, alcohol use, physical

Table 11.1 Information gathered in all three studies

Verification and consent	Lipids and lipoproteins
Informed consent	Total cholesterol
Demographic information	HDL-C
Family information	LDL-C
Interview and medical history	Lipoprotein size classes
Chronic illnesses	Triglyceride
Rose questionnaire	Apo-AI
Medications	Apo-B
Reproductive history (women)	Apo-E
Pregnancy history	Lp(a)
Menstrual history	Glucose and hormones
Hormone medications	Fasting glucose
Lifestyle and behaviors	2-h glucose
Diet food frequency questionnaire	Fasting insulin
Alcohol consumption	2-h insulin
Vitamin usage	Adiposity measures
Smoking habits	Weight
Dietary instrument	Height
Physical activity instrument	Body mass index
Inflammatory markers	Waist/hip ratio
Fibrinogen	Fat mass (bioimpedance)
Cardiac measures	Fat free mass
EKG	Blood pressure
Carotid intima media thickness (near and far walls)	Systolic blood pressure
Carotid lumen diameters (CCA and ICA)	Diastolic blood pressure

activity, and medication use. For all three studies, we assessed diet using food frequency questionnaires (Block 1998; Nobmann et al. 2005) specifically validated for the populations being studied. From the dietary surveys, we derived measures of total calories, saturated fat, cholesterol, the polyunsaturated/saturated fat ratio, and several other macro- and micronutrients, including alcohol.

A blood sample was collected under fasting conditions for measurement of glucose, insulin, lipids, and lipoproteins. A 2-h glucose tolerance test was administered. Total cholesterol, high and low density lipoprotein cholesterol (HDL-C and LDL-C), triglyceride (TG) levels, and apolipoproteins also were measured. Anthropometric measurements included height, waist and hip circumferences, and weight. Fat mass was determined by bioimpedance. Sitting blood pressure was measured three times following a 5-min rest, and the mean of the second and third measurements was used for analysis. Cardiac phenotypes were obtained from an electrocardiogram and by carotid ultrasound. Some questionnaire items were included in early phases but not later ones (e.g., family relationships and birth dates), and some phenotypes were not measured in every phase. Protocols for the SAFHS are described in Mitchell et al. (1996) and MacCluer et al. (1999); for the SHFS, in Lee et al. (1990), Howard et al. (1995), and North et al. (2003a); and for GOCADAN, in Howard et al. (2005).

Because of unique characteristics of each study population, special interests of the investigators involved, or special financial resources available, each study also has collected unique information. Investigators in the SAFHS were particularly interested in measures of oxidative stress and inflammation (Diego et al. 2007). They also had private funding that enabled them to generate microarray expression profiles (expression levels of RNA transcripts derived from lymphocytes) for 1,240 SAFHS participants, yielding data for more than 20,000 phenotypes for each individual (Göring et al. 2007). Association analyses have identified multiple transcripts that are significantly correlated with CVD risk factors (Charlesworth et al. 2010). For the SHFS, we recorded tribal enrollment and self-reported degree of Indian heritage. Both SHFS and GOCADAN measured HbA1c, and used pedometers to measure physical activity. The SHFS has cardiac and popliteal ultrasound measures. Both GOCADAN and SHFS have inflammatory markers and GOCADAN has measures of pathogen burden. Fatty acid measures are particularly important in GOCADAN (Ebbesson et al. 2008; Nobmann et al. 2005) because Alaskan Eskimos are changing from a traditional diet emphasizing fish and sea mammals to a Western diet with greater emphasis on processed foods (Nobmann et al. 1998).

For all three studies, management of pedigree, demographic, phenotypic, and genotypic data is accomplished using the computer package PedSys (Dyke 1992).

11.3.2 Genotyping

For all three studies, all family members were initially typed for approximately 400 microsatellite markers distributed across the autosomes, on average 10 centimorgans apart. These genotypic data have been used in linkage analyses as described below, to localize QTLs that influence CVD risk factors.

Fasting blood samples were collected at the clinics established for each study. Buffy coats were stored at the clinics at −80 °C and transported on dry ice to Texas Biomed for DNA isolation. To genotype SHFS and GOCADAN participants, we used the ABI PRISM Linkage Mapping Set-MD10 (version 2.5; Applied Biosystems, Foster City, CA), which consists of fluorescently-labeled PCR primer pairs that amplify dinucleotide repeat microsatellite loci (short tandem repeats) selected from the Genethon human linkage map (Dib et al. 1996). Genotyping for SAFHS used primer pairs from the MapPairs 6 and 8 Linkage Screening Set (Research Genetics Inc., Huntsville, AL). PCR amplification of DNA from study participants was performed in Applied Biosystems 9,700 thermocyclers. The products of separate PCRs for each individual were pooled, and a labeled size standard was added to each pool. The pooled PCR products were loaded into an ABI PRISM 377 or 3,100 Genetic Analyzer for laser-based automated detection and quantification, and genotypes were scored using the Genotyper software package (Applied Biosystems).

With the development of faster and less expensive methods, we moved from microsatellites to single nucleotide polymorphism (SNP) genotyping. In work aimed at identifying functional polymorphisms, single nucleotide polymorphisms (SNPs) are being typed in the regions of linkage signals. With the identification of hundreds of thousands of SNPs throughout the genome, there has been interest in performing genome-wide association studies, searching for associations between disease risk factors and specific chromosomal regions. The National Institutes of Health accepts applications to fund the generation of genome-wide SNP data, but only for projects for which the data are shared with other investigators. The American Indian communities in the SHFS and the Alaskan Eskimo communities in GOCADAN are not willing to have their genetic information broadly shared with investigators outside these two studies. Therefore, genome-wide SNP typing is not being done in SHFS or GOCADAN, although dense SNP typing is being done in regions of interest. For the SAFHS, we have used primarily private funding to generate 1 million SNPs for all family members. These data are being used for both linkage and association analyses. Whole genome sequence data also have been generated for >700 SAFHS participants.

11.3.3 Statistical Genetic Analysis

11.3.3.1 Data Cleaning

The first step in statistical genetic analysis of data for SAFHS, SHFS, and GOCADAN is the elimination of Mendelian and pedigree errors. We finalize on the pedigrees that are most supported by the genotype data, as inferred from PREST (Sun et al. 2002) statistics. Using Simwalk II (Sobel et al. 2002), we eliminated mistyping errors at blanking rates of 1.37 (SAFHS), 0.93 (SHFS), and 0.58 % (GOCADAN) of the total number of genotypes. Using Loki (Heath 1997), we computed matrices of empirical estimates of identity-by-descent allele sharing, required for our linkage analyses, at points throughout the genome for every relative pair. PREST, Simwalk II, and Loki require chromosomal maps of genotyped markers for all 22 autosomes. Across populations, we use the same sex-averaged chromosomal maps, provided by deCODE Genetics (Kong et al. 2002).

11.3.3.2 Quantitative Genetic Analysis

We obtain the estimates of heritability of CVD risk factors using maximum likelihood variance decomposition methods (e.g., Amos 1994) implemented in SOLAR (Blangero and Almasy 1997; Almasy and Blangero 1998). Among the variance terms that can be included in these analyses are the additive genetic variance, the variance due to shared household effects, random environmental variance, and additional (or alternative) components (shared spouse or sibling environments, dominance genetic effects, mitochondrial effects). Our initial heritability analyses often are done in batch mode across many phenotypes and include only a basic set of covariates: age, sex, and their higher order terms and interactions. Subsequent analyses of individual phenotypes include additional covariates that are specific for the phenotype being analyzed. These analyses enable us to estimate the relative importance of genetic, shared environmental, and random environmental effects on CVD risk factors. Phenotypes with significant heritabilities are targeted for subsequent linkage and association analyses using genotypic data, as described below.

Variance decomposition techniques, using maximum likelihood methods and implemented in SOLAR, also can be used to estimate the genetic and environmental correlations between pairs of traits (Hopper and Mathews 1982; Lange and Boehnke 1983). The genetic correlation represents the effect of shared genes (pleiotropy), and the environmental correlation, the effect of shared residual (unmeasured) environment on the phenotypic variance of two traits. Additive genetic correlations that are significantly different from zero provide evidence consistent with pleiotropy. If significant shared genetic effects are found for multiple traits, hypotheses can be generated concerning the genetic regulation of complex phenotypes.

11.3.3.3 Linkage Analysis

For phenotypes that have significant heritabilities, we perform multipoint variance component linkage analysis using SOLAR (Almasy and Blangero 1998) to determine the chromosomal locations of QTLs that influence the traits. The first analyses used approximately 400 microsatellite markers distributed across the autosomes at approximately 10 centimorgan intervals. For the SHFS and GOCADAN, we began with a screening linkage analysis of all available phenotypes, adjusted only for age, sex, and their higher order terms and interactions. As in our heritability analyses, subsequent linkage analyses of phenotypes that yield significant linkage signals include additional relevant covariates. In the SHFS, linkage analyses were conducted separately for each of the three centers and also by combining data for the three centers into a single analysis. Combined analyses included center covariates to allow for different mean trait levels in the three centers.

11.3.4 Fine Mapping and Identification of Disease-Related Variants

The initial localization of a QTL by linkage analysis tends to encompass a large genomic region, typically 10–20 centimorgans. This interval can be reduced by using methods that are based upon linkage disequilibrium, but the genomic region may still be 100 kb or greater, and the identification of functional variants can remain difficult. Because the functional polymorphism likely will not segregate in every family, we attempt to identify these polymorphisms by testing for disequilibrium in the same families in which the linkage was originally discovered. These families also are valuable for testing whether a specific set of SNPs can completely account for the linkage signal.

Using resequencing technologies, our strategy is to identify all polymorphisms within a positional candidate region by resequencing large numbers of individuals from the sample in which the linkage signal was found. Prior evidence for particular candidate genes in a linkage region may be used to prioritize the sequencing/polymorphism discovery effort. After the polymorphisms are enumerated, they are typed in the original linkage dataset using high-throughput SNP typing methods. Then they must be prioritized for molecular functional characterization. Because its large variance makes linkage disequilibrium relatively unpredictable (Abecasis et al. 2001; Terwilliger 2001), standard association methods (which exploit linkage disequilibrium) are not optimal for identifying functional polymorphic variants. We therefore use a method that effectively eliminates the correlation between a marker and a QTL that is due to linkage disequilibrium. Unfortunately, there are relatively few statistical approaches to find the main functional effects in high-dimensional SNP data (Bader 2001; Nelson et al. 2001). We are employing a Bayesian method for statistically assessing the potential functionality of observed allelic variants (Blangero et al. 2005). By using this statistical approach, we minimize the daunting task of searching through a chromosomal region using molecular techniques.

11.4 Results

Genetic studies of extended families require many years, from initial planning, to identifying and recruiting participating families, generating phenotypic data, carrying out marker genotyping, performing genetic analyses, and following up on promising leads. This is especially true in studies that are conducted in remote locations or in which identification of family members is difficult. Thus, the first quantitative genetic analyses from GOCADAN were presented at a scientific meeting in 2005, and the first publication of genetic results was in 2006. Presentation of significant linkage results began in 2006. In the SHFS, publication of quantitative genetic analyses began in 2002, and the first significant linkages were published in 2006. Although the SAFHS began much earlier (1991) and the first quantitative genetic analyses were published in 1993, the initial focus of linkage analyses was on candidate genes. It was not until 1996 that genotyping of microsatellites began, and the first QTL (for leptin on chromosome 2) was reported on by Comuzzie et al. in 1997.

The power of extended families for genetic analysis derives from the large number of relative pairs that they contain. Table 11.2 indicates the numbers of relative pairs in the SAFHS, the three field centers in the SHFS, and GOCADAN. Although the number of sib pairs is just 721–1,577, the total number of relative pairs ranges from 16,950 to 28,586. Below we summarize results of quantitative genetic analyses and linkage analyses, focusing on published papers, and with an emphasis on SAFHS, the oldest of the three studies.

11.4.1 Quantitative Genetic Analysis of CVD Risk Factors

An extensive set of phenotypes has been collected on participants in each study, from basic anthropometrics to measures related to obesity, diabetes, lipids, hormones, clotting, inflammation, oxidative stress, and the carotid arteries. As a first step in determining the chromosomal locations and identities of the genes that influence these traits, we have estimated heritabilities, i.e., the proportion of phenotypic variance that is attributable to the additive effects of genes. Correction for age, sex, and their higher order terms and interactions is done routinely, and further covariate effects often are included. The heritabilities reported here are residual heritabilities, after the effects of covariates are taken into account.

Table 11.3 lists the heritabilities for selected phenotypes in the SAFHS. The great majority of these are highly significant, indicating that a search for specific functional genes is warranted. For some phenotypes, genotype by environment interaction and/or genetic correlations with other phenotypes were detected. For example, Czerwinski et al. (2004) reported that the same gene or suite of genes affects phenotypic variation in triglycerides, LDL mean particle diameter, and to some extent, HDL-C level, and that their effects are different in smokers and nonsmokers. Bivariate analyses (Warren et al. 2005) suggest possible pleiotropic effects of genes influencing type 2 diabetes and several hemostasis-related traits. Kent et al. (2004) reported that ICAM-1 level was significantly genetically correlated with phenotypes related to obesity and to glucose homeostasis. Comuzzie et al. (2007) found significant evidence of pleiotropy between plasma levels of adiponectin and established risk factors for the metabolic syndrome and type 2 diabetes. Voruganti et al. (2008) found that a common set of genes regulating insulin resistance also regulates BMI, waist circumference, HDL cholesterol, and pulse pressure. Likewise, Voruganti et al. (2009a) found pleiotropy of genes influencing serum uric acid with waist circumference and total body fat.

Heritabilities for selected phenotypes in the SHFS are given in Table 11.4. As in SAFHS, the majority of these are highly significant. Genotype by environment interactions and genetic correlations also have been reported. For example, North and colleagues found significant genetic correlations between diabetes status and eight CVD risk factors (North et al. 2003c) and also between diabetics and nondiabetics for several obesity and lipid phenotypes (North et al. 2003b). Mosher et al. (2008) detected sex-specific genotype by diet effects on HDL-C. Mottl et al. (2009) reported genotype by diabetes and genotype by hypertension interaction for urinary albumin creatinine ratio (UACR). Franceschini et al. (2009) detected evidence for differences in genetic effects on blood pressure as a function of smoking status, alcohol intake, physical activity, and education. Melton et al. (2010) found bivariate association of a single locus on chromosome 9p21 with heart rate as measured from both echocardiogram and echocardiograph

Table 11.2 Numbers of examined relative pairs among participants in SAFHS, SHFS, and GOCADAN

Degree of relationship	Coefficient of relationship	Relationship	Number of relative pairs				
			SAFHS	SHFS-AZ	SHFS-DK	SHFS-OK	GOCADAN
First	1/2	Sibs	1,577	1,446	1,525	1,522	721
		Parent–offspring	1,333	1,320	1,256	1,205	609
Second	1/4	Avuncular	2,984	3,197	3,528	3,220	1,597
		Grandparent-grandchild	497	825	571	533	224
		Half-sibs	201	416	421	415	254
		Double first cousins	24	25	67	11	10
Third	1/8	First cousins	3,369	3,609	4,192	3,302	2,548
		Grand avuncular	798	1,711	2,058	1,609	693
		Half-avuncular	436	640	579	759	654
		Great grandparent-grandchild	34	66	22	50	10
		Other third degree	73	55	276	32	122
Fourth	1/16	First cousins once removed	3,190	4,871	6,149	4,274	4,353
		Half-first cousins	496	608	473	587	740
		Great grand avuncular	36	209	99	172	84
		Other fourth degree	91	281	473	330	394
Fifth	1/32	Second cousins	1,080	2,512	3,779	1,932	3,118
		First cousins, twice removed	359	558	509	797	1,049
		Other fifth degree	36	907	381	829	1,042
Sixth	1/64		336	1,474	1,292	1,296	3,143
Other				289	936	613	2,131
Total			16,950	25,019	28,586	23,488	23,496
Number of Participants			1,458	1,295	1,242	1,230	1,214

Doppler recordings. Wilmot et al. (2012), in analyses of serum sodium concentration in American Indians and other ethnic groups, demonstrated sex- and ethnicity-specific effects.

Table 11.5 lists published heritabilities for phenotypes in GOCADAN. The heritabilities for all but insulin and small HDL are significant at $p < 0.0001$. Pleiotropic and epistatic effects of multiple QTLs on multiple risk factors have been found. For example, SNPs in three chromosomal regions appear to have functional effects on saturated, monounsaturated, and polyunsaturated plasma fatty acids (Voruganti et al. 2010). Likewise, significant genetic correlations were found between size classes of HDL and a variety of CVD risk factors (Tejero et al. 2010).

It should be noted that comparisons of heritabilities between populations are not very meaningful: heritability measures the proportion of phenotypic variance attributed to the additive

Table 11.3 Heritabilities of CVD risk factors: San Antonio Family Heart Study

Trait	h^2	Covariates[a]	References
Lipid-related phenotypes			
HDL_{2a}-unesterified cholesterol	0.62 ± 0.09	Diabetes status, menopausal status, smoking, alcohol consumption, diabetes meds, lipid-altering meds, exogenous sex hormones	Almasy et al. (1999)
HDL_{2a}-unesterified cholesterol	0.62 ± 0.09	Diabetes status, menopausal status, smoking, alcohol consumption, diabetes meds, lipid-altering meds, exogenous sex hormones	Almasy et al. (1999)
HDL-C	0.54 ± 0.09		Arya et al. (2002)
HDL-C	0.42 ± 0.10	Plasma ApoA-I, TG, exogenous sex hormones, menopausal status	Mahaney et al. (2003)
β-lipoprotein phenotypes	0.41 ± 0.07	Diabetes status, diabetic meds, menopausal status, contraception, alcohol consumption, smoking	Rainwater et al. (2004)
Lipoprotein size phenotypes	0.30–0.45		Almasy et al. (2005)
Adiposity-Related Phenotypes			
Fat mass (bioimpedance)	0.63		Comuzzie et al. (1997)
Serum leptin level	0.71		Comuzzie et al. (1997)
Serum leptin level	0.50 ± 0.10	Diabetes status, testosterone	Martin et al. (2002)
BMI	0.54		Mitchell et al. (1999)
Abdominal skinfold average	0.38 ± 0.07		Cai et al. (2004b)
Acylation-stimulating protein	0.26 ± 0.10		Martin et al. (2004)
Diabetes-related phenotypes			
Serum insulin concentration	0.53 ± 0.09		Mitchell et al. (2000)
Insulin response to glucose index	0.13 ± 0.08		Cai et al. (2004b)
Metabolic syndrome risk			
Corrected insulin response	0.30 ± 0.07		Cai et al. (2004b)
(Lipid factor)	0.64 ± 0.09		Cai et al. (2004c)
(Adiposity factor)	0.49 ± 0.09		Cai et al. (2004c)
(Insulin–glucose factor)	0.42 ± 0.09		Cai et al. (2004c)
(BP factor)	0.26 ± 0.08		Cai et al. (2004c)
HOMA-IR	0.33	Waist circumference	Voruganti et al. (2008)
Blood pressure			
Systolic BP	0.18		Mitchell et al. (1996)
Diastolic BP	0.28		Mitchell et al. (1996)
Pulse pressure	0.21	BMI	Atwood et al. (2001b)

(continued)

Table 11.3 (continued)

Trait	h^2	Covariates[a]	References
Inflammation/oxidative stress			
Intercellular adhesion molecule-1	0.50 ± 0.06	Waist, smoking, diabetes	Kent et al. (2007)
Paraoxonase 1	0.77(f), 0.95 (m)[b]		Winnier et al. (2007)
C-reactive protein	0.17		Voruganti et al. (2008)
Hemostasis-related phenotypes	0.20–0.60		Warren et al. (2005)
Blood coagulation			
Plasminogen	0.43 ± 0.08	Smoking, contraception, alcohol, menopausal status, diabetes, diabetes Meds, lipid-altering meds, BMI, TC, HDL-C	Santamaria et al. (2007)
D-dimer	0.23 ± 0.07	Contraception, menopausal status	Diego et al. (2010)
Kidney function			
Urine albumin/creatinine ratio	0.24 ± 0.10	BMI, triglycerides, systolic BP	Arar et al. (2007)
Serum uric acid	$0.42 \pm 2 \times 10^{-7}$	BMI, waist, SBP, pulse pressure	Nath et al. (2007)
Serum uric acid	0.39	Waist/hip ratio, SBP, triglycerides, HDL-C, serum creatinine, BP meds, alcohol, smoking, diabetes status	Voruganti et al. (2009a, b)
Lifestyle			
Macronutrient intakes	0.09–0.21	(Household effects)	Cai et al. (2004a)
Cigarette and alcohol consumption (bivariate)	0.52, 0.39	Education	Viel et al. (2008)

[a] In addition to age, sex, and their higher order terms and interactions
[b] *f* females, *m* males [] included as variance component

effects of genes *within a population*, and the sources, nature, and magnitude of phenotypic variance differ between populations.

11.4.2 Mapping of Quantitative Trait Loci

11.4.2.1 San Antonio Family Heart Study

We have performed linkage screens of all quantitative phenotypes collected during the clinic visits in the SAFHS to identify loci contributing to CVD risk factors. Our application of a systematic, semi-automated approach to linkage analyses of 343 phenotypes yielded significant evidence (LOD ≥ 3.0) for 40 QTLs, far exceeding the number that would be expected by chance. Therefore, we have confidence that a substantial proportion of the significant QTLs detected in the SAFHS are reflective of genes with true, biologically important effects on our focal phenotypes. For many of these linkage signals, more extensive analyses have been done including additional covariates (see Table 11.6).

11.4.2.2 Strong Heart Family Study

We have performed a linkage screen of 125 CVD-related phenotypes in the SHFS. In analyses incorporating all three field centers, we obtained significant evidence of linkage (LOD ≥ 3.0) for 16 QTLs, again exceeding the number expected by chance. More extensive analyses of several of these signals have been published (Table 11.7). We also have significant linkage signals for heart rate, left ventricular mass, bilirubin, neutrophil cell count, and

Table 11.4 Heritabilities of CVD risk factors: Strong Heart Family Study

Trait	h^2	Covariates[a]	References
Lipid-related phenotypes			
LDL-C	0.39 ± 0.06	Estrogen, center, alcohol	North et al. (2003a)
HDL-C	0.50 ± 0.07	Estrogen, center, alcohol	North et al. (2003a)
Total cholesterol	0.39 ± 0.06	Estrogen, center	North et al. (2003a)
Ln triglyceride	0.40 ± 0.07		North et al. (2003a)
ApoA-I	0.39 ± 0.07	Estrogen, center, alcohol	North et al. (2003a)
ApoB	0.34 ± 0.07	Smoking	North et al. (2003a)
Lp(a)	0.51 ± 0.09	Center	North et al. (2003a)
VLDL-C	0.45 ± 0.09		North et al. (2003a)
VLDL-TG	0.41 ± 0.10	Smoking	North et al. (2003a)
Adiposity-related phenotypes			
Weight	0.51 ± 0.07	Center, alcohol, smoking	North et al. (2003a)
BMI	0.44 ± 0.07	Estrogen, center, alcohol, smoking	North et al. (2003a)
WHR	0.54 ± 0.07	Estrogen, center, smoking	North et al. (2003a)
Fat mass	0.52 ± 0.06	Alcohol, smoking	North et al. (2003a)
Fat free mass	0.53 ± 0.07	Alcohol, smoking	North et al. (2003a)
Diabetes-related phenotypes			
Fasting glucose	0.29 ± 0.08	Estrogen, center, alcohol	North et al. (2003a)
Ln insulin	0.44 ± 0.08	Center	North et al. (2003a
Diabetes status	0.41 ± 0.09		North et al. (2003c)
Insulin resistance syndrome factors	0.33–0.67	BP meds, lipid-lowering meds	North et al. (2003d)
Blood pressure and Heart Rate (HR)			
Systolic BP	0.23 ± 0.06		North et al. (2003a)
Diastolic BP	0.34 ± 0.07	Center	North et al. (2003a)
Echo HR	0.28 ± 0.03		Melton et al. (2010)
ECG HR	0.30 ± 0.05		Melton et al. (2010)
Inflammation/oxidative stress			
C-reactive protein	0.46 ± 0.07	Center, CVD history, physical activity,	Best et al. (2004)
Fibrinogen	0.34 ± 0.07	% Indian heritage, % body fat, BMI,	Best et al. (2004)
Paraoxonase 1	0.24 ± 0.07	Waist-hip ratio, SBP, hypertension status, alcohol, smoking, diabetes status, IGT status, TG, LDL-C, HDL-C	Best et al. (2004)
Bilirubin	0.42 ± 0.03	Center, hematocrit, SGOT, serum abumin, smoking	Melton et al. (2011)
Blood coagulation			
Ln fibrinogen	0.23 ± 0.07	Center, alcohol, smoking	North et al. (2003a)
PAI-1	0.26 ± 0.06	Center	North et al. (2003a)

(continued)

Table 11.4 (continued)

Trait	h^2	Covariates[a]	References
Carotid artery measures			
Lumen diameter	0.44 ± 0.07	Center, diabetes, impaired glucose	North et al. (2002)
Intimal–medial wall thickness	0.21 ± 0.06	Tolerance, smoking, cholesterol,	North et al. (2002)
Vascular mass	0.27 ± 0.07	Hypertension status, body surface area	North et al. (2002)
Arterial stiffness	0.23 ± 0.07		North et al. (2002)
Augmentation index	0.18 ± 0.06		North et al. (2002)
Aortic root size	0.51 ± 0.08	Center	Bella et al. (2002)
	0.44 ± 0.08	Center, height, weight, SBP, DBP	Bella et al. (2002)
Left ventricular dimensions and mass[b]			
Left ventricular mass	0.27 ± 0.08	Center	Bella et al. (2004)
LV end-diastolic chamber diameter	0.36 ± 0.08	Center	Bella et al. (2004)
Interventricular septal wall thickness	0.26 ± 0.07	Center	Bella et al. (2004)
LV posterior wall thickness	0.19 ± 0.08	Center	Bella et al. (2004)
Relative wall thickness	0.22 ± 0.07	Center	Bella et al. (2004)

[a] In addition to age, sex, and their higher order terms and interactions
[b] Further adjustment for body weight, height, SBP, heart rate, medications, and diabetes reduced heritabilities

lymphocyte cell count. In an analysis of diabetes-specific effects on weight, we identified a QTL that is a good candidate for susceptibility to fat deposition in diabetics (Franceschini et al. 2008a).

11.4.2.3 GOCADAN

Only a few of the GOCADAN linkage signals have been published (Table 11.8). We obtained significant LOD scores (≥3.0) for all adiposity-related phenotypes in women. Our linkage screens also reveal significant LOD scores for diabetes, HDL-C, mean arterial pressure, apoB, and ferritin. Suggestive LOD scores ($1.9 \leq \text{LOD} < 3.0$) were obtained for LDL size, HDL size, LDL-C, total cholesterol, apoA1, fat mass, fat-free mass, and weight.

11.5 Follow-up

The ultimate goal of our gene mapping efforts is to identify the causative functional variants responsible for our QTLs. We have begun this process in all three projects. The most extensive analyses have focused on SAFHS, the oldest of the three studies.

For the SAFHS, we have approached fine mapping in two ways. First, we have used SNP identification in strong positional candidate genes and genotyping of these novel SNPs and additional known SNPs for Bayesian quantitative trait nucleotide (BQTN) analysis (Blangero et al. 2005). For instance, we are genotyping 88 SNPs (70 novel) in the β3-adrenergic receptor gene, *ADRB3,* to follow-up our QTL for BMI on chromosome 8 (Mitchell et al. 1999) and more than 600 SNPs (250 novel) in the hepatic lipase gene, *LIPC*, for follow-up of our QTL for lipoprotein size phenotypes (Almasy et al. 2005). With the availability and practicality of whole genome SNP typing, we have turned to whole genome SNP chips to augment the fine mapping of our QTLs identified for traditional quantitative phenotypes as well as expression QTLs across the genome.

We are using gene expression analysis, bioinformatic and transcriptomic analysis, and functional analysis in SAFHS in our search for functional variants. For example, Curran et al. (2007) identified two QTLs (one nuclear and one mitochondrial) that influence mitochondrial content, and through bioinformatic and transcriptomic analyses, have identified several plausible candidates. Rutherford et al. (2007) have fine mapped a QTL influencing longitudinal change in blood pressure using SNP-association analysis within candidate genes identified from a bioinformatic search and from whole genome transcriptional expression data. Functional analyses have identified *SEPS1* as a new candidate mediator of inflammatory response (Curran et al. 2005). Using a quantitative trait linkage disequilibrium test, Bozaoglu et al. (2006) provided evidence consistent with a functional role for the *UBL5* (ubiquitin-like 5) gene in influencing traits related to metabolic syndrome. Studies such as these are allowing us to narrow in on causative functional genes. The previously mentioned whole genome sequence data also are providing new opportunities for gene identification.

In the SHFS, we have completed our linkage screen in the entire Family Study cohort. We are using panels of gene-centric SNPs to explore our QTL regions and identify candidate genes for further characterization and BQTN analyses. For our chromosome 4 obesity-related QTL (Almasy et al. 2007), we observed strong evidence of association of BMI and weight with multiple markers in two genes in the chromosome 4 region. In order to confirm our initial findings, SNPs in these two genes were genotyped in a larger sample from all three study centers. For our heart rate QTL on chromosome 9p21 (Melton et al. 2010), we observed association between SNPs in a gene encoding a hypothetical protein, *KIAA1797*. Through cross-collaboration, we subsequently showed that expression levels of the *KIAA1797* transcript were significantly associated with heart rate in SAFHS participants. We are pursuing QTLs for left ventricular mass, heart rate, presence or absence of plaque, and body composition phenotypes using the same strategy. We also have shown that the association of MYH9 polymorphisms with kidney disease phenotypes found in African Americans does not apply to American Indians (Franceschini et al. 2010).

For GOCADAN, association analysis using 1,536 gene-centric SNPs in the region of a QTL for HDL-C levels and suggestive evidence for linkage of HDL and LDL size (Cole et al. 2005) has led us to replicate several previous reports of association between SNPs and lipid levels within a 500 kb region on chromosome 19. We genotyped SNPs in candidate genes in 8p12-p21 where we have localized linkage for unsaturated fatty acids and found significant associations between fatty acids and SNPs in apolipoprotein J (*APOJ*), lipoprotein lipase (*LPL*), macrophage scavenger receptor1 (*MSR1*), and tumor necrosis factor receptor superfamily, member 10b (*TNFRSF10B*). A Bayesian association analysis based on a measured genotype model showed that SNPs in *LPL, TNFRSF10B*, and *APOJ* yielded strong statistical evidence for a functional effect on the variation in plasma fatty acid distribution (Voruganti et al. 2010).

11.6 Discussion and Conclusion

Disparities in the resources devoted to research on the causes of disease in minority populations tend to echo the inequalities in access to health care in these populations. There is an effort at the national level to address these discrepancies. Much can be learned by studying multiple racial and ethnic groups. Some genes may be more important contributors to disease susceptibility in one ethnic group than another. There may be genes that are important in all racial and ethnic groups, but because of different genetic backgrounds, environmental exposures, and lifestyles, some genes may be easier to detect in one group than another. Moreover, different ethnic groups, because of these genetic and environmental differences, may require different preventive measures and different therapies. In addition, cultural differences may mean that some preventive measures are more acceptable to the communities than others.

There are substantial similarities among the three studies. All of our study populations have a high prevalence of cardiovascular disease and are characterized by rather low physical activity levels and a high prevalence of smoking. In SAFHS and SHFS, rates of diabetes and obesity are very high. (However, in GOCADAN

Table 11.5 Heritabilities of CVD risk factors: GOCADAN

Phenotype	$h^{2\ a}$	Covariates	References	Covariate effects
Lipid-related phenotypes				
HDL-C (mg/dl)	0.51 ± 0.07	Sex, sex-specific age, age^2	Voruganti et al. (2006)	0.117
Triglycerides (mg/dl)	0.31 ± 0.08	Sex, sex-specific age, age^2	Voruganti et al. (2006)	0.020
LDL concentration (mg/dl)			Voruganti et al. (2006)	
Small	0.20 ± 0.06	Sex, sex-specific age, age^2	Voruganti et al. (2006)	0.039
Medium	0.31 ± 0.08	Sex, sex-specific age, age^2	Voruganti et al. (2006)	0.026
Large	0.30 ± 0.07	Sex, sex-specific age, age^2	Voruganti et al. (2006)	0.058
Total	0.36 ± 0.07	Sex, sex-specific age, age^2	Voruganti et al. (2006)	
Mean LDL size (nm)	0.45 ± 0.09	Sex, sex-specific age, age^2	Voruganti et al. (2006)	0.075
HDL concentration				
Small	0.20 ± 0.10	Age, sex, interactions,	Tejero et al. (2010)	
Medium	0.43 ± 0.10	Smoking, lipid-lowering meds	Tejero et al. (2010)	
Large	0.39 ± 0.10	Smoking, lipid-lowering meds	Tejero et al. (2010)	
Mean HDL size	0.89 ± 0.07	Smoking, lipid-lowering meds	Tejero et al. (2010)	
Lp(a)	0.89 ± 0.08	Smoking, lipid-lowering meds	Tejero et al. (2010)	
Total cholesterol	0.45 ± 0.09	Smoking, lipid-lowering meds	Tejero et al. (2010)	
Adiposity-related phenotypes				
Weight (kg)	0.64 ± 0.06	Sex, sex-specific age, age^2	Voruganti et al. (2006)	0.044
BMI (kg/m^2)	0.57 ± 0.07	Sex, sex-specific age, age^2	Voruganti et al. (2006)	0.037
Waist circumference (cm)	0.55 ± 0.07	Sex, sex-specific age, age^2	Voruganti et al. (2006)	0.030
Skinfold (mm)		Sex, sex-specific age, age^2	Voruganti et al. (2006)	
Subscapular	0.53 ± 0.07	Sex, sex-specific age, age^2	Voruganti et al. (2006)	
Triceps	0.47 ± 0.07	Sex, sex-specific age, age^2	Voruganti et al. (2006)	
Waist/height	0.53 ± 0.08	Sex, sex-specific age, age^2	Voruganti et al. (2011)	
Body fat (%)	0.56 ± 0.07	Sex, sex-specific age, age^2	Voruganti et al. (2011)	
Diabetes-related phenotypes				
Fasting glucose	0.31 ± 0.09	Age, sex, age x sex, lipid	Tejero et al. (2010)	
Insulin	0.24 ± 0.09	Meds, smoking	Tejero et al. (2010)	
HbA1c	0.47 ± 0.09	Meds, smoking	Tejero et al. (2010)	

(continued)

Table 11.5 (continued)

Phenotype	$h^{2\ a}$	Covariates	References	Covariate effects
Fatty acids				
Saturated FAs	0.38 ± 0.11	Age, sex, age^2, interactions	Voruganti et al. (2010)	
Monounsaturated FAs	0.48 ± 0.12	Age, sex, age^2, interactions	Voruganti et al. (2010)	
Polyunsaturated FAs	0.42 ± 0.12	Age, sex, age^2, interactions	Voruganti et al. (2010)	
Total FAs	0.55 ± 0.12	Age, sex, age^2, interactions	Voruganti et al. (2010)	
Blood pressure				
Systolic BP	0.46 ± 0.09	Age, sex, age x sex, lipid	Tejero et al. (2010)	
Diastolic BP	0.45 ± 0.09	Meds, smoking	Tejero et al. (2010)	

[a] All p values <0.0001 except for insulin ($p = 0.002$) and small HDL ($p = 0.007$)

compared to U.S. whites, the prevalence of diabetes is lower in both men and women. Prevalence of obesity is lower in men, and equal to U. S. whites in women.) All three populations are of mixed Asian and European ancestry, in varying degrees. Although each study collected unique risk factor data, a basic set of phenotypes was common to all studies. The designs of all three studies involved extended families recruited without regard to disease status. Extended families are much more powerful for statistical genetic analysis than are sib pairs or nuclear families, and the ascertainment strategy allows us to analyze multiple traits without the complications introduced by ascertainment correction.

There also are numerous differences among the three study populations. Perhaps most striking are the differences in environment and climate. The diets in the three populations are in various stages of transition from traditional to westernized. Moreover, the approaches to identifying and recruiting families differed substantially between studies because of the differing histories of the studies and the diverse environments of the populations. In identifying and recruiting families in the San Antonio Family Heart Study, we started "from scratch," but had the advantages of a circumscribed geographic area and access to city directories. The Strong Heart Family Study grew from an existing epidemiologic study (SHS) that already had recruited more than 4,500 participants in three geographic areas. Although only limited family data were available in SHS, these data provided an excellent basis for identifying extended families. GOCADAN recruitment was not focused specifically on families but rather on entire villages. A census of household members and their relationships, together with the interrelatedness of the villages, enabled us to link nearly all of the 1,200+ participants into a single large pedigree.

Genetic analyses have progressed to different degrees in the three studies, as a function of their duration and the resources available. All three studies have demonstrated significant heritabilities for a variety of CVD risk factors. Linkage analyses have revealed many significant LOD scores in the San Antonio Family Heart Study and the Strong Heart Family Study, and several in the newer GOCADAN study. In all three studies, promising signals are being followed up in an attempt to identify relevant functional polymorphisms.

As mentioned above, a growing challenge has been created by the mismatch between the cultural beliefs of some minority populations and new federal regulations concerning data sharing. We honor the conviction of the American Indians and Alaskan Eskimos in our studies that their genetic information belongs to them and must not be shared broadly (for example, by making it available through NIH). Therefore, federal

Table 11.6 San Antonio Family Heart Study selected linkage results

Trait	QTL Location	LOD	Covariates[a]	References
Lipid-related phenotypes				
HDL2a-unesterified cholesterol	8 (150 cm)	4.87	Diabetic status, postmenopausal status, smoking, alcohol consumption, diabetes meds, lipid-altering meds, exogenous sex hormones	Almasy et al. (1999)
HDL2a-unesterified cholesterol	15 (62 cm)	3.26	Diabetic status, postmenopausal status, smoking, alcohol consumption, diabetes meds, lipid-altering meds, exogenous sex hormones	Almasy et al. (1999)
HDL-C	9p (41 cm)	3.4		Arya et al. (2002)
HDL-C	16q (92 cm)	4.33	Plasma ApoA-I, TG, exogenous sex	Mahaney et al. (2003)
β-lipoprotein phenotypes	15	3.0	Diabetes status, diabetic meds, menopausal status, contraception, alcohol consumption, smoking	Rainwater et al. (2004)
Lipoprotein size phenotypes	15 (LIPC region)	1.78-3.79		Almasy et al. (2005)
HDL-EC size	5p	3.5		Almasy et al. (2005)
Adiposity-related phenotypes				
Fat mass (bioimpedance)	2p (78.3 cm)	2.75		Comuzzie et al. (1997)
Serum leptin level	2p (74.2 cm)	4.95		Comuzzie et al. (1997)
Serum leptin level	22 (D22S1685)	3.44	Diabetes status, testosterone	Martin et al. (2002)
BMI	8 (63 cm)	3.21		Mitchell et al. (1999)
Acylation-stimulating protein	17 (D17S1303)	2.7		Martin et al. (2004)
ASP and BMI (bivariate)	17	4.7		Martin et al. (2004)
ASP and HDL2a-C (bivariate)	15	3.2		Martin et al. (2004)
LpPLA2 (interacts with adiposity)	1 (153 cm)	3.39	Oral contraceptive use, menopausal	Diego et al. (2007)
			Status	
Blood pressure				
Diastolic BP	2 (D2S1790)	3.92	Systolic BP, BMI	Atwood et al. (2001a)
Systolic BP	2 (D2S1790)	1.31	Diastolic BP, BMI	Atwood et al. (2001a)
Pulse pressure	21 (D21S1440)	2.78	BMI	Atwood et al. (2001b)
ACE activity	17	4.57	BMI, menopausal status, I/D genotypes	Kammerer et al. (2004)
ACE activity	4 (D4S1548)	3.34[b]	BMI, menopausal status, I/D genotypes	Kammerer et al. (2004)
SBP rate of change	11q24.1	4.15	BMI rate of change	Rutherford et al. (2007)

(continued)

Table 11.6 (continued)

Trait	QTL Location	LOD	Covariates[a]	References
Mean arterial pressure rate of change	11q24.1	3.94	BMI rate of change	Rutherford et al. (2007)
Diabetes-related phenotypes				
Serum insulin concentration	3p (109 cm)	3.07		Mitchell et al. (2000)
Insulin response to glucose index	8 (22–26 cm)	3.09		Cai et al. (2004b)
Corrected insulin response	13q	2.98		Cai et al. (2004b)
Metabolic syndrome risk				
(lipid factor)	4p (26 cm)	3.52		Cai et al. (2004c)
(adiposity factor)	1 (30 cm)	2.53		Cai et al. (2004c)
(insulin-glucose factor)	3 (112 cm)	2.20		Cai et al. (2004c)
HOMA-IR	12 (118 cm)	3.01	Waist circumference	Voruganti et al. (2008)
Inflammation/oxidative stress				
Intercellular adhesion molecule-1	19p (33 cm)	4.95	Waist circumference, smoking, diabetes, diabetes meds	Kent et al. (2007)
Paraoxonase 1	7q (PON1 region)	31.4		Winnier et al. (2006, 2007)
Paraoxonase 1	12 (26 cm)	3.56		Winnier et al. (2006, 2007)
Plasminogen	12q14.1	2.73	Smoking, contraception, alcohol, menopausal status, diabetes, diabetes meds, lipid-altering meds, BMI, TC, HDL-C	Santamaria et al. (2007)
Kidney function				
Serum creatinine	9 (D9S922)	2.62		Arar et al. (2008)
Creatinine clearance	2 (D2S1780)	2.05		Arar et al. (2008)
Glomerular filtration rate (eGFR)	9 (D9S1122)	3.87		Arar et al. (2008)
Serum uric acid	6q (133 cm)	3.3	BMI, waist circumference, SBP, pulse pressure	Nath et al. (2007)
Serum uric acid	3p26 (D3S2387)	4.72	Waist/hip ratio, SBP, triglycerides, HDL-C, serum creatinine, BP meds, alcohol consumption, smoking, diabetes status	Voruganti et al. (2009a ,b)
Blood coagulation				
TAFI antigen	13q (D13S788)	3.09		Warren et al. (2006)
D-dimer	5p15.32-p15.2	3.32	Contraception, menopausal status	Diego et al. (2010)
Lifestyle				
Macronutrient intakes	2p22 (D2S1346)	1.0–2.62	Alcohol, smoking, diabetes status (household effects-variance component)	Cai et al. (2004a)
Cigarette and alcohol consumption (bivariate)	10 (151 cm)	3.82	Education	Viel et al. (2008)

[a] In addition to age, sex, and their higher order terms and interactions; [b] Conditional on chr 17 locus; + f = females, m = males

Table 11.7 Strong Heart Family Study selected linkage results

Trait	QTL location	LOD	Covariates[a]	References
Lipid-related phenotypes				
LDL-C (DK)	19q13.41 (93 cm)	4.3		North et al. (2006)
LDL-C (DK)	19q13.41 (93 cm)	2.7	Metabolic equiv, smoking	North et al. (2006)
HDL-C (AZ)	6p22.3-p24.3	4.4		Li et al. (2009)
Apolipoprotein A-1 (AZ)	6p22.3-p24.3	3.2		Li et al. (2009)
Apolipoprotein A-1 (DK)	9q22.2	3		Li et al. (2009)
Triglycerides (DK)	15q22.1-q22.31	3.8		Li et al. (2009)
Adiposity-related phenotypes				
BMI (DK)	2 (45 cm)	1.12		Diego et al. (2006)
Fat mass (FM) (DK)	11 (66 cm)	2.23		Diego et al. (2006)
Log FI and FM (bivariate) (DK)	2 (48 cm)	3.43		Diego et al. (2006)
Log FI and BMI (bivariate) (DK)	2 (52 cm)	2.91		Diego et al. (2006)
Weight%	1 (242 cm)	3.7		Franceschini et al. (2008a)
Weight	4q35	5.17	Center	Almasy et al. (2007)
BMI	4q35	5.08	Center	Almasy et al. (2007)
Diabetes-related phenotypes				
Insulin resistance factor scores[b]			Center	North et al. (2005)
Dyslipidemia factor	12q24.1-3 (141 cm)	2.7	Center	North et al. (2005)
Glucose-insulin-obesity factor	4q34.3 (205 cm)	2.2	Center	North et al. (2005)
Blood pressure factor	1 (237 cm)	1.6	Center	North et al. (2005)
Log fasting insulin (FI) (Dakotas)	2 (51 cm)	3.42		Diego et al. (2006)
Blood pressure and Heart Rate				
Systolic blood pressure ^ (g × sex)	17 (136 cm)	3.25	Hypertension meds	Franceschini et al. (2006)
Pulse pressure	7p15.3 (37 cm)	3.3	Center	Franceschini et al. (2008b)
Echo HR (Oklahoma)	9p21	3.67		Melton et al. (2010)
ECG HR (Oklahoma)	9p21	4.83		Melton et al. (2010)
Heart rate (EKG)	9p21 (39 cm)	3.3		Rutherford et al. (2008)
Inflammation/oxidative stress				
Bilirubin	2q37.1	6.61	Center, hematocrit, SGOT, serum albumin, smoking	Melton et al. (2011)
Blood coagulation				
Fibrinogen (DK)	7 (76 cm)	3.02	Waist circumference, diabetes status, menopausal status	Best et al. (2008)

(continued)

Table 11.7 (continued)

Trait	QTL location	LOD	Covariates[a]	References
Kidney function				
Glomerular filtration rate (AZ)	12p12.2 (39 cm)	3.5		Mottl et al. (2008)
eGFR in nonhypertensives (AZ)	12p12.2 (39 cm)	4.6		
Albumin/creatinine ratio	1q32.2 (D1S249)	2		Mottl et al. (2009)
Albumin/creatinine ratio (AZ)	1q32.2 (D1S249)	2.5		Mottl et al. (2009)
Serum uric acid	11 (71 cm)	3.56	Center, BMI, eGFR, diabetes	Voruganti et al. (2009b)
Serum uric acid	1p36 (39 cm)	3.51	Status, alcohol, medications	Voruganti et al. (2009b)

[a] In addition to age, sex, and their higher order terms and interactions; [b] females only
[c] nondiabetics only
[d] diabetes-specific

Table 11.8 GOCADAN selected linkage results

Trait	QTL location	LOD	Covariates	References
Adiposity-related phenotypes, women			Age, sex, age^2, interactions	Voruganti et al. (2011)
BMI (kg/m^2)	chr19, 61 cm	4.5	Age, sex, age^2, interactions	Voruganti et al. (2011)
Waist (inches)	chr19, 63 cm	4.8	Age, sex, age^2, interactions	Voruganti et al. (2011)
Waist/height	chr19, 65 cm	3.8	Age, sex ,age^2, interactions	Voruganti et al. (2011)
Body fat (%)	chr19, 61 cm	5.0	Age, sex, age^2, interactions	Voruganti et al. (2011)
Subscapular skinfold (cm)*	chr19, 61 cm	3.3	Age, sex ,age^2, interactions	Voruganti et al. (2011)
Triceps skinfold	chr19, 61 cm	4.0	Age, sex, age^2, interactions	Voruganti et al. (2011)
Fatty acids			Age, sex, age^2, interactions	Voruganti et al. (2010)
Saturated FAs	chr12, 146 cm	1.68	Age, sex, age^2, interactions	Voruganti et al. (2010)
Monounsaturated FAs	chr8, 50 cm	3.82	Age, sex, age^2, interactions	Voruganti et al. (2010)
Polyunsaturated FAs	chr6, 170 cm	2.93	Age, sex, age^2, interactions	Voruganti et al. (2010)
Total FAs	chr10, 97 cm	1.94	Age, sex, age^2, interactions	Voruganti et al. (2010)

funding for future gene identification efforts using NIH-funded genome-wide SNP typing or sequencing may not be an option. We will approach this problem by seeking private funding, as was done with the San Antonio Family Heart Study. (We were unsuccessful in petitioning NIH for exceptions to their broad data-sharing requirements.) We also will discuss with the study populations the sharing of data with selected researchers who are approved by them. Given the increasing importance of cardiovascular disease and related disorders in minority populations, and the promise of new genetic approaches for revealing underlying mechanisms, resolution of this issue has the highest priority.

Acknowledgments We wish to thank the participants in the San Antonio Family Heart Study, the Strong Heart Study, and GOCADAN for their generosity and for their interest in helping us as we attempt to help solve their health problems. We also wish to acknowledge the investigators, field staff, and scientific staff of these three projects, and the National Heart, Lung, and Blood Institute for providing the funding for our studies through grants P01 HL045522 (San Antonio Family Heart Study), U01 HL065529 (Strong Heart Family Study), and U01 HL082490 (GOCADAN). We also acknowledge the

Fredric C. Bartter General Clinical Research Center, supported by M01-RR01346, which provided ongoing clinical support to the San Antonio Family Heart Study. At Texas Biomed, these studies were conducted in facilities constructed with support from the Research Facilities Improvement Program Grant nos. C06 RR013556 and C06 RR017515 from the National Center for Research Resources, National Institutes of Health. The AT&T Genomics Computing Center supercomputing facilities used for statistical genetic analyses were supported in part by a gift from the SBC Foundation. The statistical genetics computer package, SOLAR, is supported by grant R01 MH059490 from the National Institute of Mental Health. The opinions expressed in this paper are those of the authors and do not necessarily reflect the views of the Indian Health Service.

References

Abecasis GR, Noguchi E, Heinzmann A, Traherne JA, Bhattacharya S, Leaves NI, Anderson GG, Zhang Y, Lench NJ, Carey A, Cardon LR, Moffatt MF, Cookson WO (2001) Extent and distribution of linkage disequilibrium in three genomic regions. Am J Hum Genet 68:191–197

Almasy L, Blangero J (1998) Multipoint quantitative-trait linkage analysis in general pedigrees. Am J Hum Genet 62:1198–1211

Almasy L, Hixson JE, Rainwater DL, Cole SA, Williams JT, Mahaney MC, VandeBerg JL, Stern MP, MacCluer JW, Blangero J (1999) Human pedigree-based quantitative trait-locus mapping: localization of two genes influencing HDL cholesterol metabolism. Am J Hum Genet 64:1686–1693

Almasy L, Rainwater DL, Cole S, Mahaney MC, VandeBerg JL, Hixson JE, Stern MP, MacCluer JW, Blangero J (2005) Joint linkage and association analysis of the hepatic lipase promotor polymorphism and lipoprotein size phenotypes. Hum Biol 77:17–25

Almasy L, Göring HH, Diego V, Cole S, Laston S, Dyke B, Howard BV, Lee ET, Best LG, Devereux R, Fabsitz RR, MacCluer JW (2007) A novel obesity locus on chromosome 4q: the Strong Heart Family Study. Obesity (Silver Spring) 15:1741–1748

Amos CI (1994) Robust variance-components approach for assessing genetic linkage in pedigrees. Am J Hum Genet 54:535–543

Arar NH, Nath S, Thameem F, Bauer R, Voruganti S, Comuzzie A, Cole S, Blangero J, MacCluer J, Abboud H (2007) Genome-wide scan for microalbuminuria in Mexican Americans: the San Antonio Family Heart Study. Genet Med 9:80–87

Arar NH, Voruganti VS, Nath SD, Thameem F, Cole S, Blangero J, MacCluer JW, Comuzzie AG, Abboud HE (2008) A genome-wide search for linkage to chronic kidney disease in a community-based sample: the SAFHS. Nephrol Dial Transplant 23:3184–3191

Arya R, Duggirala R, Almasy L, Rainwater DL, Mahaney MC, Cole S, Dyer TD, Williams K, Leach RJ, Hixson JE, MacCluer JW, O'Connell P, Stern MP, Blangero J (2002) Linkage of high density lipoprotein-cholesterol concentrations to a locus on chromosome 9p in Mexican Americans. Nat Genet 30:102–105

Atwood LD, Samollow PB, Hixson JE, Stern MP, MacCluer JW (2001a) Genome-wide linkage analysis of blood pressure in Mexican Americans. Genet Epidemiol 20:373–382

Atwood LD, Samollow PB, Hixson JE, Stern MP, MacCluer JW (2001b) Genome-wide linkage analysis of pulse pressure in Mexican Americans. Hypertension 37(part 2):42–428

Bader JS (2001) The relative power of SNPs and haplotype as genetic markers for association tests. Pharmacogenomics 2:11–24

Bella JN, MacCluer JW, Roman MJ, Almasy L, North K, Welty TK, Lee ET, Fabsitz RR, Howard BV, Devereux RB (2002) Genetic influences on aortic root size in American Indians: the Strong Heart Study. Arterioscler Thromb Vasc Biol 22:1008–1011

Bella JN, MacCluer JW, Roman MJ, Almasy L, North K, Best LG, Lee ET, Fabsitz RR, Howard BV, Devereux RB (2004) Heritability of left ventricular dimensions and mass in American Indians: the Strong Heart Study. J Hypertens 22:281–286

Best L, North KE, Tracy R, Lee ET, Howard BV, Palmieri V, MacCluer JW (2004) Genetic determination of acute phase reactant levels: the Strong Heart Study. Hum Hered 58:112–116

Best LG, North KE, Li X, Palmieri V, Umans JG, MacCluer JW, Laston S, Haack K, Göring HHH, Almas L, Lee ET, Tracy RP, Cole SA (2008) Linkage study of fibrinogen levels: the Strong Heart Family Study. BMC Med Genet 9:77

Blangero J, Almasy L (1997) Multipoint oligogenic linkage analysis of quantitative traits. Genet Epidemiol 14:959–964

Blangero J, Göring HH, Kent JW Jr, Williams JT, Peterson CP, Almasy L, Dyer TD (2005) Quantitative trait nucleotide analysis using Bayesian model selection. Hum Biol 77:541–559

Block-98 (1998) Block dietary data system. Berkeley Nutrition Services, Berkeley

Bozaoglu K, Curran JE, Elliott KS, Walder KR, Dyer TD, Rainwater DL, VandeBerg JL, Mahaney MC, Comuzzie AG, Collier GR, Zimmet P, MacCluer JW, Jowett JB, Blangero J (2006) Association of genetic variation within UBL5 with phenotypes of metabolic syndrome. Hum Biol 78:147–159

Cai G, Cole SA, Bastarrachea-Sosa RA, MacCluer JW, Blangero J, Comuzzie AG (2004a) Quantitative trait locus determining dietary macronutrient intakes is located on human chromosome 2p22. Am J Clin Nutr 80:1410–1414

Cai G, Cole SA, Freeland-Graves JH, MacCluer JW, Blangero J, Comuzzie AG (2004b) Genome-wide scans reveal quantitative trait loci on 8p and 13q related to insulin action and glucose metabolism: the

San Antonio Family Heart Study. Diabetes 53:1369–1374

Cai G, Cole SA, Freeland-Graves JH, MacCluer JW, Blangero J, Comuzzie AG (2004c) Principal component for metabolic syndrome risk maps to chromosome 4p in Mexican Americans: the San Antonio Family Heart Study. Hum Biol 76:651–665

Charlesworth JC, Curran JE, Johnson MP, Göring HHH, Dyer TD, Diego VP, Kent JW Jr, Mahaney MC, Almasy L, MacCluer JW, Moses EK, Blangero J (2010) Transcriptomic epidemiology of smoking: the effect of smoking on gene expression in lymphocytes. BMC Med Genomics 3:29

Cole SA, Laston S, Göring HHH, Diego VP, Dyer TD, Blangero J, Ebbesson S, Howard BV, MacCluer JW, Comuzzie AG (2005) Linkage of lipoprotein phenotypes to chromosome 19p in Alaska Natives from the GOCADAN Study. Obes Res 13:A100

Comuzzie AG, Hixson JE, Almasy L, Mitchell BD, Mahaney MC, Dyer TD, Stern MP, MacCluer JW, Blangero J (1997) A major quantitative trait locus determining serum leptin levels and fat mass is located on human chromosome 2. Nat Genet 15:273–276

Comuzzie AG, Tejero ME, Funahasi T, Martin LJ, Kissebah A, Takahashi M, Kihara S, Tanaka S, Rainwater DL, Matsuzawa Y, MacCluer JW, Blangero J (2007) The genes influencing adiponectin levels also influence risk factors for the metabolic syndrome and the development of type 2 diabetes. Hum Biol 79:191–200

Curran JE, Jowett JBM, Elliott KS, Gao Y, Gluschenko K, Wang J, Abel Azim DM, Cai G, Mahaney MC, Comuzzie AG, Dyer TD, Walder KR, Zimmet P, MacCluer JW, Collier GR, Kissebah AH, Blangero J (2005) Genetic variation in selenoprotein S influences inflammatory response. Nat Genet 37:1234–1241

Curran JE, Johnson MP, Dyer TD, Göring HH, Kent JW Jr, Charlesworth JC, Borg AJ, Jowett JBM, Cole SA, MacCluer JW, Kissebah AH, Moses EK, Blangero J (2007) Genetic determinants of mitochondrial content. Hum Mol Genet 16:1504–1514

Czerwinski SA, Mahaney MC, Rainwater DL, VandeBerg JL, MacCluer JW, Stern MP, Blangero J (2004) Gene by smoking interaction: evidence for effects on low-density lipoprotein size and plasma levels of triglyceride and high-density lipoprotein cholesterol. Hum Biol 76:863–876

Dib C, Faure S, Fizames C, Samson D, Drouot N, Vignal A, Millasseau P, Marc S, Hazan J, Seboun E, Lathrop M, Gyapay G, Morissette J, Weissenbach J (1996) A comprehensive genetic map of the human genome based on 5,264 microsatellites. Nature 380:152–154

Diego VP, Göring HH, Cole SA, Almasy L, Dyer TD, Blangero J, Duggirala R, Laston S, Wenger C, Cantu T, Dyke B, North KE, Schurr T, Best LG, Devereux RB, Fabsitz RR, Howard BV, MacCluer JW (2006) Fasting insulin and obesity-related phenotypes are linked to chromosome 2p: the Strong Heart Family Study. Diabetes 55:1874–1878

Diego VP, Rainwater DL, Wang XL, Cole SA, Curran JE, Johnson MP, Dyer TD, Williams JT, Moses EK, Comuzzie AG, MacCluer JW, Mahaney MC, Blangero J (2007) Genotype × adiposity interaction linkage analyses reveal a locus on chromosome 1 for lipoprotein-associated phospholipase A_2, a marker of inflammation and oxidative stress. Am J Hum Genet 80:168–177

Diego VP, Almasy L, Rainwater DL, Mahaney MC, Comuzzie AG, Cole SA, Tracy RP, Stern MP, MacCluer JW, Blangero J (2010) A quantitative trait locus on chromosome 5p influences D-dimer levels in the San Antonio Family Heart Study. Int J Vasc Med 2010:490241. doi:10.1155/2010/490241

Dyke B (1992) PEDSYS: a pedigree data management system. Population Genetics Laboratory, Southwest Foundation for Biomedical Research, San Antonio

Ebbesson SO, Laston S, Wenger CR, Dyke B, Romenesko T, Swenson M, Fabsitz RR, MacCluer JW, Devereux R, Roman M, Robbins D, Howard BV (2006) Recruitment and community interactions in the GOCADAN study. Int J Circumpolar Health 65:55–64

Ebbesson SO, Roman MJ, Devereux RB, Kaufman D, Fabsitz RR, MacCluer JW, Dyke B, Laston S, Wenger C, Comuzzie AG, Romenesko T, Ebbesson LO, Nobmann ED, Howard BV (2008) Consumption of omega-3 fatty acids is not associated with a reduction in carotid atherosclerosis: the Genetics of Coronary Artery Disease in Alaska Natives study. Atherosclerosis 199:346–353

Franceschini N, MacCluer JW, Göring HH, Cole SA, Rose KM, Almasy L, Diego V, Laston S, Lee ET, Howard BV, Best LG, Fabsitz RR, Roman MJ, North KE (2006) A quantitative trait locus-specific gene-by-sex interaction on systolic blood pressure among American Indians: the Strong Heart Family Study. Hypertension 48:266–270

Franceschini N, Almasy L, MacCluer JW, Göring HH, Cole SA, Diego VP, Laston S, Howard BV, Lee ET, Best LG, Fabsitz RR, North KE (2008a) Diabetes-specific genetic effects on obesity traits in American Indian populations: the Strong Heart Family Study. BMC Med Genet 9:90

Franceschini N, MacCluer JW, Rose KM, Rutherford S, Cole SA, Laston S, Göring HHH, Diego VP, Roman MJ, Lee ET, Best LG, Howard BV, Fabsitz RR, North KE (2008b) Genome-wide linkage analysis of pulse pressure in American Indians: the Strong Heart Study. Am J Hypertens 21:194–199

Franceschini N, Rose KM, Storti KL, Rutherford S, Voruganti VS, Laston S, Göring HH, Dyer TD, Umans JG, Lee ET, Best LG, Fabsitz RR, Cole SA, MacCluer JW, North KE (2009) Social- and behavioral-specific genetic effects on blood pressure traits: the Strong Heart Family Study. Circ Cardiovasc Genet 2:396–401

Franceschini N, Voruganti VS, Haack K, Almasy L, Laston S, Göring HH, Umans JG, Lee ET, Best LG, Fabsitz RR, MacCluer JW, Howard BV, North KE,

Cole SA (2010) The association of the MYH9 gene and kidney outcomes in American Indians: the Strong Heart Family Study. Hum Genet 127:295–301

Göring HH, Curran JE, Johnson MP, Dyer TD, Charlesworth J, Cole SA, Jowett JB, Abraham LJ, Rainwater DL, Comuzzie AG, Mahaney MC, Almasy L, MacCluer JW, Kissebah AH, Collier GR, Moses EK, Blangero J (2007) Discovery of expression QTLs using large-scale transcriptional profiling in human lymphocytes. Nat Genet 39:1208–1216

Heath SC (1997) Markov chain Monte Carlo segregation and linkage analysis for oligogenic models. Am J Hum Genet 61:748–760

Hopper JL, Mathews JD (1982) Extensions to multivariate normal models for pedigree analysis. Ann Hum Genet 46:373–383

Howard BV, Lee ET, Cowan LD, Fabsitz RR, Howard WJ, Oopik AJ, Robbins DC, Savage PJ, Yeh JL, Welty TK (1995) Coronary heart disease prevalence and its relation to risk factors in American Indians. The Strong Heart Study. Am J Epidemiol 142:254–268

Howard BV, Devereux RB, Cole SA, Davidson M, Dyke B, Ebbesson SO, Epstein SE, Robinson DR, Jarvis B, Kaufman DJ, Laston S, MacCluer JW, Okin PM, Roman MJ, Romenesko T, Ruotolo G, Swenson M, Wenger CR, Williams-Blangero S, Zhu J, Saccheus C, Fabsitz RR, Robbins DC (2005) A genetic and epidemiologic study of cardiovascular disease in Alaska Natives (GOCADAN): design and methods. Int J Circumpolar Health 64:206–221

Kammerer CM, Gouin N, Samollow PB, VandeBerg JF, Hixson JE, Cole SA, MacCluer JW, Atwood LA (2004) Two quantitative trait loci affect ACE activities in Mexican American families. Hypertension 43:466–470

Kent JW Jr, Comuzzie AG, Mahaney MC, Almasy L, Rainwater DL, VandeBerg JL, Stern MP, MacCluer JW, Blangero J (2004) Intercellular adhesion molecule-1 concentration is genetically correlated with insulin resistance, obesity and high-density lipoprotein concentration in Mexican Americans. Diabetes 53:2691–2695

Kent JW Jr, Mahaney MC, Comuzzie AG, Göring HH, Almasy L, Dyer TD, Cole SA, MacCluer JW, Blangero J (2007) Quantitative trait locus on chromosome 19 for circulating levels of intercellular adhesion molecule-1 in Mexican Americans. Atherosclerosis 195:367–373

Kong A, Gudbjartsson DF, Sainz J, Jonsdottir GM, Gudjonsson SA, Richardsson B, Sigurdardottir S, Barnard J, Hallbeck B, Masson G, Shlien A, Palsson ST, Frigge ML, Thorgeirsson TE, Gulcher JR, Stefansson K (2002) A high-resolution recombination map of the human genome. Nat Genet 31:241–247

Lange K, Boehnke M (1983) Extensions to pedigree analysis. IV. Variance components models for multivariate traits. Am J Med Genet 14:513–524

Lee ET, Welty TK, Fabsitz R, Cowan LD, Le NA, Oopik J, Cucchiara AJ, Savage PJ, Howard BV (1990) The Strong Heart Study. A study of cardiovascular disease in American Indians: design and methods. Am J Epidemiol 132:1141–1155

Li X, Monda KL, Göring HH, Haack K, Cole SA, Diego VP, Almasy L, Laston S, Howard BV, Shara NM, Lee ET, Best LG, Fabsitz RR, MacCluer JW, North KE (2009) Genome-wide linkage scan for plasma high density lipoprotein cholesterol, apolipoprotein A-1 and triglyceride variation among American Indian populations: the Strong Heart Family Study. J Med Genet 46:472–479

MacCluer JW, Stern MP, Almasy L, Atwood LA, Blangero J, Comuzzie AG, Dyke B, Haffner SM, Henkel RD, Hixson JE, Kammerer CM, Mahaney MC, Mitchell BD, Rainwater DL, Samollow PB, Sharp RM, VandeBerg JL, Williams JT (1999) Genetics of atherosclerosis risk factors in Mexican Americans. Nutr Rev 57:S59–S65

Mahaney MC, Almasy L, Rainwater DL, VandeBerg JL, Cole SA, Hixson J, Blangero J, MacCluer JW (2003) A quantitative trait locus on chromosome 16q influences normal variation in plasma HDL-C levels in Mexican Americans. Arterioscler Thromb Vasc Biol 23:339–345

Martin LJ, Mahaney MC, Almasy L, Hixson JE, Cole SA, MacCluer JW, Jaquish CE, Blangero J, Comuzzie AG (2002) A quantitative trait locus on chromosome 22 for serum leptin levels adjusted for serum testosterone. Obes Res 10:602–607

Martin LJ, Cianflone K, Zakarian R, Nagrani G, Almasy L, Rainwater DL, Cole S, Hixson JE, MacCluer JW, Blangero J, Comuzzie AG (2004) Bivariate linkage between acylation-stimulating protein and BMI and high-density lipoproteins. Obes Res 12:669–678

Melton PE, Rutherford S, Voruganti VS, Göring HH, Laston S, Haack K, Comuzzie AG, Dyer TD, Johnson MP, Kent JW Jr, Curran JE, Moses EK, Blangero J, Barac A, Lee ET, Best LG, Fabsitz RR, Devereux RB, Okin PM, Bella JN, Broeckel U, Howard BV, MacCluer JW, Cole SA, Almasy L (2010) Evidence for a gene influencing heart rate on chromosome 9p21 in American Indians: the Strong Heart Family Study. Hum Mol Genet 19:3662–3671

Melton PE, Haack K, Göring HH, Laston S, Umans JG, Lee ET, Fabsitz RR, Devereux RB, Best LG, MacCluer JW, Almasy L, Cole SA (2011) Genetic Influences on serum bilirubin in American Indians: the Strong Heart Family Study. Am J Hum Biol 23:118–125

Mitchell BD, Kammerer CM, Blangero J, Mahaney MC, Rainwater DL, Dyke B, Hixson JE, Henkel RD, Sharp RM, Comuzzie AG, VandeBerg JL, Stern MP, MacCluer JW (1996) Genetic and environmental contributions to cardiovascular risk factors in Mexican Americans: the San Antonio Family Heart Study. Circulation 94:2159–2170

Mitchell BD, Cole SA, Comuzzie AG, Almasy L, Blangero J, MacCluer JW, Hixson JE (1999) A quantitative trait locus influencing body mass index maps to the region of the b3 adrenergic receptor. Diabetes 48:1863–1867

Mitchell BD, Cole SA, Hsueh WC, Comuzzie AG, Blangero J, MacCluer JW, Hixson JE (2000) Linkage of serum insulin concentrations to chromosome 3p in Mexican Americans. Diabetes 49:513–516

Mosher MJ, Lange LA, Howard BV, Lee E, Best LG, Fabsitz RR, MacCluer JW, North KE (2008) Sex-specific interaction between APOE genotype and carbohydrate intake affects plasma HDL-C levels: the Strong Heart Family Study. Genes Nutr 3:87–97

Mottl AK, Vupputuri S, Cole SA, Almasy L, Göring HH, Diego VP, Laston S, Franceschini N, Howard BV, Lee ET, Best LG, Fabsitz RR, MacCluer JW, Umans JG, North KE (2008) Linkage analysis of glomerular filtration rate in American Indians: the Strong Heart Family Study. Kidney Int 74:1185–1191

Mottl AK, Vupputuri S, Cole SA, Almasy L, Göring HH, Diego VP, Laston S, Shara N, Lee ET, Best LG, Fabsitz RR, MacCluer JW, Umans JG, North KE (2009) Linkage analysis of albuminuria. J Am Soc Nephrol 20:1597–1606

Nath SD, Voruganti VS, Arar NH, Thameem F, Lopez-Alvarenga JC, Bauer R, Blangero J, MacCluer J, Comuzzie AG, Abboud HE (2007) Genome scan for determinants of serum uric acid variability. J Am Soc Nephrol 18:3156–3163

Nelson MR, Kardia SL, Ferrell RE, Sing CF (2001) A combinatorial partitioning method to identify multilocus genotypic partitions that predict quantitative trait variation. Genome Res 11:458–470

Nobmann ED, Ponce R, Mattil C, Devereux R, Dyke B, Ebbesson SO, Laston S, MacCluer J, Robbins D, Romenesko T, Ruotolo G, Wenger CR, Howard BV; GOCADAN study 2000–2003 (2005) Dietary intakes vary with age among Eskimo adults of Northwest Alaska in the GOCADAN Study, 2000–2003. J Nutr 135:856–862

Nobmann ED, Ebbesson SO, White RG, Schraer CD, Lanier AP, Bulkow LR (1998) Dietary intakes among Siberian Yupiks of Alaska and implications for cardiovascular disease. Int J Circumpolar Health 57:4–17

North KE, MacCluer JW, Devereux RB, Howard BV, Welty TK, Best LG, Lee ET, Fabsitz RR, Roman MJ (2002) Heritability of carotid artery structure and function: the Strong Heart Family Study. Arterioscler Thromb Vasc Biol 22:1698–1703

North KE, Howard BV, Welty TK, Best LT, Lee ET, Yeh JL, Fabsitz RR, MacCluer JW (2003a) Genetic and environmental contributions to cardiovascular disease risk in American Indians: the Strong Heart Family Study. Am J Epidemiol 157:303–314

North KE, Williams JT, Welty TK, Best LG, Lee ET, Fabsitz RR, Howard BV, MacCluer JW (2003b) Evidence for joint action of genes on diabetes status and CVD risk factors in American Indians: the Strong Heart Family Study. Int J Obes Relat Metab Discord 27:491–497

North KE, Williams K, Williams JT, Welty TK, Best LG, Lee ET, Fabsitz RR, Howard BV, Gray PS, MacCluer JW (2003c) Evidence for genetic factors underlying the insulin resistance syndrome in American Indians. Obes Res 11:1444–1448

North KE, MacCluer JW, Williams JT, Welty TK, Best LG, Lee ET, Fabsitz RR, Howard BV (2003d) Evidence for distinct genetic effects on obesity and lipid-related CVD risk factors in diabetic compared to non-diabetic American Indians: the Strong Heart Family Study. Diabetes Metab Res Rev 19:140–147

North KE, Almasy L, Göring HH, Cole SA, Diego VP, Laston S, Cantu T, Williams JT, Howard BV, Lee ET, Best LG, Fabsitz RR, MacCluer JW (2005) Linkage analysis of factors underlying insulin resistance: Strong Heart Family Study. Obes Res 13:1877–1884

North KE, Göring HH, Cole SA, Diego VP, Almasy L, Laston S, Cantu T, Howard BV, Lee ET, Best LG, Fabsitz RR, MacCluer JW (2006) Linkage analysis of LDL cholesterol in American Indian populations: the Strong Heart Family Study. J Lipid Res 47:59–66

Polk RL (1989) In: 1988–89 San Antonio City Directory. RL Polk and Co, Dallas

Rainwater DL, Mahaney MC, VandeBerg JL, Brush G, Almasy L, Blangero J, Dyke B, Hixson JE, Cole SA, MacCluer JW (2004) A quantitative trait locus influences coordinated variation in measures of ApoB-containing lipoproteins. Atherosclerosis 176:379–386

Rutherford S, Cai G, Lopez-Alvarenga JC, Kent JW Jr, Voruganti S, Proffitt JM, Curran JE, Johnson MP, Dyer TD, Jowett JB, Bastarrachea RA, Atwood LD, Göring HH, MacCluer JW, Moses EK, Blangero J, Comuzzie AG, Cole SA (2007) A chromosome 11q quantitative-trait locus influences change of blood pressure measurements over time in Mexican Americans of the San Antonio Family Heart Study. Am J Hum Genet 81:744–755

Rutherford S, Voruganti VS, Göring HH, Laston SL, Haack K, Almasy L, Comuzzie A, Lee ET, Best LG, Fabsitz RR, Devereux RB, Okin PM, Bella JN, Howard BV, MacCluer JW, Cole SA (2008) A heart rate genetic locus on chromosome 9p21 in the Strong Heart Family Study. Hypertension 52(e101):P231

Santamaria A, Diego VP, Almasy L, Rainwater DL, Mahaney MC, Comuzzie AG, Cole SA, Dyer TD, Tracy R, Stern MP, MacCluer JW, Blangero J (2007) A quantitative trait locus on chromosome 12q141 influences variation in plasma plasminogen levels in the San Antonio Family Heart Study (SAFHS). Hum Biol 79:515–523

Sobel E, Papp JC, Lange K (2002) Detection and integration of genotyping errors in statistical genetics. Am J Hum Genet 70:496–508

Sun L, Wilder K, McPeek MS (2002) Enhanced pedigree error detection. Hum Hered 54:99–110

Tejero ME, Voruganti VS, Cai G, Laston S, Wenger CR, MacCluer JW, Dyke B, Devereux R, Ebbesson SO, Fabsitz RR, Howard BV, Comuzzie AG (2010) Pleiotropic effects on subclasses of HDL, adiposity and glucose metabolism in adult Alaskan Eskimos. Am J Hum. Biol 22:444–448

Terwilliger JD (2001) On the resolution and feasibility of genome scanning approaches. Adv Genet 42:351–391

Viel K, Charlesworth J, Tejero E, Dyer T, Cole S, Haack K, MacCluer J, Blangero J, Almasy L (2008) A linkage analysis of cigarette and alcohol consumption in an unselected Mexican American population. Am J Med Genet Part B Neuropsychiatr Genet 147B:983–986

Voruganti VS, Cai G, Cole SA, Freeland-Graves JH, Laston S, Wenger CR, MacCluer JW, Dyke B, Devereux R, Ebbesson SO, Fabsitz RR, Howard BV, Comuzzie AG (2006) Common set of genes regulates low-density lipoprotein size and obesity-related factors in Alaskan Eskimos: results from the GOCADAN study. Am J Hum Biol 18:525–531

Voruganti VS, Lopez-Alvarenga JC, Nath SD, Rainwater DL, Bauer R, Cole SA, MacCluer JW, Blangero J, Comuzzie AG (2008) Genetics of variation in HOMA-IR and cardiovascular risk factors in Mexican Americans. J Mol Med (Berl) 86:303–311

Voruganti VS, Nath SD, Cole SA, Thameem F, Jowett JB, Bauer R, MacCluer JW, Blangero J, Comuzzie AG, Abboud HE, Arar NH (2009a) Genetics of variation in serum uric acid and cardiovascular risk factors in Mexican Americans. J Clin Endocrinol Metab 94:632–638

Voruganti VS, Göring HH, Mottl A, Franceschini N, Haack K, Laston S, Almasy L, Fabsitz RR, Lee ET, Best LG, Devereux RB, Howard BV, MacCluer JW, Comuzzie AG, Umans JG, Cole SA (2009b) Genetic influence on variation in serum uric acid in American Indians: the Strong Heart Family Study. Hum Genet 126:667–676

Voruganti VS, Cole SA, Ebbesson SO, Göring HH, Haack K, Laston S, Wenger CR, Tejero ME, Devereux RB, Fabsitz RR, MacCluer JW, Umans JG, Howard BV, Comuzzie AG (2010) Genetic variation in APOJ, LPL, and TNFRSF10B affects the plasma fatty acid distribution in Alaskan Eskimos. Am J Clin Nutr 91:1574–1583

Voruganti VS, Diego VP, Haack K, Cole SA, Blangero J, Göring HH, Laston S, Wenger CR, Ebbesson SO, Fabsitz RR, Devereux RB, Howard BV, Umans JG, MacCluer JW, Comuzzie AG (2011) A QTL for genotype by sex interaction for anthropometric measurements in Alaskan Eskimos (GOCADAN study) on chromosome 19q12-13. Obesity (Silver Spring) 19:1840–1846

Warren DM, Soria JM, Souto JC, Comuzzie A, Fontcuberta J, Blangero J, MacCluer JW, Almasy L (2005) Heritability of hemostasis phenotypes and their correlation with type 2 diabetes status in Mexican Americans. Hum Biol 77:1–15

Warren DM, Cole SA, Dyer TD, Soria JM, Souto JC, Fontcuberta J, Blangero J, MacCluer JW, Almasy L (2006) A locus on chromosome 13 influences levels of TAFI antigen in healthy Mexican Americans. Hum Biol 78:329–339

Wilmot B, Voruganti VS, Chang YP, Fu Y, Chen Z, Taylor HA, Wilson JG, Gipson T, Shah VO, Uman JG, Flessner MF, Hitzemann R, Shuldiner AR, Comuzzie AG, McWeeney S, Zager PG, MacCluer JW, Cole SA, Cohen DM (2012) Heritability of serum sodium concentration: evidence for sex- and ethnic-specific effects. Physiol Genomics 44:220–228

Winnier DA, Rainwater DL, Cole SA, Dyer TD, Blangero J, MacCluer JW, Mahaney MC (2006) Multiple QTLs influence variation in paraoxonase 1 (PON1) activity in Mexican Americans. Hum Biol 78:341–352

Winnier DA, Rainwater DL, Cole SA, Williams JT, Dyer TD, Blangero J, MacCluer JW, Mahaney MC (2007) Sex-specific QTL effects on variation in paraoxonase 1 (PON1) activity in Mexican Americans. Genet Epidemiol 31:66–74

Mapping of Susceptibility Genes for Obesity, Type 2 Diabetes, and the Metabolic Syndrome in Human Populations

12

Rector Arya, Sobha Puppala, Vidya S. Farook, Geetha Chittoor, Christopher P. Jenkinson, John Blangero, Daniel E. Hale, Ravindranath Duggirala, and Laura Almasy

R. Arya (✉) · V.S. Farook · C.P. Jenkinson · R. Duggirala
South Texas Diabetes and Obesity Institute, Edinburg Regional Academic Health Center, University of Texas Health Science Center at San Antonio, Edinburg, TX, USA
e-mail: arya@uthscsa.edu

V.S. Farook
e-mail: FarookV@uthscsa.edu

C.P. Jenkinson
e-mail: jenkinsonc@uthscsa.edu

R. Duggirala
e-mail: duggirala@uthscsa.edu

D.E. Hale
Division of Endocrinology and Diabetes, Department of Pediatrics, The University of Texas Health Science Center, San Antonio, TX, USA
e-mail: hale@uthscsa.edu

S. Puppala
Department of Genetics, Texas Biomedical Research Institute, San Antonio, TX, USA
e-mail: spuppala@txbiomed.org

G. Chittoor
Department of Nutrition and UNC Nutrition Research Institute, University of North Carolina at Chapel Hill, Kannapolis, NC, USA
e-mail: geetha@unc.edu

J. Blangero
South Texas Diabetes and Obesity Institute, Regional Academic Health Center, The University of Texas Health Science Center, 2102 Treasure Hills Blvd., Harlingen, TX 78550 USA
e-mail: blangero@uthsca.edu

L. Almasy
South Texas Diabetes and Obesity Institute, The University of Texas Health Science Center, San Antonio, TX, USA
e-mail: almasy@uthscsa.edu

12.1 Introduction

The epidemics of obesity, type 2 diabetes, and metabolic syndrome have become pandemics with profound public health impact worldwide, including the United States (US). The prevalence rates of these disease conditions have been increasing disturbingly in recent decades, and are associated with increased morbidity and mortality worldwide. Added to this burden, obesity, T2D, and MS prevalence rates have also increased dramatically among children and adolescents within the last few decades. In the US, these disease conditions disproportionately affect ethnic minorities, including African Americans and Mexican Americans. The disparities in children are particularly troubling because obesity is a major risk factor for future development of chronic diseases. Obesity, T2D, and MS are common complex diseases influenced by genetic, environmental factors and their interactions. The genetic contribution to obesity (Stunkard et al. 1986a; Duggirala et al. 1996, 2000; Comuzzie et al. 2001; Pankow et al. 2001; Loos and Bouchard 2003; Saunders et al. 2007), T2D (Elbein and Hasstedt 2002; Stern et al. 2002; Diamond 2003; Frayling et al. 2007), and MS and its individual components (Groop 2000; Lin et al. 2005; Biro and Wien 2010; Hinney et al. 2010) has been well established through family, twin, and adoption studies. However, progress in identification of the actual causal variants and genes contributing to these metabolic diseases

R. Duggirala et al. (eds.), *Genome Mapping and Genomics in Human and Non-Human Primates*, Genome Mapping and Genomics in Animals 5, DOI 10.1007/978-3-662-46306-2_12,

has been very limited. Three genetic mapping approaches have been widely used to localize susceptibility variants and genes that underlie the phenotypic expressions of these complex diseases, each of which has unique advantages and disadvantages (Altmuller et al. 2001; Tabor et al. 2002; Altshuler et al. 2008; Marian and Belmont 2011; Lewis and Knight 2012); please refer to the chapters on linkage and association by Almasy et al. and Hanson and Malhotra, respectively, for methodological details). The candidate gene approach examines specific genes with a potential functional role in disease pathophysiology; however, a major issue associated with this approach is nonreplication of original findings. Genome-wide linkage studies have been very successful in identifying the causal variants for single-gene disorders, but their successes in identifying genetic variants influencing complex diseases are very limited.

In recent years, however, as an alternative gene localization tool, the genome-wide association study (GWAS) method using information from common genetic variants has become a popular design. Pursuing a common disease-common variant hypothesis, GWASs have demonstrated remarkable success in localization of novel susceptibility loci for various common, complex diseases. With the advancements of the International HapMap Project and the Human Genome Project and rapid developments in high-throughput genotyping technologies, GWASs have been very successful in identifying a large number of loci for several complex diseases including obesity and T2D, implicating both known and unknown genes and new biological pathways (Frayling et al. 2007; Lango and Weedon 2008; Billings and Florez 2010; Vimaleswaran and Loos 2010; Day and Loos 2011; Fall and Ingelsson 2012). There have been efforts to replicate the original association findings or to find new association signals (e.g., *SLC16A11* [solute carrier family 16, member 11] sequence variants and T2D in Mexicans and other Latin Americans, SIGMA T2D Consortium 2014 (Williams et al. 2014)) in ethnically diverse populations, since most of the GWASs were conducted using large population- and case-control-based datasets from Europeans or populations with European ancestry (Sanghera and Blackett 2012; Saxena et al. 2012; Ng et al. 2013). One limitation of the GWAS approach is that, like linkage, it implicates a general region and an association may be due to any of a number of variants in linkage disequilibrium with a genotyped marker showing association with the phenotype of interest. Few of the GWAS signals noted above have been followed through to identification of specific functional variants.

Another drawback of these studies is that GWAS-identified common variants explain only a modest fraction of the total heritability, which is about 10 % for T2D, and less than 5 % for obesity and MS traits, suggesting that a large proportion of heritability is still unexplained (i.e., missing heritability). Therefore, there has been an increased interest in the potential role of rare variants in common complex diseases (common disease-rare variant hypothesis), which are likely to have larger effect sizes with potential functional consequences and could contribute to missing heritability (Manolio 2009; Cirulli and Goldstein 2010; Gibson et al. 2012; Agarwala et al. 2013; Zuk et al. 2014). Recent advances in next-generation sequencing technologies have made it possible to obtain complete information on rare and common sequence variants across the whole exome or genome (please refer to the chapter by Curran et al. in this volume). It is conceivable that both common and rare variants could be important contributors to complex disease risk (Gibson et al. 2012; Agarwala et al. 2013; Zuk et al. 2014). Several new collaborative national and international consortia such as the Type 2 Diabetes (T2D) Genetic Exploration by Next-generation sequencing in Ethnic Samples (T2D-GENES) and the Genetics of Type-2 Diabetes (Go-T2D) have been employing the sequencing technologies to understand the genetic architectures of complex diseases such as T2D (Flannick et al. 2014). Whole exome and whole genome sequencing studies using population-based and large pedigree-based datasets are expected to facilitate the discovery of rare and low frequency variants that influence complex phenotypes. Identification of causal variants will have great clinical significance for the development of novel preventive strategies and drug therapies for obesity, T2D, and MS, and may eventually lead to personalized medicine in the near future.

12.2 The Epidemics: Obesity, Type 2 Diabetes, and Metabolic Syndrome

12.2.1 Obesity

Obesity is a chronic disease which results from long-term positive energy balance (i.e., energy intake exceeds energy expenditure), and can be defined as an excess of body fat (Grundy 2004; Lau et al. 2007). The prevalence rates of overweight and obesity have been rapidly increasing in both developed and developing countries, and such a dramatic rise in its prevalence is attributed to overnutrition and sedentary lifestyles (Must et al. 1999; Caterson and Gill 2002; Chopra et al. 2002; Caprio 2003; Grundy 2004; Lau et al. 2007; WHO 2013). According to World Health Organization (WHO 2013) estimates, more than 1.4 billion adults (20 years and above) were estimated to be overweight in 2008; of these, over 200 million men and about 300 million women were obese. In the US, according to the National Health and Nutrition Examination Survey (NHANES) 2011–2012 data from adults aged 20 years or older, more than one-third (34.9 %) of adults were obese; and obesity rates were strikingly higher among the US minority populations: non-Hispanic (NH) European Americans [NH-EAs] = 32.6 %, NH-Asians [NH-As] = 10.8 %, Hispanics = 42.5 %, and NH-African Americans [NH-AAs] = 47.8 %] (Ogden et al. 2014). It is estimated that, by the year 2015, the occurrence of overweight and obesity in adults will be 75 % and about 44 % will be obese; with minority groups disproportionately affected (Wang and Beydoun 2007; Flegal et al. 2010). The total annual economic costs due to overweight and obesity in the US and Canada combined were estimated to be approximately $300 billion in 2009, and approximately 90 % of these total costs were related to the US (Behan and Cox 2010). Added to this burden, obesity and overweight are major risk factors for various chronic diseases such as T2D, cardiovascular disease (CVD), hypertension, MS, and certain types of cancers, leading to increased morbidity, mortality, and impaired lifestyle (Must et al. 1999; Mokdad et al. 2001; Guh et al. 2009; Schienkiewitz et al. 2012).

12.2.2 Type 2 Diabetes

Type 2 diabetes (T2D) is a complex blood glucose homeostasis disorder characterized by both insulin resistance and pancreatic β-cell dysfunction (DeFronzo 2004). The increasing prevalence of obesity parallels the increasing prevalence of T2D. Indeed, as Grubb (2002) remarks, "where obesity goes, so goes diabetes". The term "diabesity" is often used to refer to the co-occurrence of the epidemics of T2D and obesity (Caprio 2003; Kaufman 2005). Although the molecular mechanisms that underlie the association between obesity and T2D are unclear, a major feature of the pathophysiology of both obesity and T2D is insulin resistance, which manifests itself in adipose, hepatic, and skeletal muscle tissues (Goldstein 2003; DeFronzo 2004). The compound burden of an increasing global obesity epidemic together with its comorbid conditions including T2D, hypertension, and related cardiovascular complications is expected to become a major global public health problem (Sorensen 2000; Friedrich 2002; Grubbs and Brundage 2002; Zimmet 2003; Francischetti and Genelhu 2007; Apovian 2010). According to WHO (2013) estimates, globally, 347 million people have diabetes. It is estimated that about 366 million people will be afflicted with T2D by the year 2030, and the countries with the largest number of people with diabetes will be India, China, and the US by the year 2030 (Wild et al. 2004). In the US, following the Centers for Disease Control and Prevention (CDC 2011) estimates, in 2010, approximately 25.8 million people are afflicted with diabetes including about 215,000 individuals younger than 20 years (type 1 or type 2); of which, about 18.8 million have diagnosed diabetes, whereas the remaining 7 million with diabetes are unaware of their disease (see also, Cowie et al. 2009, 2010). The estimated prevalence of diagnosed diabetes based on age-adjusted data for individuals aged 20 years or older (2007–2009 National Survey Data) exhibits notable ethnic disparities (CDC 2011): NH-EAs = 7.1 %, NH-AAs = 12.6 %, NH-As = 8.4 %, and Hispanics = 11.8 % (e.g., Mexican Americans (MAs) = 13.3 %). In addition, the estimated occurrence of prediabetes in US adults aged 20 years or older is 79 million people in 2010 (CDC 2011). Using NHANES 2005–2006 data of US adults aged 20 years or older, the age and sex standardized prevalence of prediabetes [either impaired fasting glucose (IFG) or impaired glucose tolerance (IGT)] was found to be 32.0 % in MAs, compared to 25.4 % in AAs and 27.7 % in EAs (Cowie et al. 2009). Also, longstanding T2D is associated with both macrovascular (e.g., CVD) and microvascular complications such as nephropathy, neuropathy, and retinopathy and US minority populations are found to have higher rates of such complications compared to the general population (Goldstein 2003; Hunt et al. 2003; Lutale et al. 2007; Peek and Reddy 2007; Hillis et al. 2012). In the US, the estimated total economic cost of diagnosed diabetes in 2012 was $245 billion (ADA 2013).

12.2.3 Metabolic Syndrome

The clustering of cardiometabolic abnormalities such as obesity, insulin resistance, impaired glucose tolerance, dyslipidemia [elevated triglyceride and decreased high-density lipoprotein (HDL) cholesterol levels], and hypertension has been referred to as the Insulin Resistance Syndrome (IRS) or the Metabolic Syndrome (MS) (Reaven 1988; DeFronzo and Goodman 1995; Grundy 2007). The increasing prevalence of MS in global populations is mainly attributable to the rising prevalence of overweight and obesity (Grundy 2004; Deboer 2011). Obesity, insulin resistance, and their associated inflammatory processes are considered to be the major underlying mechanisms of MS (Reaven 1988; DeFronzo 1995; Cruz et al. 2005; Yang and Ming 2011). MS is associated with an increased risk for CVD, coronary heart disease (CHD), T2D, and total mortality (Lempiainen et al. 1999; Goldstein 2003; Lorenzo et al. 2006, 2007; Taylor et al. 2008). Several other metabolic disturbances have also been considered as components of MS including abnormalities of fibrinolysis, hyperuricemia, microalbuminuria, elevated markers of chronic inflammation, endothelial dysfunction, acanthosis nigricans, polycystic ovary syndrome (PCOS), and nonalcoholic fatty liver disease (NAFLD) (Isomaa 2003; Opie 2007; Brickman et al. 2010; Brown et al. 2010; Henneman et al. 2010; Deboer 2011; Lerman and Lerman 2011). A number of studies have examined the underlying structure of the MS using MS-related traits and factor analysis (Meigs 2000; Hanson et al. 2002; Arya et al. 2002a; North et al. 2003; Monda et al. 2010; Fowler et al. 2013). In general, such studies have found more than one underlying factor to explain the structure of the MS-related multivariate data, suggesting the possibility of multiple distinct or separate biological processes underlying MS. In consideration of such findings together with the fact that a large number of traits may constitute the MS, it has become difficult to precisely define MS. However, several diagnostic criteria have been proposed by expert groups to define a composite dichotomous MS phenotype in adults, using information from similar core components of MS, primarily including measures such as obesity, glucose, lipids, and blood pressure.

Aside from the debate on its clinical utility (Gordon 1998; Simmons et al. 2010; Tenenbaum and Fisman 2011; Golden et al. 2012), some commonly used MS definitions are based on guidelines proposed by the WHO, the European Group for the Study of Insulin Resistance (EGIR), the National Cholesterol Education Program—Third Adult Treatment Panel (NCEP/ATP III), the American Association of Clinical Endocrinology (AACE), the American Heart Association and National Heart, Lung and Blood Institute (AHA/NHLBI), and the International Diabetes Federation (IDF), as reviewed elsewhere (Lin et al. 2005; Ford and Li 2008; Alberti et al. 2009; Monda et al. 2010; Ford et al. 2010b; Kassi et al. 2011). However, such definitions differ in the number of traits and the population-based thresholds or cutoffs required for the diagnosis. Also, definitions of the WHO, AACE, EGIR are focused on insulin resistance, whereas the IDF focuses on waist circumference as a measure of central obesity (Monda et al. 2010; Kassi et al. 2011). Several studies have used the NCEP/ATPIII (2001) definition for its simplicity (Cruz and Goran 2004; Sung et al. 2009; Simmons et al. 2010; Henneman et al. 2010; Farook et al. 2012), since it avoids emphasis on a single cause (Grundy et al. 2005; Grundy 2005). The NCEP/ATPIII definition requires the presence of at least 3 of the following 5 factors: increased waist circumference, hypertriglyceridemia, low HDL cholesterol, hypertension, and high fasting glucose. Recently, a Joint Scientific Statement was proposed to harmonize the MS definition (Alberti et al. 2009). Using these guidelines and the NHANES 2003–2006 data, the age-adjusted prevalence of MS in US adults aged 20 years and older was estimated to range from 34.3 % (males = 36.1 % [NH-EAs = 38.4 %, NH-AAs = 25.5 %, MAs = 34.4 %], females = 32.4 % [NH-EAs = 31.3 %, NH-AAs = 38.2 %, MAs = 41.9 %]) to 38.5 % (males = 41.9 % [NH-EAs = 43.2 %, NH-AAs = 32.5 %, MAs = 44.5 %], females = 35.0 % [NH-EAs = 33.8 %, NH-AAs = 41.1 %, MAs = 44.1 %]), which corresponds to about 77 to 86 million people (Ford et al. 2010a). It is estimated that the average annual total health care costs spent by individuals with MS ($5,732) differed from those without MS ($3,581) by a magnitude of 1.6 and that the total costs increased by an average of 24 % per additional risk factor (Boudreau et al. 2009).

12.2.4 The Epidemics of Obesity, Type 2 Diabetes, and the Metabolic Syndrome in Children and Adolescents

Disturbingly, the prevalence rates of overweight and obesity among children and adolescents have also been increasing worldwide in recent decades, including in US pediatric populations, wherein certain minority ethnic groups including MAs are affected disproportionately (Mehta et al. 2007; Crocker and Yanovski 2009; Ogden and Clementi 2010; Fowler et al. 2013). Globally, more than 42 million children under the age of 5 years were overweight in 2010 (WHO 2013). In the US, based on NHANES 2011–2012 data, among children and adolescents aged 2–19 years, the rate of obesity was 16.9 % (Ogden et al. 2014). There are significant ethnic disparities in childhood obesity prevalence: NH-EAs = 14.1 %, NH-AAs = 20.2 %, NH-As = 8.6 %, and Hispanics = 22.4 %. Childhood MS-related risk factors such as obesity have been shown to be strong predictors of development of various disease conditions in adulthood including T2D, CVD, and MS (Must et al. 1999; Berenson 2002; Virdis et al. 2009; Deboer 2011). Importantly, the increasing occurrence of obesity parallels increased prevalence of MS and its correlates including T2D in children and adolescents (ADA 2000; Goran et al. 2003; Cruz et al. 2005; Pinhas-Hamiel and Zeitler 2005; Caceres et al. 2008; Deboer 2011; May et al. 2012). In the US, individuals under 20 years of age, 215,000 are affected with diabetes (type 1 or type 2) (CDC 2011). Regarding prediabetes, the unadjusted prevalences of IFG, IGT, and prediabetes in the NHANES 2005–2006 data for adolescents aged 12–19 years were 13.1, 3.4, and 16.1 %, respectively (Li et al. 2009).

There has been no fully validated definition of MS in children and adolescents, and there are concerns about pediatric MS diagnostic criteria since MS prevalence estimates vary in accordance with the definitions used (Ford and Li 2008; Huang 2008; Sumner 2009; Deboer 2011; Kassi et al. 2011). However, definitions of MS in children tend to use a similar approach to those in adults, usually the NCEP/ATPIII criteria (Ford and Li 2008). Using the NHANES 2001–2006 sample of adolescents aged 12–19 years, an estimated 8.6 % of adolescents were found to have MS, which extrapolates to approximately 2.5 million adolescents in the US population (Johnson et al. 2009). The overall prevalence was highest in Hispanic adolescents (11.2 %), followed by NH-EAs (8.9 %) and NH-AAs

(4.0 %). Several studies reported high occurrence of MS among obese children and adolescents, especially in ethnic minority groups, and MS prevalence increased with worsening obesity (Cruz et al. 2004; Weiss et al. 2004; Butte et al. 2005; Messiah et al. 2010). Cruz et al. (2004) found an inverse association between insulin sensitivity and the number of components of MS, and concluded that insulin resistance is central to the MS profile in Hispanic children with a family history of T2D. Numerous studies have identified a strong relationship between a positive parental history and/or family history of T2D and the co-occurrence of obesity and T2D related traits in children (ADA 2000; Cruz et al. 2002; Goran et al. 2003; Valdez et al. 2007). Aside from the issue of MS definition, there have been suggestions to focus on individual MS risk factors in children as they relate to obesity, insulin resistance, glucose intolerance, inflammation, endothelial dysfunction, and hypertension and their correlations (Steinberger et al. 2009; Schutte et al. 2009; Hoffman 2009).

12.2.4.1 Cardiometabolic Risk in Mexican American Children: The SAFARI Study

To address the issue of metabolic syndrome or clustering of cardiometabolic risk in MA children and adolescents, we designed the San Antonio Family Assessment of Metabolic Risk Indicators in Youth (SAFARI) study to identify signs of MS and future disease risk in MA children and adolescents in San Antonio, Texas and surrounding areas, and to examine their genetic basis. We examined 673 children and youth, aged 6–17 years old (mean age = 11.5 years and girls = 49.5 %), from large, predominantly lower income MA families at increased risk of diabetes, whose adult members had previously participated in one of three community-based genetic epidemiologic studies in San Antonio. Thus, these children represent the youngest of multiple generations from their families to have taken part in our studies. An extensive battery of clinical tests and interviews was administered to SAFARI participants to collect biomedical endpoint and covariate data on family history, demographic, phenotypic, genotypic, and environmental factors related to MS. Three of 673 children were found to have T2D, based on our clinic examinations; these children were excluded from the analyses reported here. So, our results relate to the 670 nondiabetic children and adolescents who participated in SAFARI, which were published recently (Fowler et al. 2013), and are summarized below. We defined MS as the presence of three or more of six cardiometabolic risk factors, including increased waist circumference (abdominal obesity), hyperinsulinemia, glucose intolerance (i.e., prediabetes: either impaired fasting glucose (IFG), impaired glucose tolerance (IGT), or both), hypertriglyceridemia, low HDL cholesterol (HDL-C), and elevated systolic [SBP] and/or diastolic [DBP] blood pressure. Using this definition, 18.7 % of the young people in SAFARI exhibited MS. Even very young children were affected with MS: one-third of the children with MS were less than 10 years old including three 6 year olds. While the overall prevalence of prediabetes in SAFARI children was 13.2 %, it was about 31 % in children affected with MS. In line with previous studies, the prevalence of MS increased with obesity level. In the SAFARI study, 52.7 % of children were either overweight or obese, and 33.6 % were either obese or severely obese. In SAFARI children, only about 1 % of normal-weight children had MS, but among the 65 young people in the study who were severely obese, over two-thirds had already developed MS. Since SAFARI participants came from extended families, we were able to identify strong evidence of heritability for the metabolic syndrome, its six cardiometabolic components, and MS-related risk factors. MS itself exhibited 68 % heritability, which means that 68 % of variation in the occurrence of MS is attributable to genetic factors. In addition, we found evidence for pleiotropy: genetic factors that simultaneously influence multiple, related MS traits. These findings provide insight into the complex genetic architecture underlying MS risk in these children. Insights gained through this approach may help to tailor effective dietary, physical activity, and other interventions for high-risk children and youth.

12.3 Approaches to Localize Susceptibility Genes for Obesity, Type 2 Diabetes, and Metabolic Syndrome

Obesity, T2D, MS and their related quantitative traits such as body mass index (BMI), glucose levels, and cholesterol levels have strong genetic components as revealed by twin, family, and adoption studies (Stunkard et al. 1986a, 1990; Allison et al. 1996; Edwards et al. 1997; Maes et al. 1997; Comuzzie and Allison 1998; Duggirala et al. 1999; Poulsen et al. 1999; Hanson et al. 2002; Arya et al. 2002a; North et al. 2003; Arya et al. 2004). For example, heritability estimates for obesity vary from a low of ~25 % in adoption studies (Stunkard et al. 1986b) to a medium of ~40 % in family studies (Maes et al. 1997; Duggirala et al. 1999; Poulsen et al. 1999; Arya et al. 2002a, 2004) to a high of ~70 % in twin studies (Stunkard et al. 1986a, 1990; Allison et al. 1996). The heritability of T2D was 53 %, and it was ~46 % for age at onset of T2D (Duggirala et al. 1999). Using the NCEP/ATPIII definition, the heritability of MS was found to be 51 % (Farook et al. 2012). There have been continued attempts to localize specific genetic determinants of variation in obesity, T2D, and MS and its component traits, using a variety of genetic mapping techniques including candidate gene, genome-wide linkage, and GWAS approaches (Lander and Schork 1994; Collins 1995; Lander and Kruglyak 1995; Risch and Merikangas 1996; Risch 2000; Cardon and Bell 2001; Hirschhorn and Daly 2005; Laird and Lange 2006; McCarthy 2010; Marian 2012). Earlier genetic studies of complex diseases were primarily based on a priori knowledge of potential role of a gene in the pathogenesis of a phenotype of interest, commonly known as candidate gene approach (Daly and Day 2001; Tabor et al. 2002). Subsequently, in contrast to the candidate gene approach, genome-wide linkage and association approaches have been employed as unbiased tools to screen the entire genome using information from evenly spaced genetic markers for localizing susceptibility variants/genes for a given phenotype without any consideration to the function of such markers (Goldgar 1990; Schork 1993; Kruglyak 1999; Cardon and Bell 2001; Daly and Day 2001; Tabor et al. 2002; Blangero 2004; Hirschhorn and Daly 2005; Laird and Lange 2006; McCarthy et al. 2008; Marian and Belmont 2011). A discussion on each one of the genetic mapping approaches and the corresponding findings related to obesity, T2D, and MS and its component traits is presented in the following sections.

12.3.1 Candidate Gene Association Studies

The traditional candidate gene association approach is a hypothesis-driven technique based on current understanding of the biology and pathophysiology of a disease of interest (Daly and Day 2001; Tabor et al. 2002; Bell et al. 2005; Farooqi and O'Rahilly 2005; Zhu and Zhao 2007). Candidate genes are those with known chromosomal locations and that specify molecules such as receptors, hormones, transporters, and other proteins that are part of a biochemical pathway related to the phenotype under study. This technique examines a correlation between a phenotype (a disease condition such as T2D or a quantitative trait such as insulin or glucose levels) and a genotype, using appropriate methods to deal with a discrete trait (such as differences in allele frequencies between cases and controls) or a quantitative trait (such as differences in trait mean values in accordance with different genotypes) (Almasy and MacCluer 2002; Lewis and Knight 2012). Candidate genes can be divided into two types: functional and positional (Daly and Day 2001; Bell et al. 2005; Doria et al. 2008). Functional candidates are genes that are involved in the pathogenesis of disease of interest (e.g., obesity) with a known function or role in body weight regulation and regulation of energy balance or adipose tissue biology (Clement et al. 1996; Bell et al. 2005; Choquet and Meyre 2011a). Positional candidates are genes that lie within genomic regions that have been shown to be genetically important in linkage or association studies (Bell et al. 2005; Doria et al. 2008; Choquet and Meyre 2011a). For example, signaling molecules such as the proopiomelanocortin (*POMC*) and melanocortin 4 receptor (*MC4R*) genes have been associated with obesity, and were identified through positional cloning of mouse obesity genes (Bell et al. 2005; Cai et al. 2006; Benzinou et al. 2008; Choquet and Meyre 2011a, b). Positional candidates are discussed in detail later as part of the genome-wide approaches. This approach has been effective in identifying genes responsible for extreme early-onset forms of diseases segregating as single-gene Mendelian disorders [e.g., maturity onset diabetes of the young (MODY), insulin resistance syndrome, and Wolfram syndrome] (Barroso 2005; Hansen and Pedersen 2005; Vaxillaire and Froguel 2008). However, to date, this approach has yielded disappointing results in identifying genes with measurable effects on normal variation, for example, as seen in human adiposity levels. Major reasons for the failure of traditional candidate gene studies perhaps are due to poor understanding of the pathophysiology of a phenotype of interest (e.g., obesity) (Gloyn 2003; Hansen and Pedersen 2005; Doria et al. 2008), small sample sizes, lack of replication of findings, and limitations on their ability to include all possible causative genes and polymorphisms (Daly and Day 2001; Tabor et al. 2002; Rankinen et al. 2006).

The candidate gene approach has identified several potential candidate genes for human obesity (Perusse et al. 2001; Rankinen et al. 2006). Several genes implicated in rodent models of severe or monogenic obesity have also been shown to be contributors to rare forms of early-onset obesity in humans, particularly those involved in the leptin melanocortin pathway such as leptin (*LEP*), leptin receptor (*LEPR*), *POMC*, prohormone convertase 1 (*PCSK1*), and *MC4R* (Bouatia-Naji et al. 2006; Wilson et al. 2006). Leptin plays a critical role in the regulation of body fat and body weight and leptin receptor mediates the effects of leptin; mutations in *LEP* and *LEPR* have been shown to lead to rare forms of human early-onset obesity (Tartaglia et al. 1995; Farooqi et al. 1999). The melanocortin 4 receptor plays a major role in the regulation of food intake and energy homeostasis metabolism, and mutations in the *MC4R* gene have been identified in subpopulations of morbidly obese individuals (Hinney et al. 2000, 2010); (Farooqi et al. 2003). Proopiomelanocortin is involved in feeding and pigmentation pathways and mutations in *POMC* have been reported in a few individuals with marked obesity (Krude et al. 2003). Other genes implicated in monogenic obesity in humans include *PCSK1* (i.e., regulation of energy metabolism) (Jackson et al. 1997; Farooqi et al. 2007) and brain-derived neurotrophic factor (*BDNF*) with relevance to eating behavior, body weight regulation, and hyperactivity (Rios et al. 2001; Rios 2011). Additional genes studied to identify variants influencing obesity and its related traits include: dopamine receptor D4 (*DRD4*) (Nothen et al. 1994); peroxisome proliferator-activated receptor gamma 2 (*PPARγ2*) (Ristow et al. 1998); β_3-adrenoreceptors (*ADRB3*) (Walston et al. 1995; Widen et al. 1995; Clement et al. 1996; Kurokawa et al. 2008); uncoupling protein (*UCP*) 1, 2, and 3 (Oppert et al. 1994; Norman et al. 1997; Cassell et al. 2000; Yanovski et al. 2000; Rosmond 2003); and endocannabinoid receptor 1 (*CNR1*) (Benzinou et al. 2008). Other candidate genes controlling important functions in glucose metabolism have been explored to assess their contribution to obesity. For example, a polymorphism in the tumor necrosis factor-alpha (*TNFα*) gene (G-308A) appears to improve insulin sensitivity and decrease adipocyte apoptosis. Polymorphisms in the *PPARγ-2* gene are associated with BMI, Pro115Gln with increased BMI, and Pro12Ala with decreased BMI and increased adipocyte differentiation (Fernandez-Real et al. 1997; Ristow et al. 1998; Rosmond et al. 2003). Some of the candidate genes discussed above (e.g., *MC4R*, *BDNF*, and *POMC*) have also been identified by GWASs to influence variation in common forms of obesity (see below).

Numerous candidate genes have been studied to assess their relevance to T2D risk (Gloyn 2003; Hansen and Pedersen 2005; Doria et al. 2008). For example, *PPARγ* is a strong candidate for T2D, and a common polymorphism (Pro12Ala) in

PPARγ has been associated with T2D in several European populations (Altshuler et al. 2000). The potassium inwardly rectifying channel, subfamily J, member 11 (*KCNJ11*) gene is implicated in the regulation of glucose-induced insulin secretion, and the *KCNJ11* E23 K polymorphism has been associated with T2D (Gloyn et al. 2003; Vimaleswaran and Loos 2010). These two genes harbor missense variants that are associated with T2D and encode proteins that act as targets for antidiabetes medications and management (Imamura and Maeda 2011). Insulin receptor substrate 1 (*IRS1*) is an important component of insulin action in skeletal muscle, adipose tissue, and pancreatic β-cells. *IRS1* G972R is one of the most extensively studied polymorphisms of T2D, but its association findings have been inconsistent (Kovacs et al. 2003; van Dam et al. 2004). Genetic variation at the ectonucleotide pyrophosphate phosphodiesterase (*ENPP1*) gene is associated with obesity and T2D-related traits (Jenkinson et al. 2008). Some other examples of T2D, obesity, and MS candidate genes include: Wolfram syndrome 1 (*WFS1*) (Riggs et al. 2005; Yamada et al. 2006) and adiponectin (*APM1*) (Ouchi et al. 1999; McCarthy and Froguel 2002; Arita et al. 2012). Several susceptibility genes for monogenic forms of diabetes [i.e., maturity onset diabetes of the young (MODY)] have been identified (Bell et al. 1991; Winckler et al. 2007; Vaxillaire and Froguel 2008), and recent GWASs have revealed overlap between certain loci implicated in monogenic and common forms of T2D such as hepatocyte nuclear factor 1-alpha (*HNF1A*) and *HNF4A* (Lehman et al. 2007; Voight et al. 2010; Kooner et al. 2011; Gardner and Tai 2012). Thus, despite its limitations, the candidate gene approach has contributed to an overall understanding of the mechanisms underlying the phenotypic expressions of complex diseases such as obesity, T2D, and MS.

12.3.2 Genome-Wide Linkage Studies

An alternative gene identification method is genome-wide linkage screening, discussed in detail in the Chapter by Almasy et al. This method differs from the conventional candidate gene approach in that it does not require a priori assumptions concerning the potential importance of genes or chromosomal regions and it facilitates a true search for genetic effects across the entire genome (Schork 1993; Blangero 2004). It is a hypothesis-free approach in which the entire genome is scanned to test whether certain chromosomal regions co-segregate with a trait or disease locus of interest. This approach requires related individuals such as siblings or other family members. Linkage analysis is of two types: parametric and nonparametric. Parametric, or model-based, linkage analysis involves specifying a model of inheritance for the disease (typically dominant or recessive) (Morton 1955; Elston 1992; Easton et al. 1993; Schork et al. 1993; Kruglyak et al. 1996). Nonparametric, or model-free, linkage is based on correlations between degree of allele sharing and degree of phenotypic similarity (Haseman and Elston 1972; Goldgar 1990; Schork et al. 1993; Amos 1994; Fulker et al. 1995; Kruglyak et al. 1996; Blangero and Almasy 1997; Almasy and Blangero 1998; Hauser and Boehnke 1998; Almasy et al. 1999b; Blangero et al. 2001). The parametric linkage method is more powerful when the underlying genetic model is correctly specified (Bailey-Wilson and Wilson 2011). On the other hand, given the multifactorial nature of complex traits it is difficult or impossible to specify all the required parameters of a Mendelian model of inheritance for such phenotypes. As such it has been more difficult to unravel the genetic component of complex traits using linkage approaches compared to monogenic traits (Almasy and Blangero 2008, 2009; Bailey-Wilson and Wilson 2011). However, several common disease-predisposing variants were identified in early linkage findings, for example, a widely replicated human QTL linkage for obesity-related traits (leptin) with a LOD score of 4.95 (Comuzzie et al. 1997).

Genomic regions showing statistically significant linkage are assumed to be harboring susceptibility genes. However, model-free linkage results in identification of broad chromosomal regions that harbor dozens or hundreds of genes and it is often difficult to identify a specific variant linked with the disease or phenotype of interest (Norman et al. 1997). Thus, linkage scans yield candidate chromosomal regions in which positional candidate genes are identified for further analyses. Follow-up and fine-mapping studies including combined linkage/disequilibrium analysis (Almasy et al. 1999b), are usually performed to test for association between disease phenotypes and genetic variants in the positional candidate genes. The results of linkage studies are expressed as LOD scores which are defined as the logarithm of odds (LOD) that the disease or trait locus and the genotyped marker locus are linked on the chromosome versus unlinked. An LOD score greater than 3 indicates that the null hypothesis of independent assortment (i.e., no linkage) is rejected and that there is significant evidence for linkage between the trait locus and the marker locus (Bailey-Wilson and Wilson 2011). The most important chromosomal regions exhibiting significant linkages (LOD > 3.0) with obesity, T2D, and MS-related phenotypes are presented in the following section.

12.3.2.1 Obesity Linkage Findings

The vast body of literature covering linkage and association studies of obesity and related phenotypes has been extensively reviewed (Rankinen et al. 2006). Based on five sources of evidence, i.e., single-gene mutations, Mendelian disorders, quantitative trait loci from animal studies, association and linkage studies, a human obesity gene map has been created and updated each year up to 2005 (Rankinen et al. 2006). According to this map, the number of genes, markers, and chromosomal regions that have been linked or associated with obesity-related phenotypes is currently ~253 human quantitative trait loci (single genes with large effects) from 61 genome scans (Rankinen et al. 2006). In addition, 176 human obesity cases due to single-gene mutations in 11 different genes have been reported, and 50 loci related to Mendelian syndromes relevant to obesity have been mapped to a genomic region (Rankinen et al. 2006). A few important findings from genome scans of obesity-related traits in different populations are briefly reviewed here.

The San Antonio Family Diabetes Study found evidence for linkage (LOD = 3.1) of the OB gene region (7q31.3) with the sum of five extremity skinfolds suggesting that the OB gene or a gene nearby on chromosome 7 might be involved in obesity (Duggirala et al. 1996). Strong evidence of linkage for a QTL influencing variation in plasma leptin levels in Mexican Americans on chromosome 2p21 (LOD = 5.0) comes from the San Antonio Family Heart Study (Comuzzie et al. 1997). The evidence of linkage was substantially increased (LOD = 7.5) through saturation mapping (Hixson et al. 1999). This linkage has been replicated in African-American families (Rotimi et al. 1999) and in Paris-Lille families (Hager et al. 1998). In another linkage study in Mexican Americans, significant evidence of linkage (LOD = 3.2) between the *ADRB3* locus and BMI was found on chromosome 8 (D8S1121 and *ADRB3* gene Trp64Arg polymorphism) (Mitchell et al. 1999). Strong evidence of linkage with obesity, defined as BMI > 27 kg/m^2, was found on chromosome 10p (LOD = 4.9) between markers D10S197 and D10S611, and suggestive evidence of linkage was found with plasma leptin levels on 2p21 (LOD 2.7) and 5q (LOD = 2.9) in Paris-Lille families (Hager et al. 1998). Lee et al. (Lee et al. 1999) reported evidence of linkage between markers on chromosome 20q13 and obesity phenotypes using both quantitative (BMI and percentage body fat) and qualitative traits (BMI ≥ 30 and percentage of body fat ≥40 %) in families with white ancestry. Strong evidence of linkage was found on chromosome 20 (LOD = 3.2) using BMI as a discrete trait in an affected sibpair test, while a parametric, affecteds-only analysis yielded a LOD of 3.1 ($p = 0.00009$) (Lee et al. 1999).

Genome scan results on Pima Indians include several obesity-related phenotypes such as BMI, percentage body fat, the ratio of waist-to-thigh circumference, 24-h metabolic rate, sleeping metabolic rate, 24-h respiratory quotient, and leptin levels. The strongest evidence for a QTL for BMI in the Pima was on chromosome 11q (LOD = 3.6), with suggestive evidence for linkage with other phenotypes including percentage body fat (LOD = 2.8), 24-h energy expenditure (LOD = 2.0) and diabetes status (LOD = 1.5) (Norman et al. 1997, 1998), and a bivariate LOD of 5.0 for BMI and diabetes status reported in the same region (Hanson et al. 1998). In addition, significant evidence for linkage with 24-h respiratory quotient (RQ, LOD = 3.0) was found on chromosome 20q11.2 in this population (Norman et al. 1998). Also, Walder et al. (2000) found a locus on chromosome 6p (LOD = 2.1) influencing plasma leptin concentrations in the Pimas. Importantly, Hunt et al. (2001) showed evidence for the presence of a predisposition locus with a multipoint heterogeneity LOD score of 3.5 at D20S438 for BMI on chromosome 20 in morbidly obese women from Utah pedigrees, and this region overlaps with the RQ linkage region on 20q11 in Pima Indians. In 2004, Arya et al. identified a major susceptibility locus for BMI on chromosome 4p15 in Mexican Americans (Arya et al. 2004). This 4p linkage region harbors two important positional candidate genes for obesity, *PPARGC1*, and *CCKAR*. This 4p linkage region was previously reported to be significantly linked to severe obesity in White American females (Stone et al. 2002). So far, about 10 genomic regions (chromosomes 2p, 3q, 4p, 5cen-q, 6q, 7q, 10p, 11q, 17p, and 20q) have been identified (with subsequent replications) to contain potential genes that influence obesity-related traits with measurable effects (Duggirala et al. 1996; Comuzzie et al. 2001; Duggirala et al. 2001; Loos and Bouchard 2003; Arya et al. 2004; Saunders et al. 2007). Thus, linkage studies have yielded several positive findings out of which only a few (15 QTLs) have been replicated in independent studies. Fine-mapping studies were not that successful to identify the functional variants that likely underlie these linkage signals. Based on these results from human studies, it is apparent that numerous genes are involved in influencing energy balance and fat accumulation. However, to date, the genome-wide linkage approach has shown a limited success in identifying loci for common forms of obesity. There is also very limited knowledge in regard to the genome-wide linkage screens for obesity or its related traits in children. A major locus was found on chromosome 6q that influences childhood obesity-related traits in French families (Meyre et al. 2004). Recently, major loci were identified for obesity and its related traits such as adiponectin and ghrelin in Hispanic children (Tejero et al. 2007; Voruganti et al. 2007).

12.3.2.2 Type 2 Diabetes Linkage Findings

In the past 15 years, family-based data have been examined to localize T2D susceptibility genes using genome-wide linkage or positional cloning approaches. Despite the fact that several causal mutations for the monogenic forms of diabetes, such as maturity onset diabetes of the young (MODY), have been identified using this approach, it has had limited success in mapping genes related to the common forms of T2D due to: (a) the non-Mendelian mode of inheritance of human T2D, and (b) the large chromosomal areas identified. However, based on such findings, it is increasingly apparent that T2D and related conditions are influenced by at least a few relatively common genes (i.e., oligogenes) (Duggirala et al. 1999; Comuzzie 2002). Major susceptibility loci have been identified for T2D and related phenotypes at the following chromosomal regions with claims for replication: 1q, 2q, 3q, 5q, 6q, 8p, 9q, 10q, 11q, 12q, and 20q (Stern et al. 1996; Hanis et al. 1996; Duggirala et al. 1999; Elbein and Hasstedt 2002; McCarthy and Froguel 2002; Arya et al. 2002a). Additionally, a meta-analysis of 23 T2D linkage studies from the International T2D Linkage Analysis Consortium found modest evidence for T2D susceptibility loci on chromosomes 4, 10, 14, and 16 (Guan et al. 2008).

Although several T2D linkage studies have been following up their findings with further gene discovery activities, so far, successes from such efforts have been limited to a couple of T2D studies. The localization of the NIDDM1 susceptibility gene to chromosome 2q with a LOD of 4.1 in Mexican Americans (Hanis et al. 1996), subsequently led to the discovery of Calpain 10 (*CAPN10*) on chromosome 2q37.3 as a T2D susceptibility gene (Horikawa et al. 2000). This finding was expected to provide more information toward the link between insulin resistance and insulin secretion, but could not be robustly replicated in other populations. HNF4α on chromosome 20 was identified via linkage analysis as a second putative T2D susceptibility gene (Love-Gregory et al. 2004; Silander et al. 2004a, b). Subsequently, in an Icelandic population, Reynisdottir et al. (Reynisdottir et al. 2003) identified a region on chromosome 5 (LOD = 2.9–3.4) and a region on 10 (LOD = 2.8) with suggestive linkage to T2D, and showed that the chromosome 10 region harbored the transcription factor 7-like 2 (*TCF7L2*) gene, involved in the Wnt-signaling pathway (Grant et al. 2006). The discovery of *TCF7L2* became the first major success in T2D genetics. This 10q chromosomal region was previously implicated by linkage studies (LOD: T2D = 2.9 and age at onset of T2D = 3.8) in Mexican Americans (Duggirala et al. 1999) and subsequent GWASs have shown that a linkage disequilibrium-dependent approach also would have ultimately pointed to this gene (Saxena et al. 2007; WTCCC 2007). With an overall allelic relative risk of 1.56 (Florez 2008), *TCF7L2* currently represents the best-supported T2D risk gene.

12.3.2.3 Metabolic Syndrome Linkage Findings

Genome-wide linkage analysis is a widely used approach to map susceptibility loci for metabolic syndrome and its components. A number of studies have been performed using individual components of metabolic syndrome such as T2D and obesity-related traits, dyslipidemic traits, and hypertension. Bivariate linkage analysis using combinations of these traits and linkage analyses of phenotypes derived from factor analysis of MS-related traits have also been performed (Meigs 2000; Duggirala et al. 2001; Arya et al. 2002a; Monda et al. 2010; Edwards et al. 2011; Kraja et al. 2011; Farook et al. 2012). Additionally, a few genome-wide linkage studies have been performed using yes/no definitions of metabolic syndrome as the focal phenotype. The literature on shared genetic variants for the components of metabolic syndrome is limited. Some of the interesting MS linkage findings are presented in the following discussion. Kissebah et al. (2000) performed genome-wide linkage analysis in 507 nuclear Caucasian families and found signals for two QTLs for MS-related traits. One QTL was found on chromosome 3q27 for six quantitative phenotypes related to the abdominal obesity-metabolic syndrome (LOD scores ranging from 2.4 to 3.5), in regions harboring the solute carrier family 2 of the glucose transporter (*GLUT2*) gene and the adiponectin locus. While a second QTL on 17p12 was found for plasma leptin levels [LOD = 5.0] (Kissebah et al. 2000). A German genome-wide linkage study with 250 families identified a link between MS and a locus on chromosome 1p36.13 (Hoffmann et al. 2007). This region was previously linked to gallbladder disease in Mexican Americans (Puppala et al. 2007), and also associated with an increase in body size to adipose ratio as estimated by bioelectric impedance analysis (Cai et al. 2004). In addition, this region was linked to hypertension in a Sydney sib-ship study (Benjafield et al. 2005). Bosse et al. (2007) examined 707 subjects from 264 Quebec nuclear families that showed significant evidence for linkage for MS on chromosome 15 (LOD = 3.2). Another study involving 977 Caucasians from 358 families showed that regions on chromosomes 3, 4q, and 14p were strongly linked to T2D, MS, and measures of CVD (Bowden et al. 2006). A study in African Americans showed evidence of linkage with components of metabolic syndrome: chromosomes 11q24, 13p12 for lipids and obesity, respectively, while another study on Caucasian Americans demonstrated evidence for linkage between multiple regions and various MS-related traits: 8p23 (LOD = 2.4) and lipids, 14q24 (LOD = 2.4) and obesity, and 15q15 (LOD = 3.2) and blood pressure (Kraja et al. 2005a, b). There is also evidence in Mexican Americans for major loci influencing MS-related phenotypes such as lipoprotein metabolism (Rainwater et al. 1999; Almasy et al. 1999a; Duggirala et al. 2000; Hegele 2001; Arya et al. 2002a), blood pressure (Krushkal et al. 1999; Levy et al. 2000; Rice et al. 2002), and diabetic nephropathy-related phenotypes (e.g., Krolewski et al. 2006; Arar et al. 2007; Puppala et al. 2007).

12.3.2.4 Genome-Wide Linkage Studies of Obesity, Type 2 Diabetes, and Metabolic Syndrome in Mexican Americans: Results from the San Antonio Family Diabetes/Gallbladder Study (SAFDGS)

The San Antonio Family Diabetes/Gallbladder Study (SAFDGS), an important genetic study of obesity, T2D, and MS in Mexican American families, began as the San Antonio Family Diabetes Study (SAFDS), a family-based genetic study with the primary goal of mapping susceptibility genes for type 2 diabetes and related phenotypes. The SAFDS is an extended pedigree study of 32 low-income Mexican American families ascertained on a diabetic proband. All first, second, and third degree relatives of the proband who were age 18 years or above were considered eligible for the baseline SAFDS exam, which recruited and examined 579 individuals between 1991 and 1994. Subsequently, a second follow-up and an extension of the SAFDS, called the San Antonio Family Gallbladder Study (SAFGS), examined 741 original and new individuals from 39 extended families, including 8 new families. In total, so far, 905 SAFDGS (SAFDS and SAFGS combined) participants from 40 families have taken part in at least one of the three examinations. To date, we have identified several susceptibility loci influencing obesity, T2D, and cardiovascular disease and/or its risk factors using variance components linkage analyses (Table 12.1). These include loci relating to glucose concentrations [chromosome 11, (Stern et al. 1996)], and insulin precursors and obesity-related traits [chromosome 7, (Duggirala et al. 1996)], T2D and age of diabetes onset [chromosome 10, (Duggirala et al. 1999)], fasting specific insulin concentrations, insulin resistance, and other components of the metabolic syndrome [chromosome 6, (Duggirala et al. 2001), (Arya et al. 2002a)], triglyceride levels [chromosome 15, (Duggirala et al. 2000)],

Table 12.1 SAFDGS linkage findings in Mexican Americans

Phenotype	Chromosomal location	Marker region	Position in cM	LOD	References
2-h glucose	11p	D11S899-D11S1324	62	3.4	Stern et al. 1996
32,33—split proinsulin	7q	HCPA1	163	4.2	Duggirala et al. 1996
Proinsulin	7q	HCPA1	163	3.2	Duggirala et al. 1996
Extremity skinfolds	7q	D7S514	130	3.1	Duggirala et al. 1996
T2D/age at onset	10q	D10S587	148	2.9/3.8	Duggirala et al. 1999
Triglycerides	15q	GABRB3-D15S165	25	3.9	Duggirala et al. 2000
Fasting insulin levels and insulin resistance	6q	D6S403	150	4.1 3.5	Duggirala et al. 2001
Adipo-insulin factor	6q	D6S264	179	4.9	Arya et al. 2002a, b
Adipo-insulin factor	6q	D6S403	143	4.2	Arya et al. 2002a, b
Lipid factor	7q	D7S479-D7S471	130	3.2	Arya et al. 2002a, b
HDL cholesterol	9p	D9S925-D9S741	41	3.4	Arya et al. 2002a, b
BMI	4p	D4S2912	48	4.1	Arya et al. 2004
Quantitative Martingale residual	3p	D3S2406	114	3.8	Hunt et al. 2005
GFR-CGc	2q	D2S1363	227	3.3	Puppala et al. 2007
Metabolic syndrome	7q	D7S2212-D7S821	102	3.6	Farook et al. 2012

HDL-C levels [chromosome 9, (Arya et al. 2002b)], obesity-related phenotypes [chromosome 4, (Arya et al. 2004)], Quantitative Martingale Residuals for T2D (Hunt et al. 2005), and diabetic nephropathy-related phenotypes (chromosomes 2 and 9; Puppala et al. 2007).

12.3.3 Genome-Wide Association Studies (GWASs) of Complex Disease Phenotypes

The GWAS approach is an unbiased tool to detect disease susceptibility loci, designed to identify common genetic variants with small to moderate effect sizes (e.g., Frayling et al. 2007; McCarthy and Hirschhorn 2008; Perry and Frayling 2008; Visscher et al. 2008; Manolio 2010; Grarup et al. 2014). It is an increasingly popular alternative to genome-wide linkage. Unlike linkage, GWAS does not require related individuals, making it easier to obtain large sample sizes. Most GWASs utilize case-control samples or population-based cohorts. The genome-wide association approach scans the entire genome of many hundreds to thousands of individuals using >500,000 single nucleotide polymorphisms (SNPs). This screens the whole genome at a high resolution thereby narrowing down the associated locus more precisely (Frayling and McCarthy 2007; McCarthy and Zeggini 2009). A major advantage of the GWAS approach is that it reduces the genomic area of interest to approximately 500–1,000 kb versus the 10–15 Mb usually observed in linkage screens for human disease susceptibility loci (Frayling and McCarthy 2007; McCarthy and Zeggini 2009). A simple association analysis is performed between a phenotype and each of the SNPs to identify genetic markers associated to the phenotype with a certain statistical significance. Markers used for GWAS generally have minor allele frequencies of >0.01–0.05 and are selected to tag the most common haplotypes observed in European populations followed by East Asian populations and African populations due to inherent differences in their LD patterns (Marian and Belmont 2011). Furthermore, GWAS assumes the genetic marker to be directly causal for the phenotype of interest or the marker to be in linkage disequilibrium with the causal variant elsewhere in the identified region. Correcting for the estimated number of LD blocks in the human genome and the large number of association tests in a GWAS, a p-value of $<5 \times 10^{-8}$ (0.05/1000000) is currently considered genome-wide significant or more stringently $p < 1 \times 10^{-8}$ is considered evidence of a strong association (Hoggart et al. 2008; McCarthy et al. 2008; Marian and Belmont 2011). Replication of results is considered essential for establishing the credibility of GWAS findings and a National Human Genome Research Institute (NHGRI) working group has outlined criteria for establishing a positive replication (Chanock et al. 2007). Results or data from multiple GWASs in different samples or populations may also be combined via meta-analysis, which provides additional statistical power to detect subtle genetic effects (Zeggini et al. 2008; Kraft et al. 2009; Bush and Moore 2012). More details on association analysis approaches and GWAS can be found in the Chapter by Hanson et al. GWA studies have been extremely successful for identifying susceptibility loci for various complex diseases and traits (see www.genome.gov/GWAstudies for an overview). To date, more than 150 genetic loci have been implicated in the development of monogenic, syndromic, or common forms of obesity or T2D (McCarthy 2010; Drong et al. 2012).

12.3.3.1 The GWA Studies for Obesity

In 2007, during the first wave of GWASs, the fat mass and obesity-associated (*FTO*) gene was the first novel susceptibility locus for common forms of childhood and adult obesity identified by this approach (Dina 2008; Loos and Bouchard 2008). This finding was subsequently replicated in 13 cohorts comprising more than 38,000 individuals. Each *FTO* risk allele increased BMI by 0.10–0.13 SD units, risk of overweight by 1.18-fold, and risk of obesity by 1.32-fold. Individuals who were homozygous for the risk allele weighed about 3 kg more and had a 1.67-fold increased risk for obesity than those who did not inherit a risk allele (Frayling et al. 2007; Scuteri et al. 2007). *FTO* is thought to influence T2D risk through its effects on obesity; however, there is also evidence for more direct effects. *FTO* is associated with insulin levels and insulin resistance in children even after adjustment for BMI levels (Jacobsson et al. 2008). Other genes implicated in GWASs of obesity include: insulin-induced gene 2 (*INSIG2*) (Lyon et al. 2007), melanocortin 4 receptor (*MC4R*) (Loos et al. 2008), platelet type phosphofructokinase (*PFKP*) (Andreasen et al. 2008a), catenin (cadherin-associated protein), and β-like 1 (*CTNNBLI*) (Liu et al. 2008). Of these, *FTO* and *MC4R* are the best supported and findings for *INSIG2* and *PFKP* have been inconsistent and have failed to replicate in other studies (e.g., (Andreasen et al. 2008a). There is evidence that common genetic variation near *MC4R* is associated with risk of both adiposity and insulin resistance (Chambers et al. 2008). In addition, new loci influencing common forms of obesity have been reported (Meyre et al. 2009; Thorleifsson et al. 2009).

In the second wave of obesity GWA studies, individual GWA studies were combined through collaborative efforts to increase sample size and thus power to identify additional common variants with smaller effects. For example, the Genomic Investigation of Anthropometric Traits (GIANT) consortium is such an international collaborative initiative, which brought together several research groups focusing on anthropometric traits from across Europe and the USA. Data from seven GWASs for BMI (n = 16,876) were combined in the first meta-analysis (Loos et al. 2008).

As part of the third wave of studies, a larger meta-analysis of GWA studies grew out of the GIANT consortium which consisted of 15 cohorts with an increased sample size of 32,387 adults of European ancestry yielded 35 loci (Willer et al. 2009). Of the 35 loci identified in the first GIANT meta-analysis, eight loci were strongly replicated by a series of independent studies. Important findings in this wave included two already identified loci: *FTO* and SNPs near *MC4R* and 6 other new loci: near *NEGRI1* (*neuronal growth regulator 1*), *TMEM18* (transmembrane protein 18), *SH2B1* (SH2B adapter protein1), near *KCTD15* (potassium channel tetramerisation domain containing 15), near *GNPDA2* (glucosamine-6-phosphate deaminase2), and *MTCH2* (mitochondrial carrier homologue 2). Simultaneously, deCODE Genetics performed a meta-analysis of four GWASs for BMI with 34,416 individuals including Europeans and African Americans (Thorleifsson et al. 2009). About 43 SNPs in 19 chromosomal regions were found to be associated and were subsequently replicated in 5,586 Danish individuals and confirmed in the GIANT consortium. Of the 10 loci that reached genome-wide significance, along with *FTO* and near-*MC4R*, four loci were already identified by the GIANT consortium while four more were novel and located in or near *SEC16B*, *BAT2*, between *ETV5* and *DGKG*, *BDNF*, and between *BCDUN3D* and *FAIM2*. However, variation in *BAT2* was associated with weight, but not BMI, suggesting that this locus might contribute to overall size rather than adiposity. In a recent study, Li et al. (2010a) genotyped the 12 obesity susceptibility variants identified by the GIANT consortium and deCode genetics group in 20,431 individuals in a population-based study of white Europeans. These variants showed a cumulative effect on BMI with each additional allele increasing BMI by 0.149 units, or weight by 444 g (Li et al. 2010b). However, overall, these 12 obesity susceptibility loci accounted for only 1 % of the variation in BMI, and have a very limited predictive value for obesity. GWASs exploring associations with other obesity-related traits have successfully identified 7 additional loci. For example, Meyre et al. (2009) examined association with the risk of early-onset and morbid adult obesity in 1,380 European cases and 1,416 controls. Of the 38 loci showing association with binary traits of obesity, three new risk loci in *NPC1*, near *MAF*, and near *PTER* in addition to *FTO* and *MC4R genes*, were identified and replicated in 14,186 adults and children.

Table 12.2 Some of the genome-wide association findings for obesity and related phenotypes

Study/references	Population/ethnicity	Sample size (DP/RP)	Trait	SNP	Chr loc	Gene	Effect size	*p*-value
Frayling et al. 2007	British	4,862/38,759	BMI	rs9939609	16q12.2	*FTO*	0.4	2×10^{-20}
Scuteri et al. 2007	Sardinian	4,741/3,205	BMI	rs9930506	16q12.2	*FTO*	0.52	8.6×10^{-7}
				rs1421085				
Loos et al. 2008	European	16,876/75,981	BMI	rs17782313	18q22	*MC4R*	0.13	1.5×10^{-8}
Chambers et al. 2008	South Asian	2,684/11,955	WC	rs12970134	18q22	*MC4R*	0.25	6.4×10^{-5}
Benzinou et al. 2008				rs6232	5q15-21	*PCSK1*[c]	1.34	7.3×10^{-8}
				rs6234/ rs6235		*PCSK1*[c]	1.22	2.3×10^{-12}
Benzinou et al. 2008				rs806381	6q14-15	*CNR1*[c]	1.39	3×10^{-5}
Cho et al. 2009	Asian	16,703	WHR	rs2074356	12	*C12orf51*	0.005	7.8×10^{-12}
Willer et al. 2009	European	32,387	BMI	rs2815752	1p31.1	*NEGR1*	0.15	9.3×10^{-6}
				rs6548238	2p25.2	*TMEM18*	0.24	1.2×10^{-6}
				rs10938397	4p13	*GNPDA2*	0.18	1.0×10^{-5}
				rs10838738	11p11.2	*MTCH2*	0.15	7.1×10^{-6}
				rs7498665	16p11.2	*SH2B1*	0.17	5.4×10^{-6}
				rs11084753	19q13.11	*KCTD15*	0.22	2.6×10^{-7}
Thorleifsson et al. 2009	Icelandic population	34,416/43,651	BMI	rs3101336	1p31.1	*NEGR1*	0.18	2.5×10^{-11}
				rs10913469	1q25.2	*SEC16B*	0.17	6.2×10^{-8}
				rs2867125	2p25.2	*TMEM18*	0.31	1.7×10^{-16}
				rs7647305	3q27	*ETV5*	0.23	7.2×10^{-11}
				rs4074134	11p13	*BDNF*	0.26	4.4×10^{-11}
				rs7138803	12q13	*FAIM2*	0.17	1.2×10^{-7}
				rs8049439	16p11.2	*SH2B1*	0.17	1.4×10^{-9}
				rs29941	19q13.11	*KCTD15*	0.21	7.3×10^{-12}
Meyre et al. 2009	French population	2,796/14,186	OB	rs10508503	10p12	*PTER*	0.02	6.1×10^{-1}
				rs1424233	16q22-23	*MAF*	0.03	1.3×10^{-1}
				rs1805081	18q11-12	*NPC1*	0.062	3.4×10^{-2}

(continued)

Table 12.2 (continued)

Study/references	Population/ethnicity	Sample size (DP/RP)	Trait	SNP	Chr loc	Gene	Effect size	*p*-value
Lindgren et al. 2009	European	38,580/70,639	WHR	rs2605100	1q41	*LYPLAL1*	–	1.9×10^{-4}
			WC	rs987237	6p12	*TFAP2B*	–	7.2×10^{-12}
				rs7826222	8p23.1	*MSRA*	–	2.2×10^{-3}
Heard-Costa et al. 2009	European	31,373/38,641	WC	rs10146997	14q31	*NRXN3*	0.10	7.4×10^{-6}
Scherag et al. 2010	European	2,258/31,182	OB	rs10926984	1q43-44	*SDCCAG8*	1.16	3.9×10^{-6}
				rs12145833			1.19	4.8×10^{-7}
				rs2783963			1.15	8.7×10^{-6}
				rs17150703	8p23.1	*TNKS-MSRA*	1.18	1.9×10^{-8}
				rs13278851			1.29	2.1×10^{-8}
				rs516175			1.16	9.9×10^{-8}
Speliotes et al. 2010	European	1,23,865/12,5931	BMI	rs713586	2	*RBJ-ADCY3*	0.14	6.2×10^{-22}
					2	*POMC*		
				rs1244979	16	*GPRC5B*	0.17	2.9×10^{-21}
				rs2241423	15	*MAP2K5*	0.13	1.2×10^{-18}
				rs2287019	19	*QPCTL/GIPR*	0.15	1.9×10^{-16}
				rs1514175	1	*TNN13K*	0.07	8.2×10^{-14}
				rs13107325	4	*SLC39AB*	0.19	1.5×10^{-13}
				rs2112347	5	*FLJ35779-*	0.1	2.2×10^{-13}
					5	*HMGCR*		
				rs10968576	9	*LRRN6C*	0.11	2.7×10^{-13}
				rs3810291	19	*TMEM160*	0.09	1.6×10^{-12}
				rs887912	2	*FANCL*	0.1	1.8×10^{-12}
				rs13073807	3	*CADM2*	0.1	3.9×10^{-11}
				rs11847697	14	*PRKD1*	0.17	5.8×10^{-11}
				rs2890652	2	*LRP1B*	0.09	1.4×10^{-10}
				rs1555543	1	*PTBP2*	0.06	3.7×10^{-10}
				rs4771122	13	*MTIF3*	0.09	9.5×10^{-10}
				rs4836133	5	*ZNF608*	0.07	2.0×10^{-9}
				rs4929949	11	*RPL27A*	0.06	2.8×10^{-9}
				rs206936	6	*NUTDT3-HMGA1*	0.06	3.0×10^{-8}

(continued)

Table 12.2 (continued)

Study/references	Population/ethnicity	Sample size (DP/RP)	Trait	SNP	Chr loc	Gene	Effect size	*p*-value
Heid et al. 2010	Europeans	77,167/113,636	WHR	rs9491696	6	*RSP03*	0.042	1.8×10^{-40}
				rs6905288	6	*VEGFA*	0.036	5.9×10^{-25}
				rs984222	1	*TBX15*	0.034	8.7×10^{-25}
				rs1055144	7	*WARS2*	0.04	10.0×10^{-25}
				rs10195252	2	*NFE2L3*	0.033	2.1×10^{-24}
				rs4846567	1	*GRB14*	0.032	6.9×10^{-21}
				rs1011731	1	*LYPLA1*	0.028	9.5×10^{-18}
				rs718314	12	*DNM3/PIGC*	0.03	1.1×10^{-17}
				rs1294421	6	*ITPR2/SSPN*	0.028	1.8×10^{-17}
				rs1443512	12	*LY86*	0.031	6.4×10^{-17}
				rs6795735	3	*HOXC13*	0.025	9.8×10^{-14}
				rs4823006	22	*ADAMTS9*	0.023	1.1×10^{-11}
				rs6784615	3	*ZNRF3/KREM*	0.043	3.8×10^{-10}
				rs6861681	5	*EN1*	0.022	1.9×10^{-9}
						NISCH/STAB1		
						CPEB4		
Wang et al. 2011	Non-Hispanic	1,060/2,256	Extreme obesity	rs17817449	16	*FTO*	1.6	2.5×10^{-12}
	Caucasians		WHR	rs11624704	14	*NRXN3*	NR	2.7×10^{-09}
Wen et al. 2011	East Asians	27,715	BMI (M)	rs9356744	6	*CDKAL1*	3.39	2.0×10^{-11}
		37,691		rs261967	5	*PCSk1*[c]	3.77	5.1×10^{-09}
		17,642		rs12597579	16	*GP2*	4.09	1.0×10^{-08}
				rs652722	11	*PAX6*	2.75	7.7×10^{-08}
Jiao et al. 2011	Swedish populations	4,838/5,827(M)	Morbidly	rs2116830	10	*KCNM1*	1.26	2.8×10^{-10}
			Obese	rs988712	11	*BDNF*	1.36	5.2×10^{-17}
Paternoster et al. 2011	Danish populations	2,633/2,740	Overweight	rs52376670	16	*FT0*	+	2.8×10^{-10}
				rs7132908	12	*FAIM2*	+	1.8×10^{-08}
				rs8089364	18	*MC4R*	+	3.2×10^{-08}
				rs604388	1	*SEC168*	+	5.8×10^{-07}
				rs13130484	4	*GNPDA2*	–	1.6×10^{-09}

(continued)

Table 12.2 (continued)

Study/references	Population/ethnicity	Sample size (DP/RP)	Trait	SNP	Chr loc	Gene	Effect size	*p*-value
Ng et al. 2012	African Americans	4,989	BMI	rs6794092	3	*PP13439-TMEM212*	0.167	2.4×10^{-6}
				rs268972	5	*CDH12*	0.091	5.0×10^{-5}
				rs2033195	5	*MFAP3-GALNT10*	0.094	5.6×10^{-6}
				rs815611			0.095	5.4×10^{-6}
				rs6088887	20	*FERIL4*	0.106	2.5×10^{-5}
Fox et al. 2012	European populations	10,557	VAT-SAT ratio	rs11118316	1	*LYPLAL1*	+	3.1×10^{-09}
			VAT-w	rs1659258	2	*THNSL2*	+	1.6×10^{-08}
			SAT-o	rs9922619	16	*FTO*	+	5.9×10^{-08}
Okada et al. 2012	East Asian populations	62,245	BMI	rs12149832	16q12	*FTO*	0.073	4.8×10^{-22}
				rs2030323	11	*BDNF*	0.046	3.8×10^{-16}
				rs11671664	19	*GIPR*	0.046	6.8×10^{-14}
				rs2206734	6	*CDKAL1*	0.039	1.4×10^{-11}
				rs2331841	18	*MC4R*	0.046	1.8×10^{-11}
				rs11142387	9	*KLF9*	0.040	1.3×10^{-09}
				rs516636	1	*SEC16B*	0.050	3.4×10^{-09}
Melka et al. 2012	French-Canadian adolescents	598	BMI	rs16933812	9	*PAX5*	NA	5.2×10^{-06}
								4.6×10^{-08}
			TFM	rs7638110	3	*MRPS22*	NA	9.3×10^{-09}
								2.2×10^{-06}

BMI = Body mass index, *DP* = Discovery Phase, *RP* = Replication Phase; *WC* = Waist circumference, *WHR* = Waist–hip ratio, *OB* = Obesity, *VAT* = Visceral adipose tissue, *SAT* = Subcutatneous adipose tissue, *VAT-w* = Visceral adipose tissue-women, *SAT-o* = Subcutatneous adipose tissue-overall, *TFM* = Total fat mass; d = Effect Direction; [c] loci initially identified through candidate gene approach

Subsequently, a study involving a meta-analysis of 16 GWASs ($N = 38{,}580$) from the GIANT consortium and a follow-up in 70,580 individuals for adult waist circumference (WC) and waist–hip ratio (WHR) discovered two novel loci (*TFAP2B* and *MSRA*) associated with WC and one locus *LYPLA1* associated with WHR in females (Lindgren et al. 2009). Another study of WC from the CHARGE consortium by Heard-Costa et al. (Heard-Costa et al. 2009) identified a novel locus called neurexin 3 (*NRXN3*) in addition to *FTO* and *MC4R* based on a sample of 31,373 individuals of Caucasian descent from eight cohort studies in the stage 1 and 38,641 individuals in the stage 2 analysis.

The fourth wave of obesity GWASs was dominated by two major meta-analyses from the GIANT consortium: the Speliotes et al. (2010) study was the largest BMI GWAS undertaken to date. This study examined ~2.8 million SNPs in ~124,000 individuals with targeted follow-up of 42 SNPs in ~126,000 additional individuals. They found evidence for association with 14 known obesity susceptibility loci and identified 18 new loci showing significant ($p < 5 \times 10^{-8}$) association with BMI. As shown in Table 12.2, some loci at *MC4R, POMC, SH2B1,* and *BDNF* map near key hypothalamic regulators of energy balance, while one of these loci maps close to *GIPR* which is an incretin receptor. These findings including other newly associated genes provide new insights into body weight regulation. Another meta-analysis performed by Heid et al. (2010) using WC as a focal phenotype yielded 13 novel loci as listed in Table 12.2. Scherag et al. (2010) reported two new loci *SDCCAG8* and *TNKS* for body weight regulation in a joint analysis of GWASs for extreme obesity in French and German children and adolescents, and one locus (*KCNMA1*) found to be strongly associated in an adult population (Jiao et al. 2011). In addition, findings from a few other independent GWA studies for obesity in Europeans are summarized in Table 12.2.

Obesity GWAS in Other Ethnic Populations

Two GWASs for obesity were performed in 2012 from South Asia. In the first study, Wen et al. (2012) included 27,715 East Asians (Chinese, Korean and Indonesians) in their discovery sample, which was followed by in silico and de novo replication in 37,691 and 17,642 additional East Asians, respectively. They identified three novel loci in or near the *CDKAL1, PCSK1,* and *GP2* genes, and replicated seven previously identified loci: *FTO, SEC16B, MC4R, GIPR-QPCTL, ADCY3-DNAJC27, BDNF,* and *MAP2K5* at genome-wide significance levels. In the second study, Okada et al. (2012) included a discovery sample of 26,620 Japanese individuals with replication in the East Asian sample from Wen et al. (2012) plus an additional 7,900 Japanese individuals. Association results have achieved genome-wide significance levels for five previously reported loci: *SEC16B, BDNF, FTO, MC4R, GIPR including CDKAL1* and a novel locus *KLF9.* In sum, two studies jointly identified four novel loci and replicated 7 previously reported loci. In addition, several independent GWAS studies for obesity were performed in different ethnic groups as summarized in Table 12.2.

Thus far, GWASs have successfully localized about 80 susceptibility loci associated with obesity traits (Grarup et al. 2014), hence proving that this approach has been successful in unraveling the genetic architecture of obesity and its related traits. Some of the obesity susceptibility loci so far reported that are novel, met genome-wide significance level, and replicated are summarized in Table 12.2 and Fig. 12.1. So far, of all identified loci, the genetic variation in *FTO* has the largest effect on obesity susceptibility.

12.3.3.2 The GWA Studies for Type 2 Diabetes

In the past 7 years, GWASs have greatly enhanced our understanding of the genetic architecture of complex diseases in general and T2D in particular (Frayling and McCarthy 2007; Florez 2008; McCarthy and Hirschhorn 2008; Perry and Frayling 2008; Prokopenko et al. 2008; Ridderstrale and Groop 2009). To date, three waves of T2D GWASs have been performed and several novel susceptibility genes/variants for T2D have been identified, many with subsequent replications (e.g., (Saxena et al. 2007; WTCCC 2007; Scott et al. 2007; Sladek et al. 2007; Steinthorsdottir et al. 2007; Zeggini et al. 2008). T2D genes so far reported that met genome-wide significance level and were replicated are summarized in Table 12.3. The majority of these studies have focused on the identification of association signals due to common variants (minor allele frequency >5 %).

The first wave of T2D GWA studies was marked by a study from France, composed of 661 cases and 614 controls, covering 392,935 SNPs (Sladek et al. 2007). This study identified four novel association signals at *SLC30A8* (solute carrier family 30 (zinc transporter), member 8), *LOC387761, HHEX* (hematopoietically expressed homeobox), and *EXT2* (exostoses 2). The first locus revealed a nonsynonymous polymorphism (R325W, rs13266634) in *SLC30A8* that is expressed only in insulin-producing beta cells. Two other loci contain genes that are involved in either beta-cell development or function, *IDE-KIF11-HHEX* (insulin-degrading enzyme, kinesin family member 11, and hematopoietically expressed homeobox) and *EXT2-ALX4* (exostoses 2 and ALX homeobox 4). In addition, this study confirmed the previously identified association at *TCF7L2* that was originally identified through linkage studies.

The second wave of T2D GWA studies include three major studies by the Wellcome Trust Case Control Consortium (WTCCC 2007), Diabetes Genetics Initiative [(DGI) (Saxena et al. 2007)], and Finland-United States Investigation of NIDDM Genetics [FUSION, (Scott et al. 2007)]. These studies published their independently discovered novel T2D associations at *CDKAL1* (CDK5 regulatory subunit associated protein 1-like 1), *IGF2BP2* (insulin-like growth factor 2 mRNA binding protein 2), and *CDKN2A/B* (cyclin-dependent kinase inhibitor 2A/2B) as well as replications of earlier associations of *SCL30A8* and *HHEX* with T2D (Saxena et al. 2007; Scott et al. 2007; Zeggini et al. 2008). Further replication studies confirmed associations with *SLC30A8, HHEX, CDKAL1, IGF2BP2*, and *CDKN2A/B* and three loci previously identified through linkage and candidate gene studies, *PPARG, KCNJ11*, and *TCF7L2* in both European and non-European populations. In another GWA study with 1,399 cases and 5,275 controls from Iceland, the deCODE genetics group identified an intronic variant (rs7756992) in the *CDKAL1* gene as a novel T2D locus (Steinthorsdottir et al. 2007). Furthermore, this study showed that the insulin response for rs7756992 homozygotes was ~20 % lower than for heterozygotes suggesting that this variant confers risk of T2D through reduced insulin secretion. Thus this second wave of T2D GWASs confirmed 8 T2D susceptibility loci: *TCF7L2, PPARG, KCNJ11, SCL30A8, HHEX, CDKAL1, CDKN2A/B*, and *IGF2BP2*. In addition, the WTCCC study identified a strong association between *FTO* variants and T2D, though the effect is modulated through obesity (Frayling et al. 2007).

The third wave of T2D studies was based on European GWAS with larger sample sizes so that common variants with lower effect sizes would be identified—WTCCC, FUSION, and DGI were combined to form the Diabetes Genetics Replication And Meta-analysis (DIAGRAM) consortium. Mainly, a large collaborative meta-analysis of GWA studies including previously published WTCCC, FUSION, and DG scans with a substantial sample of 4,549 cases and 5,579 controls, was performed (Zeggini et al. 2008). In the discovery stage 69 variants showed associations with T2D at $p < 10^{-4}$. In an initial attempt at replication in a sample of 22,426 individuals, of the 69 variants only 11 reached the desired significance level threshold. A subsequent replication study of these 11 variants in a larger sample of 57,366 individuals resulted in only 6 variants that reached a p-value of 5×10^{-8} for association with T2D. These six additional novel genes were *JAZF1, CDC123/CAMK1D, TSPAN8-LGR5, THADA, NOTCH2, and ADAMSTS9*. Following this study, several additional T2D genetics cohorts have been combined to form DIAGRAM + with an effective sample size of more than 22,000 European subjects. In a study conducted by Voight et al. (2010), 2,426,886 imputed and genotyped autosomal SNPs with additional interrogation of the X chromosome were examined for association with T2D as a categorical

Table 12.3 Some of the major genome-wide association study results for T2D and related phenotypes

Study/references	Population/ethnicity	Trait	Sample size (DP/RP)	Chr loc	Gene	SNP	Effect size: OR/beta	*p*-value (Study specific/ Combined *)
Sladek et al. 2007 Zeggini et al. 2007	French	T2D	1,363/5,511	10q25.3	*HHEX*	rs5015480	1.17	7.2×10^{-8}
					HHEX	rs1111875	1.14/1.13	$1.7{\times}10^{-4}/5.7{\times}10^{-10}$
				8q24.11	*SLC30A8*	rs10282940	1.15	6.1×10^{-3}
Scott et al. 2007 DGI 2007 Zeggini et al. 2007	Finnish (FUSION)	T2D	2,335/2,473	3q28	*GF2BP2*	rs4402960	1.18/1.14	$2.1 \times 10^{-8}/8.9 \times 10^{-16}$
				6p22.2	*CDKAL1*	rs7754840	1.12/1.12	$0.0095/4.1 \times 10^{-11}$
				9p21	*CDKN2B*	rs10811661	1.20/1.20	$0.0022/7.8 \times 10^{-15}$
				10q25	*TCF7L2*[l]	rs7903146	1.34/1.37	$1.3 \times 10^{-8}/1.0 \times 10^{-48}$
Zeggini et al. 2007 Scott et al. 2007 DGI 2007	British (WTCCC/UKT2D)	T2D	4,862/9,103	3q28	*GF2BP2*	rs4402960	1.17/1.14	$1.7 \times 10^{-3}/8.9 \times 10^{-16}$
				6p22.2	*CDKAL1*	rs10946398	1.20/1.12	$2.5 \times 10^{-5}/4.1 \times 10^{-11}$
				9p21	*CDKN2A*	rs7020996	1.26	1.8×10^{-7}
					CDKN2B	rs10811661	1.22/1.20	$7.6 \times 10^{-4}/7.8 \times 10^{-15}$
				10q25	*TCF7L2*[l]	rs7901695	1.37/1.37	$6.7 \times 10^{-13}/1.0 \times 10^{-48}$
DGI 2007 Scott et al. 2007 Zeggini et al. 2007	Swedish, Finnish	T2D	2,931/10,850	3q25	*PPARG*[c]	rs1801282	1.23/1.14	$1.3 \times 10^{-3}/1.7 \times 10^{-6}$
				3q28	*IGF2BP2*	rs4402960	1.17/1.14	$1.7 \times 10^{-9}/8.9 \times 10^{-16}$
				6p22.2	*CDKAL1*	rs7754840	1.08/1.12	$2.4 \times 10^{-3}/4.1 \times 10^{-11}$
				9p21	*CDKN2B*	rs10811661	1.20/1.20	$5.4 \times 10^{-8}/7.8 \times 10^{-15}$
				10q25	*TCF7L2*[l]	rs7903146	1.38/1.37	$2.3 \times 10^{-31}/1.0 \times 10^{-48}$
				11p15.1	*KCNJ11*[c]	rs5215	1.15/1.14	$1.3 \times 10^{-3}/5.0 \times 10^{-11}$
Steinthorsdottir et al. 2007	Icelandic (DECODE)	T2D	6,674/14,138	6p22.2	*CDKAL1*	rs6931514	1.20	7.7×10^{-9}
Zeggini et al. 2008	European	T2D (M)	10,128/79,792	1p12	*NOTCH2*	rs10923931	1.13	4.1×10^{-8}
				2p21	*THADA*	rs7578597	1.15	1.1×10^{-9}
				3p14	*ADAMTS9*	rs4607103	1.09	1.2×10^{-8}
				7p15	*JAZF1*	rs864745	1.10	5.0×10^{-14}
				10p13-14	*CDC123*	rs12779790	1.11	1.2×10^{-10}
				12q21	*TSPAN8*	rs7961581	1.09	1.1×10^{-9}

(continued)

Table 12.3 (continued)

Study/references	Population/ethnicity	Trait	Sample size (DP/RP)	Chr loc	Gene	SNP	Effect size: OR/beta	*p*-value (Study specific/ Combined *)
Yasuda et al. 2008	Japanese, Korean, Chinese	T2D	1,691/18,239	11p15.5	*KCNQ1*	rs2237892	1.49	6.7×10^{-13}
Unoki et al. 2008	Japanese, Singaporean	T2D	1,752/19,489	11p15.5	*KCNQ1*	rs2283228	1.26	3.1×10^{-12}
Rung et al. 2009	French, Danish	T2D	1,376/27,033	2q36	*IRS1*	rs2943641	1.19	9.3×10^{-12}
Prokopenko et al. 2009	European	T2D	36,610/82,689	11q21-q22	*MTNR1B*	rs10830963	1.09	3.3×10^{-7}
		FG					0.067	3.2×10^{-50}
Lyssenko et al. 2009	Swedish, Finnish	T2D	2,931/18,831	11q21-q22	*MTNR1B*	rs10830963	1.12	3.2×10^{-50}
		FG					0.045	0.039
Bouatia-Naji et al. 2009	French	T2D	2,151/15,464	11q21-q22	*MTNR1B*	rs1387153	1.15	6.3×10^{-5}
	European	FG					0.06	1.3×10^{-7}
Dupuis et al. 2010	European	T2D/FG	46,186/ 127,677	3q13.2-q21	*ADCY5*	rs11708067	1.12/0.027	$9.9 \times 10^{-21}/1.7 \times 10^{-14}$
		T2D/FG		1q32.2-32.3	*PROX1*	rs340874	1.07/0.013	$7.2 \times 10^{-10}/6.6 \times 10^{-6}$
		T2D/FG		7p15.3-15.1	*GCK*	rs4607517	1.07/0.062	$5.0 \times 10^{-8}/1.2 \times 10^{-44}$
		T2D/FG		2p23	*GCKR*	rs780094	1.06/0.029	$7.2 \times 10^{-10}/6.6 \times 10^{-6}$
		T2D/FG		7p21.2	*DGKB*	rs2191349	1.06/0.030	$1.1 \times 10^{-8}/5.3 \times 10^{-29}$
Tsai et al. 2010	Han Chinese	T2D	1,889/3,276	17p13.3	*SRR*	rs391300	1.28	3.06×10^{-9}
				9p24.1-23	*PTPRD*	rs17584499	1.57	8.54×10^{-10}
Qi et al. 2010	European	T2D (M)	5,643/84,605	2q24	*RBMS1* *ITGB6*	rs7593730	0.90	3.7×10^{-8}

(continued)

Table 12.3 (continued)

Study/references	Population/ethnicity	Trait	Sample size (DP/RP)	Chr loc	Gene	SNP	Effect size: OR/beta	*p*-value (Study specific/ Combined *)
Voight et al. 2010	European	T2D (M)	47,117/94.337	2	*BCL11A*	rs243021	1.08	2.9×10^{-15}
				5	*ZBED3*	rs4457053	1.08	2.8×10^{-12}
				7	*KLF14*	rs972283	1.07	2.2×10^{-10}
				8	*TP53INP1*	rs896854	1.06	9.9×10^{-10}
				9	*CHCHD9*	rs5945326	1.11	2.8×10^{-8}
				11	*CENTD2*	rs1552224	1.14	1.4×10^{-22}
				11	*KCNQ1*	rs231362	1.08	2.8×10^{-13}
				12q24	*HMGA2*	rs1531343	1.10	3.6×10^{-9}
				12	*HNF1A*	rs7957197	1.07	2.4×10^{-8}
				15	*PRC1*	rs8042680	1.07	2.4×10^{-10}
				15	*ZFAND6*	rs11634397	1.06	2.4×10^{-9}
				x	*DUSP9*	rs5945326	1.27	3.0×10^{-10}
Strawbridge et al. 2011	European	Fasting proinsulin levels	10,701	15	*LARP6*	rs1549318	0.0180	2.4×10^{-8}
				17	*SGSM2*	rs4790333	0.0179	3.0×10^{-9}
Tabassum et al. 2013	Indians	T2D	12,535	2q21	*TMEM163*	rs6723108	1.31	3.3×10^{-9}
						rs7570971	1.25	2.0×10^{-8}
Cho et al. 2012	East Asians	T2D	18,817	4	*MAEA*	rs6815464	1.13	1.57×10^{-20}
				7	*GCC1-*	rs6467136	1.11	4.96×10^{-14}
				3	*PAX4*	rs831571	1.09	8.41×10^{-11}
				6	*PSMD6*	rs9470794	1.12	2.06×10^{-10}
				6	*ZFAND3 KCNK16*	rs1535500	1.08	2.30×10^{-8}
Saxena et al. 2013	Indians	T2D	1,616	13q12	*SGCG*	rs9552911	0.67	1.82×10^{-8}
Williams et al. 2014	Mexicans and Latin Americans	T2D	3,848/4,366	17p13.1	*SLC16A11*	rs13342692	1.29	3.9×10^{-13}

T2D = Type 2 diabetes, *FG* = Fasting glucose, *DGI* = Diabetes Genetics Initiative, *DP* = Discovery phase, *RP* = Replication phase, *NR* = Not reported, *NA* = Not available, *M* = Meta-analyses, *OR* = Odds ratio, *TFM* = Total fat mass, * OR estimates and combined p-values from WTCCC (Welcome Trust Case Control Consortium)/UKT2D, DGI, and FUSION Studies. [c] loci initially identified through candidate gene approach, [1] loci were initially identified through linkage approach phenotype. As shown in Table 12.4, twelve loci were associated with T2D at genome-wide significance ($p < 5 \times 10^{-8}$). This meta-analysis confirmed associations for nine novel loci and three loci previously associated with T2D including *IRS1, MTNR1B, and KCNQ1* (Prokopenko et al. 2008; Bouatia-Naji et al. 2009; Lyssenko et al. 2009).

In 2009, three GWA studies for fasting glucose as a quantitative trait simultaneously identified *MTNR1B* as a locus influencing fasting hyperglycemia and T2D risk (Prokopenko et al. 2008; Bouatia-Naji et al. 2009; Lyssenko et al. 2009). GWAS for continuous glycemic traits particularly fasting glucose levels showed significant associations with *G6PC2* and *MTNR1B* (Prokopenko et al. 2008; Bouatia-Naji et al. 2009; Lyssenko et al. 2009). The Meta-Analysis of Glucose and Insulin-related traits Consortium (MAGIC) study examined 21 GWAS to identify loci associating with fasting glucose, fasting insulin, HOMA-beta, and HOMA-IR (Dupuis et al. 2010). In this study, following replication among 76,558 individuals from 34 additional studies, nine new loci including SNPs in or near *ADCY5, MADD, CRY2, ADRA2A, FADS1, PROX1, SLC2A2, GLIS3,* and *C2CD4B* were significantly associated with fasting glucose and one SNP near *IGF1* found to be associated with fasting insulin and HOMA-IR. This study also confirmed associations between glycemic traits and previously identified SNPs in or near *DGKB-TMEM195, GCKR, G6PC2, MTNR1B,* and *GCK*.

Thus, GWA studies of T2D and related glycemic traits have identified several candidate genes/genomic regions. As reported in Table 12.3, *TCF7L2, KCNJ11, HHEX/IDE, SLC30A8, CDKAL1, CDKN2A/2B, IGF2BP2, FTO, MC4R, PPARG, TCF2, WFS1, JAZF1, CDC123/CAMK1D, TSPAN8/LGR5, THADA, ADAMTS9, NOTCH2, KCNQ1* contain the SNPs most strongly associated with T2D (Grant et al. 2006; Salonen et al. 2007; Saxena et al. 2007; Scott et al. 2007; Sladek et al. 2007; Steinthorsdottir et al. 2007; WTCCC 2007; McCarthy and Hirschhorn 2008; Perry and Frayling 2008; Prokopenko et al. 2008; Unoki et al. 2008; Yasuda et al. 2008; Zeggini et al. 2008; Ridderstrale and Groop 2009). Most of these associations have been confirmed either by initial replication studies or by subsequent studies in other populations. It is evident from these studies that identification of common risk variants for diseases like T2D depends on large-scale genotyping of large samples of cases and controls, and that the effect sizes of most of the identified risk variants fall between 1.1 and 1.4 (Bodmer and Bonilla 2008). In fact, as detailed in Table 12.3, with an exception of *TCF7L2*, the per-allele effect sizes corresponding to most of the risk variants fall between 1.1 and 1.2. As pointed by Prokopenko et al. (2008), the T2D GWAs thus far have been concentrated on European populations, and the contribution of these genetic findings to the total phenotypic variance in susceptibility to T2D appears to be less than 10 %.

Of the identified loci, by far the strongest association with T2D has been with the *TCF7L2* gene [e.g., rs7901695] (Grant et al. 2006; Goodarzi and Rotter 2007; McCarthy and Hirschhorn 2008; Perry and Frayling 2008; Prokopenko et al. 2008; Ridderstrale and Groop 2009). Its influence on susceptibility to T2D is larger (i.e., a per-allele effect size of ~ 1.4) compared to those for other loci (Table 12.3). Lyssenko et al. (2007) examined the potential mechanisms by which genetic variants in *TCF7L2* increase susceptibility to T2D, concluding that they are associated with impaired insulin secretion, incretin effects and enhanced rate of hepatic glucose production, and that increased expression of *TCF7L2* in human islets reduced glucose-stimulated insulin secretion. Indeed, several of the susceptibility loci reported in Table 12.3 appear to be involved in insulin secretion, highlighting the potential prominent role of pancreatic beta-cell dysfunction compared to that of insulin resistance mechanisms in T2D pathogenesis (Florez 2008). Although some of the implicated loci seem to be potential biological candidates with roles in T2D pathogenesis (e.g., *IDE*), the functional relevance of them is not yet clear (see, Prokopenko et al. 2008; Ridderstrale and Groop 2009). An exception to

Table 12.4 Major genome-wide association study results for metabolic syndrome and its components

Study/ references	Population/ ethnicity	Sample size (DP/RP)	Trait	Chr. loc.	Gene or nearest gene	SNP	Effect size	*p*-value
Zabaneh and Balding (2010)	Indian Asians	2,700/2,300	Log HDL-C	16q13	*CETP*	rs3764261	0.07	1.3×10^{-48}
				16q13	*CETP*	rs9989419	−0.05	1.4×10^{-20}
				8p21.3	*LPL*	rs2083637	0.04	1.9×10^{-10}
				8p21.3	*LPL*	rs4523270	0.03	1.0×10^{-07}
				21q22.3	*FLJ41733*	rs496300	0.03	3.9×10^{-07}
				11q12.2	*FADS1*	rs174546	−0.03	6.0×10^{-07}
				11q12.2	*FADS2*	rs1535	−0.03	6.5×10^{-07}
			T2D	10q25.2	*TCF7L2*	rs7903146	1.33	6.6×10^{-07}
			DBP	9q34.11	*ENG*	rs7865146	−1.19	1.0×10^{-06}
Avery et al. (2011)	European Americans	19,486	Metabolic syndrome trait dimensions	2	*APOB*	rs1713222	NA	6.1×10^{-13}
				2	*GCKR*	rs1260326		8.1×10^{-16}
				2	*ABCB11*	rs579060		2.4×10^{-10}
				8	*LPL*	rs301		9.5×10^{-20}
				8	*TRIB1*	rs2954021		1.3×10^{-10}
				9	*ABCA1*	rs2575876		6.2×10^{-08}
				9	*ABO*	rs687621		$1.0 \text{ x} 10^{-300}$
				11	*ZNF259*	rs964184		5.5×10^{-22}
				12	*VWF*	rs216318		1.6×10^{-07}
				12	*BRAP*	rs11065987		2.9×10^{-10}
				12	*HNF1A*	rs7979473		1.1×10^{-09}
				13	*F7*	rs510335		1.0×10^{-35}
				15	*LIPC*	rs397923		1.6×10^{-15}
				16	*FTO*	rs9923233		4.9×10^{-10}
				16	*CETP*	rs247616		8.3×10^{-72}
				19	*LDLR*	rs6511720		8.3×10^{-28}
				19	*SUGP1*	rs10401969		1.1×10^{-10}
				19	*APOC1*	rs4420638		1.7×10^{-57}
				20	*PLCG1*	rs753381		4.3×10^{-08}
Avery et al. (2011)	African Americans	6,287	Metabolic syndrome trait dimension	1	*CELSR*	rs12740374	NA	3.6×10^{-13}
				1	*CRP*	rs2592887		8.4×10^{-8}
				7	*CD36*	rs3211938		4.8×10^{-10}
				8	*LPL*	rs10096633		1.8×10^{-12}
				9	*ABO*	rs8176693		6.1×10^{-75}
				12	*VWF*	rs2229446		$9.0\text{x } 10^{-9}$
				16	*CETP*	rs247616		1.9×10^{-23}
				19	*LDLR*	rs6511720		2.5×10^{-10}
				19	*PVRL2*	rs7254892		1.3×10^{-10}

(continued)

Table 12.4 (continued)

Study/ references	Population/ ethnicity	Sample size (DP/RP)	Trait	Chr. loc.	Gene or nearest gene	SNP	Effect size	*p*-value
Kraja et al. (2011)	Europeans	22,161 STAMPEED b-meta	TG-BP	2	*GCKR*	rs780093	0.18	3.0×10^{-10}
			WC-TG	2	*GCKR*	rs780093	0.19	1.9×10^{-12}
			WC-TG	2	*C2orf16*	rs1919128	−0.18	2.0×10^{-09}
			WC-TG	2	*ZNF512*	rs13022873	−0.17	5.0×10^{-09}
			WC-TG	2	*CCDC121*	rs3749147	−0.18	1.4×10^{-09}
			HDLC-GLUC	2	*ABCB11*	rs569805	0.16	8.5×10^{-08}
			WC-GLUC	6	*TFAP2B*	rs2206277	0.17	1.3×10^{-07}
			HDLC-WC	8	*LPL*	rs301	-0.22	3.2×10^{-11}
			MS	8	*LPL*	rs295	0.17	1.7×10^{-09}
			HDLC-TG	8	*LPL*	rs13702	0.29	1.0×10^{-16}
			TG-BP	8	*LPL*	rs15285	−0.27	1.3×10^{-10}
			TG-GLUC	8	*LPL*	rs2197089	0.18	1.6×10^{-09}
			BP-HDLC	8	*LPL*	rs1441756	−0.18	2.7×10^{-08}
			HDLC-WC	8	*LOC100129150*	rs9987289	0.24	3.7×10^{-08}
			HDLC-TG	8	*LOC100129150*	rs9987289	0.25	1.1×10^{-08}
			HDLC-TG	8	*TRIB1*	rs2954026	−0.16	7.9×10^{-09}
			TG-BP	8	*TRIB1*	rs2954033	0.17	8.5×10^{-09}
			BP-GLUC	11	*LOC100128354*	rs1387153	−0.19	8.1×10^{-09}
			HDLC-GLUC	11	*LOC100128354*	rs1387153	−0.21	2.4×10^{-09}
			TG-GLUC	11	*LOC100128354*	rs10830956	−0.2	4.8×10^{-11}
			TG-BP	11	*BUD13*	rs11825181	0.32	3.0×10^{-09}
			TG-GLUC	11	*BUD13*	rs11820589	0.32	5.5×10^{-09}
			HDLC-TG	11	*BUD13*	rs10790162	0.38	2.8×10^{-15}
			MS	11	*BUD13*	rs10790162	0.25	5.4×10^{-09}
			WC-TG	11	*BUD13*	rs10790162	0.39	6.6×10^{-16}
			TG-BP	11	*ZNF259*	rs11823543	0.35	2.5×10^{-09}
			TG-GLUC	11	*ZNF259*	rs12286037	−0.32	1.1×10^{-08}
			HDLC-TG	11	*ZNF259*	rs2075290	0.39	1.5×10^{-14}
			MS	11	*ZNF259*	rs2075290	0.26	2.1×10^{-09}
			WC-TG	11	*ZNF259*	rs2075290	0.41	1.1×10^{-16}
			HDLC-TG	11	*APOA5*	rs2266788	0.39	4.6×10^{-13}
			MS	11	*APOA5*	rs2266788	0.26	1.9×10^{-09}
			TG-BP	11	*APOA5*	rs2266788	0.37	3.5×10^{-08}
			WC-TG	11	*APOA5*	rs2266788	0.41	2.2×10^{-16}
			HDLC-WC	15	*LIPC*	rs10468017	0.16	5.5×10^{-08}
			HDLC-GLUC	15	*LIPC*	rs2043085	−0.17	1.3×10^{-08}
			HDLC-TG	16	*CETP*	rs173539	0.26	4.5×10^{-16}
			HDLC-WC	16	*CETP*	rs173539	0.29	1.0×10^{-16}
			MS	16	*CETP*	rs173539	0.16	9.1×10^{-09}
			BP-HDLC	16	*CETP*	rs3764261	0.29	3.3×10^{-13}
			HDLC-GLUC	16	*CETP*	rs9939224	−0.31	6.9×10^{-12}
			HDLC-TG	19	*LOC100129500*	rs439401	0.24	1.0×10^{-08}

(continued)

Table 12.4 (continued)

Study/ references	Population/ ethnicity	Sample size (DP/RP)	Trait	Chr. loc.	Gene or nearest gene	SNP	Effect size	*p*-value
Kristiansson et al. (2012)	Finnish populations	11,616	GLUC	2	*G6PC2*	rs560887	0.15	4.8×10^{-26}
			GLUC	7	*TMEM195, DGKB*	rs6947830	0.1	1.4×10^{-13}
			GLUC	7	*GCK*	rs3757840	0.1	4.4×10^{-13}
			GLUC	7	*CAMK2B*	rs1127065	0.08	8.9×10^{-11}
			GLUC	11	*MTNR1B*	rs10830962	0.12	5.0×10^{-16}
			HDL	1	*GALNT2*	rs4846922	−0.08	3.4×10^{-08}
			HDL	2	*APOB*	rs673548	−0.11	1.4×10^{-10}
			HDL	6	*HCG26, MICB*	rs3099844	−0.15	1.7×10^{-08}
			HDL	8	*LPL*	rs268	−0.38	1.9×10^{-12}
			HDL	9	*ABCA1*	rs1883025	−0.1	5.9×10^{-10}
			HDL	11	*NR1H3*	rs10838681	−0.08	1.3×10^{-09}
			HDL	15	*LIPC*	rs1532085	−0.13	4.7×10^{-24}
			HDL	16	*CETP*	rs247617	−0.25	9.3×10^{-60}
			HDL	16	*EDC4*	rs8060686	−0.11	$2.4 \text{ x} 10^{-10}$
			SBP	2	*SMEK2*	rs782590	0.09	4.0×10^{-08}
			TG	2	*APOB*	rs6711016	0.08	4.0×10^{-08}
			TG	2	*GCKR*	rs780094	0.13	5.9×10^{-20}
			TG	7	*MLXIPL*	rs13226650	0.12	1.9×10^{-11}
			TG	8	*LPL*	rs7841189	0.18	9.7×10^{-15}
			TG	11	*ZNF259*	rs964184	0.23	2.6×10^{-31}
			TG	19	*TOMM40, APOE*	rs157582	0.1	1.4×10^{-08}
			WC	16	*FTO*	rs9940128	0.09	1.7×10^{-09}

HDL-C = High-density lipoprotein—cholesterol, *T2D* = Type 2 diabetes, *DBP* = Diastolic blood pressure, *SBP* = Systolic blood pressure, *TG* = Triglycerides, *BP* = Blood pressure, *WC* = Waist circumference; *GLUC* = Glucose, *MS* = Metabolic syndrome, *NA* = Not available

this overall statement, for example, is the identification of a nonsynonymous SNP (i.e., rs13266634, Arg325Trp) at *SLC30A8* as a significant T2D risk variant (Sladek et al. 2007). *SLC30A8* gene encodes an islet-specific zinc transporter protein member 8 (*ZnT-8*), and variation in *ZnT-8* could potentially result in pancreatic β-cell dysfunction (Boesgaard et al. 2008). To help interpret the phenotypic effects of risk alleles, there also have been continued attempts to examine association between several of the T2D risk variants and other T2D related traits such as size at birth and gestational diabetes (Watanabe et al. 2007; Clausen et al. 2009; Freathy et al. 2009; Lauenborg et al. 2009).

GWAS of T2D in Non-European Populations

Independent GWA studies in the Japanese population consisting of ~3,000 individuals identified *KCNQ1* as a novel T2D susceptibility locus, in addition to the *CDKAL1* and *IGF2BP2* loci (Unoki et al. 2008; Yasuda et al. 2008). A SNP in intron 15 of *KCNQ1*, rs2237892 was found to be significantly associated with T2D in a Japanese population (Yasuda et al. 2008). Another Japanese group also independently found a strong association of a SNP in *KCNQ1* (rs2283228) with T2D in Japanese (Unoki et al. 2008). Further studies of *KCNQ1* confirmed these associations of rs2237892 in Korean, Chinese, and Europeans (Unoki et al. 2008; Hu et al. 2009; Liu et al. 2009; Qi et al. 2009), which was subsequently confirmed in Asian Indians (Been et al. 2011). Also, another SNP (rs231362) located in an intron of *KCNQ1* on chromosome 11 showed significant association with T2D in a GWAS meta-analysis in European populations (Voight et al. 2010). This intronic variant overlaps the *KCNQ1OT1* transcript, which regulates B-cell development (Been et al. 2011). These

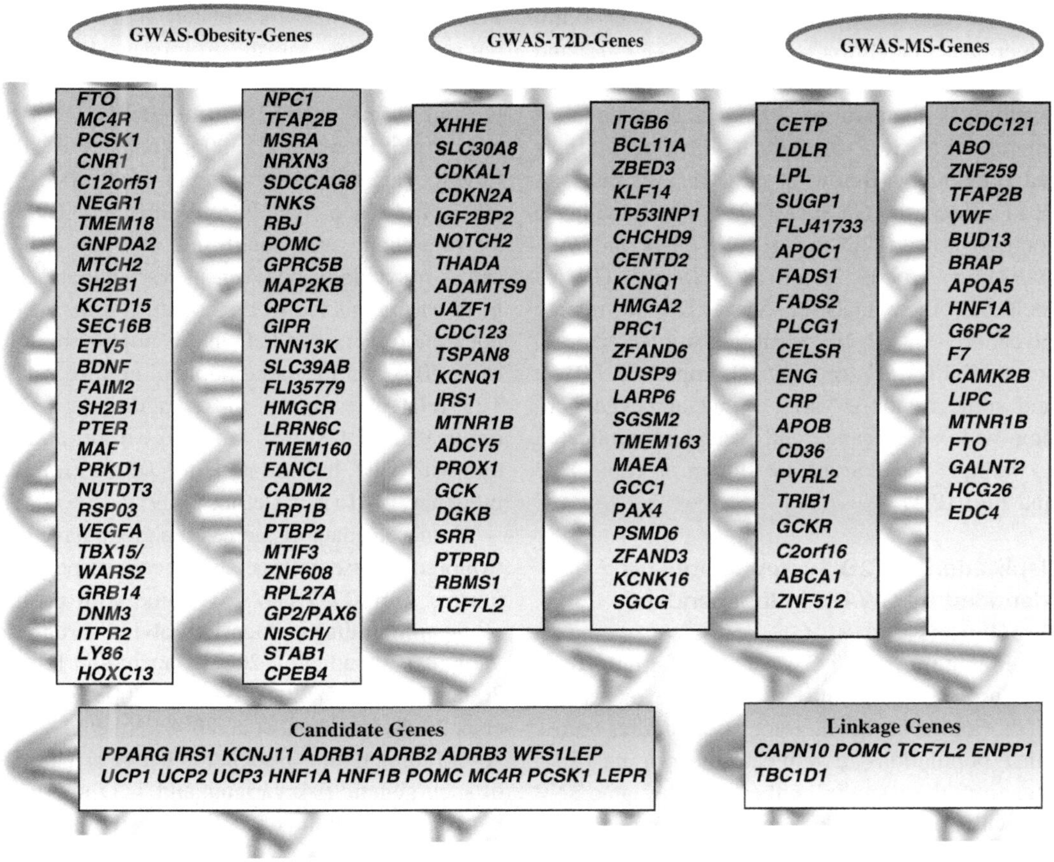

Fig. 12.1 Summary of loci identified by three mapping approaches for obesity, T2D, and MS

findings confirm *KCNQ1* as a T2D susceptibility gene in multiethnic populations. Following these genome-wide studies, a replication study in 3,210 unrelated Han Chinese has confirmed the associations of 17 previously identified common variants from genome-wide studies with T2D. Furthermore, this study indicated that common variants in *CDKAL1, CDKN2A/2B IGFBP2, SLC30A8* loci contribute to T2D risk independently or additively (Wu et al. 2008). In Han Chinese, the risk alleles of the *CDKAL1* and *CDKN2A/2B* genes increased diabetes risk by ~1.4-fold and ~1.3-fold, respectively. The allele frequencies of these risk-associated variants were also higher in the Han Chinese population compared to Europeans (Wu et al. 2008). Han Chinese are at higher risk of T2D than Europeans (WTCCC 2007; Zeggini et al. 2008). As shown in Table 12.3, a larger meta-analysis of GWAS-identified eight new loci: *GLIS3, FITM2-R3HDML-HNF4A, KCNK16, MAEA, GCC1-PAX4, PSMD6,* and *ZFAND3* for T2D in East Asians (Cho et al. 2012).

Another GWAS by Kooner et al. (2011) reported six additional new loci: *GRB14, ST6GAL1, VPS26A, HMG20A, AP3S2,* and *HNF4A* for T2D in individuals from South Asian ancestry. In addition, two T2D risk loci [*SGCG* gene, (Saxena et al. 2013) and *TMEM163*, (Tabassum et al. 2013)] were reported in Indian populations. Recently, investigators of the SIGMA consortium identified a novel locus associated with T2D in the region covering the *SLC16A11* gene ($p = 3.9 \times 10^{-13}$) on chromosome 17 in Mexican and other Latin Americans in addition to replicating *TCF7L2* and *KCNQ1* association signals (SIGMA T2D Consortium; Williams et al. 2014). T2D risk loci of interest in Mexican Americans have been reported previously (Below et al. 2011; Parra et al. 2011). Recently, Ng et al. (2014) found evidence for two novel T2D loci (i.e., *HLA-B* and *INS-IGF2*) in African Americans. So far, GWAS including meta-analyses for T2D and related glycemic traits have identified more than 90 susceptibility loci for T2D (Grarup et al. 2014), and some of those variants that achieved genome-wide significance and were well replicated in multiple populations are shown in Table 12.3 and Fig. 12.1.

Replication of T2D Susceptibility Loci Identified by GWASs in Independent Populations

Given that most T2D GWASs thus far have been in European populations, there have been a plethora of subsequent association studies across other populations to replicate the original findings from these studies, either with the same SNP or with other markers in the same genes, although results have not always been consistent across different populations, possibly due to issues such as allele frequency and LD differences. For example, as remarked by Florez (2008), association with *TCF7L2* sequence variants has been replicated in almost every population examined, with the main focus on rs7903146 and SNPs in LD with it (Goodarzi and Rotter 2007; Cauchi and Froguel 2008; Tong et al. 2009). In regard to the US populations, in addition to the ethnic groups discussed above, there were T2D GWASs performed using data from populations such as Pima Indians and Old Order Amish with limited successes (e.g., Hanson et al. 2007; Meigs et al. 2007; Rampersaud et al. 2007). But, there have been numerous studies that aimed to replicate the findings reported in Table 12.3. For example, some evidence for association between variants in *TCF7L2* and T2D or related traits has been reported in the Amish, European Americans, Mexican Americans, and African Americans [despite some inconsistent findings] (Damcott et al. 2006; Florez et al. 2006; Zhang et al. 2006; Elbein et al. 2007; Lehman et al. 2007; Sale et al. 2007; Stolerman et al. 2009; Yan et al. 2009). On the other hand, a thorough search of this gene to assess association between its variants and T2D in Pima Indians led to the conclusion that it was not a major gene for T2D in this population (Guo et al. 2007). In the same population, a search for T2D-related variants in 8 additional genes implicated by GWASs failed to exhibit association with T2D, but variation at *FTO* was found to influence BMI (Rong et al. 2009).

Numerous datasets across the globe have been explored to assess association between polymorphisms from T2D GWASs, some with positive replication findings, at least involving certain risk variants (Grarup et al. 2007; Cauchi and Froguel 2008; Herder et al. 2008; Ng et al. 2008; Sanghera et al. 2008; van Hoek et al. 2008; Wu et al. 2008). After failing to find association between a number of such genetic risk variants and T2D in African Americans (with the exception of certain variants in the *TCF7L2* gene), Lewis et al. (2008) concluded that the T2D susceptibility genes in African Americans may, in part, be different from those identified in the European-derived populations. However, Palmer et al. (2008) found modest evidence for association between variants in such genes such as *CDKAL1* and *SLC30A8* and glucose homeostasis traits in Hispanic Americans and African Americans. In recent years, there have been continued efforts to examine the transferability of the established GWAS T2D and/or glycemic trait loci in the US multiethnic populations with considerable affirmations (Chen et al. 2012; Haiman et al. 2012; Fesinmeyer et al. 2013; Ng et al. 2013). A recent international T2D trans-ethnic meta-analysis has shown the advantages of combining results from multiple ancestral groups to further understand the genetic architecture of T2D (Mahajan et al. 2014).

12.3.3.3 The GWAS of Metabolic Syndrome and Its Related Traits

Several studies have been conducted to map variants associated with components of MS, for example, examining associations with hypertension (Ehret et al. 2011), glycemic traits (Dupuis et al. 2010), and plasma lipid levels (Teslovich et al. 2010). Here we focus on studies using MS as a composite trait or studies jointly analyzing multiple MS-related traits using bivariate or multivariate approaches. GWAS findings for MS and its components that met genome-wide significance level and were replicated are summarized in Table 12.4. Zabaneh and Balding (2010) conducted a two-stage GWAS to identify common genetic variation altering risk of the MS as a composite trait, and its related traits: HDL cholesterol, plasma glucose and type 2 diabetes, abdominal obesity measured by waist-to-hip ratio, and diastolic and blood pressure in Asian Indian men, who have a high prevalence of these traits. In the discovery phase, they genotyped ~317,000 SNPs in 2,700 individuals, from which 1,500 SNPs were selected for genotyping in an additional 2,300 individuals and performed association analyses with MS (Zabaneh and Balding 2010). They found four previously reported SNPs in the genes *CETP* and *LPL*, which were associated with HDL-C ($p < 5 \times 10^{-7}$). Furthermore, they found five additional loci with SNPs that were associated with HDL-C, T2D, or DBP ($p < 10^{-6}$). Although limited number of common SNPs were found to be associated with MS traits in these Asian Indian men, they have shown high concordance with those known to be important in Europeans (Zabaneh and Balding 2010).

GWAS-based findings for MS and its related traits in European populations have been reported by Kraja et al. (2011). In this study, seven studies from the STAMPEED consortium comprising 22,161 individuals of European ancestry were subjected to bivariate GWA analyses of MS traits. MS traits were combined in all possible combinations, and individuals exceeding the thresholds for both traits were considered affected and association analyses performed. A total of 28 SNPs were associated with MS trait pairs, and these variants were located in and near fifteen genes associated with MS traits at genome-wide significance level. Since most of these bivariate associations were observed with lipid traits, the authors have concluded that these results indicate that genetic effects on lipid levels are more pronounced than for other traits (Kraja et al. 2011). The most prominent associations were in or near *LPL, CETP, APOA5, ZNF259, BUD13, TRIB1, LOC100129500,* and *LOC100128154,* of which *LPL, CETP,* and *APOA5-ZNF259-BUD13* were already known to influence lipid metabolism (Kraja et al. 2011).

Another major MS association study was conducted by Avery et al. (2011) using data from 19,486 European Americans and 6287 African Americans to identify loci associated with the clustering of metabolic phenotypes. Six phenotype domains: atherogenic dyslipidemia, vascular dysfunction, vascular inflammation, prothrombotic state, central obesity, and elevated plasma glucose, based on nineteen quantitative traits were subjected to principal component factor analysis, a data reduction approach, and eight factors were derived from the six domains. Association analysis was performed using 50 K SNP array for genotyping 49,320 SNPs and 250,000 imputed SNPs, and an additive genetic model. In European Americans, they identified SNPs reaching genome-wide significance levels ($p < 10^{-8}$) in 15 loci. Many of these were associated with one trait domain and five exhibited similar associations in African Americans. The majority of these associations were already known, for example, the association between central obesity and *FTO*. However, three new loci were identified in or near *APOC1*, *BRAP*, and *PLCG1* that were associated with multiple MS phenotype domains. In European Americans, rs4420638, located near *APOC1,* was associated with a factor phenotype ($p = 1.7 \times 10^{-57}$) and with elevated plasma glucose ($p = 8.7 \times 10^{-4}$), atherogenic dyslipidemia ($p = 1 \times 10^{-31}$), vascular inflammation ($p = 5 \times 10^{-12}$), and central obesity ($p = 1.2 \times 10^{-6}$). However, replication is needed to validate these findings. If these pleiotropic loci are confirmed in an independent population they may help characterize metabolic dysregulation and identify targets for intervention (Avery et al. 2011).

Recently, Kristiansson et al. (2012) performed a GWAS on MS and its components in four Finnish cohorts consisting of 2,637 MS cases and 7,927 controls, both with no diabetes, and also conducted a follow-up study in an independent sample with data on transcriptome and nuclear magnetic resonance-based metabolomics. In addition, they tested for loci associated with MS and its components using factor analysis. As shown in Table 12.4, twenty-two previously identified susceptibility loci for individual MS traits were replicated in their GWAS and factor analyses, and a majority of them were associated with lipid phenotypes. Importantly, a known lipid locus, the *APOA1/C3/A4/A5* gene cluster marked by the SNP rs964184, was strongly associated with MS in all four Finnish cohorts at genome-wide significance ($p = 7.23 \times 10^{-9}$) (Kristiansson et al. 2012). In a serum metabolite analysis, the same SNP rs964184 was also associated ($p = 0.024–1.88 \times 10^{-5}$) with various very low density lipoprotein, triglyceride, and high-density lipoprotein metabolites. They also found a strong association between a genetic risk score, calculated based on the number of risk alleles in loci associated with individual MS traits, and MS status in these cohorts. So far, GWAS have identified about 50 susceptibility loci for MS and its components, and those variants that achieved genome-wide significance and were well replicated in multiple studies are shown in Table 12.4 and Fig. 12.1.

GWA Replication Studies of MS and Multiethnic Populations

Most GWA studies of MS and/or its component phenotypes have been conducted using large datasets from Europeans or populations with European ancestry. Some exceptions include screens for some obesity traits, blood pressure, hypertension, renal function, and incident coronary heart disease in African Americans (AAs) (Adeyemo et al. 2009; Barbalic et al. 2011; Liu et al. 2011; Ng et al. 2012) and screens for T2D and its complications in Hispanics/Mexican Americans (MAs) (Fu et al. 2010; Parra et al. 2011), although some of the findings from these studies were only suggestive in nature partly due to the modest sample sizes available. Given such limitations, numerous studies have attempted to replicate the European-oriented GWAS findings for specific loci in non-European populations including AAs and MAs. Results have not always been consistent across different populations, possibly due to potential issues such as allele frequency and linkage disequilibrium (LD) differences across populations. Some evidence of replication for association of variants in genes such as *TCF7L2* (T2D), *KCNQ1* (T2D), and *FTO and MC4R* (obesity) has been reported in AAs and MAs (Lehman et al. 2007; Lewis et al. 2008; Palmer et al. 2011; Hester et al. 2012; Ng et al. 2014; Williams et al. 2014). In contrast, replication efforts for lipid and lipoprotein traits related to MS have been more encouraging (Chang et al. 2011; Dumitrescu et al. 2011). For example, using data from the NHANES 1991–1994 survey, Chang et al. (2011) examined association for 57 GWAS-identified or well-established susceptibility loci for lipid traits (e.g., HDL-C and triglycerides) in a multiethnic US sample. Among the examined lipid-related variants, the proportion of associations replicated in EAs (67 %) was higher than in AAs (44 %) and MAs (44 %). The search for genes that commonly influence MS-related traits is being continued. Recently, using information from 14 large epidemiological studies, several loci with pleiotropic influences on metabolic syndrome-related traits were found (Kraja et al. 2014).

12.3.3.4 GWASs and Replication Studies in Children and Youth

Thus far, a few GWA studies have been conducted for early-onset extreme obesity and other obesity-related traits in European children and adolescents (Hinney et al. 2007; Meyre et al. 2009; Scherag et al. 2010; Bradfield et al. 2012; Cousminer et al. 2013). Some of these studies found appreciable overlap of association results between children and adults, reporting associations with variants in or near genes such as *FTO*, *MC4R*, *TMEM18*, and *SDCCAG8*, for example. A GWAS of lipid traits, albeit with a very modest sample size including EA, AA, and MA children, was conducted that found some evidence for associations (e.g., a variant in *SGSM2* is associated with LDL-C levels in AAs) (Dumitrescu et al. 2011). Numerous studies have examined the relevance of GWAS-identified findings from European populations to the US pediatric or youth populations to find at least nominal evidence for association. Some examples include the replications of obesity susceptibility variants/genes in EAs (e.g., *FTO*, *TMEM18*, *MC4R*, and *BDNF*) (Zhao et al. 2011) and AAs (*FTO*) [(Bollepalli et al. 2010), but see (Klimentidis et al. 2011)]; T2D susceptibility loci (*TCF7L2*) in EAs and AAs (Dabelea et al. 2011); lipid susceptibility loci (*SORT1*) in a young EA population (Devaney et al. 2011); and, the contribution of T2D susceptibility loci (*HHEX-IDE*) to childhood obesity (Zhao et al. 2010). Some studies have examined potential genotype-by-environment (G x E) interaction influences [e.g., potential genetic influences on response to lifestyle (dietary intake and physical activity) modifications] on MS-related traits in children (Scherag et al. 2010; Garver 2011).

12.3.3.5 What Have We Learned from Genome-Wide Association Studies?

It is now well established that obesity, T2D and MS-related traits are heritable showing moderate to high heritabilities (40–70 %). Underlying causative variants are being explored using three major genetic approaches: candidate gene association, genome-wide linkage, and genome-wide association studies. As shown in Fig. 12.1, the candidate gene association approach and genome-wide linkage approach yielded a small number of common genetic variants with consistent associations for obesity, T2D, and MS. With the advent of GWASs, a new era has begun in the study of the genetic basis of common, complex diseases. To date, significant advances have been made, in particular through GWASs, in the understanding of genetic underpinnings of obesity, T2D and MS with the discovery of ~ 80 genetic loci for obesity, ~90 loci for T2D, and ~50 loci for MS and/or its components (Choquet and Meyre 2011a; Drong et al. 2012; Sandholt et al. 2012; Sanghera and Blackett 2012; Grarup et al. 2014). Most of the genetic variants identified for T2D appear to be related to B-cell dysfunction and to some extent to insulin resistance. Many of the variants for obesity appear to be involved in pathways related to energy homeostasis, and several of the identified genes appear to show associations with obesity, T2D and MS, which is suggestive of potential pleiotropic effects (Fig. 12.1). Of all identified GWAS loci for obesity, the genetic variation in *FTO* has the largest effect on obesity susceptibility. Furthermore, for most of these variants/genes, the identity of the causal genes and the functional relevance of the implicated genetic variants have yet to be established to examine their potential translation into clinical practice. In general, very few GWAS localizations have resulted in successful identification of one or more functional variants, because alleles identified in GWAS are seldom the true causative alleles but are likely in LD with them (Frazer et al. 2009). As a result, additional studies are critical to complement the results of GWAS with mechanistic studies to elucidate the biological mechanisms responsible for an observed genetic association and to identify the disease-predisposing alleles (Marian and Belmont 2011).

The effect sizes of common variants identified by GWA studies are rather small or modest. For example, the contribution of variants identified in T2D GWASs so far to the total phenotypic variance in susceptibility to T2D appears to be small (~10 %) (Billings and Florez 2010; Imamura and Maeda 2011; Drong et al. 2012). In the case of obesity, the current GWA studies account for less than 5 % of the total phenotypic variance for BMI (Drong et al. 2012). Thus, GWA studies of obesity, T2D and MS have explained only a small or modest proportion of trait specific heritabilities (i.e., missing heritability, (Manolio et al. 2009). One possibility for this is that rare variants with potentially stronger effects could be a potential source of the missing heritability, and that such variants are poorly detected by available genotyping arrays that focus on common variants (Manolio et al. 2009). Several new approaches have been proposed to identify more genetic loci, to pinpoint causal variants, and to explore the physiological mechanisms and pathways that underlie the observed associations. Furthermore, several susceptibility genes/variants have been localized by these studies for several complex diseases, and the translation of this knowledge to the prognosis and treatment of complex disease phenotype and its correlates still remains a challenge. There have been suggestions that epigenetic effects (i.e., methylation) that track underlying sequence variation could be potential contributors to heritability (McCarthy and Hirschhorn 2008; Meaburn et al. 2010; Tycko 2010). Furthermore, most GWASs have focused on studies of unrelated individuals, so there is a growing awareness that for localizing genes, family-based studies are likely to be a superior design (Thornton and McPeek 2007; Kent et al. 2007; Visscher et al. 2008).

12.4 Beyond GWA Studies, Current Research Efforts, and Future Directions

Most of the above discussed findings are association signals, and for most cases the identity of the causal genes and the functional relevance of the implicated genetic variants have yet to be established. A few exceptions, for example, include T2D associations, some involving nonsynonymous variants in the coding regions and the others involving intronic variants with potential regulatory relationships pursued through follow-up studies, related to such genes as *GCKR*, *TCF7L2*, *SLC30A8*, *KCNJ11*, and *KCNQ1* (reviews: Kato 2013; Ng and Gloyn 2013; Grarup et al. 2014; Sun et al. 2014; Thomsen and Gloyn 2014). Interestingly, we have recently shown that certain loss-of-function mutations in *SLC30A8* gene, which encodes an islet zinc transporter (ZnT8) and harbors a common variant associated with T2D risk, glucose, and proinsulin levels, are associated with T2D protection and encode unstable ZnT8 proteins (Flannick et al. 2014). Aside from these observations, the majority of GWAS-identified variants including T2D fall in noncoding regions (intronic or intergenic) of the genome, in turn highlighting their potential role in gene regulation and the associated mechanisms including transcriptional regulation, noncoding RNA function and epigenetic regulation (Elbein et al. 2012; Edwards et al. 2013; Kato 2013; Hrdličková et al. 2014).

To bridge the gap between genetic associations and disease-promoting molecular mechanisms, there have been enhanced/accelerated efforts in recent years to identify the potential molecular and biological mechanisms corresponding to the noncoding common variant T2D signals using both experimental and bioinformatic approaches. For example, a locus near *IRS1* gene was found to be associated with reduced body fat percentage and *IRS1* expression in adipose tissue in men, which was also correlated with adverse metabolic profile (e.g., increased IR and T2D risk]) (Kilpelainen et al. 2011). Using adult islet and fetal pancreas samples, Travers et al. (2013) assessed the influence of variants at the *KCNQ1* locus on regional DNA methylation and gene expression. By mapping sequence variants to open chromatin sites, Gaulton et al. (2010) found an intronic variant (rs7903146) in *TCF7L2* gene to be located in islet-selective open chromatin, suggesting its potential impact on local chromatin structure and regulatory changes. Numerous studies have been benefiting from the emerging knowledge on pancreatic islet genome regulatory mechanisms (Gaulton et al. 2010; Stitzel et al. 2010; Moran et al. 2012; Nica et al. 2013; Pasquali et al. 2014) and the publicly available databases such as ENCODE, Roadmap Epigenomics Project, and RegulomeDB for functional annotations of noncoding variants (Zhou et al. 2011; Boyle et al. 2012; Edwards et al. 2013; Fogarty et al. 2013; Lo et al. 2014). Other research activities relating to complex diseases such as obesity and T2D include: common and rare structural variation including copy number variations (CNVs) such as insertions and deletions and copy neutral variation

(Manolio et al. 2009); gene x gene interactions (Kooperberg et al. 2009); parental origin of sequence variants (Kong et al. 2009); epigenetic modifications (Gallou-Kabani and Junien 2005; Gluckman and Hanson 2008); ethnic-specific disease loci (Cho et al. 2009); and disease-associated haplotypes (Tregouet et al. 2009). In addition, the use of additional and/or refined phenotypes, both endophenotypes and intermediate phenotypes, omic-metrics (e.g., transcriptomics and metabolomics), and next-generation sequencing approaches is likely to enhance our knowledge on molecular mechanisms underlying the phenotypic expression of a disease such as T2D (Lanktree et al. 2010; Haring and Wallaschofski 2012; Robinson 2012).

12.4.1 Rare Variants and Complex Disease Phenotypes

There has been an increased interest in the potential role of "rare" variants in common complex diseases such as obesity and T2D, which are not detectable with the use of the GWAS related technologies that only focus on common variants (Cirulli and Goldstein 2010; Gibson 2011). Typically, common variants are defined as those with a minor allele frequency >5 % whereas rare variants are those with frequencies <1 % and the intermediate territory of 1–5 % frequency is often categorized as uncommon. Rare or less frequent variants with larger effects may have not been detected as the present genome-wide association scans have only limited potential to capture rarer variants. Alternatively, it is also possible that variants less common than the associated one may create "synthetic associations" by occurring more often in association with one of the alleles at the common site compared to the other (Dickson et al. 2010). In other words, synthetic association refers to a situation in which the association of a common variant with a disease is due to linkage disequilibrium between the common variant and multiple rare variants that segregate on the same haplotype (Dickson et al. 2010; Gibson 2011). However, in the absence of empirical evidence, the importance of these synthetic associations cannot be evaluated (Wray et al. 2011; Anderson et al. 2011; Goldstein 2011). In consideration of the importance of rare and low frequency variants, the data generated by projects such as the 1000 Genomes with the use of sequencing data have contributed to the development of newer arrays of SNP-chip design to include rarer (MAF ≥ 1 %) variants such as the ExomeChip [1,000 Genomes Project Consortium, (Abecasis et al. 2010; Peloso et al. 2014)].

12.4.2 The New Era of Sequencing Studies: Whole Exome Sequencing (WES) and Whole Genome Sequencing (WGS)

A new era of genome-wide sequencing has begun where we are able to generate a catalog of variants present in a DNA sequence without relying on markers and LD patterns. Using the next-generation sequencing (NGS) technologies, now it is possible to obtain complete information on rare and common sequence variants across the whole exome (WES) or genome (WGS) or a targeted region (e.g., 1,000 Genomes Project, Abecasis et al. 2010, 2012). New information on low frequency or rare variants that are associated with traits such as T2D using such technologies is emerging (Cornes et al. 2014; Estrada et al. 2014; Flannick et al. 2014; Steinthorsdottir et al. 2014). Thus genome-wide sequencing will facilitate large-scale sequence studies that will lead to identification of sequence variations, that are common, less common and rare variants that influence obesity, T2D, and MS. The limitations of the results of GWAS in explaining the heritability of complex diseases or traits may be in part due to the fact that a large number of the genetic variants in each genome are rare and private that cannot be identified using the existing technologies. Therefore, the focus has shifted toward the rare variant-common disease (RV-CD) hypothesis that implies rare and infrequent variants exert larger effect sizes (Bodmer and Bonilla 2008). Accordingly, there is a paradigm shift in genetic studies of complex traits from "common disease-common variant (CD-CV) hypothesis to rare variant-common disease hypothesis" for the identification of uncommon and rare variants with large effects. The shift has been in part accelerated by the availability of the next-generation sequencing platforms (i.e., WES and WGS) that will enable identification of the uncommon and rare variants. It is conceivable that both common and rare variants could be important contributors to complex disease risk (Gibson 2011; Agarwala et al. 2013; Zuk et al. 2014).

Next-generation sequencing (NGS) technology (Bentley et al. 2008; Wheeler et al. 2008; Metzker 2010) represents a powerful approach for studying genetic variants in the human genome or exome including rare genetic variants, efficiently. Next generation sequencing techniques, discussed in detail in the Curran et al. chapter, are emerging tools for the discovery of novel mutations underlying complex diseases or phenotypes including obesity, T2D and MS (Bamshad et al. 2011; Kilpinen and Barrett 2013). WGS enables comprehensive sequencing of the entire genome, including intronic areas that may harbor deleterious mutations while WES facilitates deeper coverage of coding regions that are important for protein function (Bamshad et al. 2011). Since there are a large number of genetic variants found in each genome, it is useful to limit multiple testing by utilizing information such as results of linkage or GWA analyses, computational predictions of the effect of a mutation on protein function, and databases of known polymorphisms to distinguish deleterious from benign variants (Chou et al. 2012). However, it is important to note that functional studies will be needed to prove the biologic effect of a variant. Thus WGS and WES are powerful screening tools with great potential to identify disease-causing mutations or variants for research and clinical applications. As sequencing costs come down, we anticipate that whole exome and whole genome sequencing will dominate genetic studies of complex diseases or traits in the near future. For example, as noted above, new information on low frequency or rare variants that are associated with T2D using such technologies has been emerging (Cornes et al. 2014; Estrada et al. 2014; Flannick et al. 2014; Steinthorsdottir et al. 2014).

12.5 Gene-by-Environment Interaction Effects on Obesity, T2D, and MS

It is well demonstrated that susceptibility to common complex diseases is determined by genetic and environmental factors and their interactions. The rapid rise in the occurrence of obesity, T2D, and MS in recent decades is attributable to changing environmental factors including socioeconomic and life style factors (e.g., poor diet and low physical activity) and the rise of obesogenic environments (e.g., fast food restaurants and television viewing time) (Trevino et al. 1999; Dehghan et al. 2005; Harper 2006; Biro and Wien 2010). However, these environmental factors do not affect all groups equally. For example, white Americans with different genetic backgrounds but living in a similar obesogenic environment are less susceptible to developing obesity ($\sim$32 %) or T2D ($\sim$8 %) compared to Pima Indians living in Arizona (Vimaleswaran and Loos 2010; Choquet and Meyre 2011a). A strong interaction between the *PPARγ* gene and nutrient environment provides a convincing example of gene–environment interaction effects on obesity and T2D. Several studies demonstrated that genetic variation in *PPARγ* is a strong modulator of physiological response to dietary fat in humans (genotype x dietary fat intake interaction), modifying lifestyle effects on obesity (Memisoglu et al. 2003; Ylonen et al. 2008; Ridderstrale and Groop 2009).

Various studies have focused on the identification of specific environmental factors affecting obesity, T2D, and MS that may interact with genetic disposition (McAllister et al. 2009). For example, using data from the Early Childhood Longitudinal Study, childhood weight status was found to be influenced by more television watching, eating fewer family meals, and unsafe neighborhoods for outdoor play (Gable et al. 2007; Steffen et al. 2009). In a sample of 422 children aged 5–10 years, an inverse association was found between sleep duration and the risk of developing overweight/obesity (Chaput et al. 2006). It has been reported that children attending public schools tend to be overweight and that the free or reduced cost food programs at private schools is positively associated with BMI levels of children (Li and Hooker 2010). In addition to behavioral traits, certain metabolic traits (e.g., insulin and leptin) have been identified to be positive predictors of weight gain (Butte et al. 2007). Another potential contributor to insulin resistance/hyperinsulinemia or to the development of T2D in children is puberty (ADA 2000; Goran et al. 2003). In Indian populations, genotype-by-diet interactions appear to play a major role in increasing the risk for diabetes (Mohan et al. 2007). It is well known that increased physical activity/exercise training is associated with favorable lipid profiles and with improvement in insulin sensitivity (Isomaa 2003; Cruz et al. 2004). Furthermore, there is strong evidence that genetic susceptibility to obesity can be altered through physical activity (Choquet and Meyre 2011a). Several studies showed a strong interaction between the *FTO* genotype and physical activity on obesity risk in adults and adolescents (Andreasen et al. 2008b; Sonestedt et al. 2009; Ahmad et al. 2011).

There is substantial evidence that dietary intake interacts with genes to modulate predisposition to complex diseases or traits. Indeed, researchers have examined gene-by-lifestyle (G × LS) interaction influences on obesity and T2D using the findings from GWA studies of T2D [e.g., *TCF7L2*] and obesity [*FTO*] (Reinehr et al. 2008; Timpson et al. 2008; Wardle et al. 2008). In European populations, gene–environment studies have reported that the association between the *FTO* gene and BMI is attenuated by physical activity levels (Rampersaud et al. 2008; Andreasen et al. 2008b; Vimaleswaran and Loos 2010). In the Old Order Amish population, physical activity was found to be inversely associated with BMI (Rampersaud et al. 2008). Although such association was only observed in individuals homozygous for a *FTO* risk allele, it was not observed in individuals with the protective allele (Rampersaud et al. 2008). There is substantial interaction between common variants in the *TCF7L2* gene and lifestyle modification in the risk of progression to T2D. For example, as seen in the Diabetes Prevention Program and the Finnish Diabetes Prevention, there was no effect of the TCF7L2 risk allele on the progression to T2D in the lifestyle intervention groups, but such an effect was found in the placebo control groups (Florez et al. 2006; Wang et al. 2007). Thus gene–environment studies show that changes in lifestyle can moderate effects of genetic susceptibility, as demonstrated by the *FTO* and physical activity and *TCF7L2* and lifestyle intervention studies. A few studies have examined the link between nutritional environment during prenatal and postnatal states and the risk to develop obesity, T2D, and MS-related traits in both childhood and adulthood (Reusens et al. 2007; Mayer-Davis 2008; Taveras et al. 2010; Dabelea and Crume 2011). In addition, intrauterine environment has been associated with the development of complex diseases in adulthood (Barker 2003; Bruce and Hanson 2010; Xita and Tsatsoulis 2010). The modification of epigenetic programming during the fetal/postnatal development due to maternal nutrition and metabolic disturbances could influence susceptibility to obesity, T2D, and MS in adulthood (Gallou-Kabani and Junien 2005; Junien and Nathanielsz 2007; Gluckman and Hanson 2008; Gluckman et al. 2008; Lusis et al. 2008; Nuyt and Alexander 2009; Dabelea and Crume 2011).

12.6 Conclusions

Obesity, type 2 diabetes, and metabolic syndrome are common complex diseases that often cluster in families and appear as comorbidities. Progression of these diseases is strongly associated with a variety of risk factors such as advancing age, genetic background, other metabolic factors such as insulin resistance, and behavioral factors (smoking, overeating, and inactivity). Genetic and environmental factors and their interactions determine the risk of developing obesity, T2D, and MS and contribute to the variation in risk profiles of various populations. To date, three major approaches have been used to identify genes influencing these diseases, better understand the disease pathogenesis, and find new therapeutic targets. However, deciphering the genetic architecture of these complex diseases or traits has been a challenging task. The candidate gene approach was largely successful in identifying variants influencing monogenic or rare disorders. But this approach has yielded disappointing results in identifying genes with measurable effects on common forms of obesity, T2D, and MS. A major limitation of the candidate gene approach is that it relies on our current knowledge of pathophysiology of a disease, though studies with larger sample sizes and meta-analyses have confirmed associations between several candidate genes and obesity, T2D, and MS-related traits.

Genome-wide linkage studies were very successful in identifying genes responsible for monogenic diseases, but few linkage studies of diseases with polygenic inheritance patterns have yielded positive findings and only a subset of these findings have been replicated in independent studies. Furthermore, fine-mapping studies were not that successful to identify the variants that likely underlie the linkage signal, partly due to limitations on follow-up of these signals from the high cost of dense genotyping and sequencing studies during this era. However, the genome-wide linkage approach has been successful to identify a small number of loci for common obesity, T2D, and MS. More recently, there have been a number of high-profile successes using the GWAS approach. So far, four waves of large-scale high-density GWAS have been conducted, which led to a series of discoveries in the fields of obesity, diabetes, and metabolic syndrome. The current number of loci reaching genome-wide significance is ~ 80 for obesity, whereas for T2D, the current number of T2D risk loci is ~ 90, and for MS the number of loci is about 50, and many of them were discovered through GWAS. However, most of these loci have not resulted in identification of functional variants and collectively they account for only a small fraction of the overall heritable risk for obesity, T2D, and MS. As such comprehensive re-sequencing and fine-mapping efforts are needed to uncover potential sources of missing heritability and unambiguously identify causal variants to start exploring the functional relevance of these loci. Next-generation sequencing technologies may lead to the discovery of rare and private variants with large effects that may yield important insights into the genetic architecture and pathophysiology of complex diseases to be utilized in clinical practice. Identification of causal variants may provide new avenues for novel therapeutic and preventive approaches for better treatment and prevention of obesity, T2D, and MS, and eventually facilitate the development of pharmacogenetics and personalized medicine.

References

Abecasis GR, Altshuler D, Auton A, Brooks LD, Durbin RM, Gibbs RA, Hurles ME, McVean GA (2010) A map of human genome variation from population-scale sequencing. Nature 467:1061–1073

Abecasis GR, Auton A, Brooks LD, Depristo MA, Durbin RM, Handsaker RE, Kang HM, Marth GT, McVean GA (2012) An integrated map of genetic variation from 1,092 human genomes. Nature 491:56–65

ADA (2000) Type 2 diabetes in children and adolescents. Am Diabetes Assoc Pediatr 105:671–680

ADA (2013) Economic costs of diabetes in the U.S. in 2012. Diabetes Care 36:1033–1046

Adeyemo A, Gerry N, Chen G, Herbert A, Doumatey A, Huang H, Zhou J, Lashley K, Chen Y, Christman M, Rotimi C (2009) A genome-wide association study of hypertension and blood pressure in African Americans. PLoS Genet 5:e1000564

Agarwala V, Flannick J, Sunyaev S, Altshuler D (2013) Evaluating empirical bounds on complex disease genetic architecture. Nat Genet 45:1418–1427

Ahmad T, Lee IM, Pare G, Chasman DI, Rose L, Ridker PM, Mora S (2011) Lifestyle interaction with fat mass and obesity-associated (FTO) genotype and risk of obesity in apparently healthy U.S. women. Diabetes Care 34:675–680

Alberti KG, Eckel RH, Grundy SM, Zimmet PZ, Cleeman JI, Donato KA, Fruchart JC, James WP, Loria CM, Smith SC Jr (2009) Harmonizing the metabolic syndrome: a joint interim statement of the

International Diabetes Federation Task Force on Epidemiology and Prevention; National Heart, Lung, and Blood Institute; American Heart Association; World Heart Federation; International Atherosclerosis Society; and International Association for the Study of Obesity. Circulation 120:1640–1645

Allison DB, Kaprio J, Korkeila M, Koskenvuo M, Neale MC, Hayakawa K (1996) The heritability of body mass index among an international sample of monozygotic twins reared apart. Int J Obes Relat Metab Disord 20:501–506

Almasy L, Blangero J (1998) Multipoint quantitative-trait linkage analysis in general pedigrees. Am J Hum Genet 62:1198–1211

Almasy L, Blangero J (2008) Contemporary model-free methods for linkage analysis. Adv Genet 60:175–193

Almasy L, Blangero J (2009) Human QTL linkage mapping. Genetica 136:333–340

Almasy L, Hixson JE, Rainwater DL, Cole S, Williams JT, Mahaney MC, VandeBerg JL, Stern MP, MacCluer JW, Blangero J (1999a) Human pedigree-based quantitative-trait-locus mapping: localization of two genes influencing HDL-cholesterol metabolism. Am J Hum Genet 64:1686–1693

Almasy L, MacCluer JW (2002) Association studies of vascular phenotypes: how and why? Arterioscler Thromb Vasc Biol 22:1055–1057

Almasy L, Williams JT, Dyer TD, Blangero J (1999b) Quantitative trait locus detection using combined linkage/disequilibrium analysis. Genet Epidemiol 17 (Suppl 1):S31–S36

Altmuller J, Palmer LJ, Fischer G, Scherb H, Wjst M (2001) Genomewide scans of complex human diseases: true linkage is hard to find. Am J Hum Genet 69:936–950

Altshuler D, Daly M, Kruglyak L (2000) Guilt by association. Nat Genet 26:135–137

Altshuler D, Daly MJ, Lander ES (2008) Genetic mapping in human disease. Science 322:881–888

Amos CI (1994) Robust variance-components approach for assessing genetic linkage in pedigrees. Am J Hum Genet 54:535–543

Anderson CA, Soranzo N, Zeggini E, Barrett JC (2011) Synthetic associations are unlikely to account for many common disease genome-wide association signals. PLoS Biol 9:e1000580

Andreasen CH, Mogensen MS, Borch-Johnsen K, Sandbaek A, Lauritzen T, Sorensen TI, Hansen L, Almind K, Jorgensen T, Pedersen O, Hansen T (2008a) Non-replication of genome-wide based associations between common variants in INSIG2 and PFKP and obesity in studies of 18,014 Danes. PLoS One 3:e2872

Andreasen CH, Stender-Petersen KL, Mogensen MS, Torekov SS, Wegner L, Andersen G, Nielsen AL, Albrechtsen A, Borch-Johnsen K, Rasmussen SS, Clausen JO, Sandbaek A, Lauritzen T, Hansen L, Jorgensen T, Pedersen O, Hansen T (2008b) Low physical activity accentuates the effect of the FTO rs9939609 polymorphism on body fat accumulation. Diabetes 57:95–101

Apovian CM (2010) The causes, prevalence, and treatment of obesity revisited in 2009: what have we learned so far? Am J Clin Nutr 91:277S–279S

Arar N, Nath S, Thameem F, Bauer R, Voruganti S, Comuzzie A, Cole S, Blangero J, MacCluer J, Abboud H (2007) Genome-wide scans for microalbuminuria in Mexican Americans: the San Antonio Family Heart Study. Genet Med 9:80–87

Arita Y, Kihara S, Ouchi N, Takahashi M, Maeda K, Miyagawa J, Hotta K, Shimomura I, Nakamura T, Miyaoka K, Kuriyama H, Nishida M, Yamashita S, Okubo K, Matsubara K, Muraguchi M, Ohmoto Y, Funahashi T, Matsuzawa Y (2012) Paradoxical decrease of an adipose-specific protein, adiponectin, in obesity. 1999. Biochem Biophys Res Commun 425:560–564

Arya R, Blangero J, Williams K, Almasy L, Dyer TD, Leach RJ, O'Connell P, Stern MP, Duggirala R (2002a) Factors of insulin resistance syndrome–related phenotypes are linked to genetic locations on chromosomes 6 and 7 in nondiabetic Mexican-Americans. Diabetes 51:841–847

Arya R, Duggirala R, Almasy L, Rainwater DL, Mahaney MC, Cole S, Dyer TD, Williams K, Leach RJ, Hixson JE, MacCluer JW, O'Connell P, Stern MP, Blangero J (2002b) Linkage of high-density lipoprotein-cholesterol concentrations to a locus on chromosome 9p in Mexican Americans. Nat Genet 30:102–105

Arya R, Duggirala R, Jenkinson CP, Almasy L, Blangero J, O'Connell P, Stern MP (2004) Evidence of a novel quantitative-trait locus for obesity on chromosome 4p in Mexican Americans. Am J Hum Genet 74:272–282

Avery CL, He Q, North KE, Ambite JL, Boerwinkle E, Fornage M, Hindorff LA, Kooperberg C, Meigs JB, Pankow JS, Pendergrass SA, Psaty BM, Ritchie MD, Rotter JI, Taylor KD, Wilkens LR, Heiss G, Lin DY (2011) A phenomics-based strategy identifies loci on APOC1, BRAP, and PLCG1 associated with metabolic syndrome phenotype domains. PLoS Genet 7: e1002322

Bailey-Wilson JE, Wilson AF (2011) Linkage analysis in the next-generation sequencing era. Hum Hered 72:228–236

Bamshad MJ, Ng SB, Bigham AW, Tabor HK, Emond MJ, Nickerson DA, Shendure J (2011) Exome sequencing as a tool for Mendelian disease gene discovery. Nat Rev Genet 12:745–755

Barbalic M, Reiner AP, Wu C, Hixson JE, Franceschini N, Eaton CB, Heiss G, Couper D, Mosley T, Boerwinkle E (2011) Genome-wide association analysis of incident coronary heart disease (CHD) in African Americans: a short report. PLoS Genet 7: e1002199

Barker DJ (2003) The developmental origins of adult disease. Eur J Epidemiol 18:733–736

Barroso I (2005) Genetics of type 2 diabetes. Diabet Med 22:517–535

Been LF, Ralhan S, Wander GS, Mehra NK, Singh J, Mulvihill JJ, Aston CE, Sanghera DK (2011) Variants in KCNQ1 increase type II diabetes susceptibility in

South Asians: a study of 3,310 subjects from India and the US. BMC Med Genet 12:18

Behan D, Cox S (2010) Obesity and its relation to mortality and morbidity costs, vol 59. The Society of Actuaries

Bell CG, Walley AJ, Froguel P (2005) The genetics of human obesity. Nat Rev Genet 6:221–234

Bell GI, Xiang KS, Newman MV, Wu SH, Wright LG, Fajans SS, Spielman RS, Cox NJ (1991) Gene for non-insulin-dependent diabetes mellitus (maturity-onset diabetes of the young subtype) is linked to DNA polymorphism on human chromosome 20q. Proc Natl Acad Sci U S A 88:1484–1488

Below JE, Gamazon ER, Morrison JV, Konkashbaev A, Pluzhnikov A, McKeigue PM, Parra EJ, Elbein SC, Hallman DM, Nicolae DL, Bell GI, Cruz M, Cox NJ, Hanis CL (2011) Genome-wide association and meta-analysis in populations from Starr County, Texas, and Mexico City identify type 2 diabetes susceptibility loci and enrichment for expression quantitative trait loci in top signals. Diabetologia 54:2047–2055

Benjafield AV, Wang WY, Speirs HJ, Morris BJ (2005) Genome-wide scan for hypertension in Sydney Sibships: the GENIHUSS study. Am J Hypertens 18:828–832

Bentley DR, Balasubramanian S, Swerdlow HP, Smith GP, Milton J, Brown CG, Hall KP, Evers DJ, Barnes CL, Bignell HR, Boutell JM, Bryant J, Carter RJ, Keira CR, Cox AJ, Ellis DJ, Flatbush MR, Gormley NA, Humphray SJ, Irving LJ, Karbelashvili MS, Kirk SM, Li H, Liu X, Maisinger KS, Murray LJ, Obradovic B, Ost T, Parkinson ML, Pratt MR, Rasolonjatovo IM, Reed MT, Rigatti R, Rodighiero C, Ross MT, Sabot A, Sankar SV, Scally A, Schroth GP, Smith ME, Smith VP, Spiridou A, Torrance PE, Tzonev SS, Vermaas EH, Walter K, Wu X, Zhang L, Alam MD, Anastasi C, Aniebo IC, Bailey DM, Bancarz IR, Banerjee S, Barbour SG, Baybayan PA, Benoit VA, Benson KF, Bevis C, Black PJ, Boodhun A, Brennan JS, Bridgham JA, Brown RC, Brown AA, Buermann DH, Bundu AA, Burrows JC, Carter NP, Castillo N, Catenazzi Chiara E, Chang S, Neil CR, Crake NR, Dada OO, Diakoumakos KD, Dominguez-Fernandez B, Earnshaw DJ, Egbujor UC, Elmore DW, Etchin SS, Ewan MR, Fedurco M, Fraser LJ, Fuentes Fajardo KV, Scott FW, George D, Gietzen KJ, Goddard CP, Golda GS, Granieri PA, Green DE, Gustafson DL, Hansen NF, Harnish K, Haudenschild CD, Heyer NI, Hims MM, Ho JT, Horgan AM, Hoschler K, Hurwitz S, Ivanov DV, Johnson MQ, James T, Huw Jones TA, Kang GD, Kerelska TH, Kersey AD, Khrebtukova I, Kindwall AP, Kingsbury Z, Kokko-Gonzales PI, Kumar A, Laurent MA, Lawley CT, Lee SE, Lee X, Liao AK, Loch JA, Lok M, Luo S, Mammen RM, Martin JW, McCauley PG, McNitt P, Mehta P, Moon KW, Mullens JW, Newington T, Ning Z, Ling NB, Novo SM, O'Neill MJ, Osborne MA, Osnowski A, Ostadan O, Paraschos LL, Pickering L, Pike AC, Pike AC, Chris PD, Pliskin DP, Podhasky J, Quijano VJ, Raczy C, Rae VH, Rawlings SR, Chiva RA, Roe PM, Rogers J, Rogert Bacigalupo MC, Romanov N, Romieu A, Roth RK, Rourke NJ, Ruediger ST, Rusman E, Sanches-Kuiper RM, Schenker MR, Seoane JM, Shaw RJ, Shiver MK, Short SW, Sizto NL, Sluis JP, Smith MA, Ernest Sohna SJ, Spence EJ, Stevens K, Sutton N, Szajkowski L, Tregidgo CL, Turcatti G, Vandevondele S, Verhovsky Y, Virk SM, Wakelin S, Walcott GC, Wang J, Worsley GJ, Yan J, Yau L, Zuerlein M, Rogers J, Mullikin JC, Hurles ME, McCooke NJ, West JS, Oaks FL, Lundberg PL, Klenerman D, Durbin R, Smith AJ (2008) Accurate whole human genome sequencing using reversible terminator chemistry. Nature 456:53–59

Benzinou M, Creemers JW, Choquet H, Lobbens S, Dina C, Durand E, Guerardel A, Boutin P, Jouret B, Heude B, Balkau B, Tichet J, Marre M, Potoczna N, Horber F, Le Stunff C, Czernichow S, Sandbaek A, Lauritzen T, Borch-Johnsen K, Andersen G, Kiess W, Korner A, Kovacs P, Jacobson P, Carlsson LM, Walley AJ, Jorgensen T, Hansen T, Pedersen O, Meyre D, Froguel P (2008) Common nonsynonymous variants in PCSK1 confer risk of obesity. Nat Genet 40:943–945

Berenson GS (2002) Childhood risk factors predict adult risk associated with subclinical cardiovascular disease. The Bogalusa heart study. Am J Cardiol 90:3L–7L

Billings LK, Florez JC (2010) The genetics of type 2 diabetes: what have we learned from GWAS? Ann N Y Acad Sci 1212:59–77

Biro FM, Wien M (2010) Childhood obesity and adult morbidities. Am J Clin Nutr 91:1499S–1505S

Blangero J (2004) Localization and identification of human quantitative trait loci: king harvest has surely come. Curr Opin Genet Dev 14:233–240

Blangero J, Almasy L (1997) Multipoint oligogenic linkage analysis of quantitative traits. Genet Epidemiol 14:959–964

Blangero J, Williams JT, Almasy L (2001) Variance component methods for detecting complex trait loci. Adv Genet 42:151–181

Bodmer W, Bonilla C (2008) Common and rare variants in multifactorial susceptibility to common diseases. Nat Genet 40:695–701

Boesgaard TW, Zilinskaite J, Vanttinen M, Laakso M, Jansson PA, Hammarstedt A, Smith U, Stefan N, Fritsche A, Haring H, Hribal M, Sesti G, Zobel DP, Pedersen O, Hansen T (2008) The common SLC30A8 Arg325Trp variant is associated with reduced first-phase insulin release in 846 non-diabetic offspring of type 2 diabetes patients—the EUGENE2 study. Diabetologia 51:816–820

Bollepalli S, Dolan LM, Deka R, Martin LJ (2010) Association of FTO gene variants with adiposity in African-American adolescents. Obesity (Silver Spring) 18:1959–1963

Bosse Y, Despres JP, Chagnon YC, Rice T, Rao DC, Bouchard C, Perusse L, Vohl MC (2007) Quantitative trait locus on 15q for a metabolic syndrome variable derived from factor analysis. Obesity (Silver Spring) 15:544–550

Bouatia-Naji N, Bonnefond A, Cavalcanti-Proenca C, Sparso T, Holmkvist J, Marchand M, Delplanque J, Lobbens S, Rocheleau G, Durand E, De GF, Chevre JC, Borch-Johnsen K, Hartikainen AL, Ruokonen A, Tichet J, Marre M, Weill J, Heude B, Tauber M, Lemaire K, Schuit F, Elliott P, Jorgensen T, Charpentier G, Hadjadj S, Cauchi S, Vaxillaire M, Sladek R, Visvikis-Siest S, Balkau B, Levy-Marchal C, Pattou F, Meyre D, Blakemore AI, Jarvelin MR, Walley AJ, Hansen T, Dina C, Pedersen O, Froguel P (2009) A variant near MTNR1B is associated with increased fasting plasma glucose levels and type 2 diabetes risk. Nat Genet 41:89–94

Bouatia-Naji N, Meyre D, Lobbens S, Seron K, Fumeron F, Balkau B, Heude B, Jouret B, Scherer PE, Dina C, Weill J, Froguel P (2006) ACDC/adiponectin polymorphisms are associated with severe childhood and adult obesity. Diabetes 55:545–550

Boudreau DM, Malone DC, Raebel MA, Fishman PA, Nichols GA, Feldstein AC, Boscoe AN, Ben Joseph RH, Magid DJ, Okamoto LJ (2009) Health care utilization and costs by metabolic syndrome risk factors. Metab Syndr Relat Disord 7:305–314

Bowden DW, Rudock M, Ziegler J, Lehtinen AB, Xu J, Wagenknecht LE, Herrington D, Rich SS, Freedman BI, Carr JJ, Langefeld CD (2006) Coincident linkage of type 2 diabetes, metabolic syndrome, and measures of cardiovascular disease in a genome scan of the diabetes heart study. Diabetes 55:1985–1994

Boyle AP, Hong EL, Hariharan M, Cheng Y, Schaub MA, Kasowski M, Karczewski KJ, Park J, Hitz BC, Weng S, Cherry JM, Snyder M (2012) Annotation of functional variation in personal genomes using RegulomeDB. Genome Res 22:1790–1797

Bradfield JP, Taal HR, Timpson NJ, Scherag A, Lecoeur C, Warrington NM, Hypponen E, Holst C, Valcarcel B, Thiering E, Salem RM, Schumacher FR, Cousminer DL, Sleiman PM, Zhao J, Berkowitz RI, Vimaleswaran KS, Jarick I, Pennell CE, Evans DM, St Pourcain B, Berry DJ, Mook-Kanamori DO, Hofman A, Rivadeneira F, Uitterlinden AG, van Duijn CM, van der Valk RJ, de Jongste JC, Postma DS, Boomsma DI, Gauderman WJ, Hassanein MT, Lindgren CM, Magi R, Boreham CA, Neville CE, Moreno LA, Elliott P, Pouta A, Hartikainen AL, Li M, Raitakari O, Lehtimaki T, Eriksson JG, Palotie A, Dallongeville J, Das S, Deloukas P, McMahon G, Ring SM, Kemp JP, Buxton JL, Blakemore AI, Bustamante M, Guxens M, Hirschhorn JN, Gillman MW, Kreiner-Moller E, Bisgaard H, Gilliland FD, Heinrich J, Wheeler E, Barroso I, O'Rahilly S, Meirhaeghe A, Sorensen TI, Power C, Palmer LJ, Hinney A, Widen E, Farooqi IS, McCarthy MI, Froguel P, Meyre D, Hebebrand J, Jarvelin MR, Jaddoe VW, Smith GD, Hakonarson H, Grant SF (2012) A genome-wide association meta-analysis identifies new childhood obesity loci. Nat Genet 44:526–531

Brickman AM, Reitz C, Luchsinger JA, Manly JJ, Schupf N, Muraskin J, DeCarli C, Brown TR, Mayeux R (2010) Long-term blood pressure fluctuation and cerebrovascular disease in an elderly cohort. Arch Neurol 67:564–569

Brown LJ, Clark PC, Armstrong KA, Liping Z, Dunbar SB (2010) Identification of modifiable chronic kidney disease risk factors by gender in an African-American metabolic syndrome cohort. Nephrol Nurs J 37(133–41):148

Bruce KD, Hanson MA (2010) The developmental origins, mechanisms, and implications of metabolic syndrome. J Nutr 140:648–652

Bush WS, Moore JH (2012) Chapter 11: genome-wide association studies. PLoS Comput Biol 8:e1002822

Butte NF, Cai G, Cole SA, Wilson TA, Fisher JO, Zakeri IF, Ellis KJ, Comuzzie AG (2007) Metabolic and behavioral predictors of weight gain in Hispanic children: the Viva la Familia Study. Am J Clin Nutr 85:1478–1485

Butte NF, Comuzzie AG, Cole SA, Mehta NR, Cai G, Tejero M, Bastarrachea R, Smith EO (2005) Quantitative genetic analysis of the metabolic syndrome in Hispanic children. Pediatr Res 58:1243–1248

Caceres M, Teran CG, Rodriguez S, Medina M (2008) Prevalence of insulin resistance and its association with metabolic syndrome criteria among Bolivian children and adolescents with obesity. BMC Pediatr 8:31

Cai G, Cole SA, Butte N, Bacino C, Diego V, Tan K, Goring HH, O'Rahilly S, Farooqi IS, Comuzzie AG (2006) A quantitative trait locus on chromosome 18q for physical activity and dietary intake in Hispanic children. Obesity (Silver Spring) 14:1596–1604

Cai G, Cole SA, Freeland-Graves JH, MacCluer JW, Blangero J, Comuzzie AG (2004) Principal component for metabolic syndrome risk maps to chromosome 4p in Mexican Americans: the San Antonio Family Heart Study. Hum Biol 76:651–665

Caprio S (2003) Obesity and type 2 diabetes: the twin epidemics. Diabetes Spectr 16:230

Cardon LR, Bell JI (2001) Association study designs for complex diseases. Nat Rev Genet 2:91–99

Cassell PG, Saker PJ, Huxtable SJ, Kousta E, Jackson AE, Hattersley AT, Frayling TM, Walker M, Kopelman PG, Ramachandran A, Snehelatha C, Hitman GA, McCarthy MI (2000) Evidence that single nucleotide polymorphism in the uncoupling protein 3 (UCP3) gene influences fat distribution in women of European and Asian origin. Diabetologia 43:1558–1564

Caterson ID, Gill TP (2002) Obesity: epidemiology and possible prevention. Best Pract Res Clin Endocrinol Metab 16:595–610

Cauchi S, Froguel P (2008) TCF7L2 genetic defect and type 2 diabetes. Curr Diab Rep 8:149–155

CDC. Centers for Disease Control and Prevention (2011) National diabetes fact sheet: national estimates and general information on diabetes and prediabetes in the United States, 2011. U.S. Department of Health and Human Services, Centers for Disease Control and Prevention, Atlanta

Chambers JC, Elliott P, Zabaneh D, Zhang W, Li Y, Froguel P, Balding D, Scott J, Kooner JS (2008) Common genetic variation near MC4R is associated with waist circumference and insulin resistance. Nat Genet 40:716–718

Chang MH, Ned RM, Hong Y, Yesupriya A, Yang Q, Liu T, Janssens AC, Dowling NF (2011) Racial/ethnic variation in the association of lipid-related genetic variants with blood lipids in the US adult population. Circ Cardiovasc Genet 4:523–533

Chanock SJ, Manolio T, Boehnke M, Boerwinkle E, Hunter DJ, Thomas G, Hirschhorn JN, Abecasis G, Altshuler D, Bailey-Wilson JE, Brooks LD, Cardon LR, Daly M, Donnelly P, Fraumeni JF Jr, Freimer NB, Gerhard DS, Gunter C, Guttmacher AE, Guyer MS, Harris EL, Hoh J, Hoover R, Kong CA, Merikangas KR, Morton CC, Palmer LJ, Phimister EG, Rice JP, Roberts J, Rotimi C, Tucker MA, Vogan KJ, Wacholder S, Wijsman EM, Winn DM, Collins FS (2007) Replicating genotype-phenotype associations. Nature 447:655–660

Chaput JP, Brunet M, Tremblay A (2006) Relationship between short sleeping hours and childhood overweight/obesity: results from the 'Quebec en Forme' Project. Int J Obes (Lond) 30:1080–1085

Chen G, Bentley A, Adeyemo A, Shriner D, Zhou J, Doumatey A, Huang H, Ramos E, Erdos M, Gerry N, Herbert A, Christman M, Rotimi C (2012) Genome-wide association study identifies novel loci association with fasting insulin and insulin resistance in African Americans. Hum Mol Genet 21:4530–4536

Cho YS, Chen CH, Hu C, Long J, Ong RT, Sim X, Takeuchi F, Wu Y, Go MJ, Yamauchi T, Chang YC, Kwak SH, Ma RC, Yamamoto K, Adair LS, Aung T, Cai Q, Chang LC, Chen YT, Gao Y, Hu FB, Kim HL, Kim S, Kim YJ, Lee JJ, Lee NR, Li Y, Liu JJ, Lu W, Nakamura J, Nakashima E, Ng DP, Tay WT, Tsai FJ, Wong TY, Yokota M, Zheng W, Zhang R, Wang C, So WY, Ohnaka K, Ikegami H, Hara K, Cho YM, Cho NH, Chang TJ, Bao Y, Hedman AK, Morris AP, McCarthy MI, Takayanagi R, Park KS, Jia W, Chuang LM, Chan JC, Maeda S, Kadowaki T, Lee JY, Wu JY, Teo YY, Tai ES, Shu XO, Mohlke KL, Kato N, Han BG, Seielstad M (2012) Meta-analysis of genome-wide association studies identifies eight new loci for type 2 diabetes in east Asians. Nat Genet 44:67–72

Cho YS, Go MJ, Kim YJ, Heo JY, Oh JH, Ban HJ, Yoon D, Lee MH, Kim DJ, Park M, Cha SH, Kim JW, Han BG, Min H, Ahn Y, Park MS, Han HR, Jang HY, Cho EY, Lee JE, Cho NH, Shin C, Park T, Park JW, Lee JK, Cardon L, Clarke G, McCarthy MI, Lee JY, Lee JK, Oh B, Kim HL (2009) A large-scale genome-wide association study of Asian populations uncovers genetic factors influencing eight quantitative traits. Nat Genet 41:527–534

Chopra M, Galbraith S, Darnton-Hill I (2002) A global response to a global problem: the epidemic of overnutrition. Bull World Health Organ 80:952–958

Choquet H, Meyre D (2011a) Genetics of obesity: what have we learned? Curr Genomics 12:169–179

Choquet H, Meyre D (2011b) Molecular basis of obesity: current status and future prospects. Curr Genomics 12:154–168

Chou J, Ohsumi TK, Geha RS (2012) Use of whole exome and genome sequencing in the identification of genetic causes of primary immunodeficiencies. Curr Opin Allergy Clin Immunol 12:623–628

Cirulli ET, Goldstein DB (2010) Uncovering the roles of rare variants in common disease through whole-genome sequencing. Nat Rev Genet 11:415–425

Clausen TD, Mathiesen ER, Hansen T, Pedersen O, Jensen DM, Lauenborg J, Schmidt L, Damm P (2009) Overweight and the metabolic syndrome in adult offspring of women with diet-treated gestational diabetes mellitus or type 1 diabetes. J Clin Endocrinol Metab 94:2464–2470

Clement K, Garner C, Hager J, Philippi A, LeDuc C, Carey A, Harris TJ, Jury C, Cardon LR, Basdevant A, Demenais F, Guy-Grand B, North M, Froguel P (1996) Indication for linkage of the human OB gene region with extreme obesity. Diabetes 45:687–690

Collins FS (1995) Positional cloning moves from perditional to traditional. Nat Genet 9:347–350

Comuzzie AG (2002) The emerging pattern of the genetic contribution to human obesity. Best Pract Res Clin Endocrinol Metab 16:611–621

Comuzzie AG, Allison DB (1998) The search for human obesity genes. Science 280:1374–1377

Comuzzie AG, Hixson JE, Almasy L, Mitchell BD, Mahaney MC, Dyer TD, Stern MP, MacCluer JW, Blangero J (1997) A major quantitative trait locus determining serum leptin levels and fat mass is located on human chromosome 2. Nat Genet 15:273–276

Comuzzie AG, Williams JT, Martin LJ, Blangero J (2001) Searching for genes underlying normal variation in human adiposity. J Mol Med 79:57–70

Cornes BK, Brody JA, Nikpoor N, Morrison AC, Dang HC, Ahn BS, Wang S, Dauriz M, Barzilay JI, Dupuis J, Florez JC, Coresh J, Gibbs RA, Kao WH, Liu CT, McKnight B, Muzny D, Pankow JS, Reid JG, White CC, Johnson AD, Wong TY, Psaty BM, Boerwinkle E, Rotter JI, Siscovick DS, Sladek R, Meigs JB (2014) Association of levels of fasting glucose and insulin with rare variants at the chromosome 11p11.2-MADD locus: Cohorts for Heart and Aging Research in Genomic Epidemiology (CHARGE) Consortium Targeted Sequencing Study. Circ Cardiovasc Genet 7:374–382

Cousminer DL, Berry DJ, Timpson NJ, Ang W, Thiering E, Byrne EM, Taal HR, Huikari V, Bradfield JP, Kerkhof M, Groen-Blokhuis MM, Kreiner-Moller E, Marinelli M, Holst C, Leinonen JT, Perry JR, Surakka I, Pietilainen O, Kettunen J, Anttila V, Kaakinen M, Sovio U, Pouta A, Das S, Lagou V, Power C, Prokopenko I, Evans DM, Kemp JP, St PB, Ring S, Palotie A, Kajantie E, Osmond C, Lehtimaki T, Viikari JS, Kahonen M, Warrington NM, Lye SJ, Palmer LJ, Tiesler CM, Flexeder C, Montgomery GW, Medland SE, Hofman A, Hakonarson H, Guxens M, Bartels M, Salomaa V, Murabito JM, Kaprio J,

Sorensen TI, Ballester F, Bisgaard H, Boomsma DI, Koppelman GH, Grant SF, Jaddoe VW, Martin NG, Heinrich J, Pennell CE, Raitakari OT, Eriksson JG, Smith GD, Hypponen E, Jarvelin MR, McCarthy MI, Ripatti S, Widen E (2013) Genome-wide association and longitudinal analyses reveal genetic loci linking pubertal height growth, pubertal timing and childhood adiposity. Hum Mol Genet 22:2735–2747

Cowie CC, Rust KF, Byrd-Holt DD, Gregg EW, Ford ES, Geiss LS, Bainbridge KE, Fradkin JE (2010) Prevalence of diabetes and high risk for diabetes using A1C criteria in the U.S. population in 1988–2006. Diabetes Care 33:562–568

Cowie L, Morgan M, White P, Gulliford M (2009) Experience of continuity of care of patients with multiple long-term conditions in England. J Health Serv Res Policy 14:82–87

Crocker MK, Yanovski JA (2009) Pediatric obesity: etiology and treatment. Endocrinol Metab Clin North Am 38:525–548

Cruz ML, Goran MI (2004) The metabolic syndrome in children and adolescents. Curr Diab Rep 4:53–62

Cruz ML, Huang TT, Johnson MS, Gower BA, Goran MI (2002) Insulin sensitivity and blood pressure in black and white children. Hypertension 40:18–22

Cruz ML, Shaibi GQ, Weigensberg MJ, Spruijt-Metz D, Ball GD, Goran MI (2005) Pediatric obesity and insulin resistance: chronic disease risk and implications for treatment and prevention beyond body weight modification. Annu Rev Nutr 25:435–468

Cruz ML, Weigensberg MJ, Huang TT, Ball G, Shaibi GQ, Goran MI (2004) The metabolic syndrome in overweight Hispanic youth and the role of insulin sensitivity. J Clin Endocrinol Metab 89:108–113

Dabelea D, Crume T (2011) Maternal environment and the transgenerational cycle of obesity and diabetes. Diabetes 60:1849–1855

Dabelea D, Dolan LM, D'Agostino R Jr, Hernandez AM, McAteer JB, Hamman RF, Mayer-Davis EJ, Marcovina S, Lawrence JM, Pihoker C, Florez JC (2011) Association testing of TCF7L2 polymorphisms with type 2 diabetes in multi-ethnic youth. Diabetologia 54:535–539

Daly AK, Day CP (2001) Candidate gene case-control association studies: advantages and potential pitfalls. Br J Clin Pharmacol 52:489–499

Damcott CM, Pollin TI, Reinhart LJ, Ott SH, Shen H, Silver KD, Mitchell BD, Shuldiner AR (2006) Polymorphisms in the transcription factor 7-like 2 (TCF7L2) gene are associated with type 2 diabetes in the Amish: replication and evidence for a role in both insulin secretion and insulin resistance. Diabetes 55:2654–2659

Day FR, Loos RJ (2011) Developments in obesity genetics in the era of genome-wide association studies. J Nutrigenet Nutrigenomics 4:222–238

Deboer MD (2011) Ethnicity, obesity and the metabolic syndrome: implications on assessing risk and targeting intervention. Expert Rev Endocrinol Metab 6:279–289

DeFronzo RA (1995) Insulin resistance and hyperinsulinemia: the link between NIDDM, CAD, hypertension and dyslipidemia. In Schwartz CLBG (ed) New horizons in diabetes mellitus and cardiovascular disease. Current Science, London, pp 11–27

DeFronzo RA (2004) Dysfunctional fat cells, lipotoxicity and type 2 diabetes. Int J Clin Pract 58(Supp l):9–21

DeFronzo RA, Goodman AM (1995) Efficacy of metformin in patients with non-insulin-dependent diabetes mellitus. The Multicenter Metformin Study Group. N Engl J Med 333:541–549

Dehghan M, Akhtar-Danesh N, Merchant AT (2005) Childhood obesity, prevalence and prevention. Nutr J 4:24

Devaney JM, Thompson PD, Visich PS, Saltarelli WA, Gordon PM, Orkunoglu-Suer EF, Gordish-Dressman H, Harmon BT, Bradbury MK, Panchapakesan K, Khianey R, Hubal MJ, Clarkson PM, Pescatello LS, Zoeller RF, Moyna NM, Angelopoulos TJ, Kraus WE, Hoffman EP (2011) The 1p13.3 LDL (C)-associated locus shows large effect sizes in young populations. Pediatr Res 69:538–543

Diabetes Genetics Initiative of Broad Institute of Harvard and MIT, LundUniversity, and Novartis Institutes of BioMedical Research, Saxena R, Voight BF, Lyssenko V, Burtt NP, de Bakker PI, Chen H, Roix JJ, Kathiresan S, Hirschhorn JN, Daly MJ, Hughes TE, Groop L, Altshuler D, Almgren P, Florez JC, Meyer J, Ardlie K, Bengtsson Boström K, Isomaa B, Lettre G, Lindblad U, Lyon HN, Melander O, Newton-Cheh C, Nilsson P, Orho-Melander M, Råstam L, Speliotes EK, Taskinen MR, Tuomi T, Guiducci C, Berglund A, Carlson J, Gianniny L, Hackett R, Hall L, Holmkvist J, Laurila E, Sjögren M, Sterner M, Surti A, Svensson M, Svensson M, Tewhey R, Blumenstiel B, Parkin M, Defelice M, Barry R, Brodeur W, Camarata J, Chia N, Fava M, Gibbons J, Handsaker B, Healy C, Nguyen K, Gates C, Sougnez C, Gage D, Nizzari M, Gabriel SB, Chirn GW, Ma Q, Parikh H, Richardson D, Ricke D,Purcell S (2007) Genome-wide association analysis identifies loci for type 2 diabetes and triglyceride levels. Science 316(5829):1331–1336

Diamond J (2003) The double puzzle of diabetes. Nature 423:599–602

Dickson SP, Wang K, Krantz I, Hakonarson H, Goldstein DB (2010) Rare variants create synthetic genome-wide associations. PLoS Biol 8:e1000294

Dina C (2008) New insights into the genetics of body weight. Curr Opin Clin Nutr Metab Care 11:378–384

Doria A, Patti ME, Kahn CR (2008) The emerging genetic architecture of type 2 diabetes. Cell Metab 8:186–200

Drong AW, Lindgren CM, McCarthy MI (2012) The genetic and epigenetic basis of type 2 diabetes and obesity. Clin Pharmacol Ther 92:707–715

Duggirala R, Blangero J, Almasy L, Arya R, Dyer TD, Williams KL, Leach RJ, O'Connell P, Stern MP (2001) A major locus for fasting insulin concentrations and insulin resistance on chromosome 6q with strong pleiotropic effects on obesity-related phenotypes in nondiabetic Mexican Americans. Am J Hum Genet 68:1149–1164

Duggirala R, Blangero J, Almasy L, Dyer TD, Williams KL, Leach RJ, O'Connell P, Stern MP (1999) Linkage of type 2 diabetes mellitus and of age at onset to a genetic location on chromosome 10q in Mexican Americans. Am J Hum Genet 64:1127–1140

Duggirala R, Blangero J, Almasy L, Dyer TD, Williams KL, Leach RJ, O'Connell P, Stern MP (2000) A major susceptibility locus influencing plasma triglyceride concentrations is located on chromosome 15q in Mexican Americans. Am J Hum Genet 66:1237–1245

Duggirala R, Stern MP, Mitchell BD, Reinhart LJ, Shipman PA, Uresandi OC, Chung WK, Leibel RL, Hales CN, O'Connell P, Blangero J (1996) Quantitative variation in obesity-related traits and insulin precursors linked to the OB gene region on human chromosome 7. Am J Hum Genet 59:694–703

Dumitrescu L, Brown-Gentry K, Goodloe R, Glenn K, Yang W, Kornegay N, Pui CH, Relling MV, Crawford DC (2011) Evidence for age as a modifier of genetic associations for lipid levels. Ann Hum Genet 75:589–597

Dupuis J, Langenberg C, Prokopenko I, Saxena R, Soranzo N, Jackson AU, Wheeler E, Glazer NL, Bouatia-Naji N, Gloyn AL, Lindgren CM, Magi R, Morris AP, Randall J, Johnson T, Elliott P, Rybin D, Thorleifsson G, Steinthorsdottir V, Henneman P, Grallert H, Dehghan A, Hottenga JJ, Franklin CS, Navarro P, Song K, Goel A, Perry JR, Egan JM, Lajunen T, Grarup N, Sparso T, Doney A, Voight BF, Stringham HM, Li M, Kanoni S, Shrader P, Cavalcanti-Proenca C, Kumari M, Qi L, Timpson NJ, Gieger C, Zabena C, Rocheleau G, Ingelsson E, An P, O'Connell J, Luan J, Elliott A, McCarroll SA, Payne F, Roccasecca RM, Pattou F, Sethupathy P, Ardlie K, Ariyurek Y, Balkau B, Barter P, Beilby JP, Ben Shlomo Y, Benediktsson R, Bennett AJ, Bergmann S, Bochud M, Boerwinkle E, Bonnefond A, Bonnycastle LL, Borch-Johnsen K, Bottcher Y, Brunner E, Bumpstead SJ, Charpentier G, Chen YD, Chines P, Clarke R, Coin LJ, Cooper MN, Cornelis M, Crawford G, Crisponi L, Day IN, de Geus EJ, Delplanque J, Dina C, Erdos MR, Fedson AC, Fischer-Rosinsky A, Forouhi NG, Fox CS, Frants R, Franzosi MG, Galan P, Goodarzi MO, Graessler J, Groves CJ, Grundy S, Gwilliam R, Gyllensten U, Hadjadj S, Hallmans G, Hammond N, Han X, Hartikainen AL, Hassanali N, Hayward C, Heath SC, Hercberg S, Herder C, Hicks AA, Hillman DR, Hingorani AD, Hofman A, Hui J, Hung J, Isomaa B, Johnson PR, Jorgensen T, Jula A, Kaakinen M, Kaprio J, Kesaniemi YA, Kivimaki M, Knight B, Koskinen S, Kovacs P, Kyvik KO, Lathrop GM, Lawlor DA, Le Bacquer O, Lecoeur C, Li Y, Lyssenko V, Mahley R, Mangino M, Manning AK, Martinez-Larrad MT, McAteer JB, McCulloch LJ, McPherson R, Meisinger C, Melzer D, Meyre D, Mitchell BD, Morken MA, Mukherjee S, Naitza S, Narisu N, Neville MJ, Oostra BA, Orru M, Pakyz R, Palmer CN, Paolisso G, Pattaro C, Pearson D, Peden JF, Pedersen NL, Perola M, Pfeiffer AF, Pichler I, Polasek O, Posthuma D, Potter SC, Pouta A, Province MA, Psaty BM, Rathmann W, Rayner NW, Rice K, Ripatti S, Rivadeneira F, Roden M, Rolandsson O, Sandbaek A, Sandhu M, Sanna S, Sayer AA, Scheet P, Scott LJ, Seedorf U, Sharp SJ, Shields B, Sigurethsson G, Sijbrands EJ, Silveira A, Simpson L, Singleton A, Smith NL, Sovio U, Swift A, Syddall H, Syvanen AC, Tanaka T, Thorand B, Tichet J, Tonjes A, Tuomi T, Uitterlinden AG, van Dijk KW, van Hoek M, Varma D, Visvikis-Siest S, Vitart V, Vogelzangs N, Waeber G, Wagner PJ, Walley A, Walters GB, Ward KL, Watkins H, Weedon MN, Wild SH, Willemsen G, Witteman JC, Yarnell JW, Zeggini E, Zelenika D, Zethelius B, Zhai G, Zhao JH, Zillikens MC, Borecki IB, Loos RJ, Meneton P, Magnusson PK, Nathan DM, Williams GH, Hattersley AT, Silander K, Salomaa V, Smith GD, Bornstein SR, Schwarz P, Spranger J, Karpe F, Shuldiner AR, Cooper C, Dedoussis GV, Serrano-Rios M, Morris AD, Lind L, Palmer LJ, Hu FB, Franks PW, Ebrahim S, Marmot M, Kao WH, Pankow JS, Sampson MJ, Kuusisto J, Laakso M, Hansen T, Pedersen O, Pramstaller PP (2010) New genetic loci implicated in fasting glucose homeostasis and their impact on type 2 diabetes risk. Nat Genet 42:105–116

Easton DF, Bishop DT, Ford D, Crockford GP (1993) Genetic linkage analysis in familial breast and ovarian cancer: results from 214 families. The Breast Cancer Linkage Consortium. Am J Hum Genet 52:678–701

Edwards KL, Newman B, Mayer E, Selby JV, Krauss RM, Austin MA (1997) Heritability of factors of the insulin resistance syndrome in women twins. Genet Epidemiol 14:241–253

Edwards KL, Wan JY, Hutter CM, Fong PY, Santorico SA (2011) Multivariate linkage scan for metabolic syndrome traits in families with type 2 diabetes. Obesity (Silver Spring) 19:1235–1243

Edwards SL, Beesley J, French JD, Dunning AM (2013) Beyond GWASs: illuminating the dark road from association to function. Am J Hum Genet 93:779–797

Ehret GB, Munroe PB, Rice KM, Bochud M, Johnson AD, Chasman DI, Smith AV, Tobin MD, Verwoert GC, Hwang SJ, Pihur V, Vollenweider P, O'Reilly PF, Amin N, Bragg-Gresham JL, Teumer A, Glazer NL, Launer L, Zhao JH, Aulchenko Y, Heath S, Sober S, Parsa A, Luan J, Arora P, Dehghan A, Zhang F, Lucas G, Hicks AA, Jackson AU, Peden JF, Tanaka T, Wild SH, Rudan I, Igl W, Milaneschi Y, Parker AN, Fava C, Chambers JC, Fox ER, Kumari M, Go MJ, van der HP, Kao WH, Sjogren M, Vinay DG, Alexander M, Tabara Y, Shaw-Hawkins S, Whincup PH, Liu Y, Shi G, Kuusisto J, Tayo B, Seielstad M, Sim X, Nguyen KD, Lehtimaki T, Matullo G, Wu Y, Gaunt TR, Onland-Moret NC, Cooper MN, Platou CG, Org E, Hardy R, Dahgam S, Palmen J, Vitart V, Braund PS, Kuznetsova T, Uiterwaal CS, Adeyemo A, Palmas W, Campbell H, Ludwig B, Tomaszewski M, Tzoulaki I, Palmer ND, Aspelund T, Garcia M, Chang YP, O'Connell JR, Steinle NI, Grobbee DE, Arking DE, Kardia SL, Morrison AC, Hernandez D, Najjar S, McArdle WL, Hadley D, Brown MJ, Connell JM, Hingorani AD, Day IN, Lawlor DA, Beilby JP, Lawrence RW, Clarke R, Hopewell JC, Ongen H, Dreisbach AW, Li Y, Young JH, Bis JC, Kahonen M, Viikari J, Adair LS, Lee NR, Chen MH, Olden M,

Pattaro C, Bolton JA, Kottgen A, Bergmann S, Mooser V, Chaturvedi N, Frayling TM, Islam M, Jafar TH, Erdmann J, Kulkarni SR, Bornstein SR, Grassler J, Groop L, Voight BF, Kettunen J, Howard P, Taylor A, Guarrera S, Ricceri F, Emilsson V, Plump A, Barroso I, Khaw KT, Weder AB, Hunt SC, Sun YV, Bergman RN, Collins FS, Bonnycastle LL, Scott LJ, Stringham HM, Peltonen L, Perola M, Vartiainen E, Brand SM, Staessen JA, Wang TJ, Burton PR, Soler AM, Dong Y, Snieder H, Wang X, Zhu H, Lohman KK, Rudock ME, Heckbert SR, Smith NL, Wiggins KL, Doumatey A, Shriner D, Veldre G, Viigimaa M, Kinra S, Prabhakaran D, Tripathy V, Langefeld CD, Rosengren A, Thelle DS, Corsi AM, Singleton A, Forrester T, Hilton G, McKenzie CA, Salako T, Iwai N, Kita Y, Ogihara T, Ohkubo T, Okamura T, Ueshima H, Umemura S, Eyheramendy S, Meitinger T, Wichmann HE, Cho YS, Kim HL, Lee JY, Scott J, Sehmi JS, Zhang W, Hedblad B, Nilsson P, Smith GD, Wong A, Narisu N, Stancakova A, Raffel LJ, Yao J, Kathiresan S, O'Donnell CJ, Schwartz SM, Ikram MA, Longstreth WT Jr, Mosley TH, Seshadri S, Shrine NR, Wain LV, Morken MA, Swift AJ, Laitinen J, Prokopenko I, Zitting P, Cooper JA, Humphries SE, Danesh J, Rasheed A, Goel A, Hamsten A, Watkins H, Bakker SJ, van Gilst WH, Janipalli CS, Mani KR, Yajnik CS, Hofman A, Mattace-Raso FU, Oostra BA, Demirkan A, Isaacs A, Rivadeneira F, Lakatta EG, Orru M, Scuteri A, Ala-Korpela M, Kangas AJ, Lyytikainen LP, Soininen P, Tukiainen T, Wurtz P, Ong RT, Dorr M, Kroemer HK, Volker U, Volzke H, Galan P, Hercberg S, Lathrop M, Zelenika D, Deloukas P, Mangino M, Spector TD, Zhai G (2011) Genetic variants in novel pathways influence blood pressure and cardiovascular disease risk. Nature 478:103–109

Elbein SC, Chu WS, Das SK, Yao-Borengasser A, Hasstedt SJ, Wang H, Rasouli N, Kern PA (2007) Transcription factor 7-like 2 polymorphisms and type 2 diabetes, glucose homeostasis traits and gene expression in US participants of European and African descent. Diabetologia 50:1621–1630

Elbein SC, Gamazon ER, Das SK, Rasouli N, Kern PA, Cox NJ (2012) Genetic risk factors for type 2 diabetes: a trans-regulatory genetic architecture? Am J Hum Genet 91:466–477

Elbein SC, Hasstedt SJ (2002) Quantitative trait linkage analysis of lipid-related traits in familial type 2 diabetes: evidence for linkage of triglyceride levels to chromosome 19q. Diabetes 51:528–535

Elston RC (1992) Segregation and linkage analysis. Anim Genet 23:59–62

Estrada K, Aukrust I, Bjorkhaug L, Burtt NP, Mercader JM, Garcia-Ortiz H, Huerta-Chagoya A, Moreno-Macias H, Walford G, Flannick J, Williams AL, Gomez-Vazquez MJ, Fernandez-Lopez JC, Martinez-Hernandez A, Centeno-Cruz F, Mendoza-Caamal E, Revilla-Monsalve C, Islas-Andrade S, Cordova EJ, Soberon X, Gonzalez-Villalpando ME, Henderson E, Wilkens LR, Le ML, Arellano-Campos O, Ordonez-Sanchez ML, Rodriguez-Torres M, Rodriguez-Guillen R, Riba L, Najmi LA, Jacobs SB, Fennell T, Gabriel S, Fontanillas P, Hanis CL, Lehman DM, Jenkinson CP, Abboud HE, Bell GI, Cortes ML, Boehnke M, Gonzalez-Villalpando C, Orozco L, Haiman CA, Tusie-Luna T, Aguilar-Salinas CA, Altshuler D, Njolstad PR, Florez JC, MacArthur DG (2014) Association of a low-frequency variant in HNF1A with type 2 diabetes in a Latino population. JAMA 311:2305–2314

Fall T, Ingelsson E (2012) Genome-wide association studies of obesity and metabolic syndrome. Mol Cell Endocrinol 382(1):740–757. doi:10.1016/j.mce.2012.08.018

Farook VS, Puppala S, Schneider J, Fowler SP, Chittoor G, Dyer TD, Allayee H, Cole SA, Arya R, Black MH, Curran JE, Almasy L, Buchanan TA, Jenkinson CP, Lehman DM, Watanabe RM, Blangero J, Duggirala R (2012) Metabolic syndrome is linked to chromosome 7q21 and associated with genetic variants in CD36 and GNAT3 in Mexican Americans. Obesity (Silver Spring) 20:2083–2092

Farooqi IS, Jebb SA, Langmack G, Lawrence E, Cheetham CH, Prentice AM, Hughes IA, McCamish MA, O'Rahilly S (1999) Effects of recombinant leptin therapy in a child with congenital leptin deficiency. N Engl J Med 341:879–884

Farooqi IS, Keogh JM, Yeo GS, Lank EJ, Cheetham T, O'Rahilly S (2003) Clinical spectrum of obesity and mutations in the melanocortin 4 receptor gene. N Engl J Med 348:1085–1095

Farooqi IS, O'Rahilly S (2005) Monogenic obesity in humans. Annu Rev Med 56:443–458

Farooqi IS, Wangensteen T, Collins S, Kimber W, Matarese G, Keogh JM, Lank E, Bottomley B, Lopez-Fernandez J, Ferraz-Amaro I, Dattani MT, Ercan O, Myhre AG, Retterstol L, Stanhope R, Edge JA, McKenzie S, Lessan N, Ghodsi M, De R, V, Perna F, Fontana S, Barroso I, Undlien DE, O'Rahilly S (2007) Clinical and molecular genetic spectrum of congenital deficiency of the leptin receptor. N Engl J Med 356:237–247

Fernandez-Real JM, Gutierrez C, Ricart W, Casamitjana R, Fernandez-Castaner M, Vendrell J, Richart C, Soler J (1997) The TNF-alpha gene Nco I polymorphism influences the relationship among insulin resistance, percent body fat, and increased serum leptin levels. Diabetes 46:1468–1472

Fesinmeyer MD, Meigs JB, North KE, Schumacher FR, Buzkova P, Franceschini N, Haessler J, Goodloe R, Spencer KL, Voruganti VS, Howard BV, Jackson R, Kolonel LN, Liu S, Manson JE, Monroe KR, Mukamal K, Dilks HH, Pendergrass SA, Nato A, Wan P, Wilkens LR, Le ML, Ambite JL, Buyske S, Florez JC, Crawford DC, Hindorff LA, Haiman CA, Peters U, Pankow JS (2013) Genetic variants associated with fasting glucose and insulin concentrations in an ethnically diverse population: results from the Population Architecture using Genomics and Epidemiology (PAGE) study. BMC Med Genet 14:98

Flannick J, Thorleifsson G, Beer NL, Jacobs SB, Grarup N, Burtt NP, Mahajan A, Fuchsberger C, Atzmon G, Benediktsson R, Blangero J, Bowden DW, Brandslund I, Brosnan J, Burslem F, Chambers J, Cho YS, Christensen C, Douglas DA, Duggirala R, Dymek Z, Farjoun Y, Fennell T, Fontanillas P, Forsen T, Gabriel S, Glaser B, Gudbjartsson DF, Hanis C, Hansen T, Hreidarsson AB, Hveem K, Ingelsson E, Isomaa B, Johansson S, Jorgensen T, Jorgensen ME, Kathiresan S, Kong A, Kooner J, Kravic J, Laakso M, Lee JY, Lind L, Lindgren CM, Linneberg A, Masson G, Meitinger T, Mohlke KL, Molven A, Morris AP, Potluri S, Rauramaa R, Ribel-Madsen R, Richard AM, Rolph T, Salomaa V, Segre AV, Skarstrand H, Steinthorsdottir V, Stringham HM, Sulem P, Tai ES, Teo YY, Teslovich T, Thorsteinsdottir U, Trimmer JK, Tuomi T, Tuomilehto J, Vaziri-Sani F, Voight BF, Wilson JG, Boehnke M, McCarthy MI, Njolstad PR, Pedersen O, Groop L, Cox DR, Stefansson K, Altshuler D (2014) Loss-of-function mutations in SLC30A8 protect against type 2 diabetes. Nat Genet 46:357–363

Flegal KM, Carroll MD, Ogden CL, Curtin LR (2010) Prevalence and trends in obesity among US adults, 1999-2008. JAMA 303:235–241

Florez JC (2008) Clinical review: the genetics of type 2 diabetes: a realistic appraisal in 2008. J Clin Endocrinol Metab 93:4633–4642

Florez JC, Jablonski KA, Bayley N, Pollin TI, de Bakker PI, Shuldiner AR, Knowler WC, Nathan DM, Altshuler D (2006) TCF7L2 polymorphisms and progression to diabetes in the Diabetes Prevention Program. N Engl J Med 355:241–250

Fogarty MP, Panhuis TM, Vadlamudi S, Buchkovich ML, Mohlke KL (2013) Allele-specific transcriptional activity at type 2 diabetes-associated single nucleotide polymorphisms in regions of pancreatic islet open chromatin at the JAZF1 locus. Diabetes 62:1756–1762

Ford ES, Li C (2008) Defining the metabolic syndrome in children and adolescents: will the real definition please stand up? J Pediatr 152:160–164

Ford ES, Li C, Zhao G (2010a) Prevalence and correlates of metabolic syndrome based on a harmonious definition among adults in the US. J Diabetes 2:180–193

Ford ES, Li C, Zhao G, Pearson WS, Tsai J, Churilla JR (2010b) Sedentary behavior, physical activity, and concentrations of insulin among US adults. Metabolism 59:1268–1275

Fox CS, Liu Y, White CC, Feitosa M, Smith AV, Heard-Costa N, Lohman K; GIANT Consortium; MAGIC Consortium; GLGC Consortium, Johnson AD, Foster MC, Greenawalt DM, Griffin P, Ding J, Newman AB, Tylavsky F, Miljkovic I, Kritchevsky SB, Launer L, Garcia M, Eiriksdottir G, Carr JJ, Gudnason V, Harris TB, Cupples LA, Borecki IB (2012) Genome-wide association for abdominal subcutaneous and visceral adipose reveals a novel locus for visceral fat in women. PLoS Genet 8(5):e1002695. doi:10.1371/journal.pgen.1002695

Fowler SP, Puppala S, Arya R, Chittoor G, Farook VS, Schneider J, Resendez RG, Upadhayay RP, Vandeberg J, Hunt KJ, Bradshaw B, Cersosimo E, VandeBerg JL, Almasy L, Curran JE, Comuzzie AG, Lehman DM, Jenkinson CP, Lynch JL, DeFronzo RA, Blangero J, Hale DE, Duggirala R (2013) Genetic epidemiology of cardiometabolic risk factors and their clustering patterns in Mexican American children and adolescents: the SAFARI Study. Hum Genet 132:1059–1071

Francischetti EA, Genelhu VA (2007) Obesity-hypertension: an ongoing pandemic. Int J Clin Pract 61:269–280

Frayling TM, McCarthy MI (2007) Genetic studies of diabetes following the advent of the genome-wide association study: where do we go from here? Diabetologia 50:2229–2233

Frayling TM, Timpson NJ, Weedon MN, Zeggini E, Freathy RM, Lindgren CM, Perry JR, Elliott KS, Lango H, Rayner NW, Shields B, Harries LW, Barrett JC, Ellard S, Groves CJ, Knight B, Patch AM, Ness AR, Ebrahim S, Lawlor DA, Ring SM, Ben Shlomo Y, Jarvelin MR, Sovio U, Bennett AJ, Melzer D, Ferrucci L, Loos RJ, Barroso I, Wareham NJ, Karpe F, Owen KR, Cardon LR, Walker M, Hitman GA, Palmer CN, Doney AS, Morris AD, Smith GD, Hattersley AT, McCarthy MI (2007) A common variant in the FTO gene is associated with body mass index and predisposes to childhood and adult obesity. Science 316:889–894

Frazer KA, Murray SS, Schork NJ, Topol EJ (2009) Human genetic variation and its contribution to complex traits. Nat Rev Genet 10:241–251

Freathy RM, Bennett AJ, Ring SM, Shields B, Groves CJ, Timpson NJ, Weedon MN, Zeggini E, Lindgren CM, Lango H, Perry JR, Pouta A, Ruokonen A, Hypponen E, Power C, Elliott P, Strachan DP, Jarvelin MR, Smith GD, McCarthy MI, Frayling TM, Hattersley AT (2009) Type 2 diabetes risk alleles are associated with reduced size at birth. Diabetes 58:1428–1433

Friedrich MJ (2002) Epidemic of obesity expands its spread to developing countries. JAMA 287:1382–1386

Fu YP, Hallman DM, Gonzalez VH, Klein BE, Klein R, Hayes MG, Cox NJ, Bell GI, Hanis CL (2010) Identification of diabetic retinopathy genes through a genome-wide association study among Mexican-Americans from Starr County, Texas. J Ophthalmol pii: 861291. doi:10.1155/2010/861291

Fulker DW, Cherny SS, Cardon LR (1995) Multipoint interval mapping of quantitative trait loci, using sib pairs. Am J Hum Genet 56:1224–1233

Gable S, Chang Y, Krull JL (2007) Television watching and frequency of family meals are predictive of overweight onset and persistence in a national sample of school-aged children. J Am Diet Assoc 107:53–61

Gallou-Kabani C, Junien C (2005) Nutritional epigenomics of metabolic syndrome: new perspective against the epidemic. Diabetes 54:1899–1906

Gardner DS, Tai ES (2012) Clinical features and treatment of maturity onset diabetes of the young (MODY). Diabetes Metab Syndr Obes 5:101–108

Garver WS (2011) Gene-diet interactions in childhood obesity. Curr Genomics 12:180–189

Gaulton KJ, Nammo T, Pasquali L, Simon JM, Giresi PG, Fogarty MP, Panhuis TM, Mieczkowski P, Secchi A, Bosco D, Berney T, Montanya E, Mohlke KL, Lieb JD, Ferrer J (2010) A map of open chromatin in human pancreatic islets. Nat Genet 42:255–259

Gibson G (2011) Rare and common variants: twenty arguments. Nat Rev Genet 13:135–145

Gibson J, Griffiths H, De Salvo G, Cole M, Jacob A, Macleod A, Yang Y, Menon G, Cree A, Ennis S, Lotery A (2012) Genome-wide association study of primary open angle glaucoma risk and quantitative traits. Mol Vis 18:1083–1092

Gloyn AL (2003) The search for type 2 diabetes genes. Ageing Res Rev 2:111–127

Gloyn AL, Weedon MN, Owen KR, Turner MJ, Knight BA, Hitman G, Walker M, Levy JC, Sampson M, Halford S, McCarthy MI, Hattersley AT, Frayling TM (2003) Large-scale association studies of variants in genes encoding the pancreatic beta-cell KATP channel subunits Kir6.2 (KCNJ11) and SUR1 (ABCC8) confirm that the KCNJ11 E23 K variant is associated with type 2 diabetes. Diabetes 52:568–572

Gluckman PD, Hanson MA (2008) Developmental and epigenetic pathways to obesity: an evolutionary-developmental perspective. Int J Obes (Lond) 32 (Suppl 7):S62–S71

Gluckman PD, Hanson MA, Beedle AS, Raubenheimer D (2008) Fetal and neonatal pathways to obesity. Front Horm Res 36:61–72

Golden SH, Brown A, Cauley JA, Chin MH, Gary-Webb TL, Kim C, Sosa JA, Sumner AE, Anton B (2012) Health disparities in endocrine disorders: biological, clinical, and nonclinical factors—an Endocrine Society scientific statement. J Clin Endocrinol Metab 97: E1579–E1639

Goldgar DE (1990) Multipoint analysis of human quantitative genetic variation. Am J Hum Genet 47:957–967

Goldstein BJ (2003) Insulin resistance: from benign to type 2 diabetes mellitus. Rev Cardiovasc Med 4(Suppl 6):S3–10

Goldstein DB (2011) The importance of synthetic associations will only be resolved empirically. PLoS Biol 9:e1001008

Goodarzi MO, Rotter JI (2007) Testing the gene or testing a variant? The case of TCF7L2. Diabetes 56:2417–2419

Goran MI, Ball GD, Cruz ML (2003) Obesity and risk of type 2 diabetes and cardiovascular disease in children and adolescents. J Clin Endocrinol Metab 88:1417–1427

Gordon DJ (1998) Factors affecting high-density lipoproteins. Endocrinol Metab Clin North Am 27:699–709, xi

Grant SF, Thorleifsson G, Reynisdottir I, Benediktsson R, Manolescu A, Sainz J, Helgason A, Stefansson H, Emilsson V, Helgadottir A, Styrkarsdottir U, Magnusson KP, Walters GB, Palsdottir E, Jonsdottir T, Gudmundsdottir T, Gylfason A, Saemundsdottir J, Wilensky RL, Reilly MP, Rader DJ, Bagger Y, Christiansen C, Gudnason V, Sigurdsson G, Thorsteinsdottir U, Gulcher JR, Kong A, Stefansson K (2006) Variant of transcription factor 7-like 2 (TCF7L2) gene confers risk of type 2 diabetes. Nat Genet 38:320–323

Grarup N, Rose CS, Andersson EA, Andersen G, Nielsen AL, Albrechtsen A, Clausen JO, Rasmussen SS, Jorgensen T, Sandbaek A, Lauritzen T, Schmitz O, Hansen T, Pedersen O (2007) Studies of association of variants near the HHEX, CDKN2A/B, and IGF2BP2 genes with type 2 diabetes and impaired insulin release in 10,705 Danish subjects: validation and extension of genome-wide association studies. Diabetes 56:3105–3111

Grarup N, Sandholt CH, Hansen T, Pedersen O (2014) Genetic susceptibility to type 2 diabetes and obesity: from genome-wide association studies to rare variants and beyond. Diabetologia 57:1528–1541

Groop L (2000) Genetics of the metabolic syndrome. Br J Nutr 83(Suppl 1):S39–S48

Grubb SR (2002) "Where obesity goes, so goes diabetes"—dual epidemics of alarming proportions. W V Med J 98:268–270

Grubbs S, Brundage SC (2002) Preconception management of chronic diseases. J S C Med Assoc 98:270–276

Grundy SM (2004) Metabolic syndrome: part II. Endocrinol Metab Clin North Am 33:xi–xiii

Grundy SM (2005) A constellation of complications: the metabolic syndrome. Clin Cornerstone 7:36–45

Grundy SM (2007) Metabolic syndrome: a multiplex cardiovascular risk factor. J Clin Endocrinol Metab 92:399–404

Grundy SM, Cleeman JI, Daniels SR, Donato KA, Eckel RH, Franklin BA, Gordon DJ, Krauss RM, Savage PJ, Smith SC Jr, Spertus JA, Costa F (2005) Diagnosis and management of the metabolic syndrome. An American Heart Association/National Heart, Lung, and Blood Institute Scientific Statement. Executive summary. Cardiol Rev 13:322–327

Guan W, Pluzhnikov A, Cox NJ, Boehnke M (2008) Meta-analysis of 23 type 2 diabetes linkage studies from the International Type 2 Diabetes Linkage Analysis Consortium. Hum Hered 66:35–49

Guh DP, Zhang W, Bansback N, Amarsi Z, Birmingham CL, Anis AH (2009) The incidence of co-morbidities related to obesity and overweight: a systematic review and meta-analysis. BMC Public Health 9:88

Guo T, Hanson RL, Traurig M, Muller YL, Ma L, Mack J, Kobes S, Knowler WC, Bogardus C, Baier LJ (2007) TCF7L2 is not a major susceptibility gene for type 2 diabetes in Pima Indians: analysis of 3,501 individuals. Diabetes 56:3082–3088

Hager J, Dina C, Francke S, Dubois S, Houari M, Vatin V, Vaillant E, Lorentz N, Basdevant A, Clement K, Guy-Grand B, Froguel P (1998) A genome-wide scan for human obesity genes reveals a major susceptibility locus on chromosome 10. Nat Genet 20:304–308

Haiman CA, Fesinmeyer MD, Spencer KL, Buzkova P, Voruganti VS, Wan P, Haessler J, Franceschini N, Monroe KR, Howard BV, Jackson RD, Florez JC, Kolonel LN, Buyske S, Goodloe RJ, Liu S, Manson JE, Meigs JB, Waters K, Mukamal KJ, Pendergrass SA, Shrader P, Wilkens LR, Hindorff LA, Ambite JL, North KE, Peters U, Crawford DC, Le ML, Pankow JS (2012) Consistent directions of effect for established type 2 diabetes risk variants across populations: the population architecture using genomics and epidemiology (PAGE) consortium. Diabetes 61:1642–1647

Hanis CL, Boerwinkle E, Chakraborty R, Ellsworth DL, Concannon P, Stirling B, Morrison VA, Wapelhorst B, Spielman RS, Gogolin-Ewens KJ, Shepard JM, Williams SR, Risch N, Hinds D, Iwasaki N, Ogata M, Omori Y, Petzold C, Rietzch H, Schroder HE, Schulze J, Cox NJ, Menzel S, Boriraj VV, Chen X, Lim LR, Lindner T, Mereu LE, Wang YQ, Xiang K, Yamagata K, Yang Y, Bell GI (1996) A genome-wide search for human non-insulin-dependent (type 2) diabetes genes reveals a major susceptibility locus on chromosome 2. Nat Genet 13:161–166

Hansen L, Pedersen O (2005) Genetics of type 2 diabetes mellitus: status and perspectives. Diabetes Obes Metab 7:122–135

Hanson RL, Bogardus C, Duggan D, Kobes S, Knowlton M, Infante AM, Marovich L, Benitez D, Baier LJ, Knowler WC (2007) A search for variants associated with young-onset type 2 diabetes in American Indians in a 100 K genotyping array. Diabetes 56:3045–3052

Hanson RL, Ehm MG, Pettitt DJ, Prochazka M, Thompson DB, Timberlake D, Foroud T, Kobes S, Baier L, Burns DK, Almasy L, Blangero J, Garvey WT, Bennett PH, Knowler WC (1998) An autosomal genomic scan for loci linked to type II diabetes mellitus and body-mass index in Pima Indians. Am J Hum Genet 63:1130–1138

Hanson RL, Imperatore G, Bennett PH, Knowler WC (2002) Components of the "metabolic syndrome" and incidence of type 2 diabetes. Diabetes 51:3120–3127

Haring R, Wallaschofski H (2012) Diving through the "-omics": the case for deep phenotyping and systems epidemiology. OMICS 16:231–234

Harper MG (2006) Childhood obesity: strategies for prevention. Fam Community Health 29:288–298

Haseman JK, Elston RC (1972) The investigation of linkage between a quantitative trait and a marker locus. Behav Genet 2:3–19

Hauser ER, Boehnke M (1998) Genetic linkage analysis of complex genetic traits by using affected sibling pairs. Biometrics 54:1238–1246

Heard-Costa NL, Zillikens MC, Monda KL, Johansson A, Harris TB, Fu M, Haritunians T, Feitosa MF, Aspelund T, Eiriksdottir G, Garcia M, Launer LJ, Smith AV, Mitchell BD, McArdle PF, Shuldiner AR, Bielinski SJ, Boerwinkle E, Brancati F, Demerath EW, Pankow JS, Arnold AM, Chen YD, Glazer NL, McKnight B, Psaty BM, Rotter JI, Amin N, Campbell H, Gyllensten U, Pattaro C, Pramstaller PP, Rudan I, Struchalin M, Vitart V, Gao X, Kraja A, Province MA, Zhang Q, Atwood LD, Dupuis J, Hirschhorn JN, Jaquish CE, O'Donnell CJ, Vasan RS, White CC, Aulchenko YS, Estrada K, Hofman A, Rivadeneira F, Uitterlinden AG, Witteman JC, Oostra BA, Kaplan RC, Gudnason V, O'Connell JR, Borecki IB, van Duijn CM, Cupples LA, Fox CS, North KE (2009) NRXN3 is a novel locus for waist circumference: a genome-wide association study from the CHARGE Consortium. PLoS Genet 5:e1000539

Hegele RA (2001) Monogenic dyslipidemias: window on determinants of plasma lipoprotein metabolism. Am J Hum Genet 69:1161–1177

Heid IM, Jackson AU, Randall JC, Winkler TW, Qi L, Steinthorsdottir V, Thorleifsson G, Zillikens MC, Speliotes EK, Magi R, Workalemahu T, White CC, Bouatia-Naji N, Harris TB, Berndt SI, Ingelsson E, Willer CJ, Weedon MN, Luan J, Vedantam S, Esko T, Kilpelainen TO, Kutalik Z, Li S, Monda KL, Dixon AL, Holmes CC, Kaplan LM, Liang L, Min JL, Moffatt MF, Molony C, Nicholson G, Schadt EE, Zondervan KT, Feitosa MF, Ferreira T, Lango AH, Weyant RJ, Wheeler E, Wood AR, Estrada K, Goddard ME, Lettre G, Mangino M, Nyholt DR, Purcell S, Smith AV, Visscher PM, Yang J, McCarroll SA, Nemesh J, Voight BF, Absher D, Amin N, Aspelund T, Coin L, Glazer NL, Hayward C, Heard-Costa NL, Hottenga JJ, Johansson A, Johnson T, Kaakinen M, Kapur K, Ketkar S, Knowles JW, Kraft P, Kraja AT, Lamina C, Leitzmann MF, McKnight B, Morris AP, Ong KK, Perry JR, Peters MJ, Polasek O, Prokopenko I, Rayner NW, Ripatti S, Rivadeneira F, Robertson NR, Sanna S, Sovio U, Surakka I, Teumer A, van Wingerden S, Vitart V, Zhao JH, Cavalcanti-Proenca C, Chines PS, Fisher E, Kulzer JR, Lecoeur C, Narisu N, Sandholt C, Scott LJ, Silander K, Stark K, Tammesoo ML, Teslovich TM, Timpson NJ, Watanabe RM, Welch R, Chasman DI, Cooper MN, Jansson JO, Kettunen J, Lawrence RW, Pellikka N, Perola M, Vandenput L, Alavere H, Almgren P, Atwood LD, Bennett AJ, Biffar R, Bonnycastle LL, Bornstein SR, Buchanan TA, Campbell H, Day IN, Dei M, Dorr M, Elliott P, Erdos MR, Eriksson JG, Freimer NB, Fu M, Gaget S, Geus EJ, Gjesing AP, Grallert H, Grassler J, Groves CJ, Guiducci C, Hartikainen AL, Hassanali N, Havulinna AS, Herzig KH, Hicks AA, Hui J, Igl W, Jousilahti P, Jula A, Kajantie E, Kinnunen L, Kolcic I, Koskinen S, Kovacs P, Kroemer HK, Krzelj V, Kuusisto J, Kvaloy K, Laitinen J, Lantieri O, Lathrop GM, Lokki ML, Luben RN, Ludwig B, McArdle WL, McCarthy A, Morken MA, Nelis M, Neville MJ, Pare G, Parker AN, Peden JF, Pichler I, Pietilainen KH, Platou CG, Pouta A, Ridderstrale M, Samani NJ, Saramies J, Sinisalo J, Smit JH, Strawbridge RJ, Stringham HM, Swift AJ, Teder-Laving M, Thomson B, Usala G, van Meurs JB, van Ommen GJ, Vatin V, Volpato CB, Wallaschofski H, Walters GB, Widen E, Wild SH, Willemsen G, Witte DR, Zgaga L, Zitting P, Beilby JP, James AL, Kahonen M, Lehtimaki T, Nieminen MS, Ohlsson C, Palmer LJ, Raitakari O, Ridker PM, Stumvoll M, Tonjes A, Viikari J, Balkau B, Ben Shlomo Y, Bergman RN, Boeing H, Smith GD, Ebrahim S, Froguel P, Hansen T, Hengstenberg C, Hveem K, Isomaa B, Jorgensen T, Karpe F, Khaw KT, Laakso M, Lawlor DA, Marre M, Meitinger T, Metspalu A, Midthjell K, Pedersen O, Salomaa V, Schwarz PE, Tuomi T, Tuomilehto J, Valle TT, Wareham NJ, Arnold AM, Beckmann JS, Bergmann S, Boerwinkle

E, Boomsma DI, Caulfield MJ, Collins FS, Eiriksdottir G, Gudnason V, Gyllensten U, Hamsten A, Hattersley AT, Hofman A, Hu FB, Illig T, Iribarren C, Jarvelin MR, Kao WH, Kaprio J, Launer LJ, Munroe PB (2010) Meta-analysis identifies 13 new loci associated with waist-hip ratio and reveals sexual dimorphism in the genetic basis of fat distribution. Nat Genet 42:949–960

Henneman P, Aulchenko YS, Frants RR, Zorkoltseva IV, Zillikens MC, Frolich M, Oostra BA, van Dijk KW, van Duijn CM (2010) Genetic architecture of plasma adiponectin overlaps with the genetics of metabolic syndrome-related traits. Diabetes Care 33:908–913

Herder C, Rathmann W, Strassburger K, Finner H, Grallert H, Huth C, Meisinger C, Gieger C, Martin S, Giani G, Scherbaum WA, Wichmann HE, Illig T (2008) Variants of the PPARG, IGF2BP2, CDKAL1, HHEX, and TCF7L2 genes confer risk of type 2 diabetes independently of BMI in the German KORA studies. Horm Metab Res 40:722–726

Hester JM, Wing MR, Li J, Palmer ND, Xu J, Hicks PJ, Roh BH, Norris JM, Wagenknecht LE, Langefeld CD, Freedman BI, Bowden DW, Ng MC (2012) Implication of European-derived adiposity loci in African Americans. Int J Obes (Lond) 36:465–473

Hillis GS, Hata J, Woodward M, Perkovic V, Arima H, Chow CK, Zoungas S, Patel A, Poulter NR, Mancia G, Williams B, Chalmers J (2012) Resting heart rate and the risk of microvascular complications in patients with type 2 diabetes mellitus. J Am Heart Assoc 1: e002832

Hinney A, Nguyen TT, Scherag A, Friedel S, Bronner G, Muller TD, Grallert H, Illig T, Wichmann HE, Rief W, Schafer H, Hebebrand J (2007) Genome wide association (GWA) study for early onset extreme obesity supports the role of fat mass and obesity associated gene (FTO) variants. PLoS One 2:e1361

Hinney A, Remschmidt H, Hebebrand J (2000) Candidate gene polymorphisms in eating disorders. Eur J Pharmacol 410:147–159

Hinney A, Vogel CI, Hebebrand J (2010) From monogenic to polygenic obesity: recent advances. Eur Child Adolesc Psychiatry 19:297–310

Hirschhorn JN, Daly MJ (2005) Genome-wide association studies for common diseases and complex traits. Nat Rev Genet 6:95–108

Hixson JE, Almasy L, Cole S, Birnbaum S, Mitchell BD, Mahaney MC, Stern MP, MacCluer JW, Blangero J, Comuzzie AG (1999) Normal variation in leptin levels in associated with polymorphisms in the proopiomelanocortin gene, POMC. J Clin Endocrinol Metab 84:3187–3191

Hoffman RP (2009) Metabolic syndrome racial differences in adolescents. Curr Diabetes Rev 5:259–265

Hoffmann K, Mattheisen M, Dahm S, Nurnberg P, Roe C, Johnson J, Cox NJ, Wichmann HE, Wienker TF, Schulze J, Schwarz PE, Lindner TH (2007) A German genome-wide linkage scan for type 2 diabetes supports the existence of a metabolic syndrome locus on chromosome 1p36.13 and a type 2 diabetes locus on chromosome 16p12.2. Diabetologia 50:1418–1422

Hoggart CJ, Clark TG, De Iorio M, Whittaker JC, Balding DJ (2008) Genome-wide significance for dense SNP and resequencing data. Genet Epidemiol 32:179–185

Horikawa Y, Oda N, Cox NJ, Li X, Orho-Melander M, Hara M, Hinokio Y, Lindner TH, Mashima H, Schwarz PE, Bosque-Plata L, Horikawa Y, Oda Y, Yoshiuchi I, Colilla S, Polonsky KS, Wei S, Concannon P, Iwasaki N, Schulze J, Baier LJ, Bogardus C, Groop L, Boerwinkle E, Hanis CL, Bell GI (2000) Genetic variation in the gene encoding calpain-10 is associated with type 2 diabetes mellitus. Nat Genet 26:163–175

Hrdličková R, Nehyba J, Bargmann W, Bose HR Jr (2014) Multiple tumor suppressor microRNAs regulate telomerase and TCF7, an important transcriptional regulator of the Wnt pathway. PLoS One 9(2):e86990. doi:10.1371/journal.pone.0086990

Hu C, Wang C, Zhang R, Ma X, Wang J, Lu J, Qin W, Bao Y, Xiang K, Jia W (2009) Variations in KCNQ1 are associated with type 2 diabetes and beta cell function in a Chinese population. Diabetologia 52:1322–1325

Huang TT (2008) Finding thresholds of risk for components of the pediatric metabolic syndrome. J Pediatr 152:158–159

Hunt KJ, Lehman DM, Arya R, Fowler S, Leach RJ, Goring HH, Almasy L, Blangero J, Dyer TD, Duggirala R, Stern MP (2005) Genome-wide linkage analyses of type 2 diabetes in Mexican Americans: the San Antonio Family Diabetes/Gallbladder Study. Diabetes 54:2655–2662

Hunt KJ, Resendez RG, Williams K, Haffner SM, Stern MP, Hazuda HP (2003) All-cause and cardiovascular mortality among Mexican-American and non-Hispanic White older participants in the San Antonio Heart Study-evidence against the "Hispanic paradox". Am J Epidemiol 158:1048–1057

Hunt SC, Abkevich V, Hensel CH, Gutin A, Neff CD, Russell DL, Tran T, Hong X, Jammulapati S, Riley R, Weaver-Feldhaus J, Macalma T, Richards MM, Gress R, Francis M, Thomas A, Frech GC, Adams TD, Shattuck D, Stone S (2001) Linkage of body mass index to chromosome 20 in Utah pedigrees. Hum Genet 109:279–285

Imamura M, Maeda S (2011) Genetics of type 2 diabetes: the GWAS era and future perspectives [Review]. Endocr J 58:723–739

Isomaa B (2003) A major health hazard: the metabolic syndrome. Life Sci 73:2395–2411

Jackson RS, Creemers JW, Ohagi S, Raffin-Sanson ML, Sanders L, Montague CT, Hutton JC, O'Rahilly S (1997) Obesity and impaired prohormone processing associated with mutations in the human prohormone convertase 1 gene. Nat Genet 16:303–306

Jacobsson JA, Klovins J, Kapa I, Danielsson P, Svensson V, Ridderstrale M, Gyllensten U, Marcus C, Fredriksson R, Schioth HB (2008) Novel genetic variant in FTO influences insulin levels and insulin resistance in severely obese children and adolescents. Int J Obes (Lond) 32:1730–1735

Jenkinson CP, Coletta DK, Flechtner-Mors M, Hu SL, Fourcaudot MJ, Rodriguez LM, Schneider J, Arya R, Stern MP, Blangero J, Duggirala R, DeFronzo RA (2008) Association of genetic variation in ENPP1 with obesity-related phenotypes. Obesity (Silver Spring) 16:1708–1713

Jiao H, Arner P, Hoffstedt J, Brodin D, Dubern B, Czernichow S, van't Hooft F, Axelsson T, Pedersen O, Hansen T, Sorensen TI, Hebebrand J, Kere J, Dahlman-Wright K, Hamsten A, Clement K, Dahlman I (2011) Genome wide association study identifies KCNMA1 contributing to human obesity. BMC Med Genomics 4:51

Johnson WD, Kroon JJ, Greenway FL, Bouchard C, Ryan D, Katzmarzyk PT (2009) Prevalence of risk factors for metabolic syndrome in adolescents: National Health and Nutrition Examination Survey (NHANES), 2001–2006. Arch Pediatr Adolesc Med 163:371–377

Junien C, Nathanielsz P (2007) Report on the IASO Stock Conference 2006: early and lifelong environmental epigenomic programming of metabolic syndrome, obesity and type II diabetes. Obes Rev 8:487–502

Kassi E, Pervanidou P, Kaltsas G, Chrousos G (2011) Metabolic syndrome: definitions and controversies. BMC Med 9:48

Kato N (2013) Insights into the genetic basis of type 2 diabetes. J Diabetes Investig 4:233–244

Kaufman F (2005) Back away from the soda! Nearly half of all children between the ages of 6 and 11 drink sweetened sodas. Are these sugar-laden beverages partly to blame for the obesity epidemic? Diabetes Forecast 58:42–45

Kent JW Jr, Dyer TD, Goring HH, Blangero J (2007) Type I error rates in association versus joint linkage/association tests in related individuals. Genet Epidemiol 31:173–177

Kilpelainen TO, Zillikens MC, Stancakova A, Finucane FM, Ried JS, Langenberg C, Zhang W, Beckmann JS, Luan J, Vandenput L, Styrkarsdottir U, Zhou Y, Smith AV, Zhao JH, Amin N, Vedantam S, Shin SY, Haritunians T, Fu M, Feitosa MF, Kumari M, Halldorsson BV, Tikkanen E, Mangino M, Hayward C, Song C, Arnold AM, Aulchenko YS, Oostra BA, Campbell H, Cupples LA, Davis KE, Doring A, Eiriksdottir G, Estrada K, Fernandez-Real JM, Garcia M, Gieger C, Glazer NL, Guiducci C, Hofman A, Humphries SE, Isomaa B, Jacobs LC, Jula A, Karasik D, Karlsson MK, Khaw KT, Kim LJ, Kivimaki M, Klopp N, Kuhnel B, Kuusisto J, Liu Y, Ljunggren O, Lorentzon M, Luben RN, McKnight B, Mellstrom D, Mitchell BD, Mooser V, Moreno JM, Mannisto S, O'Connell JR, Pascoe L, Peltonen L, Peral B, Perola M, Psaty BM, Salomaa V, Savage DB, Semple RK, Skaric-Juric T, Sigurdsson G, Song KS, Spector TD, Syvanen AC, Talmud PJ, Thorleifsson G, Thorsteinsdottir U, Uitterlinden AG, van Duijn CM, Vidal-Puig A, Wild SH, Wright AF, Clegg DJ, Schadt E, Wilson JF, Rudan I, Ripatti S, Borecki IB, Shuldiner AR, Ingelsson E, Jansson JO, Kaplan RC, Gudnason V, Harris TB, Groop L, Kiel DP, Rivadeneira F, Walker M, Barroso I, Vollenweider P, Waeber G, Chambers JC, Kooner JS, Soranzo N, Hirschhorn JN, Stefansson K, Wichmann HE, Ohlsson C, O'Rahilly S, Wareham NJ, Speliotes EK, Fox CS, Laakso M, Loos RJ (2011) Genetic variation near IRS1 associates with reduced adiposity and an impaired metabolic profile. Nat Genet 43:753–760

Kilpinen H, Barrett JC (2013) How next-generation sequencing is transforming complex disease genetics. Trends Genet 29:23–30

Kissebah AH, Sonnenberg GE, Myklebust J, Goldstein M, Broman K, James RG, Marks JA, Krakower GR, Jacob HJ, Weber J, Martin L, Blangero J, Comuzzie AG (2000) Quantitative trait loci on chromosomes 3 and 17 influence phenotypes of the metabolic syndrome. Proc Natl Acad Sci U S A 97:14478–14483

Klimentidis YC, Chen GB, Lopez-Alarcon M, Harris JJ, Duarte CW, Fernandez JR (2011) Associations of obesity genes with obesity-related outcomes in multiethnic children. Arch Med Res 42:509–514

Kong A, Steinthorsdottir V, Masson G, Thorleifsson G, Sulem P, Besenbacher S, Jonasdottir A, Sigurdsson A, Kristinsson KT, Jonasdottir A, Frigge ML, Gylfason A, Olason PI, Gudjonsson SA, Sverrisson S, Stacey SN, Sigurgeirsson B, Benediktsdottir KR, Sigurdsson H, Jonsson T, Benediktsson R, Olafsson JH, Johannsson OT, Hreidarsson AB, Sigurdsson G, Ferguson-Smith AC, Gudbjartsson DF, Thorsteinsdottir U, Stefansson K (2009) Parental origin of sequence variants associated with complex diseases. Nature 462:868–874

Kooner JS, Saleheen D, Sim X, Sehmi J, Zhang W, Frossard P, Been LF, Chia KS, Dimas AS, Hassanali N, Jafar T, Jowett JB, Li X, Radha V, Rees SD, Takeuchi F, Young R, Aung T, Basit A, Chidambaram M, Das D, Grundberg E, Hedman AK, Hydrie ZI, Islam M, Khor CC, Kowlessur S, Kristensen MM, Liju S, Lim WY, Matthews DR, Liu J, Morris AP, Nica AC, Pinidiyapathirage JM, Prokopenko I, Rasheed A, Samuel M, Shah N, Shera AS, Small KS, Suo C, Wickremasinghe AR, Wong TY, Yang M, Zhang F, Abecasis GR, Barnett AH, Caulfield M, Deloukas P, Frayling TM, Froguel P, Kato N, Katulanda P, Kelly MA, Liang J, Mohan V, Sanghera DK, Scott J, Seielstad M, Zimmet PZ, Elliott P, Teo YY, McCarthy MI, Danesh J, Tai ES, Chambers JC (2011) Genome-wide association study in individuals of South Asian ancestry identifies six new type 2 diabetes susceptibility loci. Nat Genet 43:984–989

Kooperberg C, Leblanc M, Dai JY, Rajapakse I (2009) Structures and assumptions: strategies to harness gene x gene and gene x environment interactions in GWAS. Stat Sci 24:472–488

Kovacs P, Hanson RL, Lee YH, Yang X, Kobes S, Permana PA, Bogardus C, Baier LJ (2003) The role of insulin receptor substrate-1 gene (IRS1) in type 2 diabetes in Pima Indians. Diabetes 52:3005–3009

Kraft P, Zeggini E, Ioannidis JP (2009) Replication in genome-wide association studies. Stat Sci 24:561–573

Kraja AT, Chasman DI, North KE, Reiner AP, Yanek LR, Kilpelainen TO, Smith JA, Dehghan A, Dupuis J,

Johnson AD, Feitosa MF, Tekola-Ayele F, Chu AY, Nolte IM, Dastani Z, Morris A, Pendergrass SA, Sun YV, Ritchie MD, Vaez A, Lin H, Ligthart S, Marullo L, Rohde R, Shao Y, Ziegler MA, Im HK, Schnabel RB, Jorgensen T, Jorgensen ME, Hansen T, Pedersen O, Stolk RP, Snieder H, Hofman A, Uitterlinden AG, Franco OH, Ikram MA, Richards JB, Rotimi C, Wilson JG, Lange L, Ganesh SK, Nalls M, Rasmussen-Torvik LJ, Pankow JS, Coresh J, Tang W, Linda Kao WH, Boerwinkle E, Morrison AC, Ridker PM, Becker DM, Rotter JI, Kardia SL, Loos RJ, Larson MG, Hsu YH, Province MA, Tracy R, Voight BF, Vaidya D, O'Donnell CJ, Benjamin EJ, Alizadeh BZ, Prokopenko I, Meigs JB, Borecki IB (2014) Pleiotropic genes for metabolic syndrome and inflammation. Mol Genet Metab 112:317–338

Kraja AT, Hunt SC, Pankow JS, Myers RH, Heiss G, Lewis CE, Rao DC, Province MA (2005a) Quantitative trait loci for metabolic syndrome in the Hypertension Genetic Epidemiology Network Study. Obes Res 13:1885–1890

Kraja AT, Rao DC, Weder AB, Cooper R, Curb JD, Hanis CL, Turner ST, de Andrade M, Hsiung CA, Quertermous T, Zhu X, Province MA (2005b) Two major QTLs and several others relate to factors of metabolic syndrome in the family blood pressure program. Hypertension 46:751–757

Kraja AT, Vaidya D, Pankow JS, Goodarzi MO, Assimes TL, Kullo IJ, Sovio U, Mathias RA, Sun YV, Franceschini N, Absher D, Li G, Zhang Q, Feitosa MF, Glazer NL, Haritunians T, Hartikainen AL, Knowles JW, North KE, Iribarren C, Kral B, Yanek L, O'Reilly PF, McCarthy MI, Jaquish C, Couper DJ, Chakravarti A, Psaty BM, Becker LC, Province MA, Boerwinkle E, Quertermous T, Palotie L, Jarvelin MR, Becker DM, Kardia SL, Rotter JI, Chen YD, Borecki IB (2011) A bivariate genome-wide approach to metabolic syndrome: STAMPEED consortium. Diabetes 60:1329–1339

Kristiansson K, Perola M, Tikkanen E, Kettunen J, Surakka I, Havulinna AS, Stancakova A, Barnes C, Widen E, Kajantie E, Eriksson JG, Viikari J, Kahonen M, Lehtimaki T, Raitakari OT, Hartikainen AL, Ruokonen A, Pouta A, Jula A, Kangas AJ, Soininen P, Ala-Korpela M, Mannisto S, Jousilahti P, Bonnycastle LL, Jarvelin MR, Kuusisto J, Collins FS, Laakso M, Hurles ME, Palotie A, Peltonen L, Ripatti S, Salomaa V (2012) Genome-wide screen for metabolic syndrome susceptibility Loci reveals strong lipid gene contribution but no evidence for common genetic basis for clustering of metabolic syndrome traits. Circ Cardiovasc Genet 5:242–249

Krolewski AS, Poznik GD, Placha G, Canani L, Dunn J, Walker W, Smiles A, Krolewski B, Fogarty DG, Moczulski D, Araki S, Makita Y, Ng DP, Rogus J, Duggirala R, Rich SS, Warram JH (2006) A genome-wide linkage scan for genes controlling variation in urinary albumin excretion in type II diabetes. Kidney Int 69:129–136

Krude H, Biebermann H, Gruters A (2003) Mutations in the human proopiomelanocortin gene. Ann N Y Acad Sci 994:233–239

Kruglyak L (1999) Prospects for whole-genome linkage disequilibrium mapping of common disease genes. Nat Genet 22:139–144

Kruglyak L, Daly MJ, Reeve-Daly MP, Lander ES (1996) Parametric and nonparametric linkage analysis: a unified multipoint approach. Am J Hum Genet 58:1347–1363

Krushkal J, Ferrell R, Mockrin SC, Turner ST, Sing CF, Boerwinkle E (1999) Genome-wide linkage analyses of systolic blood pressure using highly discordant siblings. Circulation 99:1407–1410

Kurokawa N, Young EH, Oka Y, Satoh H, Wareham NJ, Sandhu MS, Loos RJ (2008) The ADRB3 Trp64Arg variant and BMI: a meta-analysis of 44 833 individuals. Int J Obes (Lond) 32:1240–1249

Laird NM, Lange C (2006) Family-based designs in the age of large-scale gene-association studies. Nat Rev Genet 7:385–394

Lander E, Kruglyak L (1995) Genetic dissection of complex traits: guidelines for interpreting and reporting linkage results. Nat Genet 11:241–247

Lander ES, Schork NJ (1994) Genetic dissection of complex traits. Science 265:2037–2048

Lango H, Weedon MN (2008) What will whole genome searches for susceptibility genes for common complex disease offer to clinical practice? J Intern Med 263:16–27

Lanktree MB, Hassell RG, Lahiry P, Hegele RA (2010) Phenomics: expanding the role of clinical evaluation in genomic studies. J Investig Med 58:700–706

Lau DC, Douketis JD, Morrison KM, Hramiak IM, Sharma AM, Ur E (2007) 2006 Canadian clinical practice guidelines on the management and prevention of obesity in adults and children [summary]. CMAJ 176:S1–13

Lauenborg J, Grarup N, Damm P, Borch-Johnsen K, Jorgensen T, Pedersen O, Hansen T (2009) Common type 2 diabetes risk gene variants associate with gestational diabetes. J Clin Endocrinol Metab 94:145–150

Lee JH, Reed DR, Li WD, Xu W, Joo EJ, Kilker RL, Nanthakumar E, North M, Sakul H, Bell C, Price RA (1999) Genome scan for human obesity and linkage to markers in 20q13. Am J Hum Genet 64:196–209

Lehman DM, Hunt KJ, Leach RJ, Hamlington J, Arya R, Abboud HE, Duggirala R, Blangero J, Goring HH, Stern MP (2007) Haplotypes of transcription factor 7-like 2 (TCF7L2) gene and its upstream region are associated with type 2 diabetes and age of onset in Mexican Americans. Diabetes 56:389–393

Lempiainen P, Mykkanen L, Pyorala K, Laakso M, Kuusisto J (1999) Insulin resistance syndrome predicts coronary heart disease events in elderly nondiabetic men. Circulation 100:123–128

Lerman LO, Lerman A (2011) The metabolic syndrome and early kidney disease: another link in the chain? Rev Esp Cardiol 64:358–360

Levy D, DeStefano AL, Larson MG, O'Donnell CJ, Lifton RP, Gavras H, Cupples LA, Myers RH (2000) Evidence for a gene influencing blood pressure on chromosome 17. Genome scan linkage results for longitudinal blood pressure phenotypes in subjects from the framingham heart study. Hypertension 36:477–483

Lewis CM, Knight J (2012) Introduction to genetic association studies. Cold Spring Harb Protoc 2012:297–306

Lewis JP, Palmer ND, Hicks PJ, Sale MM, Langefeld CD, Freedman BI, Divers J, Bowden DW (2008) Association analysis in african americans of European-derived type 2 diabetes single nucleotide polymorphisms from whole-genome association studies. Diabetes 57:2220–2225

Li C, Ford ES, Zhao G, Mokdad AH (2009) Prevalence of pre-diabetes and its association with clustering of cardiometabolic risk factors and hyperinsulinemia among U.S. adolescents: National Health and Nutrition Examination Survey 2005–2006. Diabetes Care 32:342–347

Li J, Hooker NH (2010) Childhood obesity and schools: evidence from the national survey of children's health. J Sch Health 80:96–103

Li S, Zhao JH, Luan J, Ekelund U, Luben RN, Khaw KT, Wareham NJ, Loos RJ (2010a) Physical activity attenuates the genetic predisposition to obesity in 20,000 men and women from EPIC-Norfolk prospective population study. PLoS Med 7

Li S, Zhao JH, Luan J, Luben RN, Rodwell SA, Khaw KT, Ong KK, Wareham NJ, Loos RJ (2010b) Cumulative effects and predictive value of common obesity-susceptibility variants identified by genome-wide association studies. Am J Clin Nutr 91:184–190

Lin HF, Boden-Albala B, Juo SH, Park N, Rundek T, Sacco RL (2005) Heritabilities of the metabolic syndrome and its components in the Northern Manhattan Family Study. Diabetologia 48:2006–2012

Lindgren CM, Heid IM, Randall JC, Lamina C, Steinthorsdottir V, Qi L, Speliotes EK, Thorleifsson G, Willer CJ, Herrera BM, Jackson AU, Lim N, Scheet P, Soranzo N, Amin N, Aulchenko YS, Chambers JC, Drong A, Luan J, Lyon HN, Rivadeneira F, Sanna S, Timpson NJ, Zillikens MC, Zhao JH, Almgren P, Bandinelli S, Bennett AJ, Bergman RN, Bonnycastle LL, Bumpstead SJ, Chanock SJ, Cherkas L, Chines P, Coin L, Cooper C, Crawford G, Doering A, Dominiczak A, Doney AS, Ebrahim S, Elliott P, Erdos MR, Estrada K, Ferrucci L, Fischer G, Forouhi NG, Gieger C, Grallert H, Groves CJ, Grundy S, Guiducci C, Hadley D, Hamsten A, Havulinna AS, Hofman A, Holle R, Holloway JW, Illig T, Isomaa B, Jacobs LC, Jameson K, Jousilahti P, Karpe F, Kuusisto J, Laitinen J, Lathrop GM, Lawlor DA, Mangino M, McArdle WL, Meitinger T, Morken MA, Morris AP, Munroe P, Narisu N, Nordstrom A, Nordstrom P, Oostra BA, Palmer CN, Payne F, Peden JF, Prokopenko I, Renstrom F, Ruokonen A, Salomaa V, Sandhu MS, Scott LJ, Scuteri A, Silander K, Song K, Yuan X, Stringham HM, Swift AJ, Tuomi T, Uda M, Vollenweider P, Waeber G, Wallace C, Walters GB, Weedon MN, Witteman JC, Zhang C, Zhang W, Caulfield MJ, Collins FS, Davey SG, Day IN, Franks PW, Hattersley AT, Hu FB, Jarvelin MR, Kong A, Kooner JS, Laakso M, Lakatta E, Mooser V, Morris AD, Peltonen L, Samani NJ, Spector TD, Strachan DP, Tanaka T, Tuomilehto J, Uitterlinden AG, van Duijn CM, Wareham NJ, Hugh W, Waterworth DM, Boehnke M, Deloukas P, Groop L, Hunter DJ, Thorsteinsdottir U, Schlessinger D, Wichmann HE, Frayling TM, Abecasis GR, Hirschhorn JN, Loos RJ, Stefansson K, Mohlke KL, Barroso I, McCarthy MI (2009) Genome-wide association scan meta-analysis identifies three loci influencing adiposity and fat distribution. PLoS Genet 5:e1000508

Liu CT, Garnaas MK, Tin A, Kottgen A, Franceschini N, Peralta CA, de Boer IH, Lu X, Atkinson E, Ding J, Nalls M, Shriner D, Coresh J, Kutlar A, Bibbins-Domingo K, Siscovick D, Akylbekova E, Wyatt S, Astor B, Mychaleckjy J, Li M, Reilly MP, Townsend RR, Adeyemo A, Zonderman AB, de Andrade M, Turner ST, Mosley TH, Harris TB, Rotimi CN, Liu Y, Kardia SL, Evans MK, Shlipak MG, Kramer H, Flessner MF, Dreisbach AW, Goessling W, Cupples LA, Kao WL, Fox CS (2011) Genetic association for renal traits among participants of African ancestry reveals new loci for renal function. PLoS Genet 7: e1002264

Liu Y, Zhou DZ, Zhang D, Chen Z, Zhao T, Zhang Z, Ning M, Hu X, Yang YF, Zhang ZF, Yu L, He L, Xu H (2009) Variants in KCNQ1 are associated with susceptibility to type 2 diabetes in the population of mainland China. Diabetologia 52:1315–1321

Liu YJ, Liu XG, Wang L, Dina C, Yan H, Liu JF, Levy S, Papasian CJ, Drees BM, Hamilton JJ, Meyre D, Delplanque J, Pei YF, Zhang L, Recker RR, Froguel P, Deng HW (2008) Genome-wide association scans identified CTNNBL1 as a novel gene for obesity. Hum Mol Genet 17:1803–1813

Lo KS, Vadlamudi S, Fogarty MP, Mohlke KL, Lettre G (2014) Strategies to fine-map genetic associations with lipid levels by combining epigenomic annotations and liver-specific transcription profiles. Genomics 104 (2):105–112. doi:10.1016/j.ygeno.2014.04.006

Loos RJ, Bouchard C (2003) Obesity—is it a genetic disorder? J Intern Med 254:401–425

Loos RJ, Bouchard C (2008) FTO: the first gene contributing to common forms of human obesity. Obes Rev 9:246–250

Loos RJ, Lindgren CM, Li S, Wheeler E, Zhao JH, Prokopenko I, Inouye M, Freathy RM, Attwood AP, Beckmann JS, Berndt SI, Jacobs KB, Chanock SJ, Hayes RB, Bergmann S, Bennett AJ, Bingham SA, Bochud M, Brown M, Cauchi S, Connell JM, Cooper C, Smith GD, Day I, Dina C, De S, Dermitzakis ET, Doney AS, Elliott KS, Elliott P, Evans DM, Sadaf F, I, Froguel P, Ghori J, Groves CJ, Gwilliam R, Hadley D, Hall AS, Hattersley AT, Hebebrand J, Heid IM, Lamina C, Gieger C, Illig T, Meitinger T, Wichmann HE,

Herrera B, Hinney A, Hunt SE, Jarvelin MR, Johnson T, Jolley JD, Karpe F, Keniry A, Khaw KT, Luben RN, Mangino M, Marchini J, McArdle WL, McGinnis R, Meyre D, Munroe PB, Morris AD, Ness AR, Neville MJ, Nica AC, Ong KK, O'Rahilly S, Owen KR, Palmer CN, Papadakis K, Potter S, Pouta A, Qi L, Randall JC, Rayner NW, Ring SM, Sandhu MS, Scherag A, Sims MA, Song K, Soranzo N, Speliotes EK, Syddall HE, Teichmann SA, Timpson NJ, Tobias JH, Uda M, Vogel CI, Wallace C, Waterworth DM, Weedon MN, Willer CJ, Wraight, Yuan X, Zeggini E, Hirschhorn JN, Strachan DP, Ouwehand WH, Caulfield MJ, Samani NJ, Frayling TM, Vollenweider P, Waeber G, Mooser V, Deloukas P, McCarthy MI, Wareham NJ, Barroso I, Jacobs KB, Chanock SJ, Hayes RB, Lamina C, Gieger C, Illig T, Meitinger T, Wichmann HE, Kraft P, Hankinson SE, Hunter DJ, Hu FB, Lyon HN, Voight BF, Ridderstrale M, Groop L, Scheet P, Sanna S, Abecasis GR, Albai G, Nagaraja R, Schlessinger D, Jackson AU, Tuomilehto J, Collins FS, Boehnke M, Mohlke KL (2008) Common variants near MC4R are associated with fat mass, weight and risk of obesity. Nat Genet 40:768–775

Lorenzo C, Serrano-Rios M, Martinez-Larrad MT, Gonzalez-Villalpando C, Gonzalez-Sanchez JL, Martinez-Calatrava MJ, Gabriel R, Haffner SM (2007) Is waist circumference an essential component of the metabolic syndrome? Diabetes Care 30:2141–2142

Lorenzo C, Williams K, Hunt KJ, Haffner SM (2006) Trend in the prevalence of the metabolic syndrome and its impact on cardiovascular disease incidence: the San Antonio Heart Study. Diabetes Care 29:625–630

Love-Gregory LD, Wasson J, Ma J, Jin CH, Glaser B, Suarez BK, Permutt MA (2004) A common polymorphism in the upstream promoter region of the hepatocyte nuclear factor-4 alpha gene on chromosome 20q is associated with type 2 diabetes and appears to contribute to the evidence for linkage in an ashkenazi jewish population. Diabetes 53:1134–1140

Lusis AJ, Attie AD, Reue K (2008) Metabolic syndrome: from epidemiology to systems biology. Nat Rev Genet 9:819–830

Lutale JJ, Thordarson H, Abbas ZG, Vetvik K (2007) Microalbuminuria among Type 1 and Type 2 diabetic patients of African origin in Dar Es Salaam, Tanzania. BMC Nephrol 8:2

Lyon HN, Emilsson V, Hinney A, Heid IM, Lasky-Su J, Zhu X, Thorleifsson G, Gunnarsdottir S, Walters GB, Thorsteinsdottir U, Kong A, Gulcher J, Nguyen TT, Scherag A, Pfeufer A, Meitinger T, Bronner G, Rief W, Soto-Quiros ME, Avila L, Klanderman B, Raby BA, Silverman EK, Weiss ST, Laird N, Ding X, Groop L, Tuomi T, Isomaa B, Bengtsson K, Butler JL, Cooper RS, Fox CS, O'Donnell CJ, Vollmert C, Celedon JC, Wichmann HE, Hebebrand J, Stefansson K, Lange C, Hirschhorn JN (2007) The association of a SNP upstream of INSIG2 with body mass index is reproduced in several but not all cohorts. PLoS Genet 3:e61

Lyssenko V, Lupi R, Marchetti P, Del Guerra S, Orho-Melander M, Almgren P, Sjogren M, Ling C, Eriksson KF, Lethagen AL, Mancarella R, Berglund G, Tuomi T, Nilsson P, Del Prato S, Groop L (2007) Mechanisms by which common variants in the TCF7L2 gene increase risk of type 2 diabetes. J Clin Invest 117:2155–2163

Lyssenko V, Nagorny CL, Erdos MR, Wierup N, Jonsson A, Spegel P, Bugliani M, Saxena R, Fex M, Pulizzi N, Isomaa B, Tuomi T, Nilsson P, Kuusisto J, Tuomilehto J, Boehnke M, Altshuler D, Sundler F, Eriksson JG, Jackson AU, Laakso M, Marchetti P, Watanabe RM, Mulder H, Groop L (2009) Common variant in MTNR1B associated with increased risk of type 2 diabetes and impaired early insulin secretion. Nat Genet 41:82–88

Maes HH, Neale MC, Eaves LJ (1997) Genetic and environmental factors in relative body weight and human adiposity. Behav Genet 27:325–351

Mahajan A, Go MJ, Zhang W, Below JE, Gaulton KJ, Ferreira T, Horikoshi M, Johnson AD, Ng MC, Prokopenko I, Saleheen D, Wang X, Zeggini E, Abecasis GR, Adair LS, Almgren P, Atalay M, Aung T, Baldassarre D, Balkau B, Bao Y, Barnett AH, Barroso I, Basit A, Been LF, Beilby J, Bell GI, Benediktsson R, Bergman RN, Boehm BO, Boerwinkle E, Bonnycastle LL, Burtt N, Cai Q, Campbell H, Carey J, Cauchi S, Caulfield M, Chan JC, Chang LC, Chang TJ, Chang YC, Charpentier G, Chen CH, Chen H, Chen YT, Chia KS, Chidambaram M, Chines PS, Cho NH, Cho YM, Chuang LM, Collins FS, Cornelis MC, Couper DJ, Crenshaw AT, van Dam RM, Danesh J, Das D, de FU, Dedoussis G, Deloukas P, Dimas AS, Dina C, Doney AS, Donnelly PJ, Dorkhan M, Van DC, Dupuis J, Edkins S, Elliott P, Emilsson V, Erbel R, Eriksson JG, Escobedo J, Esko T, Eury E, Florez JC, Fontanillas P, Forouhi NG, Forsen T, Fox C, Fraser RM, Frayling TM, Froguel P, Frossard P, Gao Y, Gertow K, Gieger C, Gigante B, Grallert H, Grant GB, Grrop LC, Groves CJ, Grundberg E, Guiducci C, Hamsten A, Han BG, Hara K, Hassanali N, Hattersley AT, Hayward C, Hedman AK, Herder C, Hofman A, Holmen OL, Hovingh K, Hreidarsson AB, Hu C, Hu FB, Hui J, Humphries SE, Hunt SE, Hunter DJ, Hveem K, Hydrie ZI, Ikegami H, Illig T, Ingelsson E, Islam M, Isomaa B, Jackson AU, Jafar T, James A, Jia W, Jockel KH, Jonsson A, Jowett JB, Kadowaki T, Kang HM, Kanoni S, Kao WH, Kathiresan S, Kato N, Katulanda P, Keinanen-Kiukaanniemi KM, Kelly AM, Khan H, Khaw KT, Khor CC, Kim HL, Kim S, Kim YJ, Kinnunen L, Klopp N, Kong A, Korpi-Hyovalti E, Kowlessur S, Kraft P, Kravic J, Kristensen MM, Krithika S, Kumar A, Kumate J, Kuusisto J, Kwak SH, Laakso M, Lagou V, Lakka TA, Langenberg C, Langford C, Lawrence R, Leander K, Lee JM, Lee NR, Li M, Li X, Li Y, Liang J, Liju S, Lim WY, Lind L, Lindgren CM, Lindholm E, Liu CT, Liu JJ, Lobbens S, Long J, Loos RJ, Lu W, Luan J, Lyssenko V, Ma RC, Maeda S, Magi R, Mannisto S, Matthews DR, Meigs JB, Melander O, Metspalu A, Meyer J, Mirza G, Mihailov E, Moebus S, Mohan V, Mohlke KL, Morris AD, Muhleisen TW, Muller-Nurasyid M, Musk B,

Nakamura J, Nakashima E, Navarro P, Ng PK, Nica AC, Nilsson PM, Njolstad I, Nothen MM, Ohnaka K, Ong TH, Owen KR, Palmer CN, Pankow JS, Park KS, Parkin M, Pechlivanis S, Pedersen NL, Peltonen L, Perry JR, Peters A, Pinidiyapathirage JM, Platou CG, Potter S, Price JF, Qi L, Radha V, Rallidis L, Rasheed A, Rathman W, Rauramaa R, Raychaudhuri S, Rayner NW, Rees SD, Rehnberg E, Ripatti S, Robertson N, Roden M, Rossin EJ, Rudan I, Rybin D, Saaristo TE, Salomaa V, Saltevo J, Samuel M, Sanghera DK, Saramies J, Scott J, Scott LJ, Scott RA, Segre AV, Sehmi J, Sennblad B, Shah N, Shah S, Shera AS (2014) Genome-wide trans-ancestry meta-analysis provides insight into the genetic architecture of type 2 diabetes susceptibility. Nat Genet 46:234–244

Manolio TA (2009) Cohort studies and the genetics of complex disease. Nat Genet 41:5–6

Manolio TA (2010) Genomewide association studies and assessment of the risk of disease. N Engl J Med 363:166–176

Manolio TA, Collins FS, Cox NJ, Goldstein DB, Hindorff LA, Hunter DJ, McCarthy MI, Ramos EM, Cardon LR, Chakravarti A, Cho JH, Guttmacher AE, Kong A, Kruglyak L, Mardis E, Rotimi CN, Slatkin M, Valle D, Whittemore AS, Boehnke M, Clark AG, Eichler EE, Gibson G, Haines JL, Mackay TF, McCarroll SA, Visscher PM (2009) Finding the missing heritability of complex diseases. Nature 461:747–753

Marian AJ (2012) Molecular genetic studies of complex phenotypes. Transl Res 159:64–79

Marian AJ, Belmont J (2011) Strategic approaches to unraveling genetic causes of cardiovascular diseases. Circ Res 108:1252–1269

May AL, Kuklina EV, Yoon PW (2012) Prevalence of cardiovascular disease risk factors among US adolescents, 1999–2008. Pediatrics 129:1035–1041

Mayer-Davis EJ (2008) Type 2 diabetes in youth: epidemiology and current research toward prevention and treatment. J Am Diet Assoc 108:S45–S51

McAllister EJ, Dhurandhar NV, Keith SW, Aronne LJ, Barger J, Baskin M, Benca RM, Biggio J, Boggiano MM, Eisenmann JC, Elobeid M, Fontaine KR, Gluckman P, Hanlon EC, Katzmarzyk P, Pietrobelli A, Redden DT, Ruden DM, Wang C, Waterland RA, Wright SM, Allison DB (2009) Ten putative contributors to the obesity epidemic. Crit Rev Food Sci Nutr 49:868–913

McCarthy MI (2010) Genomics, type 2 diabetes, and obesity. N Engl J Med 363:2339–2350

McCarthy MI, Abecasis GR, Cardon LR, Goldstein DB, Little J, Ioannidis JP, Hirschhorn JN (2008) Genome-wide association studies for complex traits: consensus, uncertainty and challenges. Nat Rev Genet 9:356–369

McCarthy MI, Froguel P (2002) Genetic approaches to the molecular understanding of type 2 diabetes. Am J Physiol Endocrinol Metab 283:E217–E225

McCarthy MI, Hirschhorn JN (2008) Genome-wide association studies: potential next steps on a genetic journey. Hum Mol Genet 17:R156–R165

McCarthy MI, Zeggini E (2009) Genome-wide association studies in type 2 diabetes. Curr Diab Rep 9:164–171

Meaburn EL, Schalkwyk LC, Mill J (2010) Allele-specific methylation in the human genome: implications for genetic studies of complex disease. Epigenetics 5:578–582

Mehta M, Bhasin SK, Agrawal K, Dwivedi S (2007) Obesity amongst affluent adolescent girls. Indian J Pediatr 74:619–622

Meigs JB (2000) Invited commentary: insulin resistance syndrome? Syndrome X? Multiple metabolic syndrome? A syndrome at all? Factor analysis reveals patterns in the fabric of correlated metabolic risk factors. Am J Epidemiol 152:908–911

Meigs JB, Manning AK, Fox CS, Florez JC, Liu C, Cupples LA, Dupuis J (2007) Genome-wide association with diabetes-related traits in the Framingham Heart Study. BMC Med Genet 8(Suppl 1):S16

Melka MG, Bernard M, Mahboubi A, Abrahamowicz M, Paterson AD, Syme C, Lourdusamy A, Schumann G, Leonard GT, Perron M, Richer L, Veillette S, Gaudet D, Paus T, Pausova Z (2012) Genome-wide scan for loci of adolescent obesity and their relationship with blood pressure. J Clin Endocrinol Metab 97(1):E145–E150. doi:10.1210/jc.2011-1801

Memisoglu A, Hu FB, Hankinson SE, Liu S, Meigs JB, Altshuler DM, Hunter DJ, Manson JE (2003) Prospective study of the association between the proline to alanine codon 12 polymorphism in the PPAR-gamma gene and type 2 diabetes. Diabetes Care 26:2915–2917

Messiah SE, Carrillo-Iregui A, Garibay-Nieto G, Lopez-Mitnik G, Cossio S, Arheart KL (2010) Inter- and intra-ethnic group comparison of metabolic syndrome components among morbidly obese adolescents. J Clin Hypertens (Greenwich) 12:645–652

Metzker ML (2010) Sequencing technologies—the next generation. Nat Rev Genet 11:31–46

Meyre D, Delplanque J, Chevre JC, Lecoeur C, Lobbens S, Gallina S, Durand E, Vatin V, Degraeve F, Proenca C, Gaget S, Korner A, Kovacs P, Kiess W, Tichet J, Marre M, Hartikainen AL, Horber F, Potoczna N, Hercberg S, Levy-Marchal C, Pattou F, Heude B, Tauber M, McCarthy MI, Blakemore AI, Montpetit A, Polychronakos C, Weill J, Coin LJ, Asher J, Elliott P, Jarvelin MR, Visvikis-Siest S, Balkau B, Sladek R, Balding D, Walley A, Dina C, Froguel P (2009) Genome-wide association study for early-onset and morbid adult obesity identifies three new risk loci in European populations. Nat Genet 41:157–159

Meyre D, Lecoeur C, Delplanque J, Francke S, Vatin V, Durand E, Weill J, Dina C, Froguel P (2004) A genome-wide scan for childhood obesity-associated traits in French families shows significant linkage on chromosome 6q22.31-q23.2. Diabetes 53:803–811

Mitchell BD, Cole SA, Comuzzie AG, Almasy L, Blangero J, MacCluer JW, Hixson JE (1999) A quantitative trait locus influencing BMI maps to the region of the beta-3 adrenergic receptor. Diabetes 48:1863–1867

Mohan V, Jaydip R, Deepa R (2007) Type 2 diabetes in Asian Indian youth. Pediatr Diabetes 8(Suppl 9):28–34

Mokdad AH, Bowman BA, Ford ES, Vinicor F, Marks JS, Koplan JP (2001) The continuing epidemics of obesity and diabetes in the United States. JAMA 286:1195–1200

Monda KL, North KE, Hunt SC, Rao DC, Province MA, Kraja AT (2010) The genetics of obesity and the metabolic syndrome. Endocr Metab Immune Disord Drug Targets 10:86–108

Moran I, Akerman I, van de Bunt M, Xie R, Benazra M, Nammo T, Arnes L, Nakic N, Garcia-Hurtado J, Rodriguez-Segui S, Pasquali L, Sauty-Colace C, Beucher A, Scharfmann R, van AJ, Johnson PR, Berry A, Lee C, Harkins T, Gmyr V, Pattou F, Kerr-Conte J, Piemonti L, Berney T, Hanley N, Gloyn AL, Sussel L, Langman L, Brayman KL, Sander M, McCarthy MI, Ravassard P, Ferrer J (2012) Human beta cell transcriptome analysis uncovers lncRNAs that are tissue-specific, dynamically regulated, and abnormally expressed in type 2 diabetes. Cell Metab 16:435–448

Morton NE (1955) Sequential tests for the detection of linkage. Am J Hum Genet 7:277–318

Must A, Spadano J, Coakley EH, Field AE, Colditz G, Dietz WH (1999) The disease burden associated with overweight and obesity. JAMA 282:1523–1529

Ng HJ, Gloyn AL (2013) Bridging the gap between genetic associations and molecular mechanisms for type 2 diabetes. Curr Diab Rep 13:778–785

Ng MC, Hester JM, Wing MR, Li J, Xu J, Hicks PJ, Roh BH, Lu L, Divers J, Langefeld CD, Freedman BI, Palmer ND, Bowden DW (2012) Genome-wide association of BMI in African Americans. Obesity (Silver Spring) 20:622–627

Ng MC, Park KS, Oh B, Tam CH, Cho YM, Shin HD, Lam VK, Ma RC, So WY, Cho YS, Kim HL, Lee HK, Chan JC, Cho NH (2008) Implication of genetic variants near TCF7L2, SLC30A8, HHEX, CDKAL1, CDKN2A/B, IGF2BP2, and FTO in type 2 diabetes and obesity in 6,719 Asians. Diabetes 57:2226–2233

Ng MC, Saxena R, Li J, Palmer ND, Dimitrov L, Xu J, Rasmussen-Torvik LJ, Zmuda JM, Siscovick DS, Patel SR, Crook ED, Sims M, Chen YD, Bertoni AG, Li M, Grant SF, Dupuis J, Meigs JB, Psaty BM, Pankow JS, Langefeld CD, Freedman BI, Rotter JI, Wilson JG, Bowden DW (2013) Transferability and fine mapping of type 2 diabetes loci in African Americans: the Candidate Gene Association Resource Plus Study. Diabetes 62:965–976

Ng MC, Shriner D, Chen BH, Li J, Chen WM, Guo X, Liu J, Bielinski SJ, Yanek LR, Nalls MA, Comeau ME, Rasmussen-Torvik LJ, Jensen RA, Evans DS, Sun YV, An P, Patel SR, Lu Y, Long J, Armstrong LL, Wagenknecht L, Yang L, Snively BM, Palmer ND, Mudgal P, Langefeld CD, Keene KL, Freedman BI, Mychaleckyj JC, Nayak U, Raffel LJ, Goodarzi MO, Chen YD, Taylor HA Jr, Correa A, Sims M, Couper D, Pankow JS, Boerwinkle E, Adeyemo A, Doumatey A, Chen G, Mathias RA, Vaidya D, Singleton AB, Zonderman AB, Igo RP Jr, Sedor JR, Kabagambe EK, Siscovick DS, McKnight B, Rice K, Liu Y, Hsueh WC, Zhao W, Bielak LF, Kraja A, Province MA, Bottinger EP, Gottesman O, Cai Q, Zheng W, Blot WJ, Lowe WL, Pacheco JA, Crawford DC, Grundberg E, Rich SS, Hayes MG, Shu XO, Loos RJ, Borecki IB, Peyser PA, Cummings SR, Psaty BM, Fornage M, Iyengar SK, Evans MK, Becker DM, Kao WH, Wilson JG, Rotter JI, Sale MM, Liu S, Rotimi CN, Bowden DW (2014) Meta-analysis of genome-wide association studies in African Americans provides insights into the genetic architecture of type 2 diabetes. PLoS Genet 10:e1004517

Nica AC, Ongen H, Irminger JC, Bosco D, Berney T, Antonarakis SE, Halban PA, Dermitzakis ET (2013) Cell-type, allelic, and genetic signatures in the human pancreatic beta cell transcriptome. Genome Res 23:1554–1562

Norman RA, Tataranni PA, Pratley R, Thompson DB, Hanson RL, Prochazka M, Baier L, Ehm MG, Sakul H, Foroud T, Garvey WT, Burns D, Knowler WC, Bennett PH, Bogardus C, Ravussin E (1998) Autosomal genomic scan for loci linked to obesity and energy metabolism in Pima Indians. Am J Hum Genet 62:659–668

Norman RA, Thompson DB, Foroud T, Garvey WT, Bennett PH, Bogardus C, Ravussin E (1997) Genomewide search for genes influencing percent body fat in Pima Indians: suggestive linkage at chromosome 11q21-q22. Pima Diabetes Gene Group. Am J Hum Genet 60:166–173

North KE, Williams K, Williams JT, Best LG, Lee ET, Fabsitz RR, Howard BV, Gray RS, MacCluer JW (2003) Evidence for genetic factors underlying the insulin resistance syndrome in american indians. Obes Res 11:1444–1448

Nothen MM, Cichon S, Hemmer S, Hebebrand J, Remschmidt H, Lehmkuhl G, Poustka F, Schmidt M, Catalano M, Fimmers R (1994) Human dopamine D4 receptor gene: frequent occurrence of a null allele and observation of homozygosity. Hum Mol Genet 3:2207–2212

Nuyt AM, Alexander BT (2009) Developmental programming and hypertension. Curr Opin Nephrol Hypertens 18:144–152

Ogden CL, Carroll MD, Kit BK, Flegal KM (2014) Prevalence of childhood and adult obesity in the United States, 2011–2012. JAMA 311:806–814

Ogden J, Clementi C (2010) The experience of being obese and the many consequences of stigma. J Obes pii: 429098. doi:10.1155/2010/429098

Okada Y, Kubo M, Ohmiya H, Takahashi A, Kumasaka N, Hosono N, Maeda S, Wen W, Dorajoo R, Go MJ, Zheng W, Kato N, Wu JY, Lu Q, Tsunoda T, Yamamoto K, Nakamura Y, Kamatani N, Tanaka T (2012) Common variants at CDKAL1 and KLF9 are associated with body mass index in east Asian populations. Nat Genet 44:302–306

Opie LH (2007) Metabolic syndrome. Circulation 115: e32–e35

Oppert JM, Vohl MC, Chagnon M, Dionne FT, Cassard-Doulcier AM, Ricquier D, Perusse L, Bouchard C (1994) DNA polymorphism in the uncoupling protein (UCP) gene and human body fat. Int J Obes Relat Metab Disord 18:526–531

Ouchi N, Kihara S, Arita Y, Maeda K, Kuriyama H, Okamoto Y, Hotta K, Nishida M, Takahashi M, Nakamura T, Yamashita S, Funahashi T, Matsuzawa Y (1999) Novel modulator for endothelial adhesion molecules: adipocyte-derived plasma protein adiponectin. Circulation 100:2473–2476

Palmer ND, Hester JM, An SS, Adeyemo A, Rotimi C, Langefeld CD, Freedman BI, Ng MC, Bowden DW (2011) Resequencing and analysis of variation in the TCF7L2 gene in African Americans suggests that SNP rs7903146 is the causal diabetes susceptibility variant. Diabetes 60:662–668

Palmer ND, Lehtinen AB, Langefeld CD, Campbell JK, Haffner SM, Norris JM, Bergman RN, Goodarzi MO, Rotter JI, Bowden DW (2008) Association of TCF7L2 gene polymorphisms with reduced acute insulin response in Hispanic Americans. J Clin Endocrinol Metab 93:304–309

Pankow JS, Folsom AR, Cushman M, Borecki IB, Hopkins PN, Eckfeldt JH, Tracy RP (2001) Familial and genetic determinants of systemic markers of inflammation: the NHLBI family heart study. Atherosclerosis 154:681–689

Parra EJ, Below JE, Krithika S, Valladares A, Barta JL, Cox NJ, Hanis CL, Wacher N, Garcia-Mena J, Hu P, Shriver MD, Kumate J, McKeigue PM, Escobedo J, Cruz M (2011) Genome-wide association study of type 2 diabetes in a sample from Mexico City and a meta-analysis of a Mexican-American sample from Starr County, Texas. Diabetologia 54:2038–2046

Pasquali L, Gaulton KJ, Rodriguez-Segui SA, Mularoni L, Miguel-Escalada I, Akerman I, Tena JJ, Moran I, Gomez-Marin C, van de Bunt M, Ponsa-Cobas J, Castro N, Nammo T, Cebola I, Garcia-Hurtado J, Maestro MA, Pattou F, Piemonti L, Berney T, Gloyn AL, Ravassard P, Gomez-Skarmeta JL, Muller F, McCarthy MI, Ferrer J (2014) Pancreatic islet enhancer clusters enriched in type 2 diabetes risk-associated variants. Nat Genet 46:136–143

Paternoster L, Evans DM, Nohr EA, Holst C, Gaborieau V, Brennan P, Gjesing AP, Grarup N, Witte DR, Jørgensen T, Linneberg A, Lauritzen T, Sandbaek A, Hansen T, Pedersen O, Elliott KS, Kemp JP, St Pourcain B, McMahon G, Zelenika D, Hager J, Lathrop M, Timpson NJ, Smith GD, Sørensen TI (2011) Genome-wide population-based association study of extremely overweight young adults–the GOYA study. PLoS One 6(9):e24303. doi:10.1371/journal.pone.0024303

Peek R, Reddy KR (2007) Doctor Griffin Rodgers at helm of the National Institute of Diabetes and Digestive and Kidney Diseases. Gastroenterology 133:380–381

Peloso GM, Auer PL, Bis JC, Voorman A, Morrison AC, Stitziel NO, Brody JA, Khetarpal SA, Crosby JR, Fornage M, Isaacs A, Jakobsdottir J, Feitosa MF, Davies G, Huffman JE, Manichaikul A, Davis B, Lohman K, Joon AY, Smith AV, Grove ML, Zanoni P, Redon V, Demissie S, Lawson K, Peters U, Carlson C, Jackson RD, Ryckman KK, Mackey RH, Robinson JG, Siscovick DS, Schreiner PJ, Mychaleckyj JC, Pankow JS, Hofman A, Uitterlinden AG, Harris TB, Taylor KD, Stafford JM, Reynolds LM, Marioni RE, Dehghan A, Franco OH, Patel AP, Lu Y, Hindy G, Gottesman O, Bottinger EP, Melander O, Orho-Melander M, Loos RJ, Duga S, Merlini PA, Farrall M, Goel A, Asselta R, Girelli D, Martinelli N, Shah SH, Kraus WE, Li M, Rader DJ, Reilly MP, McPherson R, Watkins H, Ardissino D, Zhang Q, Wang J, Tsai MY, Taylor HA, Correa A, Griswold ME, Lange LA, Starr JM, Rudan I, Eiriksdottir G, Launer LJ, Ordovas JM, Levy D, Chen YD, Reiner AP, Hayward C, Polasek O, Deary IJ, Borecki IB, Liu Y, Gudnason V, Wilson JG, van Duijn CM, Kooperberg C, Rich SS, Psaty BM, Rotter JI, O'Donnell CJ, Rice K, Boerwinkle E, Kathiresan S, Cupples LA (2014) Association of low-frequency and rare coding-sequence variants with blood lipids and coronary heart disease in 56,000 whites and blacks. Am J Hum Genet 94:223–232

Perry JR, Frayling TM (2008) New gene variants alter type 2 diabetes risk predominantly through reduced beta-cell function. Curr Opin Clin Nutr Metab Care 11:371–377

Perusse L, Chagnon YC, Weisnagel SJ, Rankinen T, Snyder E, Sands J, Bouchard C (2001) The human obesity gene map: the 2000 update. Obes Res 9:135–169

Pinhas-Hamiel O, Zeitler P (2005) The global spread of type 2 diabetes mellitus in children and adolescents. J Pediatr 146:693–700

Poulsen P, Kyvik KO, Vaag A, Beck-Nielsen H (1999) Heritability of type II (non-insulin-dependent) diabetes mellitus and abnormal glucose tolerance—a population-based twin study. Diabetologia 42:139–145

Prokopenko I, Langenberg C, Florez JC, Saxena R, Soranzo N, Thorleifsson G, Loos RJ, Manning AK, Jackson AU, Aulchenko Y, Potter SC, Erdos MR, Sanna S, Hottenga JJ, Wheeler E, Kaakinen M, Lyssenko V, Chen WM, Ahmadi K, Beckmann JS, Bergman RN, Bochud M, Bonnycastle LL, Buchanan TA, Cao A, Cervino A, Coin L, Collins FS, Crisponi L, de Geus EJ, Dehghan A, Deloukas P, Doney AS, Elliott P, Freimer N, Gateva V, Herder C, Hofman A, Hughes TE, Hunt S, Illig T, Inouye M, Isomaa B, Johnson T, Kong A, Krestyaninova M, Kuusisto J, Laakso M, Lim N, Lindblad U, Lindgren CM, McCann OT, Mohlke KL, Morris AD, Naitza S, Orrù M, Palmer CN, Pouta A, Randall J, Rathmann W, Saramies J, Scheet P, Scott LJ, Scuteri A, Sharp S, Sijbrands E, Smit JH, Song K, Steinthorsdottir V, Stringham HM, Tuomi T, Tuomilehto J, Uitterlinden AG, Voight BF, Waterworth D, Wichmann HE, Willemsen G, Witteman JC, Yuan X, Zhao JH, Zeggini E, Schlessinger D, Sandhu M, Boomsma DI, Uda M, Spector TD, Penninx BW, Altshuler D, Vollenweider P, Jarvelin MR, Lakatta E, Waeber G, Fox CS, Peltonen L, Groop LC, Mooser V, Cupples

LA, Thorsteinsdottir U, Boehnke M, Barroso I, Van Duijn C, Dupuis J, Watanabe RM, Stefansson K, McCarthy MI, Wareham NJ, Meigs JB, Abecasis GR (2009) Variants in MTNR1B influence fasting glucose levels. Nat Genet 41(1):77–81. doi:10.1038/ng.290

Prokopenko I, McCarthy MI, Lindgren CM (2008) Type 2 diabetes: new genes, new understanding. Trends Genet 24:613–621

Puppala S, Arya R, Thameem F, Arar NH, Bhandari K, Lehman DM, Schneider J, Fowler S, Farook VS, Diego VP, Almasy L, Blangero J, Stern MP, Duggirala R, Abboud HE (2007) Genotype by diabetes interaction effects on the detection of linkage of glomerular filtration rate to a region on chromosome 2q in Mexican Americans. Diabetes 56:2818–2828

Qi L, Cornelis MC, Kraft P, Stanya KJ, Linda Kao WH, Pankow JS, Dupuis J, Florez JC, Fox CS, Paré G, Sun Q, Girman CJ, Laurie CC, Mirel DB, Manolio TA, Chasman DI, Boerwinkle E, Ridker PM, Hunter DJ, Meigs JB, Lee CH, Hu FB, van Dam RM; Meta-Analysis of Glucose and Insulin-related traits Consortium (MAGIC); Diabetes Genetics Replication and Meta-analysis (DIAGRAM) Consortium (2010) Genetic variants at 2q24 are associated with susceptibility to type 2 diabetes. Hum Mol Genet 19 (13):2706–2715. doi:10.1093/hmg/ddq156

Qi Q, Li H, Loos RJ, Liu C, Wu Y, Hu FB, Wu H, Lu L, Yu Z, Lin X (2009) Common variants in KCNQ1 are associated with type 2 diabetes and impaired fasting glucose in a Chinese Han population. Hum Mol Genet 18:3508–3515

Rainwater DL, Almasy L, Blangero J, Cole SA, VandeBerg JL, MacCluer JW, Hixson JE (1999) A genome search identifies major quantitative trait loci on human chromosomes 3 and 4 that influence cholesterol concentrations in small LDL particles. Arterioscler Thromb Vasc Biol 19:777–783

Rampersaud E, Damcott CM, Fu M, Shen H, McArdle P, Shi X, Shelton J, Yin J, Chang YP, Ott SH, Zhang L, Zhao Y, Mitchell BD, O'Connell J, Shuldiner AR (2007) Identification of novel candidate genes for type 2 diabetes from a genome-wide association scan in the Old Order Amish: evidence for replication from diabetes-related quantitative traits and from independent populations. Diabetes 56:3053–3062

Rampersaud E, Mitchell BD, Pollin TI, Fu M, Shen H, O'Connell JR, Ducharme JL, Hines S, Sack P, Naglieri R, Shuldiner AR, Snitker S (2008) Physical activity and the association of common FTO gene variants with body mass index and obesity. Arch Intern Med 168:1791–1797

Rankinen T, Zuberi A, Chagnon YC, Weisnagel SJ, Argyropoulos G, Walts B, Perusse L, Bouchard C (2006) The human obesity gene map: the 2005 update. Obesity (Silver Spring) 14:529–644

Reaven GM (1988) Banting lecture 1988. Role of insulin resistance in human disease. Diabetes 37:1595–1607

Reinehr T, Friedel S, Mueller TD, Toschke AM, Hebebrand J, Hinney A (2008) Evidence for an influence of TCF7L2 polymorphism rs7903146 on insulin resistance and sensitivity indices in overweight children and adolescents during a lifestyle intervention. Int J Obes (Lond) 32:1521–1524

Reusens B, Ozanne SE, Remacle C (2007) Fetal determinants of type 2 diabetes. Curr Drug Targets 8:935–941

Reynisdottir I, Thorleifsson G, Benediktsson R, Sigurdsson G, Emilsson V, Einarsdottir AS, Hjorleifsdottir EE, Orlygsdottir GT, Bjornsdottir GT, Saemundsdottir J, Halldorsson S, Hrafnkelsdottir S, Sigurjonsdottir SB, Steinsdottir S, Martin M, Kochan JP, Rhees BK, Grant SF, Frigge ML, Kong A, Gudnason V, Stefansson K, Gulcher JR (2003) Localization of a susceptibility gene for type 2 diabetes to chromosome 5q34-q35.2. Am J Hum Genet 73:323–335

Rice T, Rankinen T, Chagnon YC, Province MA, Perusse L, Leon AS, Skinner JS, Wilmore JH, Bouchard C, Rao DC (2002) Genomewide linkage scan of resting blood pressure: HERITAGE family study. Health, risk factors, exercise training, and genetics. Hypertension 39:1037–1043

Ridderstrale M, Groop L (2009) Genetic dissection of type 2 diabetes. Mol Cell Endocrinol 297:10–17

Riggs AC, Bernal-Mizrachi E, Ohsugi M, Wasson J, Fatrai S, Welling C, Murray J, Schmidt RE, Herrera PL, Permutt MA (2005) Mice conditionally lacking the Wolfram gene in pancreatic islet beta cells exhibit diabetes as a result of enhanced endoplasmic reticulum stress and apoptosis. Diabetologia 48:2313–2321

Rios M (2011) New insights into the mechanisms underlying the effects of BDNF on eating behavior. Neuropsychopharmacology 36:368–369

Rios M, Fan G, Fekete C, Kelly J, Bates B, Kuehn R, Lechan RM, Jaenisch R (2001) Conditional deletion of brain-derived neurotrophic factor in the postnatal brain leads to obesity and hyperactivity. Mol Endocrinol 15:1748–1757

Risch N, Merikangas K (1996) The future of genetic studies of complex human diseases. Science 273:1516–1517

Risch NJ (2000) Searching for genetic determinants in the new millennium. Nature 405:847–856

Ristow M, Muller-Wieland D, Pfeiffer A, Krone W, Kahn CR (1998) Obesity associated with a mutation in a genetic regulator of adipocyte differentiation. N Engl J Med 339:953–959

Robinson PN (2012) Deep phenotyping for precision medicine. Hum Mutat 33:777–780

Rong R, Hanson RL, Ortiz D, Wiedrich C, Kobes S, Knowler WC, Bogardus C, Baier LJ (2009) Association analysis of variation in/near FTO, CDKAL1, SLC30A8, HHEX, EXT2, IGF2BP2, LOC387761, and CDKN2B with type 2 diabetes and related quantitative traits in Pima Indians. Diabetes 58:478–488

Rosmond R (2003) Association studies of genetic polymorphisms in central obesity: a critical review. Int J Obes Relat Metab Disord 27:1141–1151

Rosmond R, Chagnon M, Bouchard C (2003) The Pro12Ala PPARgamma2 gene missense mutation is associated with obesity and insulin resistance in Swedish middle-aged men. Diabetes Metab Res Rev 19:159–163

Rotimi CN, Comuzzie AG, Lowe WL, Luke A, Blangero J, Cooper RS (1999) The quantitative trait locus on chromosome 2 for serum leptin levels is confirmed in African-Americans. Diabetes 48:643–644

Rung J, Cauchi S, Albrechtsen A, Shen L, Rocheleau G, Cavalcanti-Proença C,Bacot F, Balkau B, Belisle A, Borch-Johnsen K, Charpentier G, Dina C, Durand E, Elliott P, Hadjadj S, Järvelin MR, Laitinen J, Lauritzen T, Marre M, Mazur A,Meyre D, Montpetit A, Pisinger C, Posner B, Poulsen P, Pouta A, Prentki M, Ribel-Madsen R, Ruokonen A, Sandbaek A, Serre D, Tichet J, Vaxillaire M, Wojtaszewski JF, Vaag A, Hansen T, Polychronakos C, Pedersen O, Froguel P, Sladek R (2009) Genetic variant near IRS1 is associated with type 2 diabetes, insulin resistance and hyperinsulinemia. Nat Genet 41(10):1110–1115. doi:10.1038/ng.443. Epub 2009 Sep 6. Erratum in: Nat Genet 41(10):1156.

Sale MM, Smith SG, Mychaleckyj JC, Keene KL, Langefeld CD, Leak TS, Hicks PJ, Bowden DW, Rich SS, Freedman BI (2007) Variants of the transcription factor 7-like 2 (TCF7L2) gene are associated with type 2 diabetes in an African-American population enriched for nephropathy. Diabetes 56:2638–2642

Salonen JT, Uimari P, Aalto JM, Pirskanen M, Kaikkonen J, Todorova B, Hypponen J, Korhonen VP, Asikainen J, Devine C, Tuomainen TP, Luedemann J, Nauck M, Kerner W, Stephens RH, New JP, Ollier WE, Gibson JM, Payton A, Horan MA, Pendleton N, Mahoney W, Meyre D, Delplanque J, Froguel P, Luzzatto O, Yakir B, Darvasi A (2007) Type 2 diabetes whole-genome association study in four populations: the DiaGen consortium. Am J Hum Genet 81:338–345

Sandholt CH, Hansen T, Pedersen O (2012) Beyond the fourth wave of genome-wide obesity association studies. Nutr Diabetes 2:e37

Sanghera DK, Blackett PR (2012) Type 2 diabetes genetics: beyond GWAS. J Diabetes Metab 3

Sanghera DK, Nath SK, Ortega L, Gambarelli M, Kim-Howard X, Singh JR, Ralhan SK, Wander GS, Mehra NK, Mulvihill JJ, Kamboh MI (2008) TCF7L2 polymorphisms are associated with type 2 diabetes in Khatri Sikhs from North India: genetic variation affects lipid levels. Ann Hum Genet 72:499–509

Saunders CL, Chiodini BD, Sham P, Lewis CM, Abkevich V, Adeyemo AA, de Andrade M, Arya R, Berenson GS, Blangero J, Boehnke M, Borecki IB, Chagnon YC, Chen W, Comuzzie AG, Deng HW, Duggirala R, Feitosa MF, Froguel P, Hanson RL, Hebebrand J, Huezo-Dias P, Kissebah AH, Li W, Luke A, Martin LJ, Nash M, Ohman M, Palmer LJ, Peltonen L, Perola M, Price RA, Redline S, Srinivasan SR, Stern MP, Stone S, Stringham H, Turner S, Wijmenga C, Collier A (2007) Meta-analysis of genome-wide linkage studies in BMI and obesity. Obesity (Silver Spring) 15:2263–2275

Saxena R, Elbers CC, Guo Y, Peter I, Gaunt TR, Mega JL, Lanktree MB, Tare A, Castillo BA, Li YR, Johnson T, Bruinenberg M, Gilbert-Diamond D, Rajagopalan R, Voight BF, Balasubramanyam A, Barnard J, Bauer F, Baumert J, Bhangale T, Bohm BO, Braund PS, Burton PR, Chandrupatla HR, Clarke R, Cooper-Dehoff RM, Crook ED, Davey-Smith G, Day IN, de Boer A, de Groot MC, Drenos F, Ferguson J, Fox CS, Furlong CE, Gibson Q, Gieger C, Gilhuijs-Pederson LA, Glessner JT, Goel A, Gong Y, Grant SF, Grobbee DE, Hastie C, Humphries SE, Kim CE, Kivimaki M, Kleber M, Meisinger C, Kumari M, Langaee TY, Lawlor DA, Li M, Lobmeyer MT, Maitland-van der Zee AH, Meijs MF, Molony CM, Morrow DA, Murugesan G, Musani SK, Nelson CP, Newhouse SJ, O'Connell JR, Padmanabhan S, Palmen J, Patel SR, Pepine CJ, Pettinger M, Price TS, Rafelt S, Ranchalis J, Rasheed A, Rosenthal E, Ruczinski I, Shah S, Shen H, Silbernagel G, Smith EN, Spijkerman AW, Stanton A, Steffes MW, Thorand B, Trip M, van der HP, van der AD, van Iperen EP, van Setten J, Vliet-Ostaptchouk JV, Verweij N, Wolffenbuttel BH, Young T, Zafarmand MH, Zmuda JM, Boehnke M, Altshuler D, McCarthy M, Kao WH, Pankow JS, Cappola TP, Sever P, Poulter N, Caulfield M, Dominiczak A, Shields DC, Bhatt DL, Zhang L, Curtis SP, Danesh J, Casas JP, van der Schouw YT, Onland-Moret NC, Doevendans PA, Dorn GW, Farrall M, FitzGerald GA, Hamsten A, Hegele R, Hingorani AD, Hofker MH, Huggins GS, Illig T, Jarvik GP, Johnson JA, Klungel OH, Knowler WC, Koenig W, Marz W, Meigs JB, Melander O, Munroe PB, Mitchell BD, Bielinski SJ, Rader DJ, Reilly MP, Rich SS, Rotter JI, Saleheen D, Samani NJ, Schadt EE, Shuldiner AR, Silverstein R, Kottke-Marchant K, Talmud PJ, Watkins H, Asselbergs FW, de Bakker PI, McCaffery J, Wijmenga C, Sabatine MS, Wilson JG, Reiner A, Bowden DW, Hakonarson H, Siscovick DS, Keating BJ (2012) Large-scale gene-centric meta-analysis across 39 studies identifies type 2 diabetes loci. Am J Hum Genet 90:410–425

Saxena R, Saleheen D, Been LF, Garavito ML, Braun T, Bjonnes A, Young R, Ho WK, Rasheed A, Frossard P, Sim X, Hassanali N, Radha V, Chidambaram M, Liju S, Rees SD, Ng DP, Wong TY, Yamauchi T, Hara K, Tanaka Y, Hirose H, McCarthy MI, Morris AP, Basit A, Barnett AH, Katulanda P, Matthews D, Mohan V, Wander GS, Singh JR, Mehra NK, Ralhan S, Kamboh MI, Mulvihill JJ, Maegawa H, Tobe K, Maeda S, Cho YS, Tai ES, Kelly MA, Chambers JC, Kooner JS, Kadowaki T, Deloukas P, Rader DJ, Danesh J, Sanghera DK (2013) Genome-wide association study identifies a novel locus contributing to type 2 diabetes susceptibility in sikhs of punjabi origin from India. Diabetes 62:1746–1755

Saxena R, Voight BF, Lyssenko V, Burtt NP, de Bakker PI, Chen H, Roix JJ, Kathiresan S, Hirschhorn JN, Daly MJ, Hughes TE, Groop L, Altshuler D, Almgren P, Florez JC, Meyer J, Ardlie K, Bengtsson BK, Isomaa B, Lettre G, Lindblad U, Lyon HN, Melander O, Newton-Cheh C, Nilsson P, Orho-Melander M, Rastam L, Speliotes EK, Taskinen MR, Tuomi T, Guiducci C, Berglund A, Carlson J, Gianniny L,

Hackett R, Hall L, Holmkvist J, Laurila E, Sjogren M, Sterner M, Surti A, Svensson M, Svensson M, Tewhey R, Blumenstiel B, Parkin M, DeFelice M, Barry R, Brodeur W, Camarata J, Chia N, Fava M, Gibbons J, Handsaker B, Healy C, Nguyen K, Gates C, Sougnez C, Gage D, Nizzari M, Gabriel SB, Chirn GW, Ma Q, Parikh H, Richardson D, Ricke D, Purcell S (2007) Genome-wide association analysis identifies loci for type 2 diabetes and triglyceride levels. Science 316:1331–1336

Scherag A, Dina C, Hinney A, Vatin V, Scherag S, Vogel CI, Muller TD, Grallert H, Wichmann HE, Balkau B, Heude B, Jarvelin MR, Hartikainen AL, Levy-Marchal C, Weill J, Delplanque J, Korner A, Kiess W, Kovacs P, Rayner NW, Prokopenko I, McCarthy MI, Schafer H, Jarick I, Boeing H, Fisher E, Reinehr T, Heinrich J, Rzehak P, Berdel D, Borte M, Biebermann H, Krude H, Rosskopf D, Rimmbach C, Rief W, Fromme T, Klingenspor M, Schurmann A, Schulz N, Nothen MM, Muhleisen TW, Erbel R, Jockel KH, Moebus S, Boes T, Illig T, Froguel P, Hebebrand J, Meyre D (2010) Two new Loci for body-weight regulation identified in a joint analysis of genome-wide association studies for early-onset extreme obesity in French and german study groups. PLoS Genet 6:e1000916

Schienkiewitz A, Mensink GB, Scheidt-Nave C (2012) Comorbidity of overweight and obesity in a nationally representative sample of German adults aged 18–79 years. BMC Public Health 12:658

Schork NJ (1993) Extended multipoint identity-by-descent analysis of human quantitative traits: efficiency, power, and modeling considerations. Am J Hum Genet 53:1306–1319

Schork NJ, Boehnke M, Terwilliger JD, Ott J (1993) Two-trait-locus linkage analysis: a powerful strategy for mapping complex genetic traits. Am J Hum Genet 53:1127–1136

Schutte AE, Schutte R, Huisman HW, Rooyen JM, Malan L, Olckers A, Malan NT (2009) Classifying Africans with the metabolic syndrome. Horm Metab Res 41:79–85

Scott LJ, Mohlke KL, Bonnycastle LL, Willer CJ, Li Y, Duren WL, Erdos MR, Stringham HM, Chines PS, Jackson AU, Prokunina-Olsson L, Ding CJ, Swift AJ, Narisu N, Hu T, Pruim R, Xiao R, Li XY, Conneely KN, Riebow NL, Sprau AG, Tong M, White PP, Hetrick KN, Barnhart MW, Bark CW, Goldstein JL, Watkins L, Xiang F, Saramies J, Buchanan TA, Watanabe RM, Valle TT, Kinnunen L, Abecasis GR, Pugh EW, Doheny KF, Bergman RN, Tuomilehto J, Collins FS, Boehnke M (2007) A genome-wide association study of type 2 diabetes in Finns detects multiple susceptibility variants. Science 316:1341–1345

Scuteri A, Sanna S, Chen WM, Uda M, Albai G, Strait J, Najjar S, Nagaraja R, Orru M, Usala G, Dei M, Lai S, Maschio A, Busonero F, Mulas A, Ehret GB, Fink AA, Weder AB, Cooper RS, Galan P, Chakravarti A, Schlessinger D, Cao A, Lakatta E, Abecasis GR (2007) Genome-wide association scan shows genetic variants in the FTO gene are associated with obesity-related traits. PLoS Genet 3:e115

Silander K, Mohlke KL, Scott LJ, Peck EC, Hollstein P, Skol AD, Jackson AU, Deloukas P, Hunt S, Stavrides G, Chines PS, Erdos MR, Narisu N, Conneely KN, Li C, Fingerlin TE, Dhanjal SK, Valle TT, Bergman RN, Tuomilehto J, Watanabe RM, Boehnke M, Collins FS (2004a) Genetic variation near the hepatocyte nuclear factor-4 alpha gene predicts susceptibility to type 2 diabetes. Diabetes 53:1141–1149

Silander K, Scott LJ, Valle TT, Mohlke KL, Stringham HM, Wiles KR, Duren WL, Doheny KF, Pugh EW, Chines P, Narisu N, White PP, Fingerlin TE, Jackson AU, Li C, Ghosh S, Magnuson VL, Colby K, Erdos MR, Hill JE, Hollstein P, Humphreys KM, Kasad RA, Lambert J, Lazaridis KN, Lin G, Morales-Mena A, Patzkowski K, Pfahl C, Porter R, Rha D, Segal L, Suh YD, Tovar J, Unni A, Welch C, Douglas JA, Epstein MP, Hauser ER, Hagopian W, Buchanan TA, Watanabe RM, Bergman RN, Tuomilehto J, Collins FS, Boehnke M (2004b) A large set of Finnish affected sibling pair families with type 2 diabetes suggests susceptibility loci on chromosomes 6, 11, and 14. Diabetes 53:821–829

Simmons RK, Alberti KG, Gale EA, Colagiuri S, Tuomilehto J, Qiao Q, Ramachandran A, Tajima N, Brajkovich MI, Ben Nakhi A, Reaven G, Hama SB, Mendis S, Roglic G (2010) The metabolic syndrome: useful concept or clinical tool? Report of a WHO Expert Consultation. Diabetologia 53:600–605

Sladek R, Rocheleau G, Rung J, Dina C, Shen L, Serre D, Boutin P, Vincent D, Belisle A, Hadjadj S, Balkau B, Heude B, Charpentier G, Hudson TJ, Montpetit A, Pshezhetsky AV, Prentki M, Posner BI, Balding DJ, Meyre D, Polychronakos C, Froguel P (2007) A genome-wide association study identifies novel risk loci for type 2 diabetes. Nature 445:881–885

Sonestedt E, Roos C, Gullberg B, Ericson U, Wirfalt E, Orho-Melander M (2009) Fat and carbohydrate intake modify the association between genetic variation in the FTO genotype and obesity. Am J Clin Nutr 90:1418–1425

Sorensen TI (2000) The changing lifestyle in the world. Body weight and what else? Diabetes Care 23(Suppl 2):B1–B4

Speliotes EK, Willer CJ, Berndt SI, Monda KL, Thorleifsson G, Jackson AU, Lango AH, Lindgren CM, Luan J, Magi R, Randall JC, Vedantam S, Winkler TW, Qi L, Workalemahu T, Heid IM, Steinthorsdottir V, Stringham HM, Weedon MN, Wheeler E, Wood AR, Ferreira T, Weyant RJ, Segre AV, Estrada K, Liang L, Nemesh J, Park JH, Gustafsson S, Kilpelainen TO, Yang J, Bouatia-Naji N, Esko T, Feitosa MF, Kutalik Z, Mangino M, Raychaudhuri S, Scherag A, Smith AV, Welch R, Zhao JH, Aben KK, Absher DM, Amin N, Dixon AL, Fisher E, Glazer NL, Goddard ME, Heard-Costa NL, Hoesel V, Hottenga JJ, Johansson A, Johnson T, Ketkar S, Lamina C, Li S, Moffatt MF, Myers RH, Narisu N, Perry JR, Peters MJ, Preuss M, Ripatti S, Rivadeneira F, Sandholt C, Scott LJ, Timpson NJ, Tyrer JP, van Wingerden S, Watanabe RM, White CC, Wiklund F, Barlassina C, Chasman

DI, Cooper MN, Jansson JO, Lawrence RW, Pellikka N, Prokopenko I, Shi J, Thiering E, Alavere H, Alibrandi MT, Almgren P, Arnold AM, Aspelund T, Atwood LD, Balkau B, Balmforth AJ, Bennett AJ, Ben Shlomo Y, Bergman RN, Bergmann S, Biebermann H, Blakemore AI, Boes T, Bonnycastle LL, Bornstein SR, Brown MJ, Buchanan TA, Busonero F, Campbell H, Cappuccio FP, Cavalcanti-Proenca C, Chen YD, Chen CM, Chines PS, Clarke R, Coin L, Connell J, Day IN, den Heijer M, Duan J, Ebrahim S, Elliott P, Elosua R, Eiriksdottir G, Erdos MR, Eriksson JG, Facheris MF, Felix SB, Fischer-Posovszky P, Folsom AR, Friedrich N, Freimer NB, Fu M, Gaget S, Gejman PV, Geus EJ, Gieger C, Gjesing AP, Goel A, Goyette P, Grallert H, Grassler J, Greenawalt DM, Groves CJ, Gudnason V, Guiducci C, Hartikainen AL, Hassanali N, Hall AS, Havulinna AS, Hayward C, Heath AC, Hengstenberg C, Hicks AA, Hinney A, Hofman A, Homuth G, Hui J, Igl W, Iribarren C, Isomaa B, Jacobs KB, Jarick I, Jewell E, John U, Jorgensen T, Jousilahti P, Jula A, Kaakinen M, Kajantie E, Kaplan LM, Kathiresan S, Kettunen J, Kinnunen L, Knowles JW, Kolcic I, Konig IR, Koskinen S, Kovacs P, Kuusisto J, Kraft P, Kvaloy K, Laitinen J, Lantieri O, Lanzani C, Launer LJ, Lecoeur C, Lehtimaki T, Lettre G, Liu J, Lokki ML, Lorentzon M, Luben RN, Ludwig B, Manunta P, Marek D, Marre M, Martin NG, McArdle WL, McCarthy A, McKnight B, Meitinger T, Melander O, Meyre D, Midthjell K, Montgomery GW, Morken MA, Morris AP, Mulic R, Ngwa JS, Nelis M, Neville MJ, Nyholt DR, O'Donnell CJ, O'Rahilly S, Ong KK, Oostra B, Pare G, Parker AN, Perola M, Pichler I, Pietilainen KH, Platou CG, Polasek O, Pouta A, Rafelt S, Raitakari O, Rayner NW, Ridderstrale M, Rief W, Ruokonen A, Robertson NR, Rzehak P, Salomaa V, Sanders AR, Sandhu MS, Sanna S, Saramies J, Savolainen MJ, Scherag S, Schipf S, Schreiber S, Schunkert H, Silander K, Sinisalo J, Siscovick DS, Smit JH, Soranzo N, Sovio U, Stephens J, Surakka I, Swift AJ, Tammesoo ML, Tardif JC, Teder-Laving M, Teslovich TM, Thompson JR, Thomson B, Tonjes A, Tuomi T, van Meurs JB, van Ommen GJ (2010) Association analyses of 249,796 individuals reveal 18 new loci associated with body mass index. Nat Genet 42:937–948

Steffen LM, Dai S, Fulton JE, Labarthe DR (2009) Overweight in children and adolescents associated with TV viewing and parental weight: Project HeartBeat! Am J Prev Med 37:S50–S55

Strawbridge RJ, Dupuis J, Prokopenko I, Barker A, Ahlqvist E, Rybin D, Petrie JR, Travers ME, Bouatia-Naji N, Dimas AS, Nica A, Wheeler E, Chen H, Voight BF, Taneera J, Kanoni S, Peden JF, Turrini F, Gustafsson S, Zabena C, Almgren P, Barker DJ, Barnes D, Dennison EM, Eriksson JG, Eriksson P, Eury E, Folkersen L, Fox CS, Frayling TM, Goel A, Gu HF, Horikoshi M, Isomaa B, Jackson AU, Jameson KA, Kajantie E, Kerr-Conte J, Kuulasmaa T, Kuusisto J, Loos RJ, Luan J, Makrilakis K, Manning AK, Martínez-Larrad MT, Narisu N, Nastase Mannila M, Ohrvik J, Osmond C, Pascoe L, Payne F, Sayer AA, Sennblad B, Silveira A, Stancáková A, Stirrups K, Swift AJ, Syvänen AC, Tuomi T, van 't Hooft FM, Walker M, Weedon MN, Xie W, Zethelius B; DIAGRAM Consortium; GIANT Consortium; MuTHER Consortium; CARDIoGRAM Consortium; C4D Consortium, Ongen H, Mälarstig A, Hopewell JC, Saleheen D, Chambers J, Parish S, Danesh J, Kooner J, Ostenson CG, Lind L, Cooper CC, Serrano- Ríos M, Ferrannini E, Forsen TJ, Clarke R, Franzosi MG, Seedorf U, Watkins H, Froguel P, Johnson P, Deloukas P, Collins FS, Laakso M, Dermitzakis ET, Boehnke M, McCarthy MI, Wareham NJ, Groop L, Pattou F, Gloyn AL, Dedoussis GV, Lyssenko V, Meigs JB, Barroso I, Watanabe RM, Ingelsson E, Langenberg C, Hamsten A, Florez JC (2011) Genome-wide association identifies nine common variants associated with fasting proinsulin levels and provides new insights into the pathophysiology of type 2 diabetes. Diabetes 60 (10):2624–2634. doi:10.2337/db11-0415

Steinberger J, Daniels SR, Eckel RH, Hayman L, Lustig RH, McCrindle B, Mietus-Snyder ML (2009) Progress and challenges in metabolic syndrome in children and adolescents: a scientific statement from the American Heart Association Atherosclerosis, Hypertension, and Obesity in the Young Committee of the Council on Cardiovascular Disease in the Young; Council on Cardiovascular Nursing; and Council on Nutrition, Physical Activity, and Metabolism. Circulation 119:628–647

Steinthorsdottir V, Thorleifsson G, Reynisdottir I, Benediktsson R, Jonsdottir T, Walters GB, Styrkarsdottir U, Gretarsdottir S, Emilsson V, Ghosh S, Baker A, Snorradottir S, Bjarnason H, Ng MC, Hansen T, Bagger Y, Wilensky RL, Reilly MP, Adeyemo A, Chen Y, Zhou J, Gudnason V, Chen G, Huang H, Lashley K, Doumatey A, So WY, Ma RC, Andersen G, Borch-Johnsen K, Jorgensen T, Vliet-Ostaptchouk JV, Hofker MH, Wijmenga C, Christiansen C, Rader DJ, Rotimi C, Gurney M, Chan JC, Pedersen O, Sigurdsson G, Gulcher JR, Thorsteinsdottir U, Kong A, Stefansson K (2007) A variant in CDKAL1 influences insulin response and risk of type 2 diabetes. Nat Genet 39:770–775

Steinthorsdottir V, Thorleifsson G, Sulem P, Helgason H, Grarup N, Sigurdsson A, Helgadottir HT, Johannsdottir H, Magnusson OT, Gudjonsson SA, Justesen JM, Harder MN, Jorgensen ME, Christensen C, Brandslund I, Sandbaek A, Lauritzen T, Vestergaard H, Linneberg A, Jorgensen T, Hansen T, Daneshpour MS, Fallah MS, Hreidarsson AB, Sigurdsson G, Azizi F, Benediktsson R, Masson G, Helgason A, Kong A, Gudbjartsson DF, Pedersen O, Thorsteinsdottir U, Stefansson K (2014) Identification of low-frequency and rare sequence variants associated with elevated or reduced risk of type 2 diabetes. Nat Genet 46:294–298

Stern MP, Duggirala R, Mitchell BD, Reinhart LJ, Shivakumar S, Shipman PA, Uresandi OC, Benavides E, Blangero J, O'Connell P (1996) Evidence for linkage of regions on chromosomes 6 and 11 to plasma glucose concentrations in Mexican Americans. Genome Res 6:724–734

Stern MP, Williams K, Haffner SM (2002) Identification of persons at high risk for type 2 diabetes mellitus: do we need the oral glucose tolerance test? Ann Intern Med 136:575–581

Stitzel ML, Sethupathy P, Pearson DS, Chines PS, Song L, Erdos MR, Welch R, Parker SC, Boyle AP, Scott LJ, Margulies EH, Boehnke M, Furey TS, Crawford GE, Collins FS (2010) Global epigenomic analysis of primary human pancreatic islets provides insights into type 2 diabetes susceptibility loci. Cell Metab 12:443–455

Stolerman ES, Manning AK, McAteer JB, Fox CS, Dupuis J, Meigs JB, Florez JC (2009) TCF7L2 variants are associated with increased proinsulin/insulin ratios but not obesity traits in the Framingham Heart Study. Diabetologia 52:614–620

Stone S, Abkevich V, Hunt SC, Gutin A, Russell DL, Neff CD, Riley R, Frech GC, Hensel CH, Jammulapati S, Potter J, Sexton D, Tran T, Gibbs D, Iliev D, Gress R, Bloomquist B, Amatruda J, Rae PM, Adams TD, Skolnick MH, Shattuck D (2002) A major predisposition locus for severe obesity, at 4p15-p14. Am J Hum Genet 70:1459–1468

Stunkard AJ, Foch TT, Hrubec Z (1986a) A twin study of human obesity. JAMA 256:51–54

Stunkard AJ, Harris JR, Pedersen NL, McClearn GE (1990) The body-mass index of twins who have been reared apart. N Engl J Med 322:1483–1487

Stunkard AJ, Sorensen TI, Hanis C, Teasdale TW, Chakraborty R, Schull WJ, Schulsinger F (1986b) An adoption study of human obesity. N Engl J Med 314:193–198

Sumner AE (2009) Ethnic differences in triglyceride levels and high-density lipoprotein lead to underdiagnosis of the metabolic syndrome in black children and adults. J Pediatr 155:S7–S11

Sun X, Yu W, Hu C (2014) Genetics of type 2 diabetes: insights into the pathogenesis and its clinical application. Biomed Res Int 2014:926713

Sung KC, Kim BJ, Kim BS, Lee WY, Park JB, Wilson AM (2009) A comparison of the prevalence of the MS and its complications using three proposed definitions in Korean subjects. Am J Cardiol 103:1732–1735

Tabassum R, Chauhan G, Dwivedi OP, Mahajan A, Jaiswal A, Kaur I, Bandesh K, Singh T, Mathai BJ, Pandey Y, Chidambaram M, Sharma A, Chavali S, Sengupta S, Ramakrishnan L, Venkatesh P, Aggarwal SK, Ghosh S, Prabhakaran D, Srinath RK, Saxena M, Banerjee M, Mathur S, Bhansali A, Shah VN, Madhu SV, Marwaha RK, Basu A, Scaria V, McCarthy MI, Venkatesan R, Mohan V, Tandon N, Bharadwaj D (2013) Genome-wide association study for type 2 diabetes in Indians identifies a new susceptibility locus at 2q21. Diabetes 62:977–986

Tabor HK, Risch NJ, Myers RM (2002) Candidate-gene approaches for studying complex genetic traits: practical considerations. Nat Rev Genet 3:391–397

Tartaglia LA, Dembski M, Weng X, Deng N, Culpepper J, Devos R, Richards GJ, Campfield LA, Clark FT, Deeds J (1995) Identification and expression cloning of a leptin receptor, OB-R. Cell 83:1263–1271

Taveras EM, Gillman MW, Kleinman K, Rich-Edwards JW, Rifas-Shiman SL (2010) Racial/ethnic differences in early-life risk factors for childhood obesity. Pediatrics 125:686–695

Taylor H, Liu J, Wilson G, Golden SH, Crook E, Brunson CD, Steffes M, Johnson WD, Sung JH (2008) Distinct component profiles and high risk among African Americans with metabolic syndrome: the Jackson Heart Study. Diabetes Care 31:1248–1253

Tejero ME, Cai G, Goring HH, Diego V, Cole SA, Bacino CA, Butte NF, Comuzzie AG (2007) Linkage analysis of circulating levels of adiponectin in Hispanic children. Int J Obes (Lond) 31:535–542

Tenenbaum A, Fisman EZ (2011) "The metabolic syndrome… is dead": these reports are an exaggeration. Cardiovasc Diabetol 10:11

Teslovich TM, Musunuru K, Smith AV, Edmondson AC, Stylianou IM, Koseki M, Pirruccello JP, Ripatti S, Chasman DI, Willer CJ, Johansen CT, Fouchier SW, Isaacs A, Peloso GM, Barbalic M, Ricketts SL, Bis JC, Aulchenko YS, Thorleifsson G, Feitosa MF, Chambers J, Orho-Melander M, Melander O, Johnson T, Li X, Guo X, Li M, Shin CY, Jin GM, Jin KY, Lee JY, Park T, Kim K, Sim X, Twee-Hee OR, Croteau-Chonka DC, Lange LA, Smith JD, Song K, Hua ZJ, Yuan X, Luan J, Lamina C, Ziegler A, Zhang W, Zee RY, Wright AF, Witteman JC, Wilson JF, Willemsen G, Wichmann HE, Whitfield JB, Waterworth DM, Wareham NJ, Waeber G, Vollenweider P, Voight BF, Vitart V, Uitterlinden AG, Uda M, Tuomilehto J, Thompson JR, Tanaka T, Surakka I, Stringham HM, Spector TD, Soranzo N, Smit JH, Sinisalo J, Silander K, Sijbrands EJ, Scuteri A, Scott J, Schlessinger D, Sanna S, Salomaa V, Saharinen J, Sabatti C, Ruokonen A, Rudan I, Rose LM, Roberts R, Rieder M, Psaty BM, Pramstaller PP, Pichler I, Perola M, Penninx BW, Pedersen NL, Pattaro C, Parker AN, Pare G, Oostra BA, O'Donnell CJ, Nieminen MS, Nickerson DA, Montgomery GW, Meitinger T, McPherson R, McCarthy MI, McArdle W, Masson D, Martin NG, Marroni F, Mangino M, Magnusson PK, Lucas G, Luben R, Loos RJ, Lokki ML, Lettre G, Langenberg C, Launer LJ, Lakatta EG, Laaksonen R, Kyvik KO, Kronenberg F, Konig IR, Khaw KT, Kaprio J, Kaplan LM, Johansson A, Jarvelin MR, Janssens AC, Ingelsson E, Igl W, Kees HG, Hottenga JJ, Hofman A, Hicks AA, Hengstenberg C, Heid IM, Hayward C, Havulinna AS, Hastie ND, Harris TB, Haritunians T, Hall AS, Gyllensten U, Guiducci C, Groop LC, Gonzalez E, Gieger C, Freimer NB, Ferrucci L, Erdmann J, Elliott P, Ejebe KG, Doring A, Dominiczak AF, Demissie S, Deloukas P, de Geus EJ, de Faire U, Crawford G, Collins FS, Chen YD, Caulfield MJ, Campbell H, Burtt NP, Bonnycastle LL, Boomsma DI, Boekholdt SM, Bergman RN, Barroso I, Bandinelli S, Ballantyne CM, Assimes TL, Quertermous T, Altshuler D, Seielstad M, Wong TY, Tai ES, Feranil AB, Kuzawa CW, Adair LS, Taylor HA Jr, Borecki IB, Gabriel SB, Wilson JG, Holm H, Thorsteinsdottir U, Gudnason V, Krauss RM, Mohlke KL, Ordovas

JM, Munroe PB, Kooner JS, Tall AR, Hegele RA, Kastelein JJ, Schadt EE, Rotter JI, Boerwinkle E, Strachan DP, Mooser V, Stefansson K, Reilly MP, Samani NJ, Schunkert H, Cupples LA, Sandhu MS, Ridker PM, Rader DJ, van Duijn CM, Peltonen L, Abecasis GR, Boehnke M, Kathiresan S (2010) Biological, clinical and population relevance of 95 loci for blood lipids. Nature 466:707–713

Thomsen SK, Gloyn AL (2014) The pancreatic beta cell: recent insights from human genetics. Trends Endocrinol Metab 25:425–434

Thorleifsson G, Walters GB, Gudbjartsson DF, Steinthorsdottir V, Sulem P, Helgadottir A, Styrkarsdottir U, Gretarsdottir S, Thorlacius S, Jonsdottir I, Jonsdottir T, Olafsdottir EJ, Olafsdottir GH, Jonsson T, Jonsson F, Borch-Johnsen K, Hansen T, Andersen G, Jorgensen T, Lauritzen T, Aben KK, Verbeek AL, Roeleveld N, Kampman E, Yanek LR, Becker LC, Tryggvadottir L, Rafnar T, Becker DM, Gulcher J, Kiemeney LA, Pedersen O, Kong A, Thorsteinsdottir U, Stefansson K (2009) Genome-wide association yields new sequence variants at seven loci that associate with measures of obesity. Nat Genet 41:18–24

Thornton T, McPeek MS (2007) Case-control association testing with related individuals: a more powerful quasi-likelihood score test. Am J Hum Genet 81:321–337

Timpson NJ, Emmett PM, Frayling TM, Rogers I, Hattersley AT, McCarthy MI, Davey SG (2008) The fat mass- and obesity-associated locus and dietary intake in children. Am J Clin Nutr 88:971–978

Tong Y, Lin Y, Zhang Y, Yang J, Zhang Y, Liu H, Zhang B (2009) Association between TCF7L2 gene polymorphisms and susceptibility to type 2 diabetes mellitus: a large Human Genome Epidemiology (HuGE) review and meta-analysis. BMC Med Genet 10:15

Travers ME, Mackay DJ, Dekker NM, Morris AP, Lindgren CM, Berry A, Johnson PR, Hanley N, Groop LC, McCarthy MI, Gloyn AL (2013) Insights into the molecular mechanism for type 2 diabetes susceptibility at the KCNQ1 locus from temporal changes in imprinting status in human islets. Diabetes 62:987–992

Tregouet DA, Konig IR, Erdmann J, Munteanu A, Braund PS, Hall AS, Grosshennig A, Linsel-Nitschke P, Perret C, DeSuremain M, Meitinger T, Wright BJ, Preuss M, Balmforth AJ, Ball SG, Meisinger C, Germain C, Evans A, Arveiler D, Luc G, Ruidavets JB, Morrison C, van der HP, Schreiber S, Neureuther K, Schafer A, Bugert P, El Mokhtari NE, Schrezenmeir J, Stark K, Rubin D, Wichmann HE, Hengstenberg C, Ouwehand W, Ziegler A, Tiret L, Thompson JR, Cambien F, Schunkert H, Samani NJ (2009) Genome-wide haplotype association study identifies the SLC22A3-LPAL2-LPA gene cluster as a risk locus for coronary artery disease. Nat Genet 41:283–285

Trevino RP, Marshall RM Jr, Hale DE, Rodriguez R, Baker G, Gomez J (1999) Diabetes risk factors in low-income Mexican-American children. Diabetes Care 22:202–207

Tsai FJ, Yang CF, Chen CC, Chuang LM, Lu CH, Chang CT, Wang TY, Chen RH, Shiu CF, Liu YM, Chang CC, Chen P, Chen CH, Fann CS, Chen YT, Wu JY (2010) A genome-wide association study identifies susceptibility variants for type 2 diabetes in Han Chinese. PLoS Genet 6(2):e1000847. doi:10.1371/journal.pgen.1000847

Tycko B (2010) Mapping allele-specific DNA methylation: a new tool for maximizing information from GWAS. Am J Hum Genet 86:109–112

Unoki H, Takahashi A, Kawaguchi T, Hara K, Horikoshi M, Andersen G, Ng DP, Holmkvist J, Borch-Johnsen K, Jorgensen T, Sandbaek A, Lauritzen T, Hansen T, Nurbaya S, Tsunoda T, Kubo M, Babazono T, Hirose H, Hayashi M, Iwamoto Y, Kashiwagi A, Kaku K, Kawamori R, Tai ES, Pedersen O, Kamatani N, Kadowaki T, Kikkawa R, Nakamura Y, Maeda S (2008) SNPs in KCNQ1 are associated with susceptibility to type 2 diabetes in East Asian and European populations. Nat Genet 40:1098–1102

Valdez R, Greenlund KJ, Khoury MJ, Yoon PW (2007) Is family history a useful tool for detecting children at risk for diabetes and cardiovascular diseases? A public health perspective. Pediatrics 120(Suppl 2):S78–S86

van Dam RM, Hoebee B, Seidell JC, Schaap MM, Blaak EE, Feskens EJ (2004) The insulin receptor substrate-1 Gly972Arg polymorphism is not associated with Type 2 diabetes mellitus in two population-based studies. Diabet Med 21:752–758

van Hoek M, Dehghan A, Witteman JC, van Duijn CM, Uitterlinden AG, Oostra BA, Hofman A, Sijbrands EJ, Janssens AC (2008) Predicting type 2 diabetes based on polymorphisms from genome-wide association studies: a population-based study. Diabetes 57:3122–3128

Vaxillaire M, Froguel P (2008) Monogenic diabetes in the young, pharmacogenetics and relevance to multifactorial forms of type 2 diabetes. Endocr Rev 29:254–264

Vimaleswaran KS, Loos RJ (2010) Progress in the genetics of common obesity and type 2 diabetes. Expert Rev Mol Med 12:e7

Virdis A, Ghiadoni L, Masi S, Versari D, Daghini E, Giannarelli C, Salvetti A, Taddei S (2009) Obesity in the childhood: a link to adult hypertension. Curr Pharm Des 15:1063–1071

Visscher PM, Andrew T, Nyholt DR (2008) Genome-wide association studies of quantitative traits with related individuals: little (power) lost but much to be gained. Eur J Hum Genet 16:387–390

Voight BF, Scott LJ, Steinthorsdottir V, Morris AP, Dina C, Welch RP, Zeggini E, Huth C, Aulchenko YS, Thorleifsson G, McCulloch LJ, Ferreira T, Grallert H, Amin N, Wu G, Willer CJ, Raychaudhuri S, McCarroll SA, Langenberg C, Hofmann OM, Dupuis J, Qi L, Segre AV, van Hoek M, Navarro P, Ardlie K, Balkau B, Benediktsson R, Bennett AJ, Blagieva R, Boerwinkle E, Bonnycastle LL, Bengtsson BK, Bravenboer B, Bumpstead S, Burtt NP, Charpentier G, Chines PS, Cornelis M, Couper DJ, Crawford G, Doney AS, Elliott KS, Elliott AL, Erdos MR, Fox CS, Franklin

CS, Ganser M, Gieger C, Grarup N, Green T, Griffin S, Groves CJ, Guiducci C, Hadjadj S, Hassanali N, Herder C, Isomaa B, Jackson AU, Johnson PR, Jorgensen T, Kao WH, Klopp N, Kong A, Kraft P, Kuusisto J, Lauritzen T, Li M, Lieverse A, Lindgren CM, Lyssenko V, Marre M, Meitinger T, Midthjell K, Morken MA, Narisu N, Nilsson P, Owen KR, Payne F, Perry JR, Petersen AK, Platou C, Proenca C, Prokopenko I, Rathmann W, Rayner NW, Robertson NR, Rocheleau G, Roden M, Sampson MJ, Saxena R, Shields BM, Shrader P, Sigurdsson G, Sparso T, Strassburger K, Stringham HM, Sun Q, Swift AJ, Thorand B, Tichet J, Tuomi T, van Dam RM, van Haeften TW, van Herpt T, Vliet-Ostaptchouk JV, Walters GB, Weedon MN, Wijmenga C, Witteman J, Bergman RN, Cauchi S, Collins FS, Gloyn AL, Gyllensten U, Hansen T, Hide WA, Hitman GA, Hofman A, Hunter DJ, Hveem K, Laakso M, Mohlke KL, Morris AD, Palmer CN, Pramstaller PP, Rudan I, Sijbrands E, Stein LD, Tuomilehto J, Uitterlinden A, Walker M, Wareham NJ, Watanabe RM, Abecasis GR, Boehm BO, Campbell H, Daly MJ, Hattersley AT, Hu FB, Meigs JB, Pankow JS, Pedersen O, Wichmann HE, Barroso I, Florez JC, Frayling TM, Groop L, Sladek R, Thorsteinsdottir U, Wilson JF, Illig T, Froguel P, van Duijn CM, Stefansson K, Altshuler D, Boehnke M, McCarthy MI (2010) Twelve type 2 diabetes susceptibility loci identified through large-scale association analysis. Nat Genet 42:579–589

Voruganti VS, Goring HH, Diego VP, Cai G, Mehta NR, Haack K, Cole SA, Butte NF, Comuzzie AG (2007) Genome-wide scan for serum ghrelin detects linkage on chromosome 1p36 in Hispanic children: results from the Viva La Familia study. Pediatr Res 62:445–450

Walder K, Hanson RL, Kobes S, Knowler WC, Ravussin E (2000) An autosomal genomic scan for loci linked to plasma leptin concentration in Pima Indians. Int J Obes Relat Metab Disord 24:559–565

Walston J, Silver K, Bogardus C, Knowler WC, Celi FS, Austin S, Manning B, Strosberg AD, Stern MP, Raben N (1995) Time of onset of non-insulin-dependent diabetes mellitus and genetic variation in the beta 3-adrenergic-receptor gene. N Engl J Med 333:343–347

Wang J, Kuusisto J, Vanttinen M, Kuulasmaa T, Lindstrom J, Tuomilehto J, Uusitupa M, Laakso M (2007) Variants of transcription factor 7-like 2 (TCF7L2) gene predict conversion to type 2 diabetes in the Finnish Diabetes Prevention Study and are associated with impaired glucose regulation and impaired insulin secretion. Diabetologia 50:1192–1200

Wang K, Li WD, Zhang CK, Wang Z, Glessner JT, Grant SF, Zhao H, Hakonarson H, Price RA (2011) A genome-wide association study on obesity and obesity-related traits. PLoS One 6(4):e18939. doi:10.1371/journal.pone.0018939. Erratum in: PLoS One. 2012; 7 (2). doi:10.1371/annotation/a34ee94e-3e6a-48bd-a19e-398a4bb88580

Wang Y, Beydoun MA (2007) The obesity epidemic in the United States—gender, age, socioeconomic, racial/ethnic, and geographic characteristics: a systematic review and meta-regression analysis. Epidemiol Rev 29:6–28

Wardle J, Carnell S, Haworth CM, Plomin R (2008) Evidence for a strong genetic influence on childhood adiposity despite the force of the obesogenic environment. Am J Clin Nutr 87:398–404

Watanabe RM, Allayee H, Xiang AH, Trigo E, Hartiala J, Lawrence JM, Buchanan TA (2007) Transcription factor 7-like 2 (TCF7L2) is associated with gestational diabetes mellitus and interacts with adiposity to alter insulin secretion in Mexican Americans. Diabetes 56:1481–1485

Weiss R, Dziura J, Burgert TS, Tamborlane WV, Taksali SE, Yeckel CW, Allen K, Lopes M, Savoye M, Morrison J, Sherwin RS, Caprio S (2004) Obesity and the metabolic syndrome in children and adolescents. N Engl J Med 350:2362–2374

Wen W, Cho YS, Zheng W, Dorajoo R, Kato N, Qi L, Chen CH, Delahanty RJ, Okada Y, Tabara Y, Gu D, Zhu D, Haiman CA, Mo Z, Gao YT, Saw SM, Go MJ, Takeuchi F, Chang LC, Kokubo Y, Liang J, Hao M, Le Marchand L, Zhang Y, Hu Y, Wong TY, Long J, Han BG, Kubo M, Yamamoto K, Su MH, Miki T, Henderson BE, Song H, Tan A, He J, Ng DP, Cai Q, Tsunoda T, Tsai FJ, Iwai N, Chen GK, Shi J, Xu J, Sim X, Xiang YB, Maeda S, Ong RT, Li C, Nakamura Y, Aung T, Kamatani N, Liu JJ, Lu W, Yokota M, Seielstad M, Fann CS, Wu JY, Lee JY, Hu FB, Tanaka T, Tai ES, Shu XO (2012) Meta-analysis identifies common variants associated with body mass index in east Asians. Nat Genet 44:307–311

Wheeler DA, Srinivasan M, Egholm M, Shen Y, Chen L, McGuire A, He W, Chen YJ, Makhijani V, Roth GT, Gomes X, Tartaro K, Niazi F, Turcotte CL, Irzyk GP, Lupski JR, Chinault C, Song XZ, Liu Y, Yuan Y, Nazareth L, Qin X, Muzny DM, Margulies M, Weinstock GM, Gibbs RA, Rothberg JM (2008) The complete genome of an individual by massively parallel DNA sequencing. Nature 452:872–876

WHO (2013) Obesity and overweight. Fact sheet. No 311. WHO Media Centre, World Health Organization (WHO)

Widen E, Lehto M, Kanninen T, Walston J, Shuldiner AR, Groop LC (1995) Association of a polymorphism in the beta 3-adrenergic-receptor gene with features of the insulin resistance syndrome in Finns. N Engl J Med 333:348–351

Wild S, Roglic G, Green A, Sicree R, King H (2004) Global prevalence of diabetes: estimates for the year 2000 and projections for 2030. Diabetes Care 27:1047–1053

Willer CJ, Speliotes EK, Loos RJ, Li S, Lindgren CM, Heid IM, Berndt SI, Elliott AL, Jackson AU, Lamina C, Lettre G, Lim N, Lyon HN, McCarroll SA, Papadakis K, Qi L, Randall JC, Roccasecca RM, Sanna S, Scheet P, Weedon MN, Wheeler E, Zhao JH, Jacobs LC, Prokopenko I, Soranzo N, Tanaka T, Timpson NJ, Almgren P, Bennett A, Bergman RN, Bingham SA, Bonnycastle LL, Brown M, Burtt NP,

Chines P, Coin L, Collins FS, Connell JM, Cooper C, Smith GD, Dennison EM, Deodhar P, Elliott P, Erdos MR, Estrada K, Evans DM, Gianniny L, Gieger C, Gillson CJ, Guiducci C, Hackett R, Hadley D, Hall AS, Havulinna AS, Hebebrand J, Hofman A, Isomaa B, Jacobs KB, Johnson T, Jousilahti P, Jovanovic Z, Khaw KT, Kraft P, Kuokkanen M, Kuusisto J, Laitinen J, Lakatta EG, Luan J, Luben RN, Mangino M, McArdle WL, Meitinger T, Mulas A, Munroe PB, Narisu N, Ness AR, Northstone K, O'Rahilly S, Purmann C, Rees MG, Ridderstrale M, Ring SM, Rivadeneira F, Ruokonen A, Sandhu MS, Saramies J, Scott LJ, Scuteri A, Silander K, Sims MA, Song K, Stephens J, Stevens S, Stringham HM, Tung YC, Valle TT, van Duijn CM, Vimaleswaran KS, Vollenweider P, Waeber G, Wallace C, Watanabe RM, Waterworth DM, Watkins N, Witteman JC, Zeggini E, Zhai G, Zillikens MC, Altshuler D, Caulfield MJ, Chanock SJ, Farooqi IS, Ferrucci L, Guralnik JM, Hattersley AT, Hu FB, Jarvelin MR, Laakso M, Mooser V, Ong KK, Ouwehand WH, Salomaa V, Samani NJ, Spector TD, Tuomi T, Tuomilehto J, Uda M, Uitterlinden AG, Wareham NJ, Deloukas P, Frayling TM, Groop LC, Hayes RB, Hunter DJ, Mohlke KL, Peltonen L, Schlessinger D, Strachan DP, Wichmann HE, McCarthy MI, Boehnke M, Barroso I, Abecasis GR, Hirschhorn JN (2009) Six new loci associated with body mass index highlight a neuronal influence on body weight regulation. Nat Genet 41:25–34

Williams AL, Jacobs SB, Moreno-Macias H, Huerta-Chagoya A, Churchhouse C, Marquez-Luna C, Garcia-Ortiz H, Gomez-Vazquez MJ, Burtt NP, Aguilar-Salinas CA, Gonzalez-Villalpando C, Florez JC, Orozco L, Haiman CA, Tusie-Luna T, Altshuler D (2014) Sequence variants in SLC16A11 are a common risk factor for type 2 diabetes in Mexico. Nature 506:97–101

Wilson SG, Adam G, Langdown M, Reneland R, Braun A, Andrew T, Surdulescu GL, Norberg M, Dudbridge F, Reed PW, Sambrook PN, Kleyn PW, Spector TD (2006) Linkage and potential association of obesity-related phenotypes with two genes on chromosome 12q24 in a female dizygous twin cohort. Eur J Hum Genet 14:340–348

Winckler W, Weedon MN, Graham RR, McCarroll SA, Purcell S, Almgren P, Tuomi T, Gaudet D, Bostrom KB, Walker M, Hitman G, Hattersley AT, McCarthy MI, Ardlie KG, Hirschhorn JN, Daly MJ, Frayling TM, Groop L, Altshuler D (2007) Evaluation of common variants in the six known maturity-onset diabetes of the young (MODY) genes for association with type 2 diabetes. Diabetes 56:685–693

Wray NR, Purcell SM, Visscher PM (2011) Synthetic associations created by rare variants do not explain most GWAS results. PLoS Biol 9:e1000579

WTCCC (2007) Genome-wide association study of 14,000 cases of seven common diseases and 3,000 shared controls. Nature 447:661–678

Wu Y, Li H, Loos RJ, Yu Z, Ye X, Chen L, Pan A, Hu FB, Lin X (2008) Common variants in CDKAL1, CDKN2A/B, IGF2BP2, SLC30A8, and HHEX/IDE genes are associated with type 2 diabetes and impaired fasting glucose in a Chinese Han population. Diabetes 57:2834–2842

Xita N, Tsatsoulis A (2010) Fetal origins of the metabolic syndrome. Ann N Y Acad Sci 1205:148–155

Yamada T, Ishihara H, Tamura A, Takahashi R, Yamaguchi S, Takei D, Tokita A, Satake C, Tashiro F, Katagiri H, Aburatani H, Miyazaki J, Oka Y (2006) WFS1-deficiency increases endoplasmic reticulum stress, impairs cell cycle progression and triggers the apoptotic pathway specifically in pancreatic beta-cells. Hum Mol Genet 15:1600–1609

Yan Y, North KE, Ballantyne CM, Brancati FL, Chambless LE, Franceschini N, Heiss G, Kottgen A, Pankow JS, Selvin E, West SL, Boerwinkle E (2009) Transcription factor 7-like 2 (TCF7L2) polymorphism and context-specific risk of type 2 diabetes in African American and Caucasian adults: the Atherosclerosis Risk in Communities study. Diabetes 58:285–289

Yang Z, Ming XF (2011) CD36: the common soil for inflammation in obesity and atherosclerosis? Cardiovasc Res 89:485–486

Yanovski JA, Diament AL, Sovik KN, Nguyen TT, Li H, Sebring NG, Warden CH (2000) Associations between uncoupling protein 2, body composition, and resting energy expenditure in lean and obese African American, white, and Asian children. Am J Clin Nutr 71:1405–1420

Yasuda K, Miyake K, Horikawa Y, Hara K, Osawa H, Furuta H, Hirota Y, Mori H, Jonsson A, Sato Y, Yamagata K, Hinokio Y, Wang HY, Tanahashi T, Nakamura N, Oka Y, Iwasaki N, Iwamoto Y, Yamada Y, Seino Y, Maegawa H, Kashiwagi A, Takeda J, Maeda E, Shin HD, Cho YM, Park KS, Lee HK, Ng MC, Ma RC, So WY, Chan JC, Lyssenko V, Tuomi T, Nilsson P, Groop L, Kamatani N, Sekine A, Nakamura Y, Yamamoto K, Yoshida T, Tokunaga K, Itakura M, Makino H, Nanjo K, Kadowaki T, Kasuga M (2008) Variants in KCNQ1 are associated with susceptibility to type 2 diabetes mellitus. Nat Genet 40:1092–1097

Ylonen SK, Salminen I, Lyssenko V, Virtanen SM, Groop L, Aro A, Saloranta C (2008) The Pro12Ala polymorphism of the PPAR-gamma2 gene affects associations of fish intake and marine n-3 fatty acids with glucose metabolism. Eur J Clin Nutr 62:1432–1439

Zabaneh D, Balding DJ (2010) A genome-wide association study of the metabolic syndrome in Indian Asian men. PLoS One 5:e11961

Zeggini E, Scott LJ, Saxena R, Voight BF, Marchini JL, Hu T, de Bakker PI, Abecasis GR, Almgren P, Andersen G, Ardlie K, Bostrom KB, Bergman RN, Bonnycastle LL, Borch-Johnsen K, Burtt NP, Chen H, Chines PS, Daly MJ, Deodhar P, Ding CJ, Doney AS, Duren WL, Elliott KS, Erdos MR, Frayling TM, Freathy RM, Gianniny L, Grallert H, Grarup N, Groves CJ, Guiducci C, Hansen T, Herder C, Hitman GA, Hughes TE, Isomaa B, Jackson AU, Jorgensen T, Kong A, Kubalanza K, Kuruvilla FG, Kuusisto J, Langenberg C, Lango H, Lauritzen T, Li Y, Lindgren

CM, Lyssenko V, Marvelle AF, Meisinger C, Midthjell K, Mohlke KL, Morken MA, Morris AD, Narisu N, Nilsson P, Owen KR, Palmer CN, Payne F, Perry JR, Pettersen E, Platou C, Prokopenko I, Qi L, Qin L, Rayner NW, Rees M, Roix JJ, Sandbaek A, Shields B, Sjogren M, Steinthorsdottir V, Stringham HM, Swift AJ, Thorleifsson G, Thorsteinsdottir U, Timpson NJ, Tuomi T, Tuomilehto J, Walker M, Watanabe RM, Weedon MN, Willer CJ, Illig T, Hveem K, Hu FB, Laakso M, Stefansson K, Pedersen O, Wareham NJ, Barroso I, Hattersley AT, Collins FS, Groop L, McCarthy MI, Boehnke M, Altshuler D (2008) Meta-analysis of genome-wide association data and large-scale replication identifies additional susceptibility loci for type 2 diabetes. Nat Genet 40:638–645

Zeggini E, Weedon MN, Lindgren CM, Frayling TM, Elliott KS, Lango H, Timpson NJ, Perry JR, Rayner NW, Freathy RM, Barrett JC, Shields B, Morris AP, Ellard S, Groves CJ, Harries LW, Marchini JL, Owen KR, Knight B, Cardon LR, Walker M,Hitman GA, Morris AD, Doney AS; Wellcome Trust Case Control Consortium (WTCCC),McCarthy MI, Hattersley AT (2007) Replication of genome-wide association signals in UK samples reveals risk loci for type 2 diabetes. Science 316(5829):1336–1341. Epub 2007 Apr 26. Erratum in: Science 317(5841):1035-6

Zhang C, Qi L, Hunter DJ, Meigs JB, Manson JE, van Dam RM, Hu FB (2006) Variant of transcription factor 7-like 2 (TCF7L2) gene and the risk of type 2 diabetes in large cohorts of U.S. women and men. Diabetes 55:2645–2648

Zhao J, Bradfield JP, Zhang H, Annaiah K, Wang K, Kim CE, Glessner JT, Frackelton EC, Otieno FG, Doran J, Thomas KA, Garris M, Hou C, Chiavacci RM, Li M, Berkowitz RI, Hakonarson H, Grant SF (2010) Examination of all type 2 diabetes GWAS loci reveals HHEX-IDE as a locus influencing pediatric BMI. Diabetes 59:751–755

Zhao J, Bradfield JP, Zhang H, Sleiman PM, Kim CE, Glessner JT, Deliard S, Thomas KA, Frackelton EC, Li M, Chiavacci RM, Berkowitz RI, Hakonarson H, Grant SF (2011) Role of BMI-associated loci identified in GWAS meta-analyses in the context of common childhood obesity in European Americans. Obesity (Silver Spring) 19:2436–2439

Zhou X, Maricque B, Xie M, Li D, Sundaram V, Martin EA, Koebbe BC, Nielsen C, Hirst M, Farnham P, Kuhn RM, Zhu J, Smirnov I, Kent WJ, Haussler D, Madden PA, Costello JF, Wang T (2011) The Human Epigenome Browser at Washington University. Nat Methods 8:989–990

Zhu M, Zhao S (2007) Candidate gene identification approach: progress and challenges. Int J Biol Sci 3:420–427

Zimmet P (2003) The burden of type 2 diabetes: are we doing enough? Diabetes Metab 29:6S9–6S18

Zuk O, Schaffner SF, Samocha K, Do R, Hechter E, Kathiresan S, Daly MJ, Neale BM, Sunyaev SR, Lander ES (2014) Searching for missing heritability: designing rare variant association studies. Proc Natl Acad Sci U S A 111:E455–E464

Genetic Influence on the Human Brain 13

D. Reese McKay, Anderson M. Winkler, Peter Kochunov, Emma E.M. Knowles, Emma Sprooten, Peter T. Fox, John Blangero, and David C. Glahn

D.R. McKay (✉) · A.M. Winkler · E.E.M. Knowles · E. Sprooten · D.C. Glahn
Department of Psychiatry, Yale University School of Medicine, 300 George Street, New Haven CT 06511, USA
e-mail: mckay.reese@gmail.com

A.M. Winkler
e-mail: anderson.winkler@yale.edu

E.E.M. Knowles
e-mail: emma.knowles@yale.edu

E. Sprooten
e-mail: emma.sprooten@yale.edu

D.C. Glahn
e-mail: david.glahn@yale.edu

D.R. McKay · A.M. Winkler · E.E.M. Knowles · E. Sprooten · D.C. Glahn
Olin Neuropsychiatry Research Center, Institute of Living—Whitehall Building, Hartford Hospital, 200 Retreat Avenue, Hartford, CT 06106, USA

D.R. McKay · P.T. Fox
Research Imaging Institute, University of Texas Health Science Center San Antonio, 8403 Floyd Curl Drive, San Antonio, TX 78229, USA
e-mail: fox@uthscsa.edu

A.M. Winkler
Centre for Functional MRI of the Brain, University of Oxford, Oxford OX3 9DU, UK

P. Kochunov
Maryland Psychiatric Research Center, Department of Psychiatry, University of Maryland School of Medicine, 655 West Baltimore St., Baltimore, MD 21201, USA
e-mail: pkochunov@mprc.umaryland.edu

J. Blangero
Department of Genetics, Texas Biomedical Research Institute, 7620 NW Loop 410, San Antonio, TX 78227, USA
e-mail: blangero@uthscsa.edu

The current form of a species reflects advantageous behaviors ingrained and selected by the interaction of genes and environment. That the human mind is the apex of this process is a principle stated and restated by philosophers and scientists throughout history. Now more than ever science is on the cusp of directly linking genes to human brain function.

Imaging genetics—the combination of imaging and genetic information to map gene effects in the brain—enjoys an embarrassment of data riches and equally abundant unrealized discovery. In the late 1980s, the field of human genetics was revolutionized by the discovery of copious molecular markers, advances in fast and cost-effective genotyping methods, and the development of powerful statistical methods. The emergence of human brain mapping nearly paralleled this timeline. Functional magnetic resonance imaging (FMRI) in the 1990s, on the heels of discoveries spawned by positron emission tomography (PET) in the 1980s, pushed knowledge of the brain's inner workings to unprecedented levels. That these frontiers of scientific discovery could inform one another was demonstrated in 2001 (Thompson et al. 2001), just months after the completion of the first working drafts of the human genome sequence were published (Venter et al. 2001; Lander et al. 2001).

Today, we are in the midst of another chapter in the genomic revolution, driven by the development of massively parallel gene sequencing technology that is capable of rapidly genotyping

R. Duggirala et al. (eds.), *Genome Mapping and Genomics in Human and Non-Human Primates*, Genome Mapping and Genomics in Animals 5, DOI 10.1007/978-3-662-46306-2_13,

hundreds of thousands of polymorphic markers per sample. As a result, the power of whole genome sequence data is outstripping biometric image-based discovery. Defining brain phenotypes that represent the action of genes is the challenge of our time. In particular, there is urgent need for programmatic criteria to extract meaningful phenotypes from neuroimages. Research to properly define traits from an endless possibility of image-based metrics has been shortsighted. This represents a fundamental dilemma that must be overcome. In turn, a systematic program for discovery will be established and our understanding of gene-brain interaction will embrace topics that we cannot yet envision. Indeed, the time is now and the potential for discovery is ripe!

13.1 Traits and Subjects

To date, an over-reliance on obsolete study designs has limited the progress of imaging-genetics and led to a minefield of inconsistent findings. As research of complex brain pathology progressed into the genomic age, investigators naturally gravitated toward methods that were successful for studying affected populations; notably, phenotype and subject criteria related to diagnostic status. Because the degree of impairment and presentation of symptoms in brain-related disorders vary widely among affected individuals (including subclinical impairment), diagnostic categorizations are problematic. This has motivated a more powerful alternative strategy, namely the use of quantitative traits as phenotypes (Blangero 2004; Gottesman and Gould 2003). To date, quantitative traits are applied in three general study design classes: Case-control, twin/sibling pair, and extended pedigree.

Studies utilizing large extended pedigrees have multiple benefits compared to twin designs, including increased power to detect heritable effects, less confounding of genetic effects with shared environmental effects because of the inclusion of multiple households within pedigrees, and greater mathematical power to localize and identify causal quantitative trait loci, and far more power to examine the effects of rare variation (Blangero 2004; for in depth discussion of the rare variation strategy, see Chap. 16 by Curran et al. this volume). However, these advantages are not without added burden. Familial studies typically require more participants than twin studies. Recruiting large families to participate in imaging-genetics studies requires that many family members live in close proximity. As is the case in all quantitative genetic studies, extremely reliable, nonlabile, phenotypes are required. An added benefit of focusing on randomly selected large extended pedigrees is that many different image-based phenotypes can be analyzed in a single study.

13.1.1 Normal Brain Variation

Early large-scale brain-imaging research focused on young, healthy, normal adult subjects (Mazziotta et al. 1995). In the past decade, normative studies of brain structure and function have been extended to the entire human lifespan, from childhood through senescence (Biswal et al. 2010; Glahn et al. 2010; Gogtay et al. 2004; Mazziotta et al. 2001; Sowell et al. 2003; Thompson et al. 2005). Going forward, these streams of research should be the foundation for image-based gene discovery instead of unfounded metrics in clinical populations. Additionally, it is highly likely that the genes involved in normal phenotypic variation are also involved in pathological variation. This further mandates research of genetic influence on normal brain structure and function, as truly understanding pathology may require a better understanding of normal variation.

In vivo MRI data is inherently quantitative and is capable of depicting an immense number of potential phenotypes. Image-based metrics can be drawn from any source of contrast including tissue type, anisotropy level, blood flow, and oxygenation level, among many others. Brain volume, total gray matter, and other global measurements were shown to be highly heritable (Bartley et al. 1997), lobar measurements followed (Geschwind et al. 2002), then measurements of Brodmann areas and specific gyri (Peper et al. 2007; Winkler et al. 2010), and most recently voxelwise analysis of the whole image space (Stein et al. 2010). Unfortunately, the power to choose has been a double-edged sword.

Phenotypes are often tested in abundance, as there is no established method for selecting phenotypes and data driven techniques provide un-biased perspective. Yet, mapping genes or sets of genes to structure–function relationships has remained elusive. An alternative approach is selecting, modeling, and evaluating potential phenotypes based on our ability to test neuroscience driven hypotheses. Though seemingly apparent, this notion is a drastic deviation from modern high-profile methods, such as testing every voxel in an image for genome-wide association. Not only does such a broad net increase type 1 error, but it also undermines decades of neuroscience-imaging research with a moot question: Do genetic variants influence voxels in MR images? Instead of addressing an arbitrary aspect of image processing (voxels), phenotypes used for gene identification analyses should reflect our understanding of the brain.

Herein, we share the results and conclusions drawn from testing and applying candidate phenotypic measurements in an extended pedigree MRI study. Subjects were randomly ascertained and phenotypes represent normal variation. Extended pedigree designs are more powerful than twin designs for localizing the effects of genes, but also require automated, quantitative, and robust metrics. Because the actions of genes are unknown, phenotypes should represent image-based neuroscientific truth, as we presume. We place focus on basic neuroanatomy. This will develop a foundation for understanding how genetic influence is reflected in the brain structure and function that we quantify using MRI.

While many genetic studies of mental disorders focus on the presence of a particular disease, this diagnostic endpoint is often distant from determinant etiology (Plomin et al. 2009). Conversely, quantitative phenotypes that are genetically correlated with disease liability can be measured in all individuals (both affected and unaffected) and provide greater power to detect disease-related genetic factors than affection status alone (Blangero 2004; Glahn et al. 2012; Gottesman and Gould 2003). Extended pedigrees provide an ideal framework to exploit these advantages, amongst others. Keeping with this strategy, MRI data was obtained from participants in the "Genetics of Brain Structure and Function" (GOBS) study. GOBS is a pseudorandom ascertainment of extended Mexican–American families in the San Antonio area. In 1991, initial investigations were designed to identify risk factors for diabetes, hypertension, and obesity. Since then, first, second, and third degree relatives of original probands and spouses have been recruited. The diversity of biological relationships and large number of informative pairs is indicative of the multigenerational depth and expanse of these large pedigrees (Table 13.1).

13.2 Background and Significance

13.2.1 Why Is Structural MRI Appropriate for Studying Genetic Underpinnings of the Brain?

Statistical genetics quantifies covariance between phenotypic and genetic variability. The statistical power of such analysis is strongly dependent on the precision of phenotypic measurements. Modern MRI technology is capable of providing phenotypic measurements with both high precision and reproducibility. The intersession, scan–rescan variability of MRI-based phenotypes such as global brain volume is less than 1 % (Lemieux et al. 1999). The intersession variability of more localized structural phenotypes such as hemispheric, lobar, and tissue volumes or gray matter thickness is estimated to be in the 3–10 % range (Agartz et al. 2001; Julin et al. 1997; Lerch and Evans 2005).

13.2.2 Is a Trait Influenced by Genetic Factors?

Quantitative genetic analysis partitions trait covariance among related individuals into genetic and environmental components. For the univariate case (a single trait, such as total brain volume), the covariance matrix (Ω) in a family (pedigree) of n members can be modeled as $\Omega = 2\Phi\sigma_a^2 + I_n\sigma_e^2$, where Φ is the $n \times n$ kinship matrix for the pedigree (Table 13.1), σ_a^2 is the variance in the trait due to additive genetic effects, I_n is an $n \times n$ identity matrix, and σ_e^2 is the variance due to random environmental effects. The most fundamental genetic parameter is the heritability (h^2) of a trait $h^2 = \sigma_a^2/(\sigma_a^2 + \sigma_e^2)$. While this model is for the simplest case of only two variance components (additive genetic and environmental), it is readily extendable via the addition of variance terms in the denominator to allow for additional variance components such as those including dominance genetic variance, X-linked genetic variance, mitochondrial effects, and maternal effects (Almasy and Blangero 1998). Covariates such as sex, age, and their interaction (age × sex) are routinely included in these genetic models. Regression terms are estimated for each covariate, and the likelihood of a model in which the covariate effect is estimated is compared to the likelihood of a model in which the covariate effects are constrained to zero.

Table 13.1 A sample of the pair-wise relationships within Mexican–American pedigrees of participants in the GOBS study

Number of relative pairs	Familial relationship	Coefficient of relationship
2	Monozygotic twins	1
1,004	Parent-offspring	1/2
1,192	Siblings	1/2
352	Grandparent-grandchild	1/4
2,407	Avuncular	1/4
175	Half-siblings	1/4
7	Great grandparent-grandchild	1/8
675	Grand-avuncular	1/8
361	Half-avuncular	1/8
2,783	1st cousins	1/8
34	Great grand-avuncular	1/16
19	Half grand-avuncular	1/16
2,797	1st cousins, once removed	1/16
402	Half 1st cousins	1/32
343	1st cousins, twice removed	1/32
10	Half 1st cousins, once removed	1/32
955	2nd cousins	1/32
321	2nd cousins, once removed	1/64

13.2.3 Are Two Traits Influenced by the Same Genes?

Using the information contained in the kinship matrix and maximum likelihood variance decomposition techniques, the phenotypic correlation between any two traits can be partitioned into additive genetic and random environmental components. This is often referred to as multivariate genetic analysis. The phenotypic correlation (ρ_p) between two traits (x and y), the additive genetic (ρ_a) and random environmental correlations (ρ_e) between the two traits, and their heritabilities (denoted as h_x^2 and h_y^2) are related as follows:

$$\rho_p(x, y) = \left[h_x^2 h_y^2\right]^{1/2} \rho_a(x, y) + \left[\left(1 - h_x^2\right)\left(1 - h_y^2\right)\right]^{1/2} \rho_e(x, y) \tag{13.1}$$

The additive genetic correlation ranges from −1 to 1 and is a measure of the shared genetic basis of the two traits. An absolute additive genetic correlation of 1.0 indicates complete pleiotropy, where the same genes or sets of genes affect both traits (Almasy et al. 1997). Alternatively, a genetic correlation between 1 and 0 indicates incomplete pleiotropy, meaning that the two traits are influenced to some extent by the same genes, but each trait also has a unique genetic basis. A genetic correlation between −1 and 0 indicates a slightly more complicated circumstance where the two phenotypes are divergent. Similarly, the random environmental correlation is estimated and serves as a measure of the strength of the correlated response of the traits to nongenetic factors. In the maximum likelihood framework, the likelihoods of models that constrain the genetic correlation (or environmental correlation) between the traits to zero are compared to the likelihood of models that allow the genetic correlation (or environmental correlation) between the traits to be estimated.

This method of genetic correlation analysis allows the determination of (prior to gene mapping or QTL studies) whether two or more brain-related phenotypes are: (1) Influenced by the same sets of genes, (2) by partly overlapping sets of genes, or (3) have no genetic effects in common. These analyses can be used to test a wide variety of hypotheses concerning the genetic architecture of brain-related phenotypes. For example, a series of tests can evaluate whether genes that influence brain structure also influence brain function (as measured by neurocognitive testing).

13.3 Genetic Analysis of Brain-Based Phenotypes

13.3.1 Heritability of the Human and Nonhuman Primate Brain

The size, shape, and internal structure of the primate brain vary considerably between individuals within a species and a significant portion of this intrasubject variability is influenced by genetic factors. While very early stages of primate brain development are predominately mediated by genetic programs (Rubenstein et al. 1999; Rubenstein and Rakic 1999), later stages of development, organization, and brain maturation result from a complex interaction of genetic and environmental influences (Rakic 1988). Studies in nonhuman primates have provided heritability estimates for brain weight ranging between 0.42 and 0.75 (Cheverud et al. 1990a, b; Rogers et al. 2007). Human imaging studies have expanded upon these initial findings. Phenotypes based on lobar measurements are less heritable than global phenotypes and have been shown to vary by lobe (Geschwind et al. 2002). Brodmann areas or specific gyri, though widely variable, are slightly less heritable than lobar phenotypes (Winkler et al. 2010; Wright et al. 2002). Together, these reports demonstrate an indirect relationship between estimated genetic influence and phenotype spatial resolution. Reduced heritability estimates for smaller structures might be associated with the reliability of image analyses rather than an intrinsic reduction in the genetic influences of these regions. However, it is more likely that whole brain phenotypes reflect the action of many genes and are more readily transmitted. Therefore, high heritability values do not convey gene-finding feasibility.

13.3.2 Genetic Influence on Gray Matter

Gray matter primarily consists of neuronal cell bodies. Gray matter is distributed across the surface of the cerebral hemispheres (cerebral cortex) and of the cerebellum (cerebellar cortex). Large collections of gray matter are also present in the thalamic nuclei and basal ganglia and cerebellar nuclei.

The most thorough demonstration of genetic influence on gray matter was provided independently by Panizzon et al. (2009) and Winkler et al. (2010). Specifically, these efforts sought the fundamental actions of genes by investigating the relationship between gray matter volume, surface area and thickness in brain regions similar to Brodmann Areas. Using different samples and designs, both studies concluded that variability of both cortical surface area and thickness were influenced by independent genetic factors, indicating that measurements of gray matter volume confound these effects. Furthermore, focusing on cortical surface area or thickness rather than volume places the investigator closer to the theoretical action of genes.

Since these studies, investigators have increased the resolution of genetic investigations of gray matter by moving from Brodmann Areas to pointwise cortical reconstructions (Figs. 13.1

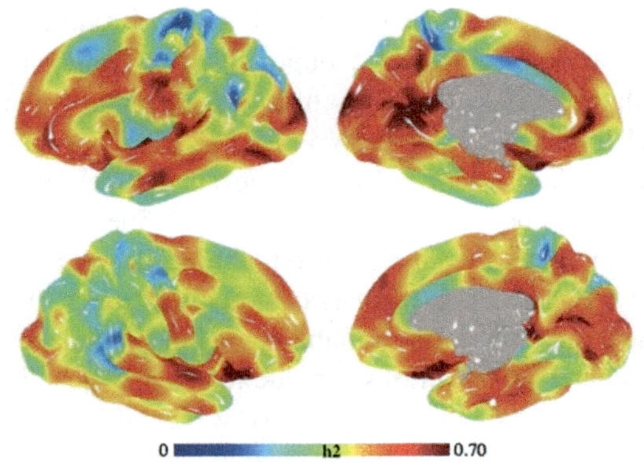

Fig. 13.1 Heritability of pointwise cortical surface area. Phenotypes were defined by parceling each hemisphere into 40,962 vertices using the Freesurfer image analysis suite, which is documented and freely available (Dale et al. 1999; Fischl et al. 1999, http://surfer.nmr.mgh.harvard.edu/). Genetic analyses were performed using the SOLAR software package, which is also freely available (Almasy and Blangero 1998, http://www.txbiomed.org/departments/genetics/genetics-detail?p=37)

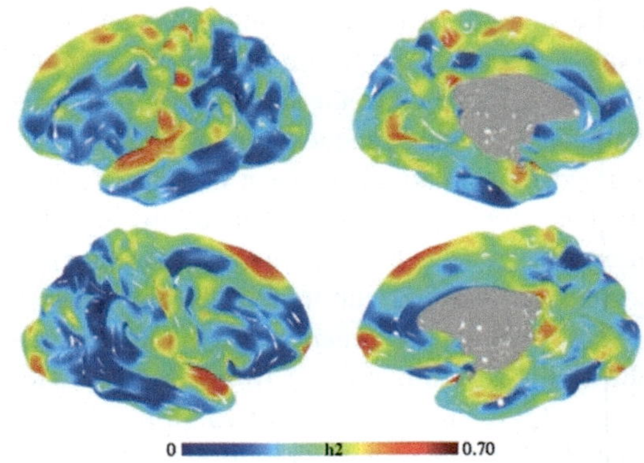

Fig. 13.2 Heritability of pointwise cortical thickness. Image and genetic analyses were performed analogously to those of Fig. 13.1

and 13.2, Winkler et al. 2012). Doing so, alleviates any undue influence of assuming the genetic underpinnings of the cortex correspond to Brodmann Areas.

The conscientious student may draw similarities between this pointwise approach and the voxelwise genome-wide association approach that was criticized in Sect. 13.1. It is important to note that the goal of the analytic techniques used to create Figs. 13.1 and 13.2 is not to identify genes, but to identify heritable traits (i.e. brain regions) that cluster genetically and will therefore have more power for subsequent gene discovery. Such a pointwise approach contributed to the search for genetic roots of the brain by providing phenotypes for the first cortical atlas constructed entirely from genetic information (Chen et al. 2012). In this extremely elegant work, Chen and colleagues used a fuzzy clustering technique in 406 twins to parcel cortical surface area into genetic subdivisions. Boundaries of the cortical map corresponded to meaningful structural and functional organization. Therefore, the Chen subdivisions represent traits that will have greater statistical power for gene identification studies than phenotypes that are nothing more than products of image processing (e.g., individual voxels or vertices).

13.3.3 Genetic Influence on White Matter

Cerebral white matter tracts, or fasciculi, consist primarily of glial cells, myelin, and axons that transmit signals from one region of the cerebrum to another and between the cerebrum and lower brain centers.

Kochunov et al. (2010) demonstrated a significant genetic influence on cerebral white matter in 467 subjects from extended pedigrees. White matter heritability for fractional anisotropy [FA, a measure of white matter integrity (Beaulieu 2002)] averaged across the whole brain was 0.53, $p = 2 \times 10^{-7}$. Figure 13.3 depicts voxel-level heritability estimates projected onto the white matter skeleton. Evidence for genetic control was relatively higher in the inferior fronto-occipital fasciculus ($h^2 = 0.74$), the anterior corona radiate ($h^2 = 0.84$), genu ($h^2 = 0.73$), and the superior longitudinal fasciculus ($h^2 = 0.81$). Heritability estimates were consistently higher for left hemisphere regions than their contralateral area, inline with observations that left hemisphere FA-values are less variable than those on the right (Hua et al. 2009).

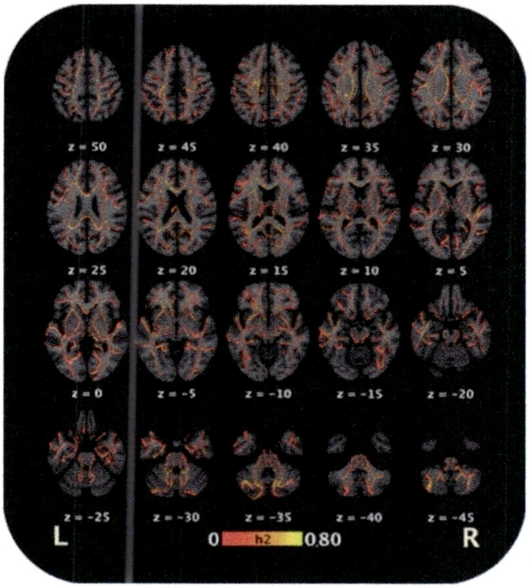

Fig. 13.3 Voxel-level heritability estimates of white matter tract microstructure are presented in standard brain space. Heritability estimates varied from 0 to 0.80 and indicate genetic control of FA-values throughout cortex

Genetic correlations for analogous tracts in left and right hemispheres were high, indicating that common genes influence contralateral tracts. Genetic correlations between the corpus callosum and the other white matter tracts were significant, with the exception of the internal capsule and the cingulum. However, the internal capsule and the cingulum were genetically correlated with each other and other tracts, providing evidence for pleiotropy between different tracts.

This data suggests that voxel-level FA-values are influenced by genetic factors, the microstructure of the major white matter tracts is heritable and that partially overlapping, but not completely common, genetic factors control axonal anatomy of these tracts. These findings are consistent with the notion that a relatively large set of genes influence white matter microstructure and that these genes are not common to all observed white matter tracts. Rather, some tracts are influenced by relatively unique genetic factors. These findings imply that diffusion-based genetic studies of brain-related illnesses should focus on the tract or tracts implicated in disorders, rather than genes that may influence white matter more generally.

13.3.4 Genetic Influence on Functional Connectivity in the Default-Mode Network

When the brain is not engaged in specific tasks, spontaneous fluctuations in neuronal activity give rise to coherent and structured connectivity networks (Biswal et al. 1995; Fox and Raichle 2007, Beckmann et al. 2005), as identified through connectivity analyses with functional MRI and PET. One network, termed the default mode (Raichle et al. 2001), is believed to support self-referential or nondirected cognitive processing (Gusnard et al. 2001) and thought to characterize basal neural activity. Aberrant default-mode connectivity has been reported in individuals with a host of neurological and psychiatric illnesses, suggesting that this intrinsic network is sensitive to pathophysiologic alterations in brain function and structure (Broyd et al. 2009).

Glahn et al. (2010) demonstrated significant genetic influence ($h^2 = 0.42$, $p = 4.6 \times 10^{-3}$) over default-mode functional connectivity independent of genetic influence on regional gray matter density in 333 subjects from extended pedigrees (Glahn et al. 2010). Establishing the heritability of default-mode functional connectivity authorizes the use of resting-state networks as phenotypes in the search for the genetic roots of illnesses that have been associated with altered default-mode connectivity. Furthermore, identification of the genes that influence the intrinsic functional architecture of the human brain would represent a significant advance for basic neuroscience, independent of the ramifications for brain disorders.

Because the default-mode reflects a "baseline" system, it is plausible that the genes that influence default-mode connectivity also contribute to general regulation of brain metabolism, cerebral blood flow, or other aspects of basic neuronal activity. Identification of these genes will provide an important vantage point for understanding the brain's intrinsic architecture and the influence that those systems have on a host of neurological and psychiatric illnesses. Future studies mapping and identifying the actual quantitative trait loci will provide insight into the genes that influence default-mode functional connectivity.

13.3.5 Cognitive Ability: Genetic Influence on Intelligence

Evidence in favor of pleiotropic effects on various anatomic phenotypes and neurocognitive function has been reported, however, there is little work examining pleiotropic influences on brain morphology, network activity, and neurocognitive variation in triad. Thompson et al. (2001) provided preliminary evidence that prefrontal gray matter density and general cognitive ability covary in healthy twins. These findings were extended by Posthuma et al. (2002) who applied a formal bivariate correlational analysis, concluding that gray and white volume matter and intelligence are mediated by a common set of genes. Since, this same group reported that a single underlying genetic factor mediates working memory ability and global gray and white matter volumes. In contrast processing speed was genetically related only to white matter volume (Posthuma et al. 2003). More recently, performance on a spatial delayed response task and integrity of the superior longitudinal fasciculus were found to share common genetic factors (Karlsgodt et al. 2010). Together, these reports provide strong evidence for overlap between neurocognitive and neuroanatomic phenotypes.

13.4 Initial Conclusions

A decade after the decade of the brain and a decade after the sequencing of the human genome, many thought more would have been discovered. The most glaring nonevent, given the emphasis and allocation of resources, is the general lack of early diagnosis, treatment or prevention of complex disorders. Indeed, the density and combinatorial nature of the two fields has proven immense. Yet and still, many efforts are underway to define measurements from cortex, subcortical nuclei, white matter, and functional connectivity for use as phenotypes in imaging-genetics studies. In time, a systematic program for discovery will yield genetic roots of neuroanatomy and basic brain function.

The remainder of the chapter includes our prospective on the directions that imaging genomics should move in the next decade, pointing out several pitfalls and limitations of the current field.

13.4.1 Over-Reliance on Association and Dysfunction

Human brain mapping relied solely on association of lesion location and neurological deficit for a century after Broca (and others) first made clear associations between structure and function in the 1860s. Investigators observed behavioral deficits, formed hypotheses, and awaited a postmortem autopsy to hunt for lesions in the brain. Due to over-reliance on this method, brain mapping lapsed into a scientific backwater, lasting well into the 1900s. Swapping the brain of that era with the genome of the 2000s, the fields of brain mapping and imaging-genetics employed similar strategies: associations were drawn between dichotomous behavioral traits and a poorly understood entity. Often, a genome wide association study from thousands of case-control subjects was used to nominate candidate genes. Thereafter, functional imaging was used to associate brain traits with a specific variant of those candidate genes. Such a "double association" approach has failed to establish a foundation for further discovery and frequently caused more muddle than clarity in attempted replication studies.

13.4.2 Under-Reliance on Quantitative Traits and Function

Priority and focus must sway from categories of illness toward indices of normal variation. Understanding the genetic influences that determine variation in neuroanatomic structure and connectivity among normal healthy subjects are likely to elucidate how those processes are disrupted in brain illnesses. Because brain measures vary within the normal population, it is possible to localize influential genes in samples of healthy individuals. Such samples could significantly improve our ability to find genes associated with neuroanatomic variability. Identifying such genes would constitute a significant step forward in understanding the biological mechanisms that govern brain anatomy, providing prospective a priori hypotheses for testing in clinical populations. With properly defined quantitative traits, this will lead to superior gene discovery efforts.

13.4.3 Relation to Gene Discovery

Most of the studies discussed herein do not provide information concerning the identity of causal genes. However, they do provide substantial evidence that there are genes involved in the variability of brain structure and function, and that image-based biometrics are sensitive to genetic mediation. Identification of the underlying genes will provide an important vantage point for understanding the brain's intrinsic structural architecture and the influence that it has on other domains of neuroscience, including clinical impairment.

Showing significant heritability provides critical information necessary before these methods can be appropriately used in studies designed to identify or functionally characterize genes. The identification of one or more genes that influence gross anatomy should provide a causal point in the biological chain that governs variation in anatomical features across individuals. The discovery of such genes could dramatically improve our understanding of how molecular processes influence structure–function relationships throughout the brain. This, in turn, should provide important leads for how these processes are disrupted in illnesses associated with aberrant anatomical traits. The characterization of normal genetic influence in phenotypes relevant to fundamental neuroscience is the initial step toward this vast discovery process (Glahn et al. 2007).

13.5 Implications for the Immediate Future

13.5.1 Lessons Learnt

Some parallels between the fields of brain mapping and imaging-genetics are unavoidable. Others, particularly those that have proven detrimental for brain mapping, should be avoided at all costs by imaging-genetics researchers. Already discussed was an over-reliance on dysfunction, and the lesion method in particular; unfortunately, it is too late to avoid this wave of influence. Another parallel is over-localization of function. Neuroscientists, to some degree, still suffer from the "Grandmother cell" dogma where the sole function of a hypothetical neuron was theorized to identify one's grandmother (Konorski 1967). More fashionably, recent reports have adopted the term "Jennifer Aniston neuron" (Quiroga 2012). From this unfounded line of thought, imaging-geneticists must take caution in implicating single genes or SNPs for highly complex (and conceptualized) function. Rather, the field should take note of the breakthrough that has taken place in many fields and embrace the network-of-genes concept over the single-gene concept.

13.5.2 FMRI

FMRI is slowly becoming a one-stop-shop in brain mapping research. Limitations for use in imaging-genetics research must be considered. As our goal is to characterize phenotypes that will eventually lead to the discovery of causal gene sets, the extraction of highly stable traits is a prime directive. Paradigm-based FMRI is intrinsically state-dependent and less stable than structural and resting-state MRI. Typically, functional imaging data is averaged across subjects to improve signal to noise ratio because individual subject data can be sporadic. Furthermore, each block of FMRI data is only indicative of a single paradigm, meaning separate scans would have to be acquired in every subject for every task of interest. To guide gene discovery with task-based FMRI, it will become obligatory to model results from published activation studies to identify the most stable and consistent paradigm-induced activation patterns.

13.5.3 Meta-Analysis

Meta-analytic uses of functional imaging data are more reliable. Recently, the BrainMap database (www.brainmap.org) was used to guide a study seeking genetic influence of general cognitive ability. Specifically, regions corresponding to activations induced by working memory tasks were defined meta-analytically. The boundaries of these regions were then exported to a separate cohort for subsequent analyses (Karlsgodt et al. 2010). This work provides proof-of-concept that the spatial extent of paradigm task activations predicted by models of published results can be used to lessen the search space in studies conducted in independent populations. However, it remains to be seen whether the results of Karlsgodt and colleagues would have been improved had FMRI data been acquired on a per subject basis.

Recently, independent component analysis on the entire BrainMap database was used to extract functional connectivity networks (FCNs). The same FCNs were then shown to closely correspond to resting-state networks extracted from thirty subjects, entirely independent of BrainMap (Smith et al. 2009). This groundbreaking finding provides compelling evidence for the coherence of FCNs extracted from resting state data and networks activated by behavioral and cognitive challenges. Because meta-analytic results pool information from many studies, they can be used to guide genetic analysis of structural MRI perhaps with more stability and power than traditional functional MRI. Furthermore, using resting state data in conjunction with meta-analytic results to investigate genetic influence of networks that correspond to task activations is a powerful, cutting edge construct.

13.6 Implications for the Distant Future

13.6.1 Epigenetics

Neuroplasticity is partially modulated by genetic factors and partially modulated by epigenetics, which are dynamic changes that influence the expression of genes without changing the DNA sequence. Epigenetic processes are of particular clinical interest because their external triggers (e.g., diet, drug abuse, and stress) can affect a person's vulnerability to many diseases, including psychiatric disorders. This fledgling field is a natural progression of genetic and environment influence that will gain momentum as our knowledge of gene function improves.

13.6.2 Social Science

The human brain is particularly sensitive to social stimuli. Some feel this has accelerated the rate of human brain evolution in that humans have complex neuronal circuitry for processing interactive social information (i.e. predicting others' reactions and emotions and responding appropriately). Research has revealed that parenting style and early-life stress can epigenetically modify the expression of genes that influence brain morphology and function (Weaver et al. 2004). Such findings may seem far-fetched, considering we do not fully understand the function(s) of genes whose expression levels are reportedly influenced. However, we should not expect the diversity of implications to have bounds.

References

Agartz I, Okuguwa G, Nordstrom M, Greitz D, Magnotta V, Sedvall G (2001) Reliability and reproducibility of brain tissue volumetry from segmented MR scans. Eur Arch Psychiatry Clin Neurosci 251:255–261

Almasy L, Blangero J (1998) Multipoint quantitative-trait linkage analysis in general pedigrees. Am J Hum Genet 62:1198–1211

Almasy L, Dyer T, Blangero J (1997) Exploiting pleiotropy to map genes for oligogenic phenotypes using extended pedigree data. Genet Epidemiol 14:975–980

Bartley AJ, Jones DW, Weinberger DR (1997) Genetic variability of human brain size and cortical gyral patterns. Brain 120:257–269

Beaulieu C (2002) The basis of anisotropic water diffusion in the nervous system—a technical review. NMR Biomed 15:435–455

Beckmann CF, DeLuca M, Devlin JT, Smith SM (2005) Investigations into resting-state connectivity using independent component analysis. Philos Trans R Soc Lond B Biol Sci 360:1001–1013

Biswal B, Yetkin FZ, Haughton VM, Hyde JS (1995) Functional connectivity in the motor cortex of resting human brain using echo-planar MRI. Magn Reson Med 34:537–541

Biswal BB, Mennes M, Zuo XN et al (2010) Toward discovery science of human brain function. Proc Natl Acad Sci USA 107:4734–4739. doi:10.1073/pnas.0911855107

Blangero J (2004) Localization and identification of human quantitative trait loci: king harvest has surely come. Curr Opin Genet Dev 14:233–240

Broyd SJ, Demanuele C, Debener S, Helps SK, James CJ, Sonuga-Barke EJ (2009) Default-mode brain dysfunction in mental disorders: a systematic review. Neurosci Biobehav Rev 33:279–296

Chen CH, Gutierrez ED, Thompson W, Panizzon MS, Jernigan TL, Eyler LT, Fennema-Notestine C, Jak AJ, Neale MC, Franz CE, Lyons MJ, Grant MD, Fischl B, Seidman LJ, Tsuang MT, Kremen WS, Dale AM (2012) Hierarchical genetic organization of human cortical surface area. Science 335:1634–1636

Cheverud JM, Falk D, Hildebolt C, Moore AJ, Helmkamp RC, Vannier M (1990a) Heritability and association of cortical petalias in rhesus macaques (*Macaca mulatta*). Brain Behav Evol 35:368–372

Cheverud JM, Falk D, Vannier M, Konigsberg L, Helmkamp RC, Hildebolt C (1990b) Heritability of brain size and surface features in rhesus macaques (*Macaca mulatta*). J Hered 81:51–57

Dale AM, Fischl B, Sereno MI (1999) Cortical surface-based analysis. I: Segmentation and surface reconstruction. Neuroimage 9:179–194

Day-Williams AG, Blangero J, Dyer TD, Lange K, Sobel EM (2011) Linkage analysis without defined pedigrees. Genet Epidemiol 35:360–370

Fischl B, Sereno MI, Dale AM (1999) Cortical surface-based analysis. II: Inflation., flattening., and a surface-based coordinate system. Neuroimage 9:195–207

Fox MD, Raichle ME (2007) Spontaneous fluctuations in brain activity observed with functional magnetic resonance imaging. Nat Rev Neurosci 8:700–711

Geschwind DH, Miller BL, DeCarli C, Carmelli D (2002) Heritability of lobar brain volumes in twins supports genetic models of cerebral laterality and handedness. Proc Natl Acad Sci USA 99:3176–3181

Glahn DC, Paus T, Thompson PM (2007) Imaging genomics: mapping the influence of genetics on brain structure and function. Hum Brain Mapp 28:461–463

Glahn DC, Winkler AM, Kochunov P, Almasy L, Duggirala R, Carless MA, Curran JC, Olvera RL, Laird AR, Smith SM, Beckmann CF, Fox PT, Blangero J (2010) Genetic control over the resting brain. Proc Natl Acad Sci 107:1223–1228

Glahn DC, Curran JE, Winkler AM, Carless MA, Kent JW Jr, Charlesworth JC, Johnson MP, Göring HH, Cole SA, Dyer TD, Moses EK, Olvera RL, Kochunov P, Duggirala R, Fox PT, Almasy L, Blangero J (2012) High dimensional endophenotype ranking in the search for major depression risk genes. Biol Psychiatry 71:6–14

Gogtay N, Giedd JN, Lusk L, Hayashi KM, Greenstein D, Vaituzis AC, Nugent TF 3rd, Herman DH, Clasen LS, Toga AW, Rapoport JL, Thompson PM (2004) Dynamic mapping of human cortical development during childhood through early adulthood. Proc Natl Acad Sci USA 101:8174–8179

Gottesman II, Gould TD (2003) The endophenotype concept in psychiatry: etymology and strategic intentions. Am J Psychiatry 160:636–645

Gusnard DA, Akbudak E, Shulman GL, Raichle ME (2001) Medial prefrontal cortex and self-referential mental activity: relation to a default mode of brain function. Proc Natl Acad Sci USA 98:4259–4264

Hua K, Oishi K, Zhang J, Wakana S, Yoshioka T, Zhang W, Akhter KD, Li X, Huang H, Jiang H, van Zijl P, Mori S (2009) Mapping of functional areas in the human cortex based on connectivity through association fibers. Cereb Cortex 19:1889–1895

Julin P, Melin T, Andersen C, Ispberg B, Svensson L, Wahlund LO (1997) Reliability of interactive three-dimensional brain volumetry using MP-RAGE magnetic resonance imaging. Psychiatry Res 76:41–49

Karlsgodt KH, Kochunov P, Winkler AM, Laird R, Almasy L, Duggirala R, Olvera RL, Fox PT, Blangero J, Glahn DC (2010) A multimodal assessment of the genetic control over working memory. J Neuroscience 30:8197–8202

Kochunov P, Glahn DC, Lancaster JL, Winkler AM, Smith S, Thompson PM, Almasy L, Duggirala R, Fox PT, Blangero J (2010) Genetics of microstructure of cerebral white matter using diffusion tensor imaging. Neuroimage 53:1109–1116

Konorski J (1967) Integrative activity of the brain; an interdisciplinary approach. University of Chicago Press, USA

Lander ES, International Human Genome Sequencing Consortium: Eric S Lander et al (2001) Initial sequencing and analysis of the human genome. Nature 409:860–921

Lemieux L, Hagemann G, Krakow K, Woermann FG (1999) Fast, accurate and reproducible automatic segmentation of the brain in T1-weighted volume MRI data. Magn Reson Med 42:127–135

Lerch JP, Evans AC (2005) Cortical thickness analysis examined through power analysis and a population simulation. Neuroimage 24:163–173

Mazziotta J, Toga A, Evans A, Fox P, Lancaster J, Zilles K, Woods R, Paus T, Simpson G, Pike B, Holmes C, Collins L, Thompson P, MacDonald D, Iacoboni M, Schormann T, Amunts K, Palomero-Gallagher N, Geyer S, Parsons L, Narr K, Kabani N, Le Goualher G, Boomsma D, Cannon T, Kawashima R, Mazoyer B (2001) A probabilistic atlas and reference system for the human brain: International Consortium for Brain Mapping (ICBM). Phil Trans R Soc Lond 356:1293–1322

Mazziotta JC, Toga AW, Evans A, Fox P, Lancaster J (1995) A probabilistic atlas of the human brain: theory and rationale for its development. NeuroImage 2:89–101

Panizzon MS, Fennema-Notestine C, Eyler LT, Jernigan TL, Prom-Wormley E, Neale M, Jacobson K, Lyons MJ, Grant MD, Franz CE, Xian H, Tsuang M, Fischl B, Seidman L, Dale A, Kremen WS (2009) Distinct genetic influences on cortical surface area and cortical thickness. Cereb Cortex 19:2728–2735

Peper JS, Brouwer RM, Boomsma DI, Kahn RS, Hulshoff Pol HE (2007) Genetic influences on human brain structure: a review of brain imaging studies in twins. Hum Brain Mapp 28:464–473

Plomin R, Haworth C, Davis O (2009) Common disorders are quantitative traits. Nature 10:872–878

Posthuma D, Baare WF, Hulshoff Pol HE, Kahn RS, Boomsma DI, De Geus EJ (2003) Genetic correlations between brain volumes and the WAIS-III dimensions of verbal comprehension, working memory, perceptual organization, and processing speed. Twin Res 6:131–139

Posthuma D, De Geus EJ, Baare WF, Hulshoff Pol HE, Kahn RS, Boomsma DI (2002) The association between brain volume and intelligence is of genetic origin. Nat Neurosci 5:83–84

Quiroga RQ (2012) The Jennifer Aniston Neuron. Borges and memory—encounters with the human brain. Massachusetts Institute of Technology, Cambridge, pp 159–180

Raichle ME, MacLeod AM, Snyder AZ, Powers WJ, Gusnard DA, Shulman GL (2001) A default mode of brain function. Proc Natl Acad Sci USA 98:676–682

Rakic P (1988) Specification of cerebral cortical areas. Science 241:170–176

Rogers J, Kochunov P, Lancaster J, Shelledy W, Glahn D, Blangero J, Fox P (2007) Heritability of brain volume, surface area and shape: an MRI study in an extended pedigree of baboons. Hum Brain Mapp 28:576–583

Rubenstein JLR, Rakic P (1999) Genetic control of cortical development. Cereb Cortex 9:521–523

Rubenstein JLR, Anderson S, Shi L, Miyashita-Lin E, Bulfone A, Hevner R (1999) Genetic control of cortical regionalization and connectivity. Cereb Cortex 9:524–532

Smith SM, Fox PT, Miller KL, Glahn DC, Fox PM, Mackay CE, Filippini N, Watkins KE, Toro R, Laird AR, Beckmann CF (2009) Correspondence of the brain's functional architecture during activation and rest. Proc Natl Acad Sci USA 106:13040–13045

Sowell ER, Peterson BS, Thompson PM, Welcome SE, Henkenius AL, Toga AW (2003) Mapping cortical change across the human life span. Nat Neurosci 6:309–315

Stein JL, Hua X, Lee S, Ho AJ, Leow AD, Toga AW, Saykin AJ, Shen L, Foroud T, Pankratz N, Huentelman MJ, Craig DW, Gerber JD, Allen AN, Corneveaux JJ, Dechairo BM, Potkin SG, Weiner MW, Thompson P (2010) Alzheimer's Disease Neuroimaging Initiative: Voxelwise genome-wide association study (vGWAS). Neuroimage 53:1160–1174. doi:10.1016/j.neuroimage.02.032

Thompson PM, Cannon TD, Narr KL, van Erp T, Poutanen VP, Huttunen M, Lönnqvist J, Standertskjöld-Nordenstam CG, Kaprio J, Khaledy M, Dail R, Zoumalan CI, Toga AW (2001) Genetic influences on brain structure. Nat Neurosci 4:1253–1258

Thompson PM, Sowell ER, Gogtay N, Giedd JN, Vidal CN, Hayashi KM, Leow A, Nicolson R, Rapoport JL, Toga AW (2005) Structural MRI and brain development. Int Rev Neurobiol 67:285–323

Venter JC, Adams MD, Myers EW et al (2001) The sequence of the human genome. Science 291:1304–1351

Weaver ICG, Cervoni N, Champagne FA, D'Alessio AC, Sharma S, Seckl JR et al (2004) Epigenetic programming by maternal behavior. Nat Neurosci 7:847–854

Winkler AM, Kochunov P, Blangero J, Almasy L, Zilles K, Fox PT, Duggirala R, Glahn DC (2010) Cortical thickness or grey matter volume? The importance of selecting the phenotype for imaging genetics studies. Neuroimage 53:1135–1146

Winkler AM, Sabuncu MR, Yeo BT, Fischl B, Greve DN, Kochunov P, Nichols TE, Blangero J, Glahn DC (2012) Measuring and comparing brain cortical surface area and other areal quantities. Neuroimage 61:1428–1443

Wright IC, Sham P, Murray RM, Weinberger DR, Bullmore ET (2002) Genetic contributions to regional variability in human brain structure: methods and preliminary results. Neuroimage 17:256–271

14 Variation, Genetics, and Evolution of the Primate Craniofacial Complex

Richard J. Sherwood and Dana L. Duren

14.1 Introduction

A remarkable discovery was recently announced regarding the genetic influence on the vertebrate craniofacial complex (Abzhanov et al. 2006; Campas et al. 2010). The subject of the study was the genus *Geospiza*, better known as Darwin's finches, the poster genus for evolutionary adaptation. It is well known that the beaks of the various species of these finches vary in depth, width, and length, and that the resulting shapes correspond with the ecological niche of the particular bird. In 2006, Abzhanov and colleagues described how different levels of expression of calmodulin (CaM), a calcium mediator, account for the variation in beak length (Abzhanov et al. 2006). Following previous work demonstrating that variance in beak depth and width was similarly described by levels of bone morphogenetic proteins-4 (BMP4) (Abzhanov et al. 2004), this work provides an elegant description of the genetic mechanism of morphological differentiation of craniofacial structures. While, in one sense, a beak is a discrete anatomical unit, it is also true that it is a complex of multiple hard and soft tissues with geometric properties extending beyond length, depth, and width. The significance of this work lies in the identification of the relationship between, and relative independent action of, CaM and BMP4 with respect to specific metric traits.

In contrast to the advances in avian cranial genetics, the genetic mechanisms responsible for variation of the primate craniofacial complex are still poorly understood. The current understanding of the genetic underpinnings of the primate craniofacial complex comes primarily from three sources, extrapolation from developmental studies of fish or avian animal models, analysis of dysmorphic syndromes in humans, or from the application of modern quantitative genetic approaches including genome-wide linkage analyses. In this chapter, we explore the genetic influences on primate craniofacial morphology and examine the relevance to diverse fields from evolutionary biology to biomedicine.

14.2 Primate Craniofacial Diversity

The order *Primates* is represented by roughly 400 species exhibiting great diversity in body size, locomotor habit, and environmental

R.J. Sherwood (✉)
Division of Morphological Sciences and Biostatistics, Department of Community Health, Department of Pediatrics, Boonshoft School of Medicine, Wright State University, 3171 Research Blvd, Kettering, OH 45420, USA
e-mail: Richard.sherwood@wright.edu

D.L. Duren
Division of Morphological Sciences and Biostatistics, Department of Community Health, Department of Orthopaedic Surgery, Boonshoft School of Medicine, Wright State University, Kettering, OH 45420, USA
e-mail: Dana.duren@wright.edu

R. Duggirala et al. (eds.), *Genome Mapping and Genomics in Human and Non-Human Primates*, Genome Mapping and Genomics in Animals 5, DOI 10.1007/978-3-662-46306-2_14,

adaptation. Craniofacial trends in primate evolution have included changes in orbital morphology and orientation related to an increased emphasis on visual cues, and a relative increase in cranial capacity (Ross and Ravosa 1993; Ross 1996). Figure 14.1 provides examples of craniofacial form in two cercopithecoid primates (vervet monkey and baboon), a ceboid (spider monkey), and a prosimian (indri). Most primates exhibit a generalized mammalian cranial form, although there are interesting exceptions such as the beaver-like aye–aye (*Daubentonia*).

Some of the most dramatic evolutionary changes in primate craniofacial form are seen among the *Hominini,* the tribe including humans and their ancestors. These include significant changes in each of the craniofacial components, most notably the dramatic expansion of the brain and neurocranium, the concomitant increase in flexion of the basicranium at the pituitary fossa (the craniometric point known as sella), and a reduction in dimensions of the splanchnocranium with a resulting orthognathic disposition of the face. Figure 14.2 presents a comparison of bisected human and chimpanzee crania where these differences are readily apparent.

14.3 Background

14.3.1 Structure and Development

The skull (cranium and mandible) is a complex anatomical structure, with a developmental history that includes osteogenic precursors derived from both neural crest cells and mesoderm, and a

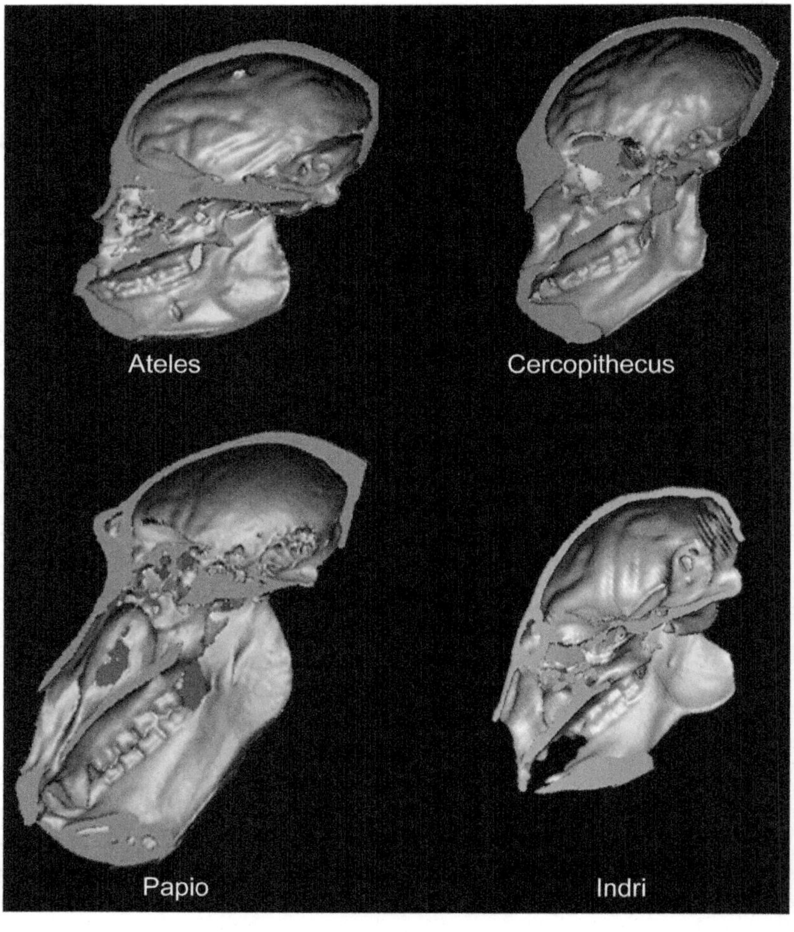

Fig. 14.1 CT reconstructions of the internal aspect of four primate taxa. All images are scaled to the distance from sella (the pituitary fossa) to nasion (the intersection of the nasal and frontal bones)

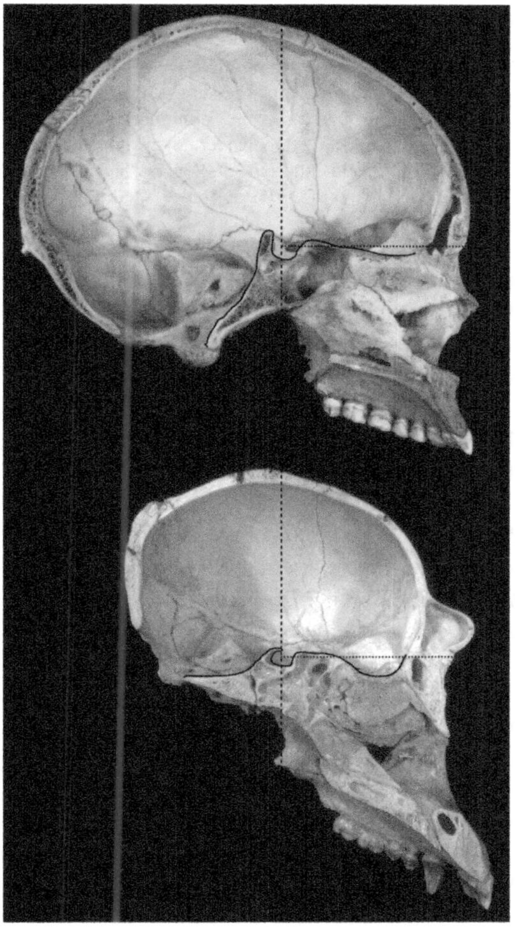

Fig. 14.2 Internal aspect of bisected human (*top*) and chimpanzee crania. Crania are aligned at sella (the pituitary fossa indicated by *vertical line*) and scaled to the distance from sella to nasion (*horizontal line*). Superior margin of basicranium is outlined in *black*

functional constituency including the housing of a diverse array of sensory and mechanical components. The craniofacial complex is frequently discussed in terms of developmental and functional components. The basicranium, which includes the sphenoid, ethmoid, and portions of the occipital and temporal bones, is phylogenetically the oldest component and is dominated by endochondral ossification during development. The neurocranium is identified as those bones surrounding the brain, such as the parietals and the squamous portions of the frontal, temporal, and occipital bones. The splanchnocranium is dominated by the zygomatics, maxillae, and mandible but also includes numerous small bones such as the nasals, the lacrimals, and the unpaired midline vomer. Of these components, the basicranium largely undergoes endochondral ossification; the neurocranium and splanchnocranium are predominately formed through intramembranous ossification, although several bones of the splanchnocranium demonstrate both forms of ossification (e.g., the mandible). It has been suggested that growth of bones derived from these two processes differs, with intramembranous bone largely governed by the surrounding mechanical environment, and endochondral bone regulated by the intrinsic genetic program within cartilaginous precursors (Enlow 1990; Lieberman et al. 2000). This suggested dichotomy, however, was problematic from the start because several cranial bones, such as the sphenoid, occipital, and mandible, utilize both forms of ossification for specific regions (Langille and Hall 1993).

The functional environment of the skull and surrounding soft tissues are also complex. Externally, the bones of the skull are subjected to biomechanical stresses imposed by nuchal, masticatory, and facial musculature and their associated tendons and fasciae. Internally, neurocranial growth has been hypothesized to be directed by brain size as well as fiber orientation of the meninges (Moss and Young 1960). In addition, the epithelium of the paranasal sinuses and the air spaces of the temporal bone may ultimately play a role in configuration of the associated bones and the distribution of mechanical strains within them (Sherwood 1999; Witmer 1997).

14.3.2 Paradigms for Genetic Research of Craniofacial Morphology

The past two decades have seen a considerable transition in the biological sciences largely as a result of the advances in genomic research. Craniofacial research, and most notably research into craniofacial anomalies, has moved from categorization of syndromes based on phenotypic patterns to the identification of specific gene mutations responsible for these syndromes.

While crude surgical approaches to the cranium and intracranial structures appear as early as 6500 BC, systematic interest in the anatomy of the craniofacial complex more likely began with the work of Herophilos (third century BC) or with the comparative anatomical approach of Galen (second century BC). The descriptive nature of anatomical observation was the dominant paradigm well into the nineteenth century, when quantitative analysis of cranial form began with the work of anatomists such as Blumenbach, Retzius, Broca, Morton, and Lombroso. This quantitative work was largely designed to describe differences between racial groups or, as with the case of the biological determinism of Lombroso, to predict potential criminal tendencies in individuals. The first studies in hereditary transmission of craniofacial features began with the pioneering work of Sir Francis Galton in 1875 (e.g., Galton 1885, 1876a, b), who was able to demonstrate heritable aspects of craniofacial form by examining sets of twins. Investigation of the growth of basicranial and intracranial structures began in 1931 with the application of the Bolton method standardizing radiographic technique allowing for consistent quantification of internal cranial structures (Broadbent 1931).

The descriptive paradigm that had dominated craniofacial research began to shift with the landmark paper of Moss and Young (1960), describing a functional approach to craniofacial biology (craniology in their terminology). This approach considered that cranial form closely reflects the functional demands of the associated hard and soft tissues and focused on the physical constraints placed upon the growing cranium. Importantly, functional craniology formalized the concept of the skull as a complex of both integrated and independent components.

As genetic methodology improved, the genetic basis for craniofacial form began to emerge as the dominant research topic. By the early 1980s Slavkin (1983) described the "genetic paradigm," as forming the basis for research into congenital defects. He defines this paradigm as recognizing the interaction between the gene and the environment in producing a phenotype. Importantly he stressed that

> not all traits that appear multiple times in the same family or pedigree are "genetic" in origin, and possible contributions from "non-genetic" factors (like mutagens, carcinogens, teratogens, nutritional status, environmental insults) must always be considered Slavkin 2001, p. 466).

Not surprisingly, with the rapid growth in genetic data, the perceived role of the environment began to diminish shortly thereafter. By the late 1990s, Moss (1997a), the father of functional craniology, was clearly concerned by the lack of consideration of nongenetic influences on craniofacial growth, identifying the "genomic thesis" as the dominant paradigm of morphogenesis. He suggested that the role of the environment was being overlooked in favor of genetic deterministic models, despite significant evidence for epigenetic/genomic interactions throughout development. Such decided shifts in thought are not uncommon following significant technological advances and, over time, there is typically a return to more synthetic approaches incorporating all available lines of evidence. This is currently evident in the increased attempts at a systems biology approach (Ideker et al. 2001a, b; Ideker and Krogan 2012), which is again advocating a more holistic approach integrating environmental, gene, and gene network data to provide a comprehensive view of the system under investigation.

The application of, and the need for, a systems biology approach to craniofacial biology was described in a recent review of gene discovery advances in craniofacial biology. Handrigan et al. (2007, p. 110) noted that current research is characterized by a piecemeal approach "focusing on one stage of development, one part of the face, or on just a few signaling pathways." The multifactorial basis of many syndromes, ranging from craniosynostosis to tooth agenesis, is becoming clear with new genetic components identified on a regular basis. Handrigan et al. note, "These manifold etiologies reflect the overriding integration and complexity of molecular regulation in craniofacial development and emphasize the need for exhaustive surveying of the involved genes and gene pathways." (Handrigan et al. 2007, p. 109–110).

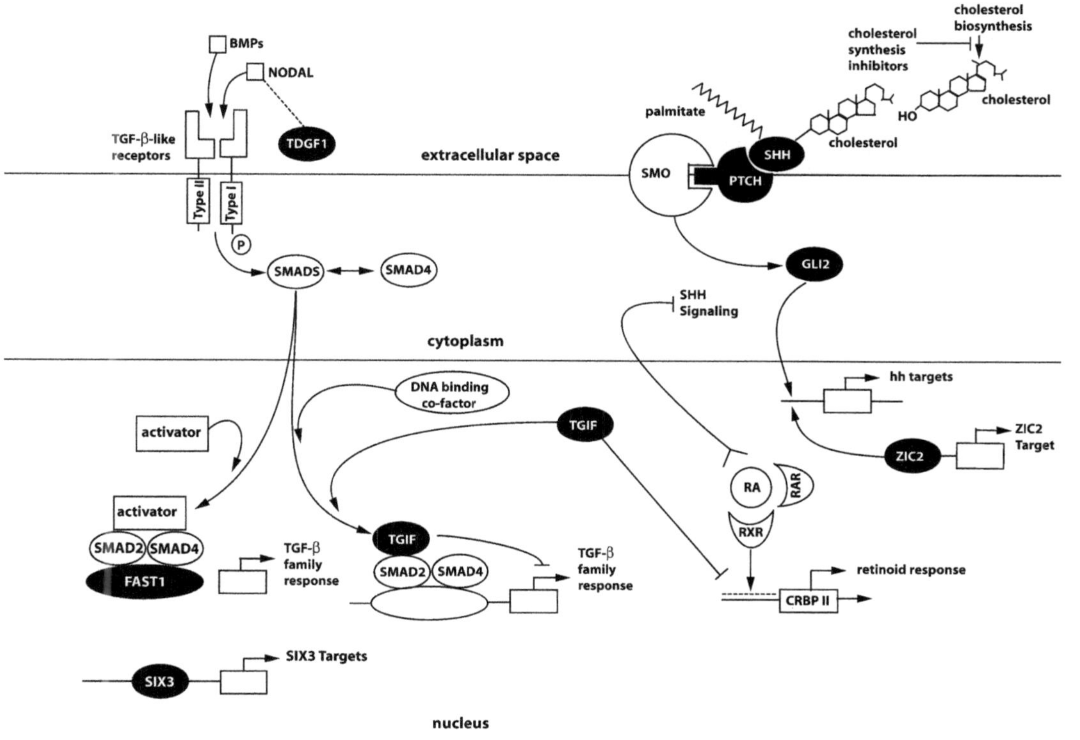

Fig. 14.3 Signaling pathway associated with Holoprosencephaly (HPE). Genes in *black* have been implicated in HPE in humans (after Ming and Muenke 2002)

The systems biology approach stresses the hierarchical nature of biological information and prioritizes the elucidation of gene networks to characterize the system under investigation. With regard to craniofacial biology, a well-developed pathway model has been developed relative to the disorder holoprosencephaly (HPE) (Fig. 14.3). This disorder had been characterized as genetically heterogeneous with at least eight genes being identified as etiologic factors. Additional research has identified the signaling pathways linking these genes, thus identifying the basis for the range of phenotypes seen and the genetic heterogeneity (Gripp et al. 2000; Ming et al. 2002; Ming and Muenke 2002; Orioli et al. 2001; Roessler et al. 2003). Identification of additional such signaling pathways and gene networks is critical for a complete understanding of normal craniofacial development and the etiology of dysmorphologies.

14.4 Genetics of the Craniofacial Complex

14.4.1 Developmental Genetics of the Craniofacial Complex

The genetic contributions to early craniofacial development have been the subject of study for many decades and significant findings are frequent. Not surprisingly, as much of craniofacial development relies upon the proper formation of the underlying skeletal substructure, many of the genes involved in craniofacial development are those that contribute to general skeletal development throughout the body. These include a number of fibroblast growth factors or their receptors (*Fgf* or *Fgfr*), bone morphogenetic proteins (*Bmp*), or signaling molecules such as sonic hedgehog (*Shh*) or the *Wnt* family (Handrigan et al. 2007;

Havens et al. 2008; Helms and Schneider 2003; Hu and Helms 1999). Craniofacial anomalies associated with mutations in these genes frequently occur alongside other skeletal anomalies. For instance, mutations in fibroblast growth factor receptors are the cause of several craniosynostotic disorders (Apert syndrome, Crouzon syndrome), which along with the craniofacial symptoms of premature suture closure, are also characterized by limb anomalies such as syndactyly. Even a relatively discrete craniofacial disorder, such as cleft lip and palate, may be part of a syndrome with multiple postcranial skeletal and soft-tissue symptoms. Phenotypes within the cleft lip and palate spectrum have been associated with *Bmp* signaling, specifically *Bmp4* deficiency, which is also linked to other alterations in facial form (particularly in mandibular morphology) and postcranial dysmorphologies such as syndactyly and polydactyly (Bonilla-Claudio et al. 2012, Murray and Schutte 2004; Naruse et al. 2007; Zhang et al. 2002).

The question then becomes, if the genes above are responsible for large-scale skeletal morphogenesis, what are the factors dictating the intricate details of craniofacial morphogenesis? Part of the answer lies in the action of these genes along spatial or temporal gradients. For instance, variation in *Bmp4* expression has been shown to correlate with variation in beak morphology in *Geospiza* as noted above, and also with differences in cichlid jaw morphology (Albertson et al. 2003; Albertson and Kocher 2006, reviewed in Helms et al. 2005) and in tooth and palate development in mice (Feng et al. 2002; Gong and Guo 2003). The other part of the answer may lie in additional, currently unknown, genes with smaller, more localized effects.

14.4.2 Genetic Heterogeneity in Dysmorphic Syndromes

When examining the current literature one cannot help but be impressed by the wealth of detailed genetic information that is rapidly becoming available for cranial disorders (e.g., Cohen 2002; Hennekam et al. 2010; Mulliken 2002). It is also clear, however, that the advances made in the genetics of craniofacial disorders do not provide unambiguous answers to questions of causation. For instance, Cohen (2002) lists at least six disorders resulting from mutations in the *FGFR3* gene, at least five disorders associated with mutations in *FGFR2*, and at least nine separate mutations associated with holoprosencephaly (Ming and Muenke 2002). In other words, disorders such as Crouzon or Pfeiffer syndromes, along with other craniosynostotic syndromes, are not distinct entities but rather variable manifestations along a continuous scale. This heterogeneity has made some researchers suggest that, instead of numerous individual distinct syndromes, there are only a handful of syndromes each with considerable variation along a continuum. This idea has largely been rejected, as syndromes do tend to present a definable set of symptoms that breed true in families. Cohen and MacLean (1999) suggest several ways to integrate phenotypic and genotypic nomenclature that are likely to become standard practice as we continue to elucidate these relationships. While their system may be a bit cumbersome (e.g., the simple Crouzon syndrome would be replaced by "Crouzon syndrome, FGFR2, Cys278Phe"), such a system may become necessary for clarity.

In discussing the problems associated with this genetic heterogeneity, Cohen (2002, p. 9) states that "other factors are involved that are not understood at the present time." There are two clear candidates for these other factors: (1) the environment; or (2) other, currently unknown, genes. Environmental insults resulting in growth perturbation or gross anatomical deformities are relatively commonly encountered in utero and range from mechanical disruptions, such as amniotic bands, to complications based on placental-cord insufficiencies, to the introduction of teratogenic substances (Cohen 1990; Cox 2004; Moss 1997b; Sherwood et al. 1992, 1997). The subtle effects of a "normal" environment (acknowledging the extreme heterogeneity of any individual's environment) on variability are less easily characterized.

The other potential confounding factor in understanding the genetics of dysmorphology is the relationship of mutated genes with other genes. While it is readily acknowledged that complex traits are often oligogenic in nature (i.e., a few genes with pronounced and identifiable effects of varying degrees are together responsible for most of the genetic contribution to the phenotypic variance of a trait), there still persists an expectation that a given mutation will produce a singular outcome. Even if the (nongenetic) environment were held constant, this expectation would not be warranted. The cumulative pleiotropic effects of genes and gene-by-gene interactions would be expected to produce a wide range of phenotypes proportional to the number of genes involved. In other words, variability among normal genes would be expected to produce variable phenotypes when acting in concert with a mutated gene. The basic genetics underlying normal variation of the craniofacial complex are not well defined but clearly important for continued progress.

14.4.3 Animal Models for Human Craniofacial Genetics

A number of animal models have been used to explore the genetic underpinnings of craniofacial structures. Zebrafish and chicks have been used extensively to study the genetics influencing early development of important structures such as the pharyngeal arch system (Helms and Schneider 2003; Yelick et al. 1996; Yelick and Schilling 2002). Murine models have also proven important especially for understanding the genetics of the dentition and palate (Jernvall et al. 1998; Jernvall and Thesleff 2000; Miettinen et al. 1999; Vaahtokari et al. 1996). Mammalian models are important for understanding aspects of human craniofacial genetics such as the integration or modularity of the cranium (e.g., Cheverud 1995).

Nonhuman primates, given their phylogenetic proximity to humans, would serve as the best model. The craniofacial complex of nonhuman primates has been the subject of numerous anatomical studies (e.g., Hylander 1979, 1986; Ravosa et al. 2000; Ross and Hylander 2000; Ross 2001; Vinyard et al. 2003; Washburn 1947). Much of this research has been aimed at elucidating the evolutionary history of the order by understanding how craniofacial components, the basicranium, neurocranium, and splanchnocranium, are integrated in both a developmental and evolutionary sense. Within primates, a number of associations between the basicranium and other structures have been suggested. As the basicranium serves as the floor to the neurocranium, the most obvious association is between the skeletal elements of the base and the brain. Scientists have long considered brain size and the extent of basicranial flexion to be related in primates. Humans possess both a large brain (relative to body mass) and a strongly flexed cranial base (Lieberman et al. 2000). Within non-human primates, a significant correlation between relative encephalization and cranial base angle has also been demonstrated (Ross and Ravosa 1993). However, not all brain/base relationships are consistent throughout primates. For example, Lieberman et al. (2000) report significant correlations between brain stem volume and cranial base flexion in strepsirrhines (lemurs and lorises) but not in haplorhines (tarsiers, monkeys, apes, and humans).

Associations have also been suggested between basicranial and facial structures such as the orientation of the orbits and the anterior cranial base (Ravosa 1991; Ross and Ravosa 1993). Again, a difference exists in correlations between haplorhines and strepsirrhines with the former being characterized by significant correlations between orbit orientation and the anterior cranial base, most likely due to the close approximation of the orbits below the olfactory tract (Lieberman et al. 2000); McCarthy and Lieberman (2001) have also identified an integrated region they term the "facial block" defined by the superoposterior portions of the face. The facial block is said to rotate about an axis loosely defined by the greater wings of the sphenoid bone during ontogeny. In haplorhines, the orientation of the block is correlated with the cranial base angle. Strepsirrhines do not show this correlation.

14.4.4 Quantitative Genetic Studies of the Craniofacial Complex in Animals

Despite these acknowledged correlations between craniofacial components, it is not clear what elements are the primary determinants driving craniofacial variation in primates. While experimental approaches to primate craniofacial morphology are not practical, quantitative genetic techniques are proving fruitful in elucidating the genetic architecture underlying craniofacial variation. Quantitative genetic analysis of craniofacial traits has primarily focused on two families of primates: *Callitrichidae* represented by the saddle-back tamarin (*Saguinus fuscicollis*) (Cheverud 1995), and *Cercopithecidae* represented by the rhesus macaque (*Macaca mulatta*) (Cheverud and Buikstra 1981a, b, 1982; Cheverud 1982; Cheverud et al. 1990a, b; McGrath et al. 1984) and baboon (*Papio hamadryas ssp.*) (Hlusko et al. 2002; Hlusko and Mahaney 2003). These studies focused on facial, mandibular, and dental traits. Sherwood et al. (2006a, b, c, 2008c, d, 2011) broadened this perspective and included internal measures of the basicranium along with neurocranial and splanchnocranial phenotypes in the baboon.

The first step in quantitative genetic analysis of complex traits is to establish the relative genetic influence on traits. Narrow-sense heritability provides such a measure. Narrow-sense heritability is expressed as

$$h^2 = \sigma_A^2 / \sigma_P^2 \quad (14.1)$$

where σ_A^2 refers to the additive genetic variance and σ_P^2 refers to the total phenotypic variance. In a study by Cheverud (1982) of macaque facial metrics, heritability estimates, calculated using mother-offspring pairs, were moderate ($\sim$0.33). The sample available for this study was drawn from 297 positively identified individuals containing 51 mother–offspring sets with a total of 134 mother–offspring pairs. While the analysis resulted in a number of significant heritabilities, the small sample size may explain why approximately 52 % of the estimates were not significant. A similar study of craniofacial traits in tamarins found heritabilities averaging 0.37 with a range of 0.04–0.94. While the number of related individuals, 134 animals, was slightly less than in the macaque study, extended genealogies were available and the heritabilities were calculated using a maximum-likelihood approach with pedigree data. With this methodology, the number of significant heritabilities increased to 67 %. In a study of dental metrics, using the pedigreed population of baboons at the Texas Biomedical Research Institute (formerly the Southwest Foundation for Biomedical Research), Hlusko and colleagues (Hlusko et al. 2002; Hlusko and Mahaney 2003) report heritabilities ranging from 0.38 to 0.85 for dental metrics of baboons (*Papio*) with all heritabilities significant.

Genetic correlations (ρ_G) provide a means to examine the shared effects of genes on traits. As noted, a number of associations have been described for the primate craniofacial complex at the phenotypic level and these have been further explored at the genetic level using the concept of morphological integration. The concept of morphological integration was formalized in 1958 (Olson and Miller 1958) and is used to describe how the interdependent nature of traits relates to the total complex form of an organism.

In several classic papers Cheverud explored the integration of the primate cranium from phenotypic and genetic perspectives (Cheverud 1982, 1995). In an analysis of the macaque skull, 56 measures were partitioned into function sets (F-sets) based on existing research. Two primary functional matrices, neurocranial and orofacial, were identified with three and four submatrices, respectively (frontal, parietal, occipital in the neurocranial matrix, orbital, nasal, oral, and masticatory in the orofacial matrix). Theoretically, there would be a hierarchical pattern of correlations with the measures in each submatrix and matrix being more correlated than measures spanning submatrices or matrices.

Phenotypically, the expectation of a hierarchical relationship is met. That is, Cheverud reports that the average coefficient of determination (r^2) for traits within the same F-set is more than five times higher than average r^2 values among traits from different F-sets (Cheverud 1982). The relationship was somewhat different, however, when the genetic correlations were examined. The average r^2 for traits within and among F-sets was more similar than that seen with phenotypic correlations, indicating that "F-sets are not necessarily independently evolving entities" (Cheverud 1982, p. 508). When analyzed separately there was a difference between the neurocranial and orofacial sets. Neurocranial traits showed greater correlations within submatrices than between neurocranial submatrices. In contrast, traits within orofacial submatrices tended to show roughly equivalent correlations independent of whether they were within or among other orofacial submatrices.

Using a slightly different design where traits were assigned to one of six sets (oral, nasal, cranial vault, orbit, zygomatic, cranial base), a similar study was conducted on a small New World monkey, the saddle-back tamarin (Cheverud 1995). In this study, cranial vault and oral traits showed higher average levels of genetic correlation (0.49 and 0.66, respectively) to traits within their respective sets than with traits in other sets. Nasal, orbital, cranial base, and zygomatic traits showed no tendency for higher genetic correlations within sets relative to those between sets.

14.4.4.1 Dentition

Within comparative and evolutionary anatomy, the dentition has frequently served as the focus of much research. The reasons for this are multiple. First, teeth are essential to the procurement and mastication of food, as well as for inter- and intraspecific communication. Teeth are discrete elements that are relatively easy to examine in living animals (high-resolution dental casting methods are readily available). The morphology of the teeth varies greatly within primates. Finally, teeth are among the most durable of biological structures and are, therefore, more prone to fossilization than many other elements. As a result, the dentition and jaws provide an excellent source of information regarding adaptations to a given environment and may even provide detailed information on the niche occupied by an animal or even behavioral aspects. It is true that many primate taxa are known largely, if not entirely, by dentition alone.

Primates are heterodontic animals with up to four different tooth types with each type having been described as evolving as "largely independent units" (Weiss et al. 1998, p. 369). Primitive mammalian dental formulas, seen in early primates, consisted of three incisors, one canine, four premolars, and three molars in each quadrant of the jaw. Most modern mammals have reduced the number of teeth within each jaw, with many eliminating some types (e.g., the lack of canines and premolars in rodents).

The development of the dentition is complex with precursors derived from ectoderm (ameloblasts) and neural crest cells (odontoblasts, cementum). The dentition begins development as a series of epithelial ingrowths into the subjacent ectomesenchyme. The presumptive tooth progresses through three well-characterized phases, the bud, cap, and bell stages. It is during the last of these stages, the bell stage, where substantial histo- and morphodifferentiation occurs. By late bell stage, the hard tissue components of the tooth, dentin, and enamel have begun to form and the nerve and vascular supply are beginning to develop (Ten Cate 1989). Permanent dentition begins as successional tooth germs arising from the dental lamina adjacent to the dental organ of the incisors through premolars. Permanent molars have no deciduous precursors and arise from a posterior extension of the dental lamina. Within each stage a number of genes have been identified, which, when disrupted, can result in dental agenesis (e.g., *PAX9* or *MSX1*), dentin dysgenesis (e.g., *COL1A1*, *COL1A2*), or amelogenesis imperfecta (e.g., *AMELX*, *ENAM*) (Hu and Simmer 2007).

Morphogenesis of individual teeth has been studied in the mouse and is largely directed by two signaling centers, the primary and secondary enamel knots (Jernvall et al. 1994; Jernvall and Thesleff 2000; Vaahtokari et al. 1996). The primary enamel knot develops during the transition from the bud to cap stage at the point where epithelial folding begins to define tooth shape

(Cho et al. 2007; Jernvall and Thesleff 2000). In multicusped teeth, the primary enamel knot is removed apoptotically and the secondary enamel knots appear at the site of individual cusps. As noted, the enamel knots are signaling centers and Jernvall and Thesleff (2000) have identified reiterative patterns of expression, particularly in reciprocal *FGF* signaling between the primary and secondary knots and the underlying mesenchyme.

14.4.4.2 Quantitative Genetic Studies of Primate Dentition

In an effort to elucidate the genetic architecture of primate dentition, Hlusko and colleagues have explored the quantitative genetics of dentition in the baboon (Hlusko et al. 2004a, b, 2006). In an analysis of genetic correlations among dental traits there is an expectation of hierarchical relationships similar to those discussed for cranial components. For the dentition, it is hypothesized that antimeric teeth (e.g., left and right first molars) will show a high degree of genetic correlation (with ρ_G approaching or equaling 1.00 indicating complete pleiotropy). Because of the developmental relationship, serial pairs of teeth (e.g., M_1, M_2, etc.) would also be expected to exhibit high levels of genetic correlation, followed by occluding pairs of teeth with slightly lower expectations for genetic correlations.

On examination of molar cusp patterning and cingular remnant expression, the expectations of complete pleiotropy for traits from antimeric teeth were met (Hlusko et al. 2004a; Hlusko and Mahaney 2003). Genetic correlations for cingular remnant traits also showed the expected pattern with a reduction in magnitude from antimeric pairs, to serial pairs, to occluding pairs. Molar cusp patterning showed a slight deviation from the expected patterns wherein many of the serially homologous traits in mandibular molars demonstrated genetic correlations equal to one, the same traits in serial maxillary molars demonstrate incomplete pleiotropy (genetic correlations different from one).

14.4.4.3 Current Work on the Quantitative Genetics of the Human and Nonhuman Craniofacial Complex

We have undertaken three studies designed to elucidate the genetic architecture of the craniofacial complex. Two of these studies are designed to be parallel complementary studies: one examining craniofacial structure in humans (Sherwood et al. 2005, 2008a, 2011), the other in a nonhuman primate, the baboon (Sherwood et al. 2006a, 2008b, c). The third study focuses on the dentognathic complex in humans (Duren et al. 2006, Sherwood et al. 2007). In the first studies, each craniofacial developmental component is characterized by a series of metric traits derived from lateral cephalographs, while the third study uses high-resolution dental casts.

The first study involves participants in the Fels Longitudinal Study (Roche 1992), the largest and longest running study of human growth and development. Throughout the study there has been a concentration on aspects of skeletal growth, most notably on methods of assessing skeletal maturation from hand-wrist and knee radiographs (Roche et al. 1988a, b; Roche 1989; Xi and Roche 1990). Cranial radiography of Fels Longitudinal Study participants was conducted between 1931 and 1982. In keeping with the general focus of the study, primary attention has been on growth and development of cranial components in participants. Several key papers focused on the growth of specific bones or anatomical units, for example, early work by Young (1957) on the frontal and parietal bones, or Garn and Lewis' work on the mandible (Garn et al. 1963; Lewis et al. 1982, 1985). A series of papers also detailed growth of cranial base structures (Lewis and Roche 1972; Lewis et al. 1985) including a classic paper investigating changes in basicranial flexion (i.e., saddle angle) (Lewis and Roche 1977). Significant findings from this work include the identification of subtle but distinct pubertal spurts in basicranial dimensions in both males and females. Most growth studies restrict analysis to ages 18 years

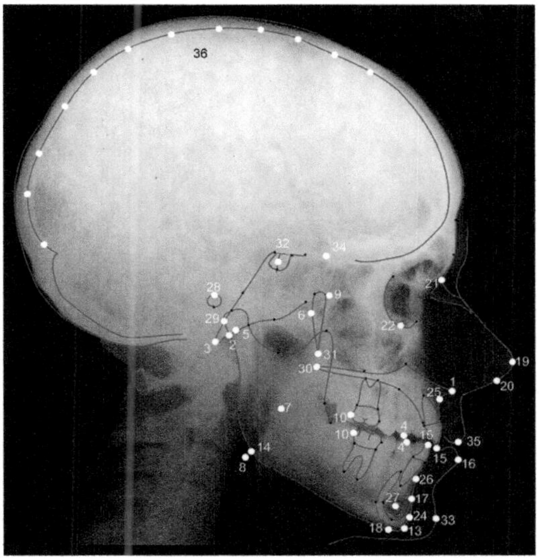

Fig. 14.4 Lateral cephalograph showing the 47 craniometric points used for measurements (For details of methodology, see Sherwood et al. 2011)

or below. Because of the unique quality of the Fels Longitudinal Study data, some studies have investigated the changes that continue throughout the lifetime (Garn et al. 1967; Lewis and Roche 1988). Both of these studies note a small, but significant, growth in skull dimensions past attainment of adulthood.

Recent work using the original craniofacial data collected from the Fels Longitudinal Study examined the genetic architecture of 80 traits based on 47 craniometric points (Fig. 14.4) derived from lateral cephalographs. All traits were significantly heritable (Table 14.1 provides data for basicranial traits as an example). Examination of genetic correlations between traits identified a subset of traits exhibiting shared genetic effects. While our initial work revolved around analysis of the original data collected by Lewis (Sherwood et al. 2008a), subsequent efforts focused on reanalysis of the entire lateral cephalographic collection have recently been published (Sherwood et al. 2011; Sherwood and McNulty 2011). This collection of numerous phenotypes drawn from standard craniometric or orthodontic analyses allows a full characterization of all components of the craniofacial complex.

The parallel study of the baboon craniofacial complex uses the pedigreed population from the Texas Biomedical Research Institute/Southwest National Primate Research Center, San Antonio, Texas. These animals are a mixture of two subspecies, *Papio hamadryas anubis* and *Papio hamadryas cynocephalus* and their hybrids. Following the protocol established in the Fels Longitudinal Cranial Study, lateral cephalographs were taken of 830 baboons. These were phenotyped in the same manner as the human cephalographs although a portion of the phenotypes do not translate onto the shape of the baboon skull; therefore, there are fewer traits measured in this sample. Recent work (Sherwood et al. 2008c, 2011) has shown that the craniofacial traits in the baboon, similar to those in the human study, are all significantly heritable.

Both the human and baboon studies successfully identified QTLs influencing variation in craniofacial traits. Ten significant QTLs were identified for human craniofacial traits (Sherwood et al. 2004, 2011), and 14 QTLs were identified for baboon craniofacial traits (Sherwood et al. 2008c). Many of the regions identified in both species contain genes known to influence craniofacial features (e.g., *SIX3, OTCS, BMP6*, or several members of the WNT family). Future work will seek to systematically interrogate the QTLs, prioritize the genes within, and conduct functional assessment of sequence variation in those candidate genes. The goal of this work is not only to identify genes with a potential to result in dysmorphologies when mutated, but to better characterize the variation in the background genetic matrix with which mutated genes interact.

14.5 Implications

As with *Geospiza*, the Darwin's finches described at the start of this chapter, the diversity of craniofacial forms across primate species raises questions of the interplay between genetic control, functional adaptation, and architectural byproducts of those processes (i.e., spandrels, Gould and Lewontin 1979). The magnitude of

Table 14.1 Heritability estimates (h^2) and standard errors for basicranial traits

					Covariates					
Variable	N	$h^{2\,a}$	S.E.	p	Sex	Age	Sex × Age	Age^2	Sex × Age^2	% Var[b]
Sella to sphenoethmoidal junction (mm)	975	0.32	0.07	2.51E-08	●			●		7.3
Sella to posterior nasal spine (mm)	969	0.42	0.06	9.87E-17	●	●	●	●	●	31.5
Sella to nasion (mm)	974	0.45	0.06	6.49E-19	●			●		27.3
Posterior condylion to S–N (mm)	974	0.47	0.06	1.39E-18	●					4.0
Porion location (mm)	953	0.22	0.07	4.75E-05	●	●				10.9
Nasion to sella to basion (degrees)	964	0.58	0.06	4.47E-25	●					2.2
Cranial deflection (degrees)	946	0.16	0.06	5.43E-04	●	●				2.0
Basion to sella (mm)	965	0.43	0.06	1.30E-13	●					28.7
Basion to posterior nasal spine (mm)	962	0.36	0.06	2.16E-15	●	●				10.6
Basion to nasion (mm)	965	0.42	0.06	5.41E-16	●			●		31.0

Significant covariates, and the percent variance explained by those covariates, are indicated

[a] h_0: $h^2 = 0$

[b] Percent variation of trait explained by significant covariate effects

these interactions, and the effect on evolutionary trajectories, will be increasingly understood as the genetic influences on primate craniofacial variation are revealed. Clinical applications, in the form of tissue engineering and gene therapies, will benefit from detailed analysis of the genetic underpinnings for craniofacial variation in humans and in closely related animal species.

14.5.1 Evolutionary Implications

The evolutionary history of the order *Primates* is of great interest for a variety of reasons, not the least of which is that humans belong to the order. Phylogenetic reconstruction of fossil primates, and notably the *Hominoidea* (apes and humans), have relied almost exclusively on analyses of craniofacial remains. These analyses often incorporate extensive trait lists enumerating hundreds of characters that are frequently treated as independent.

Phylogenetic analyses can benefit from genetic research in three ways. First, identification of a heritable component to craniofacial morphology is necessary to demonstrate that traits have evolutionary relevance. While, to some, it may seem obvious that traits of the craniofacial complex are under genetic influence, it is important to point out that previous studies failed to identify significant heritabilities for craniofacial traits in humans and nonhuman primates. It is also reasonable to suggest that, for some traits, there are significant environmental influences (in the largest sense) that may limit the ability to detect the genetic influences on variation.

Second, characterization of the levels of integration and modularity in the cranium will help determine the levels of independence between traits used in phylogenetic analyses. Given the complex three-dimensional architecture of the primate skull, it is difficult to imagine that changes in one structure will not be associated with concomitant changes in other structures. Phenotypic integration of the craniofacial complex has been discussed. Numerous studies have also demonstrated levels of pleiotropy between traits, including traits from different developmental components. As phylogenetic analyses of fossil remains essentially employ morphological traits as surrogate measures of underlying genetic similarity and differences, the use of genetically correlated traits may bias phylogenetic assessments by effectively reducing the genetic signal being analyzed (Sherwood et al. 2008a).

Finally, the localization of QTL and genes influencing variation in the craniofacial complex allows us to begin to identify the true traits upon which evolutionary forces act. This enables the expansion of current genetic techniques aimed at determining the timing of evolutionary events and may answer some long-standing questions within paleoanthropology, such as the rapid expansion of the hominin brain approximately 2 million years ago.

14.5.2 Biomedical Implications

Few modern scientific endeavors have enjoyed the publicity, and concomitant controversies, as has the explosion of genetic research in the past two decades. While many people are familiar with the Human Genome Project, they may not realize that genome maps for a wide variety of animals and plants ranging from beetles to pigs to the platypus are becoming available. Harold Slavkin, the former director of the National Institute of Dental and Craniofacial Research, described the potential impact of this research as including "understanding fundamental basics of diseases and disorders, targeting research to the fundamental root causes of disease processes, risk assessment for preclinical interventions, diagnostics, and tailoring treatment and therapeutics to individual risk and responses" (Slavkin 2001, p. 476). In the decade since that statement was written, a number of advances have been made into the research and application of clinically relevant genetic techniques.

The craniofacial and dentognathic complexes comprise one of the primary foci for research into areas of gene therapy and tissue engineering (Wan et al. 2006). The clinical reasons for this focus are numerous; even small craniofacial defects (whether congenital or acquired) can influence multiple aspects of physical and mental health. Additionally, for the dentition, discrete elements such as the teeth provide an easily managed object for manipulation, and the "normality" of the engineered structures is relatively easy to assess. Current approaches to regenerative medicine are examining the potential of restoring specific tissues in the pulp chamber of teeth (Murray et al. 2007; Nakashima 2005), periodontal ligaments (Jin et al. 2004; Nakahara 2006), complete teeth (Duailibi et al. 2006; Hu et al. 2006), or the supporting bone (Dunn et al. 2005; Nussenbaum and Krebsbach 2006, Rutherford et al. 2003; Young et al. 2005a, b). Gene therapy has even been investigated as a means to accelerate orthodontic treatment (Kanzaki et al. 2006). Increased characterization of the genetic architecture of the human craniofacial and dentognathic complexes will facilitate application of gene therapy and tissue engineering approaches.

14.6 Conclusions

Significant advances to understanding craniofacial biology have been made since the days of pure descriptive anatomy. Just as the formalization of functional craniology opened new avenues of research resulting in a new understanding of craniofacial form, the genomic revolution is providing new insights on a regular basis. While bird and rodent models have proven extremely valuable in elucidating developmental determinants, use of an animal in close phylogenetic proximity to humans, such as the baboon or other nonhuman primates, will become increasingly important, most notably in development of new therapeutic techniques. New approaches in quantitative genetics may prove particularly valuable in these endeavors.

Acknowledgments The craniofacial research program at Wright State University is funded by the National Institute for Dental and Craniofacial Research, National Institutes of Health (DE016692, DE016408, and DE018497 to RJ Sherwood, NIH grant P51 RR13986 to the Southwest National Primate Research Center).

References

Abzhanov A, Protas M, Grant BR, Grant PR, Tabin CJ (2004) Bmp4 and morphological variation of beaks in Darwin's finches. Science 305:1462–1465

Abzhanov A, Kuo WP, Hartmann C, Grant BR, Grant PR, Tabin CJ (2006) The calmodulin pathway and evolution of elongated beak morphology in Darwin's finches. Nature 442:563–567

Albertson RC, Streelman JT, Kocher TC (2003) Directional selection has shaped the oral jaws of Lake Malawi cichlid fishes. Proc Natl Acad Sci USA 100:5252–5257

Albertson RC, Kocher TC (2006) Genetic and developmental basis of cichlid trophic diversity. Heredity 97:211–221

Bonilla-Claudio M, Wang J, Bai Y, Klysik E, Selever J, Martin JF (2012) Bmp signaling regulates a dose-dependent transcriptional program to control facial skeletal development. Development 139:709–719. doi:10.1242/dev.073197

Broadbent BH Jr (1931) A new x-ray technique and its application to orthodontic practice. Angle Orthod 1:45

Campas O, Mallarino R, Herrel A, Abzhanov A, Brenner MP (2010) Scaling and shear transformations capture beak shape variation in Darwin's finches. Proc Natl Acad Sci USA 107:3356–3360. doi:10.1073/pnas.0911575107

Cheverud JM, Buikstra JE (1981a) Quantitative genetics of skeletal nonmetric traits in the rhesus macaques on Cayo Santiago. I. Single trait heritabilities. Am J Phys Anthropol 54:43–49

Cheverud JM, Buikstra JE (1981b) Quantitative genetics of skeletal nonmetric traits in the rhesus macaques on cayo santiago. II. Phenotypic, genetic, and environmental correlations between traits. Am J Phys Anthropol 54:51–58

Cheverud JM (1982) Phenotypic, genetic, and environmental morphological integration in the cranium. Evolution 36:499–516. doi:10.2307/2408096

Cheverud JM, Buikstra JE (1982) Quantitative genetics of skeletal nonmetric traits in the rhesus macaques of Cayo Santiago. III. Relative heritability of skeletal nonmetric and metric traits. Am J Phys Anthropol 59:151–155

Cheverud JM, Falk C, Hildebolt C, Moore AJ, Helmkamp RC, Vannier M (1990a) Heritability and association of cortical petalias in rhesus macaques (*Macaca mulatta*). Brain Behav Evol 35:368–372

Cheverud JM, Falk C, Vannier M, Konigsberg L, Helmkamp RC, Hildebolt C (1990b) Heritability of brain size and surface features in rhesus macaques (*Macaca mulatta*). J Hered 81:51–57

Cheverud JM (1995) Morphological integration in the saddle-back tamarin (*Saguinus fuscicollis*) cranium. Am Nat 145:63–89. doi:10.1086/285728

Cho SW, Lee HA, Cai J, Lee MJ, Kim JY, Ohshima H, Jung HS (2007) The primary enamel knot determines the position of the first buccal cusp in developing mice molars. Differentiation 75:441–451

Cohen MM (1990) Anomalies, syndromes, and dysmorphic growth and development. In: Enlow DH (ed) Facial Growth. WB Saunders Co, Philadelphia, pp 331–345

Cohen MM Jr, MacLean RE (1999) Should syndromes be defined phenotypically or molecularly? Resolution of the dilemma. Am J Med Genet 86:203–204

Cohen MM (2002) Perspectives on craniofacial anomalies, syndromes, and other disorders. In: Lin KY, Ogle RC, Jane JA (eds) Craniofacial Surgery: Science and Surgical Technique. WB Saunders Co, Philadelphia, pp 3–38

Cox TC (2004) Taking it to the max: the genetic and developmental mechanisms coordinating midfacial morphogenesis and dysmorphology. Clin Genet 65:163–176

Duailibi SE, Duailibi MT, Vacanti JP, Yelick PC (2006) Prospects for tooth regeneration. Periodontol 2000 (41):177–187

Dunn CA, Jin P, Taba M Jr, Franceschi RT, Bruce RR, Giannobile WV (2005) BMP gene delivery for alveolar bone engineering at dental implant defects. Mol Ther 11:294–299

Duren CL, Williams-Blangero S, Subedi J, Shrestha R, Jha B, Towne B, Sherwood RJ (2006) Genetic architecture of the human dentognathic complex. http://www.craniofacialgenetics.org/. Accessed 13 Nov 2006

Enlow CH (1990) Control processes in facial growth. In: Enlow CH (ed) facial growth. WB Saunders Co, Philadelphia, pp 229–248

Feng JP, Zhang J, Tan X, Lu Y, Guo C, Harris SE (2002) Identification of cis-DNA regions controlling Bmp4 expression during tooth morphogenesis *in vivo*. J Dent Res 81:6–10

Galton F (1876a) A theory of heredity. J Anthropol Inst 5:329–348

Galton F (1876b) The history of twins, as a criterion of the relative powers of nature and nurture. J Anthropol Inst 5:391–406

Galton F (1885) Types and their inheritance. Science 6:268–275

Garn SM, Lewis AB, Vicinus JH (1963) The inheritance of symphyseal size during growth. Angle Orthod 33:222–231

Garn SM, Rohmann CG, Wagner B, Ascoli W (1967) Continuing bone growth throughout life: a general phenomenon. Am J Phys Anthropol 26:313–317

Gong SG, Guo C (2003) Bmp4 gene is expressed at the putative site of fusion in the midfacial region. Differentiation 71:228–236

Gould SJ, Lewontin RJ (1979) The spandrels of san marco and the panglossian paradigm: a critique of the adaptationist programme. Proc R Soc Lond 205:581–598

Gripp KW, Wotton C, Edwards MC, Roessler E, Ades L, Meinecke P, Richieri-Costa A, Zackai EH, Massague J, Muenke M, Elledge SJ (2000) Mutations in TGIF cause holoprosencephaly and link NODAL signalling to human neural axis determination. Nat Genet 25:205–208

Handrigan GR, Buchtova M, Richman JM (2007) Gene discovery in craniofacial development and disease–cashing in your chips. Clin Genet 71:109–119

Havens BA, Velonis C, Kronenberg MS, Lichtler AC, Oliver B, Mina M (2008) Roles of FGFR3 during morphogenesis of Meckel's cartilage and mandibular bones. Dev Biol 316:336–349

Helms JA, Schneider RA (2003) Cranial skeletal biology. Nature 423:326–331

Helms JA, Cordero C, Tapadia MC (2005) New insights into craniofacial morphogenesis. Development 132:851–861

Hennekam RCM, Krantz IC, Allanson JE (2010) Gorlin's Syndromes of the Head and Neck, 5th edn. Oxford University Press, Oxford

Hlusko LJ, Weiss KM, Mahaney MC (2002) Statistical genetic comparison of two techniques for assessing molar crown size in pedigreed baboons. Am J Phys Anthropol 117:182–189

Hlusko LJ, Mahaney MC (2003) Genetic contributions to expression of the baboon cingular remnant. Arch Oral Biol 48:663–672

Hlusko LJ, Maas ML, Mahaney MC (2004a) Statistical genetics of molar cusp patterning in pedigreed baboons: implications for primate dental development and evolution. J Exp Zool B Mol Dev Evol 302:268–283

Hlusko LJ, Suwa G, Kono RT, Mahaney MC (2004b) Genetics and the evolution of primate enamel thickness: a baboon model. Am J Phys Anthropol 124:223–233

Hlusko LJ, Lease LR, Mahaney MC (2006) Evolution of genetically correlated traits: tooth size and body size in baboons. Am J Phys Anthropol 131:420–427

Hu B, Nadiri A, Kuchler-Bopp S, Perrin-Schmitt F, Peters H, Lesot H (2006) Tissue engineering of tooth crown, root, and periodontium. Tissue Eng 12:2069–2075

Hu C, Helms JA (1999) The role of sonic hedgehog in normal and abnormal craniofacial morphogenesis. Development 126:4873–4884

Hu JC, Simmer J (2007) Developmental biology and genetics of dental malformations. Orthod Craniofac Res 10:45–52

Hylander WL (1979) Experimental analysis of temporomandibular joint reaction force in macaques. Am J Phys Anthropol 51:433–456

Hylander WL (1986) *In vivo* bone strain as an indicator of masticatory bite force in *Macaca fascicularis*. Archs Oral Biol 31:149–157

Ideker T, Galitski T, Hood L (2001a) A new approach to decoding life: systems biology. Annu Rev Genomics Hum Genet 2:343–372

Ideker T, Thorsson V, Ranish JA, Christmas R, Buhler J, Eng JK, Bumgarner R, Goodlett CR, Aebersold R, Hood L (2001b) Integrated genomic and proteomic analyses of a systematically perturbed metabolic network. Science 292:929–934

Ideker T, Krogan NJ (2012) Differential network biology. Mol Syst Biol 8:565. doi:10.1038/msb.2011.99

Jernvall J, Kettunen P, Karavanova I, Martin LB, Thesleff I (1994) Evidence for the role of the enamel knot as a control center in mammalian tooth cusp formation: non-dividing cells express growth stimulating Fgf-4 gene. Int J Dev Biol 38:463–469

Jernvall J, Aberg T, Kettunen P, Keranen S, Thesleff I (1998) The life history of an embryonic signaling center: BMP-4 induces p21 and is associated with apoptosis in the mouse tooth enamel knot. Development 125:161–169

Jernvall J, Thesleff I (2000) Reiterative signaling and patterning during mammalian tooth morphogenesis. Mech Dev 92:19–29

Jin P, Anusaksathien O, Webb SA, Printz MA, Giannobile WV (2004) Engineering of tooth-supporting structures by delivery of PDGF gene therapy vectors. Mol Ther 9:519–526

Kanzaki H, Chiba M, Arai K, Takahashi I, Haruyama N, Nishimura M, Mitani H (2006) Local RANKL gene transfer to the periodontal tissue accelerates orthodontic tooth movement. Gene Ther 13:678–685

Langille RM, Hall BK (1993) Pattern formation and the neural crest. In: Hanken J, Hall BK (eds) The skull, vol 1. University of Chicago Press, Chicago, pp 77–111

Lewis AB, Roche AF (1972) Elongation of the cranial base in girls during pubescence. Angle Orthod 42:358–367

Lewis AB, Roche AF (1977) The saddle angle: constancy or change? Angle Orthod 47:46–54

Lewis AB, Roche AF, Wagner B (1982) Growth of the mandible during pubescence. Angle Orthod 52:325–342

Lewis AB, Roche AF, Wagner B (1985) Pubertal spurts in cranial base and mandible. Comparisons within individuals. Angle Orthod 55:17–30

Lewis AB, Roche AF (1988) Late growth changes in the craniofacial skeleton. Angle Orthod 58:127–135

Lieberman CE, Ross CF, Ravosa MJ (2000) The primate cranial base: ontogeny, function, and integration. Am J Phys Anthropol Suppl 31:117–169

McCarthy RC, Lieberman CE (2001) Posterior maxillary (PM) plane and anterior cranial architecture in primates. Anat Rec 264:247–260

McGrath JW, Cheverud JM, Buikstra JE (1984) Genetic correlations between sides and heritability of asymmetry for nonmetric traits in rhesus macaques on Cayo Santiago. Am J Phys Anthropol 64:401–411

Miettinen PJ, Chin JR, Shum L, Slavkin HC, Shuler CF, Derynck R, Werb Z (1999) Epidermal growth factor receptor function is necessary for normal craniofacial development and palate closure. Nat Genet 22:69–73

Ming JE, Kaupas ME, Roessler E, Brunner HG, Golabi M, Tekin M, Stratton RF, Sujansky E, Bale SJ, Muenke M (2002) Mutations in PATCHED-1, the receptor for SONIC HEDGEHOG, are associated with holoprosencephaly. Hum Genet 110:297–301

Ming JE, Muenke M (2002) Multiple hits during early embryonic development: digenic diseases and holoprosencephaly. Am J Hum Genet 71:1017–1032

Moss ML, Young RW (1960) A functional approach to craniology. Am J Phys Anthropol 18:281–292

Moss ML (1997a) The functional matrix hypothesis revisited. 3. The genomic thesis. Am J Orthod Dentofac Orthop 112:338–342

Moss ML (1997b) The functional matrix hypothesis revisited. 4. The epigenetic antithesis and the resolving synthesis. Am J Orthod Dentofac Orthop 112:410–417

Mulliken JB (2002) The craniofacial surgeon as amateur geneticist. J Craniofac Surg 13:3–17

Murray JC, Schutte BC (2004) Cleft palate: players, pathways and pursuits. J Clin Invest 113:1676–1678

Murray PE, Garcia-Godoy F, Hargreaves KM (2007) Regenerative endodontics: a review of current status and a call for action. J Endod 33:377–390

Nakahara T (2006) A review of new developments in tissue engineering therapy for periodontitis. Dent Clin North Am 50:265–276, ix–x

Nakashima M (2005) Bone morphogenetic proteins in dentin regeneration for potential use in endodontic therapy. Cytokine Growth Factor Rev 16:369–376

Naruse T, Takahara M, Takagi M, Oberg KC, Ogino T (2007) Busulfan-induced central polydactyly, syndactyly and cleft hand or foot: a common mechanism of disruption leads to divergent phenotypes. Dev Growth Differ 49:533–541

Nussenbaum B, Krebsbach PH (2006) The role of gene therapy for craniofacial and dental tissue engineering. Adv Drug Deliv Rev 58:577–591

Olson E, Miller R (1958) Morphological integration. University of Chicago Press, Chicago

Orioli IM, Castilla EE, Ming JE, Nazer J, Burle de Aguiar MJ, Llerena JC, Muenke M (2001) Identification of novel mutations in SHH and ZIC2 in a South American (ECLAMC) population with holoprosencephaly. Hum Genet 109:1–6

Ravosa MJ (1991) Interspecific perspective on mechanical and nonmechanical models of primate circumorbital morphology. Am J Phys Anthropol 86:369–396

Ravosa MJ, Vinyard CJ, Hylander WL (2000) Stressed out: masticatory forces and primate circumorbital form. Anat Rec 261:173–175

Roche AF, Chumlea WC, Guo SS (1988a) The assessment of hand-wrist skeletal maturity. Clin Res 36:896

Roche AF, Chumlea WC, Thissen C (1988b) Assessing the skeletal maturity of the hand-wrist. In: Thomas CC (ed) Fels method. Thomas Publisher, Springfield

Roche AF (1989) Relative utility of carpal skeletal ages. Am J Hum Biol 1:479–482

Roche AF (1992) Growth, maturation and body composition: the fels longitudinal study 1929-1991. Cambridge University Press, Cambridge

Roessler E, Du YZ, Mullor JL, Casas E, Allen WP, Gillessen-Kaesbach G, Roeder ER, Ming JE, Altaba A, Muenke M (2003) Loss-of-function mutations in the human GLI2 gene are associated with pituitary anomalies and holoprosencephaly-like features. Proc Natl Acad Sci USA 100:13424–13429

Ross CF, Ravosa MJ (1993) Basicranial flexion, relative brain size, and facial kyphosis in nonhuman primates. Am J Phys Anthropol 91:305–324

Ross C (1996) Adaptive explanation for the origins of the anthropoidea (Primates). Am J Primatol 40:205–230

Ross CF, Hylander WL (2000) Electromyography of the anterior temporalis and masseter muscles of owl monkeys (*Aotus trivirgatus*) and the function of the postorbital septum. Am J Phys Anthropol 112:455–468

Ross CF (2001) *In vivo* intraorbital bone strain from the lateral orbital wall of Macaca and the functioning of the craniofacial haft. Am J Phys Anthropol Suppl 32:128

Rutherford RB, Nussenbaum B, Krebsbach PH (2003) Bone morphogenetic protein 7 ex vivo gene therapy. Drug News Perspect 16:5–10

Sherwood RJ, May RL, Meindl RS, Robinson HB (1992) Growth alteration in the pathological human fetus. Am J Phys Anthropol Suppl 14:150

Sherwood RJ, Robinson HB, Meindl RS, May RL (1997) Pattern and process of growth of the abnormal human fetus. Hum Biol 69:849–871

Sherwood RJ (1999) Pneumatic processes in the temporal bone of chimpanzee (*Pan troglodytes*) and gorilla (*Gorilla gorilla*). J Morphol 241:127–137

Sherwood RJ, Duren CL, Blangero J, Dyer T, Cole SA, Siervogel RM, Towne B (2004) A genome wide-linkage scan for quantitative trait loci influencing the craniofacial complex. http://www.ashg.org/cgi-bin/ashg04s/ashg04 Accessed Dec 2004

Sherwood RJ, Duren CL, Blangero J, Dyer T, Cole SA, Siervogel RM, Towne B (2005) Craniofacial genetics and the Fels Longitudinal Study. http://www.craniofacialgenetics.org/ Accessed 2 Dec 2005

Sherwood RJ, Duren CL, Blangero J, Mahaney MC, Towne B (2006a) Face value: comparative quantitative genetics of the human (*Homo sapiens*) and baboon (*Papio hamadryas*) craniofacial complex. Paleoanthropol PAS 2006a Abstracts:A01

Sherwood RJ, Duren CL, Mahaney MC, Havill LM, Towne B (2006b) Integration and modularity in the baboon craniofacial complex. http://www.craniofacialgenetics.org/ Accessed 15 Nov 2006

Sherwood RJ, Duren CL, Mahaney MC, Towne B (2006b) Quantitative genetics of modern baboon

(*Papio hamadryas*) craniofacial variation. Am J Phys Anthropol Suppl 42:164–165

Sherwood RJ, Duren CL, Blangero J, Subedi J, Shrestha R, Jha B, Towne B, Williams-Blangero S (2007) Genetic influence and integration of dental traits. Paleoanthropol PAS 2007 Abstracts:A28

Sherwood RJ, Duren CL, Demerath EW, Czerwinski SA, Siervogel RM, Towne B (2008a) Quantitative genetics of modern human cranial variation. J Hum Evol 54:909–914

Sherwood RJ, Duren CL, Havill LM, Rogers J, Cox LA, Towne B, Mahaney MC (2008b) A genome-wide linkage scan for quantitative trait loci influencing the craniofacial complex in baboons (*Papio hamadryas spp.*). Genetics 180:619–628

Sherwood RJ, Duren CL, Havill LM, Rogers J, Cox LA, Towne B, Mahaney MC (2008c) A genomewide linkage scan for quantitative trait loci influencing the craniofacial complex in baboons (*Papio hamadryas spp.*). Genetics 180:619–628

Sherwood RJ, Mahaney MC, Duren CL, Havill LM, Cox LA, Rogers J, Towne B (2008d) Variation, genetics, and evolution of the primate craniofacial complex. Am J Phys Anthropol Suppl 43:192

Sherwood RJ, Duren CL, Mahaney MC, Blangero J, Dyer TC, Cole SA, Czerwinski SA, Chumlea WC, Siervogel RM, Choh AC, Nahhas RW, Lee M, Towne B (2011) A genome-wide linkage scan for quantitative trait loci influencing the craniofacial complex in humans (*Homo sapiens sapiens*). Anat Rec (Hoboken) 294:664–675

Sherwood RJ, McNulty KP (2011) Dissecting the genetic architecture of craniofacial shape. In: Lestrel PE (ed) Biological Shape Analysis: Proceedings of the 1st International Symposium. World Scientific Singapore, pp 145–171

Slavkin HC (1983) Research on craniofacial genetics and developmental biology: implications for the future of academic dentistry. J Dent Educ 47:231–238

Slavkin HC (2001) The human genome, implications for oral health and diseases, and dental education. J Dent Educ 65:463–479

Ten Cate AR (1989) Oral histology: development, structure, and function. In: The CV. Mosby Company, St. Louis

Vaahtokari A, Aberg T, Jernvall J, Keranen S, Thesleff I (1996) The enamel knot as a signaling center in the developing mouse tooth. Mech Dev 54:39–43

Vinyard CJ, Wall CE, Williams SH, Hylander WL (2003) Comparative functional analysis of skull morphology of tree-gouging primates. Am J Phys Anthropol 120:153–170

Wan CC, Nacamuli RP, Longaker MT (2006) Craniofacial bone tissue engineering. Dent Clin North Am 50:175–190 vii

Washburn SL (1947) The relation of the temporal muscle to the form of the skull. Anat Rec 99:239–248

Weiss KM, Stock CW, Zhao Z (1998) Dynamic interactions and the evolutionary genetics of dental patterning. Crit Rev Oral Biol Med 9:369–398

Witmer LM (1997) The evolution for the antorbital cavity of archosaurs: a study in soft-tissue reconstruction in the fossil record with an analysis of the function of pneumaticity. J Vert Paleontol 17:1–74

Xi HJ, Roche AF (1990) Differences between the hand-wrist and the knee in assigned skeletal ages. Am J Phys Anthropol 83:95–102

Yelick PC, Driever W, Neuhauss S, Stashenko P (1996) Craniofacial cartilage development in zebrafish. Ann NY Acad Sci 785:360–361

Yelick PC, Schilling TF (2002) Molecular dissection of craniofacial development using zebrafish. Crit Rev Oral Biol Med 13:308–322

Young RW (1957) Postnatal growth of the frontal and parietal bones in white males. Am J Phys Anthropol 15:367–386

Young CS, Kim SW, Qin C, Baba O, Butler WT, Taylor RR, Bartlett JC, Vacanti JP, Yelick PC (2005a) Developmental analysis and computer modelling of bioengineered teeth. Arch Oral Biol 50:259–265

Young CS, Abukawa H, Asrican R, Ravens M, Troulis MJ, Kaban LB, Vacanti JP, Yelick PC (2005b) Tissue-engineered hybrid tooth and bone. Tissue Eng 11:1599–1610

Zhang Z, Song Y, Zhao X, Zhang X, Fermin C, Chen Y (2002) Rescue of cleft palate in Msx1-deficient mice by transgenic Bmp4 reveals a network of BMP and Shh signaling in the regulation of mammalian palatogenesis. Development 129:4135–4146

Genetic Influences on Behavior in Nonhuman Primates

15

Julia N. Bailey, Christopher Patterson, and Lynn A. Fairbanks

15.1 Introduction

The genetic basis for behaviors has been shown in a wide range of species, from singular cellular protozoans to human beings. The development of behaviors is driven by both nature and nurture: behavioral phenotypes are caused by the expression of genes within environments, and these genes change their expression patterns throughout the life of the organism in response to environmental stimuli. Experiences (especially during early ontogenic stages of life) can have long-lasting effects on the behavior patterns of that organism (Breed and Sanchez 2012).

Nonhuman primates are useful for the study of genetic influences of behavior because they have complex behaviors and social structures comparable to humans since they are closely related genetically (Blomquist and Brent 2013). Pedigreed populations have the added advantage in that confounding effects which might obscure the genetic control of behaviors such as diet and environment can be tightly controlled. Another advantage of working with nonhuman primates is that biological samples can be collected more frequently, and from tissues that can be difficult to collect from human subjects (e.g., spinal fluid) (Jasinska et al. 2012). Because of their genetic similarities to humans, nonhuman primates act as model organisms for studying human diseases, many of which—like anxiety, alcoholism and drug addiction—fall within the purview of behavioral genetics.

However, there are major difficulties inhibiting the study of behavior genetics in nonhuman primates. There are few suitable populations of captive or semi-captive animals which have known genetic relations. The majority of behavioral genetic investigations in nonhuman primates have involved studying a very limited set of species such as rhesus macaques (*Macaca mulatta*), baboons (*Papio hamadryas* and *Papio anubis*), vervets (*Chlorocebus aethiops sabaeus*), and chimpanzees (*Pan troglodytes*). These species mimic many aspects of human behavior in that they live in complex societies with defined social roles. They experience frequent social stressors; hence, biological adaptive measures have evolved, many of which mirror the adaptive measures that have evolved in humans. However, this focus on a limited number of species may have contributed to significant bias when attempting to generalize behavioral similarities and differences across all nonhuman primates. This is particularly affected by the fact that two phylogenic branches of the primate evolutionary tree (the prosimians and New World monkeys)

J.N. Bailey (✉) · C. Patterson · L.A. Fairbanks
Department of Epidemiology, University of California, Los Angeles, Los Angeles
CA 90095, USA
e-mail: jbailey@mednet.ucla.edu

C. Patterson
e-mail: hrafnfaedhir@gmail.com

L.A. Fairbanks
e-mail: lfairbanks@mednet.ucla.edu

R. Duggirala et al. (eds.), *Genome Mapping and Genomics in Human and Non-Human Primates*, Genome Mapping and Genomics in Animals 5, DOI 10.1007/978-3-662-46306-2_15,

have been broadly ignored. Thus, the research presented is limited to the species studied, which do not include representatives from all species available for comparison.

This chapter begins with a discussion of heritability in some of the published genes that have been demonstrated to have an effect on the behavior of nonhuman primates, and the mechanisms (if understood) by which the variants within those genes produce observable differences in primate behaviors. While this chapter is focused specifically on the genetic control of nonhuman primate behavior, evidence from other animal models such as mice and humans will be discussed to provide evolutionary context to the discussion. The chapter includes guidelines and recommendations to improve behavioral genetic research, and provides new tools and methods that will take the field into the future.

This is an exciting time for behavioral genetics on nonhuman primates, as the field is in its infancy and there is still much to discover.

15.2 Heritability: Is It Genetic?

Heritability (h^2) is a measure of the amount of a trait that may be genetic, and it is often used to determine whether there is sufficient genetic signal to be used to localize genes. A measure with a larger heritability score may give a higher chance of success of localizing a gene for that trait in that particular population. However, the number of genes involved, the trait architecture, and the effect size of each gene on the behavioral measure are important factors in determining the locus of the genes responsible for the behavioral variation. Linkage-based mapping techniques may have difficulty mapping the genetic loci of a trait, even if that trait has a large h^2, especially if the trait has a polygenic or oligogenic genetic architecture, and those genes each carry only a small effect size (Anderson et al. 2010; Göring et al. 2007).

More specifically, heritability is the proportion of variation attributed to genetics compared to the total variation seen in the phenotypic trait. The remainder of the variation in a given trait can then be explained by individual differences, and by differences in unique and shared environments. Heritability can be quantifiably estimated by decomposing the phenotypic variation using statistical methods such as variance component analyses. There are currently only a few nonhuman primate populations that have been studied for heritability of behavioral traits, due to the requirement of knowing the relatedness or the pedigree status of individuals. There are few 'pedigreed' colonies suitable for studies on the genetic effect of behaviors, so each population is studied for specific behavioral traits of interest to the researchers, using different methods and study populations for each study. Heritability results must be taken in context of the cohort under study, the complexity of the pedigree, and the number of individuals.

One of the colonies that has been studied for genetic inheritance on behavior in nonhuman primates is the Vervet Research Colony (VRC). The VRC pedigree consists of a 16 mutigenerational pedigree, matrilineal colony of vervet monkeys (*C. aethiops sabaeus*). All subjects are raised in social groups that are managed to reflect the natural social composition of vervet groups in the wild. Variance component analyses utilizing the genetic relatedness of each colony member has been used in this population to estimate heritability of several behavioral traits. Novelty-seeking was measured by using a novel (but unthreatening) object in the home enclosure, and is significantly heritable ($h^2 = 0.47 \pm 0.01$, $p < 0.0001$) implying that 47 % of the variation of the trait is under genetic control. (Bailey et al. 2007) Impulsivity and impulsive aggression, measured using the intruder challenge test developed at the VRC, uses the resident-intruder paradigm to assess the behavioral response of an individual to an unfamiliar conspecific on the periphery of the subjects' home enclosure. This challenge elicits species-typical reactions of interest, arousal, and aggression toward a social stranger (the 'intruder') for both males and females, and it amplifies individual differences in characteristic reaction tendencies. Animals scoring high on social impulsivity rush over to the intruder immediately without taking the time to assess the situation. Social impulsivity is also found to be significantly heritable, ($h^2 = 0.35 \pm 0.11$, $p < 0.0001$) (Fairbanks 2001). Subscales of the index are independently heritable: both impulsive approaching ($h^2 = 0.25 \pm 0.10$, $p = 0.0008$) and aggressiveness (0.61 ± 0.12, $p < 0.0001$) (Fairbanks et al. 2004).

At the Harlow Primate Laboratory, alcohol consumption was studied using animals drawn from a large ongoing longitudinal study investigating genetic and environmental factors affecting neurobiology, including the behavior of alcohol consumption. In one study of 156 rhesus macaques belonging to a single pedigree who receive identical early rearing backgrounds, the heritability to consume alcohol was also significant, and 19.8 % of the variance was attributable to additive genetic effects (Lorenz et al. 2006).

Two hundred and eighty-five pedigreed rhesus monkeys (*Macaca mulatta*) from both the Harlow Primate Laboratory and the Wisconsin National Primate Research Center (Madison, WI) were studied for heritability of specific behavioral traits. Of the five behaviors studied, two were significant 'freezing duration' (behavioral inhibition) ($h^2 = 0.38$, $p = 0.0120$) and 'orienting to the intruder' (vigilance) ($h^2 = 0.91$, $p < 0.0001$). The other traits—'duration of locomotion', 'hostility', and 'frequency of cooing'—were not significantly heritable (Rogers et al. 2008).

Responses to novel and stressful environments were studied in 85 rhesus monkeys (*M. mulatta*) at the Oregon National Primate Research Center, which has a standard matriarchal colony (Williamson et al. 2006). The traits were tested using a set of temperament-testing paradigms, and heritabilities were estimated using variance component-based quantitative genetic analyses with much of the genetic information arising from paternal half-sibs. Significant heritabilities include latency to leave the mother during the initial 5-min observation period ($h^2 = 1.00$, $p \leq 0.05$), explore ($h^2 = 1.00$, $p \leq 0.01$, and movement ($h^2 = 1.00$, $p \leq 0.05$) during the alone-1 period and movement during the alone-2 period ($h^2 = 1.00$, $p \leq 0.05$). A factor analysis was also performed on the behaviors and seven factors emerged from the analyses. The only one statistically significant was factor 2, which consisted of movement during the test, and it was highly heritable ($h^2 = 1.00$, $p \leq 0.01$). Other factors (Factor 7, explore novelty) reached a heritability of 1 but were not significant. Nonsignificant factors included factor 1—distress vocalization, factor 3—distress cues, factor 4—delayed independence, factor 5—early independence, and factor 6—explore familiar environment. The nonsignificance of the higher heritabilities may be a reflection of the small sample size not being large enough for power, resulting in imprecise estimates.

Zoo populations were used to study the heritability of personality on 145 chimpanzees from 13 zoos that participated in the ChimpanZoo Program of the Jane Goodall Institute. Heritability was estimated using the symmetric differences squared (SDS) technique (Weiss et al. 2000). SDS incorporates phenotypic differences among all possible pairs of subjects in the sample, whether related or unrelated of all the traits studied. Only dominance showed significant heritability ($h^2 = 0.63$, $p = 0.0000$). Shared zoo effects accounted for only a negligible proportion of the variance for all factors.

Genetic origins of social networks and social behaviors were studied in 107 of the free-ranging rhesus macaques on the island of Cayo Santiago (Brent et al. 2013). The animals were assessed for relatedness, and variance component analyses methods were used to assess heritability. When controlling for age, sex, dominance, rank, and social household effects, a significant heritability ($h^2 = 0.84$, $p = 0.0250$) was found with grooming betweenness, which is an index of affiliate social positioning (Brent et al. 2013).

Though calculated with different measures in different populations, these heritabilities show us that many of the behaviors have strong heritable genetic underpinnings and are under genetic control. These results also demonstrate that for many traits, the genetics is not the most important factor driving the variance, and that the environment is also significant. This would make gene detection difficult.

15.3 Understanding the Genetic Control of Behaviors

Genes that control behavioral traits adaptively tend to have broad systemic similarities: they tend to belong to diverse multigene families; they are expressed in the cells of the brain, sensory organs, or other nervous tissue; and they are involved in either the processing of environmental stimuli, the mediation of internal states such as hunger, affiliation, and emotions, or are associated with neuronal development and neuroplasticity necessary for learning (reviewed in Bendesky and Bargmann 2011). Since there are so many genes that participate in each pathway, polymorphisms or variants in several genes can be associated with similar phenotypes and can contribute additively to their severity. It is also probable that different polymorphisms or variants in the same gene can have differing effects on phenotypes. While an exhaustive list of these signaling molecules and the host of genes that interact with them is beyond the scope of this chapter, these three classes of genes neuroreceptors, and transporters will be discussed in the context of two neurotransmitters well-studied in nonhuman primates (dopamine and serotonin) and how they relate to behaviors in nonhuman primates.

15.3.1 The Dopamine and Serotonin System

Both dopamine (*DA*) and serotonin (*5-HT* or 5-hydroxytryptophan) are in a class of neurotransmitters called monoamines, which are simple organic molecules synthesized via enzymatic action from amino acids (tyrosine and tryptophan, respectively). These molecules conduct the action potential of the neuron across the synapse to other connected neurons, exciting or inhibiting their own potential to fire.

Dopamine has been broadly conserved in its role coordinating motor function and reward-based learning across most phyla of the animal kingdom (the arthropods appear to be the sole exemption to this rule). In vertebrates, dopaminergic neurons connect regions of the brain associated with reward-based learning, such as the ventral tegmental area and the nucleus accumbens. In response to primary rewards or stimuli that have become associated with rewards through conditioned learning, these neurons experience phasic activation (increased bursts of action potentials). Dopamine appears to encode a reward prediction error: dopamine release and the phasic activation of dopaminergic neurons strongly increase in respect to rewards that exceed expectation, and drop below the baseline level of activity if the expectations of rewards are not met. Furthermore the release of dopamine drives reward-seeking behavior, increasing the likelihood that the individual will repeat behaviors that have become associated with greater reward expectations (reviewed in Barron et al. 2010).

Increased dopamine signaling is associated with many interrelated behaviors, including reward-seeking, conditioned learning, social dominance and extroversion, aggression, voluntary physical activity and motor control, working memory and focus, and addictive and compulsive behaviors. Many stimulants, such as cocaine and amphetamines, act by increasing the level of dopamine available for signaling in the synaptic cleft, hence why the psychoactive effects of these drugs (arousal, confidence, extroversion, aggression, etc.) are similar to the effects of dopamine signaling. Meanwhile, drugs that reduce dopamine activity, such as neuroleptics, impair concentration, reduce motivation, and cause anhedonia (the inability to experience pleasure).

While dopamine is highly associated with the rewards centers of the brain, serotonergic projections are especially dense in the limbic system, a set of structures that are responsible for the regulation of mood, emotional learning, memory and fear response. Serotonin also plays an important role in development, and many studies into early-life adversity and stress have demonstrated long lasting effects on serotonin signaling in the brain (reviewed in Nordquist and Oreland 2010).

It is also an important modulator of appetite and sleep cycles, mood, and inhibitory control. Low levels of serotonin are associated with depression, anxiety, stress-reactivity, and aggression, as well as increased risk-taking in gambling tasks. All of these conditions share components of increased emotional-reactivity and impulse-control.

Cerebrospinal fluid (CSF) concentrations of the terminal metabolites of dopamine (*HVA*, homovanillic acid) and serotonin (*5-HIAA*, 5-hydroxyindoleacetic acid) have been used as proxy biomarkers (endophenotypes) of overall levels of dopaminergic and serotonergic metabolism. These studies have demonstrated that the variance observed in the concentration of these metabolites are highly heritable and are stable over time and across environment (Freimer et al. 2007; Kaplan et al. 2002).

15.3.2 Neuroreceptor Proteins

Neuroreceptor genes play an especially important role in the modification of behavior. This family of proteins localizes at the synapse of the neuron, and when they bind to their target ligands (e.g., a neurotransmitter, or an odor molecule), they stimulate a molecular cascade, exciting or inhibiting the action potential of the neuron. Most of these receptors are coupled to similar complexes of G-Proteins, and it is through this shared mechanism that both serotonin and dopamine activate or inhibit the activity of the neuron (reviewed in Barnes and Sharp 1999; Bendesky and Bargmann 2011; Callier et al. 2003). However, in the presence of too much of their specific ligand, the receptors can become desensitized to the synaptic signal. Over extended periods, this can lead to a long-term down-regulation of neuroreceptors available at the synapse and a significant decrease in synaptic efficiency, especially with the serotonin receptors, as we will see with the examples of early-life adversity models.

There are five known receptor proteins that bind to dopamine and translate the dopaminergic signal into neural activity, however only two of them, *DRD1* (dopamine receptor *D1*) and *DRD4* (dopamine receptor *D4*) have been studied in respect to nonhuman primate behavior.

DRD1 is the most highly expressed of the five dopamine receptors, and upon binding with its agonist, acts to increase the action potential of the neuron (reviewed in Callier et al. 2003). A single nucleotide polymorphism (SNP) in the 5′ UTR of *DRD1* has been associated with the alcohol consumption of adolescent, male rhesus macaques that had been maternally deprived and peer-reared (Newman et al. 2005). Maternal deprivation is frequently used as a model condition for early-life stress and adversity, and tends to produce anxious and impulsive behaviors during adolescence. Female rhesus macaques and maternally-reared male carriers of this allele did not show increased propensity to consume alcohol, and this allele exemplifies the effect confounding factors such as gender and early rearing experience have on behavior (Newman et al. 2005).

While *DRD1* excites the neuron, *DRD4* is an inhibitory receptor. Its expression throughout the brain is much lower than DRD1, but its binding affinity and selectivity to dopamine is much higher (reviewed in Callier et al. 2003). A VNTR (variable number of tandem repeats) in exon III of the *DRD4* gene has been linked to novelty-seeking behavior in vervet monkeys (*C. aethiops sabaeus*). Carriers of the rare, 5-repeat variant displayed significantly shorter latencies to approach a large and potentially threatening object with which they had no prior experience than were carriers of the more common 6-repeat variant. The variance observed was consistent across age-groups, the only other demographic factor that was shown to significantly account for the variance in observed novelty-seeking scores (Bailey et al. 2007). In addition, juvenile carriers of the 5-repeat variant also scored higher on the Social Impulsivity Index, as measured by the shortness of latency to approach an unfamiliar conspecific with risky, assertive, and aggressive behavior. Social-impulsivity scores were also influenced by age and sex factors, but also by the genotype of the juvenile's mother, finding that the highest scores occurred in variant-carrying juveniles with variant-carrying mothers (Fairbanks et al. 2012). This illustrates two points: (1) the *DRD4* variant is a risk factor that is influenced by the developmental environment, and (2) almost everyone with any risk genotype also has one or both parents with the risk genotype, so they are likely to have both genetic and environmental influences operating. Similar repeat variations have been detected in humans, dogs, horses, and chimpanzees. In humans, variants have been associated with novelty seeking, risk taking behavior, and Attention Deficit Hyperactivity Disorder (Ptácek et al. 2011).

While the other dopamine receptors have not been fully investigated in the behaviors of nonhuman primates, they have been studied in respect to other mammalian species. Polymorphisms in *DRD2* (dopamine receptor *D2*, also an inhibitory receptor) have been associated with increased risk for alcoholism (Noble et al. 1998), pathological gambling (Lobo et al. 2010), and other addictive/impulse-control behaviors in humans (Ariza et al. 2012).

While dopamine has five known neuroreceptors in primates, serotonin has fourteen. The sheer number demonstrates the complexity of studying candidate genes. All but one of the receptors operate through the same molecular mechanism of G-protein complexes as the dopamine receptors (reviewed in Barnes and Sharp 1999).

A couple of studies have investigated the impact that rearing-history has on the expression of these receptors in both rhesus macaques (*Macaca mulatta*) and marmoset monkeys (*Callithrix jacchus*). Parental deprivation during infancy in marmosets produces a pro-depressive state, increased stress-reactivity, and general anhedonia that can persist until adolescence. (Law et al. 2009) That same study found that peer-reared marmosets had decreased *5-HTR1A* mRNA (serotonin receptor 1A, as measured by in situ hybridization—ISH—and real-time polymerase chain reaction—RT-PCR–) and binding (via positron emission tomography—PET—imaging techniques) in the hippocampus, a region associated with memory formation that is disproportionately affected by long-term stress, and that *5-HTR1A* mRNA was correlated with cerebrospinal fluid (CSF) concentrations of cortisol, a biomarker for stress-response. Another study (Spinelli et al. 2010) duplicated the same study using rhesus macaques with both magnetic resonance imaging (MRI) and PET scans. They found that peer-reared monkeys had an overall decrease of *5-HTR1A* density and binding throughout the brain. In females, the receptor density in the dorso-medial prefrontal cortex (associated with cognitive decision-making and emotional control) was significantly higher in peer-reared subjects versus their maternally-reared counterparts.

15.3.3 Transporter Proteins

Another important category of neuromodular genes that affect behavior is a class of solute carrier proteins (SLC) or transporter genes associated with each neurotransmitter. These proteins moderate the signal transmission by reuptaking excess dopamine back into the presynaptic neuron and repackaging the neurotransmitter into synaptic vesicles, functionally terminating the neurotransmitter signal and resetting the neuron for the next time it needs to fire. These proteins are the target of several drugs, both therapeutic and illicit. Cocaine competitively binds with the dopamine transporter (*DAT*), preventing the re-uptake of dopamine into the presynaptic neuron, while amphetamines reverse DAT activity, pumping dopamine back out into the synaptic cleft. Likewise, the serotonin transporter (*5-HTT* or SERT) is the target of a class of antidepressants called SSRIs (selective serotonin reuptake inhibitors), which functionally increases the amount of serotonin available for signaling. In macaques, both the genes that encode both the dopamine (*DAT*) and serotonin (*5-HTT* or SERT) transporters have alleles that differentially alter the pattern of expression in the brain.

In the *5'UTR* promoter region of *SLC6A3* (the gene that encodes the *DAT* protein), two single nucleotide polymorphisms (SNPs) associated with transcription factor binding sites have been associated with social rank in cynomolgus macaques (Miller-Butterworth et al. 2007). One of those SNPs was also found in rhesus macaques, and is also associated with social dominance behaviors. While it is not clear how these variants alter the expression of the dopamine transporter, the reduced transcription of *DAT* mRNA correspondingly reduces the density of *DAT* within dopaminergic neurons. This would presumptively increase the concentration of synaptic dopamine available for signaling.

The serotonin transporter (*5-HTT*, encoded by the *SLC6A4* gene [Solute Carrier, family 6, member 4]) together with its linked polymorphic region (*5-HTTLPR*) is probably the most studied gene in nonhuman primate behavior. In both humans and rhesus macaques, there is an analogous 21 bp length variant *rh5-HTTLPR* which is located in the same region as the serotonin transporter gene promoter polymorphism identified in humans. However, it is not in the same precise location. Therefore, two major alleles segregate in both species, descriptively called *Long* (*L*) and *Short* (*S*) for their relative sizes. The core sequence is similar, (*C*)*7 AGCAT*(*C*)6, but there is difference in the variation in allele length for the *L* allele that is usually attributed to the association with human behaviors; humans have 17 repeat units while rhesus macaques have 24 repeat units (Trefilov et al. 2000). Functional studies show similar effects between humans and rhesus polymorphisms in that the *S*-allele results in decreased transcriptional efficiency of the serotonin transporter. Variation in the serotonin transporter gene promoter has been shown to be related to several behavioral traits in humans including anxiety and depression (Goenjian et al. 2012).

In nonhuman primates, these two variants have been associated with observable differences in development and reproductive timing. Studies performed on the free-ranging macaques of Cayo Santiago in Puerto Rico, found that the number of *S*-alleles carried by each monkey was predictive of the age at which male macaques left their natal group. Homozygous S-carriers dispersed approximately 6 months earlier than heterozygotes and 14 months earlier than homozygous L-carriers (Krawczak et al. 2005; Trefilov et al. 2000).

The serotonin transporter has also been implicated in several social behaviors, notably the construction of social dominance hierarchies. Social dominance hierarchies represent a collection of behaviors observed in many species of captive and free-ranging nonhuman primates. Dominant individuals tend to be more aggressive, initiate agonistic encounters, display attack gestures and vocalizations, and consistently defeat lower ranking conspecifics. Subordinates display gestures and vocalizations associated with submission and flight, and tended to flee or cower when placed in agonist encounters with a dominant individual (Miller-Butterworth et al. 2007). While many factors can influence social dominance hierarchies such as personality, early-life history, physiological traits such as size, and the immediate social environment, both male and female macaques tended to retain the same relative dominance status even when assigned to different social groups (reviewed in Miller-Butterworth et al. 2008). As stated before, CSF *5-HIAA* concentrations (a biomarker of overall serotonin metabolism) have been positively associated with increased social status in female cynomolgus macaques and negatively associated with social status in males, and *5-HTTLPR* genotypes and early-rearing experience have been shown to affect CSF *5-HIAA* concentrations. (Bennett et al. 2002) In a study of psychosocial stress in the form of social reorganization and subordinate social status, 40 females were drawn from middle ranking genealogies of several large social grounds and reorganized into groups: those dependent on *5-HTTLPR* genotypes; those with only *LL*-homozygote individuals; and those in which all individuals had at least one *S* allele. Most of the measures (morning cortisol concentrations, glucocorticoid negative feedback, weight loss, and abdominal fat loss) were not significantly associated with genotype. There appeared to be an interaction with social status, genotype, and changes in serum concentrations of leptin and triiodiothyronine. Dominant *LL*-homozygote females had the highest levels while subordinate

S-variant females had the lowest level (Jarrell et al. 2008).

Watson et al. (2009) found that male rhesus *S*-allele carriers spent less time looking at the eye region of faces, and had larger pupil diameter when gazing at photographs of familiar high-status males from the same cohort. They also experienced higher risk-aversion on gambling tasks when presented in conjunction with another high-status individual. In the same activity, *LL*-homozygotes demonstrated increased risk-seeking behavior.

In the study of social networks in free-ranging rhesus macaques on the island of Cayo Santiago (Brent et al. 2013), one measure of sociality was associated with serotonergic genes profiles. Specifically the 'grooming eigenvector', which represented the tendency of individuals to spend a lot of time in grooming behaviors, was associated with an interaction of the *5-HTTLPR* and *TPH2* genotypes (tryptophan hydroxylase 2 is the rate-limiting enzyme required for serotonin biosynthesis in the brain).

A study on the prevalence of social dominance behaviors in respect to seven different species of macaque showed that the relation to social organization may be more controlled by genetic factors than by environmental ones. They found that species which displayed relaxed patterns of dominance, open relationships, and higher levels of conciliatory behavior tended to be monomorphic in the upstream promoter region of the *rh5-HTT* gene. Rhesus macaques (*Macaca mulatta*), the most stratified species of macaques, had three variants. This relationship of hierarchical social dominance to the amount of allelic variation was also linked to the polymorphic region in the monoamine oxidase A (MAO-A) gene, an enzyme required for the degradation of several monoamine neurotransmitters including both dopamine and serotonin (Wendland et al. 2006).

The last group of related behavioral traits associated with the serotonin transporter has to do with stress-reactivity and anxiety, two traits which show a strong gene-environment interaction between the *5-HTTLPR* and the prior experience of stress (usually modeled in experiments by maternal-deprivation and peer-rearing) (Barr et al. 2003, 2004). The limbic-hypothalamus-pituitary-adrenal (LHPA) axis is the central mechanism by which the nervous system and endocrine system modulate the reaction to stress and the fight-or-flight response. Because the LHPA axis is so well understood, studies frequently use serum concentrations of many of the hormones described above as endophenotypes for these behaviors. The reaction to stress is quantified by a baseline (nonstressed) measure, and by another reading following the exposure to stress. Researchers have used several methods to model ethologically-relevant stressors including social separation, threat (usually introducing a plastic snake or a fake predator), intrusion by an unknown conspecific, intrusion by a human researcher (nonthreatening), and relocation. Each of these stressors produce distinct behavioral responses organized through different parts of the limbic system, and produce similar interactions with the LHPA axis.

The *5-HTTLPR* variants have been associated with differences in the serum concentrations of the stress hormones, and this interacted with a history of environmental stressors (peer-reared vs. mother-reared). Rhesus macaques with the *LS*-genotype and peer-reared macaques each showed increased adrenocorticotropic hormone (ACTH) release in response to stress, and together these conditions increased the release of ACTH synergistically. (Barr et al. 2004) Another study found that mother macaques with the *LL*-genotype had consistent serum cortisol levels over the course of 6 months of study, while mothers with the *LS*-genotype showed significantly greater fluctuations in this trait over the same period. These *LS* mothers were also found to be more likely to be abusive to their infants (McCormack et al. 2009).

In addition, studies using PET scans and fMRI (functional magnetic resonance imaging) have found that *rh5-HTTLPR* S-carriers demonstrated increased limbic reactivity in response to specific aversive stimuli. For example, S-carriers displayed increased metabolic activity (measured using fluorodeoxyglucose PET imaging) in the amygdala in response to relocation stress, and the bed nucleus of the stria terminalis (BNST) in response to threat (Kalin et al. 2008). The metabolic activity of the BNST has been shown to be highly predictive of the "freezing" response of monkeys in response to threats (Oler et al. 2010; Rogers et al. 2008). Oler et al. (2009) used PET imaging to demonstrate that *5-HTT* availability (an index of its density and binding affinity) in the amygdala, hippocampal, and BNST regions correlated positively with several behavioral and neuroendocrine measures of anxious temperaments.

15.3.4 Interaction of Dopamine and Serotonin Signaling

The story becomes even more complex when considering the fact that many genes affect both dopamine and serotonin. There are many other genes important in the dopamine pathway, such as *COMT* (*catechol-O-methyl transferase*), *DBH* (*dopamine β hydroxylase*), and Tyrosine hydroxylase (*TH*), which is an enzyme responsible for catalyzing the rate-controlling step in dopamine biosynthesis. Most genes and variants have not been thoroughly studied in humans, let alone in nonhuman primates.

In cynomolgus macaques (*Macaca fascicularis*), social dominance rank illustrates the interaction between these two system. Both dominant males and females had significantly higher HVA concentrations than subordinates. Dominant males (but not females) had significantly lower CSF 5-HIAA concentrations (Kaplan et al. 2002; Riddick et al. 2009). In a study of free-ranging rhesus macaques, low CSF *5-HIAA* concentrations early in life were associated with delayed migration from the natal group, and increased aggression and premature death, but the individuals who survived were more likely to attain higher social ranks (Howell et al. 2007).

15.3.5 Caveats and Guidelines

While studying genetic influences on behaviors in nonhuman primates is very promising, there are several caveats of which we must be aware. Primary among them is the question of whether the comparison of phenotypes between species is valid and meaningful.

There are issues with cross-species analyses and analogous behavioral traits. Some traits, such as aggression, dominance, and extroversion, may be more comparable than others like anxiety. It is unclear if the anxieties found amongst different nonhuman primates are comparable. Is the anxiety of a social primate the same as the anxiety of a nonsocial primate? Does it matter if the trait is measured directly or is a composite or endophenotype? It is not inconceivable to think there would be differences at least in the manifestation of social anxiety. In a genetic sense, these traits are phenocopies of each other, and have different genetic mechanisms. So this leads us to question whether we can expect replication in genetics of primate behaviors if the phenotypes under study are not similar.

Many of these traits are obviously present in humans; however, it is unclear exactly how to map them back for specific comparison. This would be necessary to detect similar genetic mechanisms. For example, there are several types of anxiety defined by psychiatrists in the DSM (Diagnostic and Statistical Manual): generalized anxiety; social anxiety; and 'anxiety disorders' such as posttraumatic stress disorder (PTSD) and obsessive compulsive disorder (OCD). It is suspected that in humans, these anxieties have unique and shared genes (Domschke and Deckert 2012), though it is unclear how to map or to correlate the excitability/anxiety found in nonhuman primates to any or all of these anxiety disorders.

As mentioned previously, the current results in this field are biased because of species and phenotypes studied. Species bias is an issue because most research is performed on a very small subset of primate species, as there are few controlled populations where genetic relations between individuals are known.

Associated with this is another issue of variant detection bias. Most variants that are tested are genotyped using specific PCR primers, such as for the *5-HTTLPR* variants that have already been associated with some easily quantified behavioral phenotype. This is the "low-hanging fruit", and it only tests for one particular variant within a given gene of interest, ignoring other variants that might also contribute to the phenotype. Also the PCR'd genotype does not necessarily mean that the specific variant that was studied is disease-causing. It could be possible that the PCR'd genotype and the actual-disease causing variant are close together and are in high degree of linkage disequilibrium.

Association studies require large, out-bred populations (such as humans) to detect possibly causal variants. Most of the primate populations used for research are controlled, and are thus at least somewhat inbred, which can dramatically reduce the power of the method to detect phenotype affecting variants, unless methods such as linkage analysis that exploit the relatedness in pedigrees are used.

Another way to discover genes is to perform linkage analyses, either of specific genes or chromosomal regions or a 'genome scan' that searches markers over all the chromosomes. An advantage of a 'genome scan' over candidate gene studies is that it allows for the discovery of novel genes, since in a candidate gene study one has to first have a candidate gene. In a genome scan one can detect new candidate genes. In order to do linkage genome scans, however, one needs a genetic marker map of the species under study.

15.4 Conclusions and Future Directions

The study of Genetic Influences on Behavior in Nonhuman Primates is in its infancy, and there is much room to grow. Up to this point, most of the work has been in calculation of determining traits, and in calculation of heritabilities, and in examining a limited number of candidate genes. In order for the field to develop, there is a need to have reliable, valid phenotypes, and to thoroughly test each gene hypothesis, including all potential variants. Researchers in the field need to be cognizant that the phenotypes and genotypes may not be comparable between species (or even subspecies) and that this problem is probably even more complex than hypothesized.

There are many new tools emerging from the human genome project (as described elsewhere in this volume) which will prove to be very useful in the study of the genetic influences on behaviors in nonhuman primates. Chief among these are the databases of comparative genomics, of proteomics, and of all the sequenced species. New maps will be developed, including new sequenced maps. New techniques like Next Generation or Deep Sequencing will allow us to study the genomes in depth and to more exactly determine genetic variants responsible for observable behavioral variation in nonhuman primates.

References

Anderson TJC, Williams JT, Nair S, Sudimack D, Barends M, Jaidee A, Price RN, Nosten F (2010) Inferred relatedness and heritability in malaria parasites. Proc Biol Sci 277:2531–2540

Ariza M, Garolera M, Jurado MA, Garcia-Garcia I, Hernan I, Sánchez-Garre C, Vernet-Vernet M, Sender-Palacios MJ, Marques-Iturria I, Pueyo R, Segura B, Narberhaus A (2012) Dopamine genes (*DRD2/ANKK1-TaqA1* and *DRD4-7R*) and executive function: their interaction with obesity. PLoS ONE 7:e41482

Bailey JN, Breidenthal SE, Jorgensen MJ, McCracken JT, Fairbanks LA (2007) The association of DRD4 and novelty seeking is found in a nonhuman primate model. Psychiatr Genet 17:23–27

Barnes NM, Sharp T (1999) A review of central 5-HT receptors and their function. Neuropharmacology 38:1083–1152

Barr CS, Newman TK, Becker ML, Parker CC, Champoux M, Lesch KP, Goldman D, Suomi SJ, Higley JD (2003) The utility of the nonhuman primate; model for studying gene by environment interactions in behavioral research. Genes Brain Behav 2:336–340

Barr CS, Newman TK, Shannon C, Parker C, Dvoskin RL, Becker ML, Schwandt M, Champoux M, Lesch KP, Goldman D, Suomi SJ, Higley JD (2004) Rearing condition and rh5-HTTLPR interact to influence limbic-hypothalamic-pituitary-adrenal axis response to stress in infant macaques. Biol Psychiatry 55:733–738

Barron AB, Søvik E, Cornish JL (2010) The role of dopamine and related copounds in reward-seeking behavior across animal phyla. Front Behav Neurosci 4:163

Bendesky A, Bargmann CI (2011) Genetic contributions to behavioural diversity at the gene-environment interface. Nat Rev Genet 12:809–820

Bennett AJ, Lesch KP, Heils A, Long JC, Lorenz JG, Shoaf SE, Champoux M, Suomi SJ, Linnoila MV, Higley JD (2002) Early experience and serotonin transporter gene variation interact to influence primate CNS function. Mol Psychiatry 7:118–122

Blomquist GE, Brent LNJ (2013) Applying quantitative genetic methods to primate social behavior. Int J Primatol. doi:10.1007/s10764-013-9709-5

Breed M, Sanchez L (2012) Both environment and genetic makeup influence behavior. Nat Educ Knowl 3:68

Brent LJ, Heilbronner SR, Horvath JE, Gonzallez-Martinex J, Ruiz-Lambides A, Robinson AG, Pate Skene JH, Platt ML (2013) Genetic origins of social networks in rhesus macaques. Sci Rep 3:1042

Callier S, Snapyan M, Le Crom S, Prou D, Vincent JD, Vernier P (2003) Evolution and cell biology of dopamine receptors in vertebrates. Biol Cell 95:489–502

Domschke K, Deckert J (2012) Genetics of anxiety disorders-status quo and quo vadis. Curr Pharm Des 18(35):5691–5698. Review. PMID:22632468

Fairbanks LA (2001) Individual differences in response to a stranger: social impulsivity as a dimension of temperament in vervet monkeys (Cercopithecus aethiops sabaeus). J Comp Psychol 115(1):22–28. PMID:11334215

Fairbanks LA, Newman TK, Bailey JN, Jorgensen MJ, Breidenthal SE, Ophoff RA, Comuzzie AG, Martin LJ, Rogers J (2004) Genetic contributions to social impulsivity and aggressiveness in vervet monkeys. Biol Psychiatry 55:642–647

Fairbanks LA, Way BM, Breidenthal SE, Bailey JN, Jorgensen MJ (2012) Maternal and offspring dopamine D4 receptor genotypes interact to influence juvenile impulsivity in vervet monkeys. Psychol Sci 23:1099–1104

Freimer NB, Service SK, Ophoff RA, Jasinska AJ, McKee K, Villeneuve A, Belisle A, Bailey JN, Breidenthal SE, Jorgensen MJ, Mann JJ, Cantor RM, Dewar K, Fairbanks LA (2007) A quantitative trait locus for variation in dopamine metabolism mapped in a primate model using reference sequences from related species. Proc Natl Acad Sci USA 104:15811–15816

Goenjian AK, Bailey JN, Walling DP, Steinberg AM, Schmidt D, Dandekar U, Noble EP (2012) Association of *TPH1*, *TPH2*, and *5HTTLPR* with PTSD and depressive symptoms. J Affect Disord 140:244–252

Göring HH, Curran JE, Johnson MP, Dyer TD, Charlesworth J, Cole SA, Jowett JB, Abraham LJ, Rainwater DL, Comuzzie AG, Mahaney MC, Almasy L, MacCluer JW, Kissebah AH, Collier GR, Moses EK, Blangero J (2007) Discover of expression QTLs using large-scale transcriptional profiling in human lymphocytes. Nat Genet 39:1208–1216

Howell S, Westergaard G, Hoos B, Chavanne TJ, Shoaf SE, Cleveland A, Snoy PJ, Suomi SJ, Dee Higley J (2007) Serotonergic influences on life-history outcomes in free-ranging male rhesus macaques. Am J Primatol 69:851–865

Jarrell H, Hoffman JB, Kaplan JR, Berga S, Kinkead B, Wilson ME (2008) Polymorphisms in the serotonin reuptake transporter gene modify the consequences of social status on metabolic health in female rhesus monkeys. Physiol Behav 93:807–819

Jasinska AJ, Lin MK, Service S, Choi OW, DeYoung J, Grujic O, Kong SY, Jung Y, Jorgensen MJ, Fairbanks LA, Turner T, Cantor RM, Wasserscheid J, Dewar K, Warren W, Wilson RK, Weinstock G, Jentsch JD, Freimer NB (2012) A non-human primate system for large-scale genetic studies of complex traits. Hum Mol Genet 21:3307–3316

Kalin NH, Shelton SE, Fox AS, Rogers J, Oakes TR, Davidson RJ (2008) The serotonin transporter genotype is associated with intermediate brain phenotypes that depend on the context of eliciting stressor. Mol Psychiatry 13:1021–1027

Kaplan JR, Manuck SB, Fontenot MB, Mann JJ (2002) Central nervous system monoamine correlates of social dominance in cynomolgus monkeys (*Macaca fascicularis*). Neuropsychopharmacology 26:431–443

Krawczak M, Trefilov A, Berard J, Bercovitch F, Kessler M, Sauermann U, Croucher P, Nürnberg P, Widdig A, Schmidtke J (2005) Male reproductive timing in Rhesus macaques is influenced by the 5HTTLPR promoter polymorphism of the serotonin transporter gene. Biol Reprod 72:1109–1113

Law AJ, Pei Q, Walker M, Gordon-Andrews H, Weickert CS, Feldon J, Pryce CR, Harrison PJ (2009) Early parental deprivation in the marmoset monkey produces long-term changes in hippocampal expression of genes involved in synaptic plasticity and implicated in mood disorder. Neuropsychopharmacology 34:1381–1394

Lobo DS, Souza RP, Tong RP, Casey DM, Hodgins DC, Smith GJ, Williams RJ, Schopflocher DP, Wood RT, el-Guebaly N, Kennedy JL (2010) Association of functional variants in the dopamine D2-like receptors with risk for gambling behaviour in healthy Caucasian subjects. Biol Psychol 85:33–37

Lorenz JG, Long JC, Linnoila M, Goldman D, Suomi SJ, Higley JD (2006) Genetic and other contributions to alcohol intake in rhesus macaques (*Macaca mulatta*). Alcohol Clin Exp Res 30:389–398

McCormack K, Newman TK, Higley JD, Maestripieri D, Sanchez MM (2009) Serotonin transporter gene variation, infant abuse, and responsiveness to stress in rhesus macaque mothers and infants. Horm Behav 55:538–547

Miller-Butterworth CM, Kaplan JR, Barmada MM, Manuck SB, Ferrell RE (2007) The serotonin transporter: sequence variation in Macaca fascicularis and its relationship to dominance. Behav Genet 37:678–696

Miller-Butterworth CM, Kaplan JR, Shaffer J, Devlin B, Manuck SB, Ferrell RE (2008) Sequence variation in the primate dopamine transporter gene and its relationship to social dominance. Mol Biol Evol 25:18–28

Newman TK, Syagailo YV, Barr CS, Wendland JR, Champoux M, Graessle M, Suomi SJ, Higley JD, Lesch KP (2005) Monoamine oxidase A gene promoter variation and rearing experience influences aggressive behavior in rhesus monkeys. Biol Psychiatry 57:167–172

Noble EP, Zhang X, Ritchie T, Lawford BR, Grosser SC, Young RM, Sparkes RS (1998) D2 dopamine receptor and GABA(A) receptor beta3 subunit genes and alcoholism. Psychiatry Res 81:133–147

Nordquist N, Oreland L (2010) Serotonin, genetic variability, behaviour, and psychiatric disorders–a review. J Med Sci 115:2–10

Oler JA, Fox AS, Shelton SE, Christian BT, Murali D, Oakes TR, Davidson RJ, Kalin NH (2009) Serotonin transporter availability in the amygdala and bed nucleus of the stria terminalis predicts anxious temperament and brain glucose metabolic activity. J Neurosci 29:9961–9966

Oler JA, Fox AS, Shelton SE, Rogers J, Dyer TD, Davidson RJ, Shelledy W, Oakes TR, Blangero J, Kalin NH (2010) Amygdalar and hippocampal substrates of anxious temperament differ in their heritability. Nature 466:864–868

Ptácek R, Kuzelová H, Stefano GB (2011) Dopamine D4 receptor gene DRD4 and its association with psychiatric disorders. Med Sci Monit 17:RA215–RA220

Riddick NV, Czoty PW, Gage HD, Kaplan JR, Nader SH, Icenhower M, Pierre PJ, Bennett A, Garg PK, Garg S, Nader MA (2009) Behavioral and neurobiological characteristics influencing social hierarchy formation in female cynomolgus monkeys. Neuroscience 158:1257–1265

Rogers J, Shelton SE, Shelledy W, Garcia R, Kalin NH (2008) Genetic influences on behavioral inhibition and anxiety in juvenile rhesus macaques. Genes Brain Behav 7:463–469

Spinelli S, Chefer S, Carson RE, Jagoda E, Lang L, Heilig M, Barr CS, Suomi SJ, Higley JD, Stein EA (2010) Effects of early-life stress on serotonin(1A) receptors in juvenile Rhesus monkeys measured by positron emission tomography. Biol Psychiatry 67:1146–1153

Trefilov A, Berard J, Krawczak M, Schmidtke J (2000) Natal dispersal in rhesus macaques is related to serotonin transporter gene promoter variation. Behav Genet 30:295–301

Watson KK, Ghodasra JH, Platt ML (2009) Serotonin transporter genotype modulates social reward and punishment in rhesus macaques. PLoS ONE 4:e4156

Weiss A, King JE, Figueredo AJ (2000) The heritability of personality factors in chimpanzees (*Pan troglodytes*). Behav Genet 30:213–221

Wendland JR, Lesch KP, Newman TK, Timme A, Gachot-Neveu H, Thierry B, Suomi SJ (2006) Differential functional variability of serotonin transporter and monoamine oxidase a genes in macaque species displaying contrasting levels of aggression-related behavior. Behav Genet 36:163–172

Williamson DE, Coleman K, Bacanu SA, Devlin BJ, Rogers J, Ryan ND, Cameron JL (2006) Heritability of fearful-anxious endophenotypes in infant rhesus macaques: a preliminary report. Biol Psychiatry 53:284–291

16 Genomic Studies of Human Populations: Resequencing Approaches to the Identification of Human Quantitative Loci

Joanne E. Curran, Claire Bellis, Laura Almasy, and John Blangero

16.1 Introduction

The primary goal of the complex disease genomics field is to identify loci influencing disease susceptibility. The field has progressed substantially in recent years with the development of new methodologies for genome-wide assessment of sequence variation. Data is rapidly accumulating that rare variants have a large cumulative effect on normal phenotypic variation and are extremely important to disease (Blangero 2004; De La Vega et al. 2011; Li and Leal 2008; Pelak et al. 2010). Pedigree-based studies represent an implicit enrichment strategy for identifying such rare variants. Mendelian transmissions from parents to offspring maximize the chance that multiple copies of rare variants exist in the pedigree. These variants can then be identified by direct resequencing and statistical tests that minimize the influence of spurious linkage disequilibrium (Blangero et al. 2005; Kent et al. 2007). The key factor in the identification of rare variants then becomes resequencing sufficient numbers of chromosomes to capture all existing sequence variation. Given the likely importance of rare variation, a comprehensive sequencing strategy is the best means for detecting all such variants with sufficient copies. Whole genome sequence represents the "holy grail" for genetic studies. Prior approaches have only sampled partial variation from the genome. Unlike most other sciences, the causal state space for genetics is finite and we now have the tools available to comprehensively examine it.

J.E. Curran (✉) · C. Bellis · L. Almasy · J. Blangero
Department of Genetics, Texas Biomedical Research Institute, PO Box 760549, San Antonio, TX 78245-0549, USA
e-mail: jcurran@txbiomedgenetics.org

C. Bellis
e-mail: cbellis@txbiomedgenetics.org

L. Almasy
e-mail: almasy@txbiomedgenetics.org

J. Blangero
e-mail: john@txbiomedgenetics.org

16.2 Rationale for Next-Generation Sequencing Studies

16.2.1 Human Genetic Variation

Human genetic variation manifests itself in all aspects of human phenotypic variation, and has direct implications for the discovery of the underlying genes responsible for the observed heritability (the proportion of the total variance of a phenotype that is attributable to the additive effects of alleles) of any given trait. Localization and identification of these causal genes is the main goal of any genetic study of human disease. The heritability of a given trait tells us how important a role genetic variation is likely to play in the causation of variation. For many complex traits, known genetic variants only account for a small proportion of the total heritability (using plasma HDL-C levels as an example, known

R. Duggirala et al. (eds.), *Genome Mapping and Genomics in Human and Non-Human Primates*, Genome Mapping and Genomics in Animals 5, DOI 10.1007/978-3-662-46306-2_16,

variants account for less than 10 % of its total heritability) (Chasman et al. 2009), indicating that most of the genes/genetic variants have yet to be identified. This problem, termed "missing heritability", holds for many other complex human traits (Manolio et al. 2009).

What is the biological source of heritability in humans? Ultimately, it comes from observable functional genetic variation at the sequence level. A functional variant is one that influences the focal phenotype via some molecular mechanism. Thus, functional variants can be considered to be phenotype-specific in this context. It is the primary goal of complex disease genetics to identify such directly functional variants since they also will directly implicate the causal genes involved in the disease process.

16.2.2 Identification of Human Genetic Variation

For a given trait with significant heritability, how does one go about localizing and identifying these causal genetic factors that ultimately determine heritability?

Many recent advances in analysis of human quantitative traits have been made in the context of genetically complex diseases (Blangero 2004). In the absence of complete sequence information for all study participants, gene localization depends either on the random effect of known genetic markers assessed via linkage (for details, see the Chap. 3 by Almasy et al., in this volume), or the main effect of the markers via association (for details, see the Chapter by Hanson and Malhotra in this volume).

16.2.2.1 Gene Localization by Linkage

Before the era of high-throughput genotyping and next-generation sequencing, complex disease genetics in pedigrees was concentrated on genome scanning using a high-density map of genetic markers evenly distributed throughout the genome. Such genome scan, or linkage, information was used to identify chromosomal regions that contain variants influencing disease risk factors. Classical penetrance model-free linkage analysis is biased against rare functional variants. This bias is the result of the usual practice of estimating a single residual heritability and single quantitative trait loci (QTL)-specific heritability for the entire set of pedigrees examined. However, if rare functional variation is important, we would expect the magnitude of correlation between relatives to vary across pedigrees reflecting variance in pedigree-specific heritability. In fact, evidence of heterogeneity in heritabilities across pedigrees is expected under a model in which there are rare variants of moderate effect segregating. The assumption of heritability (both total and QTL-specific) homogeneity will lead to many missed QTL signals if rare variation is important.

To better search for QTLs due to rare functional variation, pedigree- and lineage-specific linkage analyses are required. These analyses may be done using a simple extension of the variance component model and simultaneously accounting for potential covariates. One approach for pedigree-specific linkage analysis is analogous to that long utilized for Mendelian disorders of dichotomous diseases. Basically, a search for linkage within each pedigree is performed using the usual variance component model (for details, please refer to Chap. 3 by Almasy et al., in this volume) with the added constraint that mean parameters must be held constant to that estimated from all of the data. This constraint is conceptually similar to ascertainment correction and guarantees that all phenotype deviations are referenced to the total population rather than to the specific pedigree being considered. Although each pedigree can be argued to represent a set of unique localization hypotheses, it may also be prudent to address the increased number of parameters being investigated (one additional QTN-specific heritability for each pedigree). Therefore, a mixture model analogous to heterogeneity logarithm of odds (LOD) testing performed in parametric linkage analysis, such as that being implemented in our computer package, SOLAR should also be employed. Some of the pedigrees may segregate functional variants at a given location, while others will not. All testing should be performed in the standard variance component framework. Using this approach in an example and performing a formal test of heritability heterogeneity, we determined that about 15 % of lipid-related traits in the San Antonio Family Study show strong evidence for differences in total heritability across pedigrees that is consistent with the presence of rare functional variants; and pedigree-specific linkage analyses revealed additional genome-wide significant QTLs for 140 traits that were missed using conventional linkage analysis, suggesting a large underlying source of rare functional variation (Unpublished data). These results highlight the importance of such an approach for the identification of QTLs due to rare variation.

16.2.2.2 Gene Localization by Association

Association studies are limited to detecting the effects of relatively common (with allele frequencies greater than 0.10) genetic variants and largely have not led to causal gene identification. Genome-wide association studies (GWAS) exploit correlations between closely spaced markers that are a function of linkage disequilibrium. These correlations are limited by the differences in the allele frequencies of the markers. Thus, common variants cannot be strongly correlated with rare ones. GWAS panels of single nucleotide polymorphisms (SNPs) are selected to represent common variation across the genome with modern human GWAS panels including a million or more markers that are correlated with a large portion of SNPs with allele frequencies of 0.05 or greater at an r^2 of at least 0.8. Each SNP on the panel represents not only itself but also serves as a proxy for some number of ungenotyped markers. An association signal could be due to any of the variants in disequilibrium with the genotyped SNP. Although widely misunderstood, these associations do not represent gene identifications but are bona fide QTL localizations that must be deeply sequenced in order to identify the underlying causal genes and followed up with functional work. Unfortunately, to date, few causal genes have resulted from these types of studies and the ones that have been identified were generally known as prior candidate genes. Most functional genetic variation is likely to be much less frequent. Thus, these classical localization approaches will miss most of the functional genetic signal, resulting in the missing heritability problem described above. Therefore, in order to be able to detect the effects of rare functional variants, a high-throughput next-generation sequencing approach needs to be employed to exhaustively search for variants.

16.2.2.3 Gene Localization by Sequencing

The identification of causal genes, using a genome sequencing approach, will obligately generate information on the pathways of these genes and will directly identify novel drug targets. Unlike epidemiological approaches, causal inference is possible using genetic strategies. Because DNA sequence variation is not influenced by other biological or environmental factors, genetic variation that correlates with disease risk must obligately reflect causation. Of course, identifying the exact causal sequence variants is difficult and represents one of the main challenges of modern human genetics. The ability to identify genes that are causally involved with disease risk provides an unparalleled opportunity to quickly determine biological pathways that are involved in pathology. Modern genomic technologies that allow the unbiased examination of all genes simultaneously can be exploited to rapidly identify genes involved in disease susceptibility. Given this information, each gene in an empirically identified molecular network that is proven to be involved in disease risk becomes a potential drug target. Whole genome sequencing allows a comprehensive search for functional sequence variation, to identify novel genes with alterations that have a substantially higher likelihood of representing functional variants of relevance for human physiological variation.

More recently, rare variants have been receiving increased attention in an attempt to explain the "missing heritability" problem observed from GWAS for many common traits. Rare variants are likely to have larger effect sizes and could contribute significantly to missing heritability; and it is postulated that these variants are also likely to have obvious functional consequences (Cirulli and Goldstein 2010; Manolio et al. 2009; Pelak et al. 2010). The primary technology for identifying rare variants is sequencing, either of target regions or entire genomes; however the sample set selected for sequencing will be of particular importance. These topics will be addressed in the following sections.

16.3 Next-Generation Sequencing Applications

16.3.1 Targeted Sequencing

16.3.1.1 Promoter Sequencing

Gene transcription in complex organisms is controlled by the intricate balance of proteins binding to promoter, enhancer and repressor sites within DNA sequences, and is regulated by both cis-acting factors in the flanking gene sequence and trans-acting external modulators regulated by other cellular characteristics. In recent years, there has been considerable interest in the effect of cis-acting variants on gene transcription. Several cis-acting elements exist that are involved in regulating gene expression; however the simplest and potentially most important is the 5′ promoter region. Promoter sequences are the most precisely defined of the many cis-acting regulatory regions of a gene and are of critical importance for their role in initiating gene transcription (Buckland et al. 2005; Coleman et al. 2002a, b; Rockman and Wray 2002). The role of promoters in initiating gene transcription highlights them as a potential source of genetic variation that may affect the expression level of a gene, and given their fixed location, promoters are also an ideal region for both genomic and functional analyses of genetic variation.

Several studies have investigated the functional relevance of variants within promoter regions. Buckland et al. (2005) performed a meta-analysis of approximately 700 gene promoters, interrogating the first 500 bp upstream of the transcription start site. Of the variants investigated, their results showed strong bias towards a promoter location for functional SNPs. Of all SNPs, 50 % were within the first 100 bp and 75 % within 200 bp of the transcription start site (Buckland et al. 2005). A second study performed by Rockman and Wray (2002) investigated 141 promoter variants involved in regulating over 100 genes, spread over the autosomes and X chromosome. They found that 63 % of the 107 genes studied had allelic differences of twofold or greater in their rates of transcription. Similar to the study of Buckland et al. as described above, 58.9 % of the functional variants were located within the first 500 bp upstream of the transcription start site. An additional 12.8 % fell 3′ to the start of transcription and another 12.8 % were more than 1 kb upstream of the transcription start. Only 1.4 % of functional variants were more than 10 kb upstream of their start sites (Rockman and Wray 2002). A more recent study by Sinnett et al. (2006) analyzed the promoter region (defined as 2 kb upstream of transcription initiation) of 197 genes in a multi-ethnic panel of 40 individuals. Their analysis identified 1,838 promoter variants for assessment and results showed that 75 % of the variants predicted functional roles, modifying putative transcription factor binding sites (Sinnett et al. 2006). A number of specific functional genes influencing complex phenotypes in humans have been successfully identified, and in our own work, we identified functional promoter variants in selenoprotein S (SELS), a gene involved in inflammation and in presenilins-associated rhomboid-like (PARL), a gene involved in mitochondrial integrity (Curran et al. 2005, 2010).

16.3.1.2 Candidate Gene Sequencing

Traditionally candidate genes have been identified through a variety of different methods including the comprehensive searching of the publicly available genomic databases, dense SNP mapping or genome-wide transcriptional profiling in the sample population. No matter the method of identification, the next step for a positional candidate gene is to comprehensively resequence the gene in sufficient individuals to maximize the probability of identifying all genetic variation. Selection of the most informative sample set for sequencing is highly dependent on the population being assessed (i.e., families or unrelated individuals). In a large extended pedigree, key members, such as founders, will impart the most information. For samples of unrelated individuals, it will be impossible to capture all variation, though the use of phenotypic extremes for sequencing will likely identify variants of larger effect sizes. The relative position of the variant to the gene's structure strongly influences the probability that the variation affects the function of the gene product. Thus, for resequencing in the sample, the most comprehensive strategy is to include all exons, intronic regions shown to be evolutionarily conserved as identified by comparative genomics (if the total intronic region is too large), 2 kb of the 3′UTR region and up to 5 kb of the putative promoter region, for each gene. Sanger sequencing, the most common method for such sequencing is now too costly and is being surpassed by next-generation sequencing applications on the smaller instruments including the Illumina MiSeq and the Life Technologies Ion Torrent.

16.3.1.3 QTL Sequencing

Linkage analyses identify chromosomal regions, of varying size, that contain QTLs influencing disease risk factors. This information significantly reduces the genomic search space, but still requires further effort to localize the specific genes and variants contributing to the signal. In comparison to a QTL region identified by association ($\sim$500 kb), a QTL region identified by linkage is typically 10–15 megabases (Mb) in size. To identify the underlying genes influencing this linkage signal, deep comprehensive sequencing is required. Many studies have performed fine mapping, with SNP markers, across linkage regions to try and narrow the search space, however this has been met with limited success as the variants assessed have only been common. Until recently, the sequencing of a QTL region by Sanger sequencing has been too costly, though with the release of the small scale next-generation sequencing instruments mentioned above, such sequencing is rapid and relatively inexpensive.

16.3.2 Exome Sequencing

Traditional complex phenotype research has focused on the analysis of protein coding variants that directly impact the protein structure and function, often dominant in simple disorders. We are now able to do this on a genome-wide scale, using a whole exome sequencing approach. Whole exome sequencing represents a currently accessible technology that enables the rapid identification of functional protein coding variation influencing phenotypic variation. The exome constitutes approximately 1 % of the human genome. This represents roughly 30 Mb that is split across 200,000 exons. Exome sequencing allows the identification of all coding variants, including those non-synonymous variants that alter protein sequence, which are most likely to have direct functional consequences. Modern sequencing technology allows us to entertain such a comprehensive approach.

Recent studies suggest that exome sequencing can be very powerful and that many rare potentially functional coding variants are likely to be found (Choi et al. 2009; Ng et al. 2010). Using targeted whole exome sequencing of 12 individuals, Ng et al. (2009) found approximately 6,000 non-synonymous variants per individual and predicted that it would have been about 8,500 with better sequencing coverage. These variants also were primarily rare (Ng et al. 2009). Using a less sensitive technique, Hedges et al. (2009) sequenced 8 independent exomes and found an average of 3,847 non-synonymous variants per individual with 683 being novel

(Hedges et al. 2009). All of these sequencing studies suggest a large number of relatively rare protein coding variants lurk within human populations. A recent study by Bowden et al. (2010) identified a rare variant (of ~1 % frequency) that accounts for 17 % of the variance in plasma adiponectin in a large Hispanic American sample, using a family-based whole exome sequencing approach (Bowden et al. 2010).

16.3.3 Whole Genome Sequencing

Whole genome sequencing (WGS) allows a comprehensive search for functional sequence variation, to identify novel genes with alterations that have a substantially higher likelihood of representing functional variants of relevance for human physiological variation. While identifying the exact causal variants influencing a trait represents one of the main challenges of human genetics, the potential to identify such variants using WGS is significantly increased and provides an unparalleled opportunity to quickly determine biological pathways involved. Each gene in an empirically identified network becomes a potential drug target. WGS represents the "holy grail" for genetic studies. Prior approaches have only sampled partial variation from the genome. However, the first studies employing WGS are now being performed in sufficiently large samples to likely produce benefit. One of the earliest applications of WGS in human gene identification has been to severe disorders that are thought to be possible single gene, Mendelian conditions. Sequencing in small samples of such patients has identified putative functional mutations in a large proportion of cases. WGS in a sample of six patients with severe early onset epilepsy and their parents was successful in all six cases, identifying de novo mutations in four individuals, parental isodisomy in one, and a recessive mutation in another (Martin et al. 2014). Among the first published WGS studies for a complex human phenotype is an examination of bipolar disorder in a large Old Order Amish pedigree (Georgi et al. 2014). This study identified multiple chromosomal regions shared among affected family members, each with multiple potential deleterious variants, suggesting a complex and potentially heterogeneous genetic architecture underlying bipolar disorder even in this population isolate.

16.3.4 Other Sequencing Applications

16.3.4.1 RNA Sequencing (RNA-Seq)

The transcriptome is defined as the complement of RNA molecules (transcripts) in a cell. Using modern genomic technology, it is now possible to discover, profile and quantify RNA transcripts (for details, please refer to Chap. 5 by Göring in this volume). Characterization of the transcriptome is essential for identifying and interpreting functional elements of the genome, and understanding disease development. Compared to array based transcript analyses, RNA-Seq provides several advantages over array based assays including a more precise quantification of transcripts and their isoforms than other methods; it is not limited to detecting transcripts that correspond to known genomic sequences; junctions between exons can be assayed; allele-specific expression differences and alternative splicing events can be detected. RNA profiling tools have been around for decades, though tremendous progress has been made in advancing the technology. From the days of Northern blots and serial analysis of gene expression (SAGE) analysis we have moved to gene-expression microarrays and now deep sequencing. With these tremendous advances in technology, the information content obtained from RNA analysis has also significantly increased, and like that of genome sequence, a substantial computational framework is essential.

16.3.4.2 Methylation Sequencing

The methylation pattern of DNA has been shown to influence gene expression patterns.

The implementation of next-generation sequencing has made it possible to study methylation patterns on a genome wide scale, rather than on a gene by gene basis. There are two common forms of methylation sequencing: whole genome bisulfite sequencing and MeDIP-Seq. In whole genome bisulfite sequencing, genomic DNA is bisulfite treated and all unmethylated cytosine bases are converted to uracil. Methylated cytosine bases (those containing a 5′ methyl group; 5′-methylcytosine) are not affected and once sequenced, the methylation status of each allele can be determined (Callinan and Feinberg 2006; Pomraning et al. 2009). This requires whole genome sequencing and is still somewhat cost prohibitive. The second method, MeDIP-Seq is an attempt to reduce the sequenced material, increasing throughput and reducing the cost. In this method, methylated DNA is selected for prior to sequencing using an antibody against 5′-methylcytosine. The unmethylated DNA is then washed away, leaving only the highly enriched methylated material for sequencing (Down et al. 2008; Pomraning et al. 2009). Both of these methods have their own benefits and one best suited to the study design should be chosen.

16.3.4.3 Mitochondrial Sequencing

The mitochondria are essential to life, being the major cellular site of energy production and respiration. A great deal of research has implicated mitochondrial dysfunction in a variety of human diseases including cancer, obesity, multiple sclerosis, several psychiatric disorders and a wide range of age related disorders (Begriche et al. 2006; Dakubo et al. 2006; Dutta et al. 2006; Fattal et al. 2006; Wallace 2005; Weissig et al. 2004). The mitochondrial genome is very small, consisting of about 16.5 kb and encodes genes for the biochemical reactions of respiration, and specific molecules involved in protein synthesis. The genome however only encodes a small number of mitochondrial functioning proteins; most of the proteins found in the mitochondria are nuclear encoded.

Mitochondrial sequencing is very popular among anthropologists and genealogists to investigate human evolution and diversity; however it has been gaining more interest from geneticists given the essential role of the mitochondria in maintaining cellular homeostasis. Given the small size of the mitochondrial genome, Sanger sequencing methods are still feasible; though sequencing is also possible using next-generation sequencing technology and mitochondrial variant information is captured when performing whole genome sequencing.

16.4 Next-Generation Sequencing Study Design

16.4.1 Return of the Family Study

Given the cumulative effect of rare variants on normal phenotypic variation and their importance to disease, different strategies are required to identify such variants than those that have been employed to assess common genetic variation. Most obvious, optimal capture and detection of rare functional variants will require a return to pedigree-based studies. Pedigree studies are one of these implicit designs that provide several advantages to the identification of rare variation, the main advantage being that rarer variants will be present at a much higher frequency than in the general population. Mendelian transmissions from parents to offspring maximize the chance that multiple copies of rare variants exist in the pedigree. These variants can then be identified by direct resequencing and statistical tests that minimize the influence of spurious linkage disequilibrium (Blangero et al. 2005; Kent et al. 2007). The key factor in the identification of rare variants then becomes resequencing sufficient numbers of chromosomes to capture all existing sequence variation. Given the likely importance of rare variation, a comprehensive sequencing strategy is the best means for detecting all such variants with sufficient copies.

Deep sequencing for functional variation in large pedigrees offers many benefits over studies of unrelated individuals, predominantly a greater number of copies of private variants. Additionally, for rare but non-private variants, extended pedigrees lead to substantially increased variance in allele frequency which permits a much wider potential for large variant-specific heritabilities (genetic signals). Given the growing awareness that rare functional variation appears to be responsible for observable phenotypic genetic variation, it is clear that individual pedigrees can provide significant evidence for gene identification for even complex quantitative phenotypes such as lipids. With the vast extent of private functional variation, any pedigree may hold an overt key to a disease-relevant gene. A pedigree-specific rare functional variant with small relative effect size (in relation to population attributable risk or QTN-specific heritability), but with a larger absolute effect size that is further enriched by Mendelian statistical mechanics within an extended pedigree, can be sufficient to verify that a given gene is involved in endophenotype variation.

16.4.2 Unrelated Individual Study Designs

Much of this chapter has focused on the importance of pedigree-based study designs for the identification of rare variation, but what options are possible for unrelated individuals?

For less rare, but still uncommon, variants with minor allele frequencies greater than 0.005, large studies of highly selected unrelated individuals such as those employed in the pioneering work by the research group at the University of Texas Southwestern Medical Center (Cohen et al. 2004, 2005, 2006) have led to the identification of rarer functional variants influencing lipids. However, such studies are inefficient and not suited to the largest functional class involving effectively private functional variants. By definition, any set of unrelated individuals could never capture more than a single copy of such a variant. However, due to the large case/control samples accumulated in various biorepositories during the era of GWAS studies, it is likely that interest in gene discovery in samples of unrelated individuals will continue and these sample collections may prove useful in generalizing family-based gene identification results to population-based samples. In particular, it may be useful to examine genes nominated by family studies in large collections of unrelated individuals to identify and characterize other functional variants in these genes.

16.5 Next-Generation Sequencing (NGS) Technologies

NGS approaches presently available, and that have been implemented in the new wave of sequencing projects, provide DNA read data generated using a high-throughput methodology that employs substantially different underlying chemistry dynamics. While there are several different technologies on the market, in this review we focus on the three companies that have led the revolution: Illumina, Life Technologies (previously Applied Biosystems) and Roche.

16.5.1 Illumina

Illumina recently added the HiSeq 2,500 platform to its family of sequencing instruments, which also includes the HiSeq 2,000, the Genome Analyzer IIx and the MiSeq. The HiSeq 2,500 incorporates the HiSeq 2,000 architecture with the onboard cluster generation of the MiSeq to switch between high output run mode and a rapid turnaround run mode. The rapid run mode is capable of sequencing a 30x genome in a day or fast multiplexed applications, such as exomes or RNA-Seq. Up to 120 Gb of sequence (1.2 billion reads), using 2 × 150 bp read lengths, is generated in ~27 h during this rapid run phase. In high output mode, the HiSeq 2,000 & 2,500 are identical with 600 Gb output (6 billion paired-end reads) in ~11 days for a 2 × 100 bp read length. Given their output, the HiSeq systems are most widely used for whole genome and whole exome sequencing. Additional applications include de novo sequencing, RNA-Seq, small RNA discovery, DNA methylation sequencing and cytogenetic analysis.

The Genome Analyzer (GA) IIx is the most widely published and adopted next-generation sequencing technology. The GAIIx is capable of outputting 95 Gb of data per run (640 million paired-end reads) in ~14 days at a 2 × 150 bp read length. The GAIIx is most widely used for RNA-seq, ChIP-Seq for gene regulation analysis, small genome sequencing, targeted resequencing, de novo sequencing, and amplicon sequencing.

The MiSeq is a fully integrated sequencing system that performs cluster generation, sequencing and data analysis all onboard the instrument within a 24 h period for 2 × 150 bp paired-end reads. The MiSeq platform is capable of outputting up to 5 Gb of sequence per 27 h run, for 2 × 150 bp reads and up to 8.5 Gb of data in 39 h for 2 × 250 bp runs. Applications of the MiSeq include highly multiplexed amplicon sequencing, targeted sequencing, small genome sequencing, ChIP-Seq and small RNA sequencing.

16.5.2 Life Technologies

The Life Technologies Support Oligonucleotide Ligation Detection (SOLiD) system performs massively parallel sequencing by stepwise ligation (all DNA is sequenced at the same time). The unique assay uses a 2 base encoding to distinguish between SNPs and errors. The SOLiD 4 instrument can generate 100 Gb of sequence data, or 1.4 billion reads per 16 day run, with 2 × 50 bp mate-paired reads. As with the other instruments, the SOLiD 4 can be used for whole genome sequencing, de novo sequencing, targeted sequencing, methylation sequencing, ChIP-Seq, small RNA sequencing and transcriptome sequencing.

The Ion Torrent system is a single instrument that uses semiconductor chips and semiconductor technology for a variety of sequencing throughputs. The Chip is the machine and can scale in density for pretty much any application. The throughput of the instrument is fast with a 2 h sequencing run for 200–400 bp reads, generating a between 10 Mb and 1 Gb of sequencing depending on the Chip selected. The Ion Torrent is best suited to small genome sequencing, targeted sequencing, target capture, RNA-Seq and miRNA-Seq, library assessment and ChIP-Seq.

16.5.3 Roche

The Roche 454 GS FLX+ is an ultra high-throughput automated DNA sequencing system.

The major advantage of the GS FLX + approach to sequencing besides the throughput is its ability to achieve read lengths in the order of 1,000 bp, the longest of any of the NGS technologies. The technology is also flexible enough to combine both long shotgun reads and paired end reads for complex genomes. The instrument can generate 700 Mb of sequence, 1 million shotgun reads, per 23 h instrument run. Applications of the GS FLX+ include genome sequencing, de novo sequencing, targeted sequencing, and transcriptome sequencing.

The Roche 454 GS Junior, like the MiSeq, is a compact integrated system capable of sample prep to analysis in a single run. The instrument generates 35 Mb of sequence (100,000 shotgun or 70,000 amplicon reads at an average of 400 bp) per 10 h run. The GS Junior is best suited to amplicon sequencing, genome and de novo sequencing of microbes, and transcriptome sequencing.

16.5.4 Genetic Analysis Services

It is not always feasible or possible for laboratories to perform sequencing in-house, and for this situation, there are many companies available that will perform sequencing, alignment and some data interpretation services. A web search of sequencing services will alert you to the many companies that are available, but here are some better known services.

16.6 Illumina and Certified Service Providers

The Illumina Genome Network is a network of sequencing teams at different institutions using Illumina sequencing platforms and providing whole genome sequencing services.

Illumina FastTrack Services provides whole genome sequencing services performed at Illumina by their scientists.

16.7 Complete Genomics

CGI provides a complete end-to-end sequencing service for human genomes, from sample preparation to analysis, providing ready called sequence data. The turnaround time is 3–4 months and a minimum 40x mapper coverage is guaranteed, sequenced on their own proprietary technology.

Other service providers include the National Center for Genome Resources, Beckman Coulter Genomics, BGI, SeqWright, HudsonAlpha and CIDR. Data interpretation services are provided by Knome, DNASTAR, Broad Institute and the Sanger Institute.

16.8 Conclusion

Looking back to the advent of the Human Genome Project back in 1990, it then took 13 years to complete the sequencing of 8 human genomes and cost billions of dollars. If we look at the technological advances that have occurred since then and assume this is the way of the future, we can only imagine that we will be able to obtain sequence information much more rapidly than we do now and the price will continue to decrease. The field of human genetics will be dominated by complete genome sequencing. In light of these times, we now need to think ahead and pay some focus to the computational burden that these projects will pose and furthermore, look towards biology and making sense of the genetic information we will soon be overwhelmed with.

References

Begriche K, Igoudjil A, Pessayre D, Fromenty B (2006) Mitochondrial dysfunction in NASH: causes, consequences and possible means to prevent it. Mitochondrion 6:1–28

Blangero J (2004) Localization and identification of human quantitative trait loci: king harvest has surely come. Curr Opin Genet Dev 14:233–240

Blangero J, Göring HHH, Kent JWJ, Williams JT, Peterson CP, Almasy L, Dyer TD (2005) Quantitative trait nucleotide analysis using Bayesian Model Selection. Hum Biol 77:541–559

Bowden DW, An SS, Palmer ND, Brown WM, Norris JM, Haffner SM, Hawkins GA, Guo X, Rotter JI, Chen YD, Wagenknecht LE, Langefeld CD (2010) Molecular basis of a linkage peak: exome sequencing and family-based analysis identify a rare genetic variant in the ADIPOQ gene in the IRAS Family Study. Hum Mol Genet 19:4112–4120

Buckland PR, Hoogendoorn B, Coleman SL, Guy CA, Smith SK, O'Donovan MC (2005) Strong bias in the location of functional promoter polymorphisms. Hum Mutat 26:214–223

Callinan PA, Feinberg AP (2006) The emerging science of epigenomics. Hum Mol Genet 15 Spec 1:R95–101

Chasman DI, Paré G, Mora S, Hopewell JC, Peloso G, Clarke R, Cupples LA, Hamsten A, Kathiresan S, Mälarstig A, Ordovas JM, Ripatti S, Parker AN, Miletich JP, Ridker PM (2009) Forty-three loci associated with plasma lipoprotein size, concentration, and cholesterol content in genome-wide analysis. PLoS Genet 5:e1000730

Choi M, Scholl UI, Ji W, Liu T, Tikhonova IR, Zumbo P, Nayir A, Bakkaloğlu A, Ozen S, Sanjad S, Nelson-Williams C, Farhi A, Mane S, Lifton RP (2009) Genetic diagnosis by whole exome capture and massively parallel DNA sequencing. Proc Natl Acad Sci USA 106:19096–19101

Cirulli ET, Goldstein DB (2010) Uncovering the roles of rare variants in common disease through whole-genome sequencing. Natl Rev Genet 11:415–425

Cohen JC, Kiss RS, Pertsemlidis A, Marcel YL, McPherson R, Hobbs HH (2004) Multiple rare alleles contribute to low plasma levels of HDL cholesterol. Science 305:869–872

Cohen JC, Pertsemlidis A, Fahmi S, Esmail S, Vega GL, Grundy SM, Hobbs HH (2006) Multiple rare variants in NPC1L1 associated with reduced sterol absorption and plasma low-density lipoprotein levels. Proc Natl Acad Sci USA 103:1810–1815

Cohen J, Pertsemlidis A, Kotowski IK, Graham R, Garcia CK, Hobbs HH (2005) Low LDL cholesterol in individuals of African descent resulting from frequent nonsense mutations in PCSK9. Natl Genet 37:161–165

Coleman SL, Buckland PR, Hoogendoorn B, Guy CA, Smith K, O'Donovan MC (2002a) Experimental

analysis of the annotation of promoters in the public database. Hum Mol Genet 11:1817–1821
Coleman SL, Hoogendoorn B, Guy CA, Smith SK, O'Donovan MC, Buckland PR (2002b) Streamlined approach to functional analysis of promoter-region polymorhisms. Biotechniques 33:412–418
Curran JE, Jowett JB, Elliott KS, Gao Y, Gluschenko K, Wang J, Abel Azim DM, Cai G, Mahaney MC, Comuzzie AG, Dyer TD, Walder KR, Zimmet P, MacCluer JW, Collier GR, Kissebah AH, Blangero J (2005) Genetic variation in selenoprotein S influences inflammatory response. Natl Genet 37:1234–1241
Curran JE, Jowett JB, Abraham LJ, Diepeveen LA, Elliott KS, Dyer TD, Kerr-Bayles LJ, Johnson MP, Comuzzie AG, Moses EK, Walder KR, Collier GR, Blangero J, Kissebah AH (2010) Genetic variation in PARL influences mitochondrial content. Hum Genet 127:183–190
Dakubo GD, Parr RL, Costello LC, Franklin RB, Thayer RE (2006) Altered metabolism and mitochondrial genome in prostate cancer. J Clin Pathol 59:10–16
De La Vega FM, Bustamante CD, Leal SM (2011) Genome-wide association mapping and rare alleles: from population genomics to personalized medicine. Pac Symp Biocomput 74–75
Down TA, Rakyan VK, Turner DJ, Flicek P, Li H, Kulesha E, Gräf S, Johnson N, Herrero J, Tomazou EM, Thorne NP, Bäckdahl L, Herberth M, Howe KL, Jackson DK, Miretti MM, Marioni JC, Birney E, Hubbard TJ, Durbin R, Tavaré S, Beck S (2008) A Bayesian deconvolution strategy for immunoprecipitation-based DNA methylome analysis. Natl Biotechnol 26:779–785
Dutta R, McDonough J, Yin X, Peterson J, Chang A, Torres T, Gudz T, Macklin WB, Lewis DA, Fox RJ, Rudick R, Mirnics K, Trapp BD (2006) Mitochondrial dysfunction as a cause of axonal degeneration in multiple sclerosis patients. Ann Neurol 59:478–489
Fattal O, Budur K, Vaughan AJ, Franco K (2006) Review of the literature on major mental disorders in adult patient with mitochondrial diseases. Psychosomatics 47:1–7
Georgi B, Craig D, Kember RL, Liu W, Lindquist I, Nasser S, Brown C, Egeland JA, Paul SM, Bućan M (2014) Genomic view of bipolar disorder revealed by whole genome sequencing in a genetic isolate. PLoS Genet 10:e1004229
Hedges D, Burges D, Powell E, Almonte C, Huang J, Young S, Boese B, Schmidt M, Pericak-Vance MA, Martin E, Zhang X, Harkins TT, Züchner S (2009) Exome sequencing of a multigenerational human pedigree. PLoS ONE 4:e8232
Kent JW Jr, Dyer TD, Göring HHH, Blangero J (2007) Type I error rates in association versus joint linkage/association tests in related individuals. Genet Epidemiol 31:173–177
Li B, Leal SM (2008) Methods for detecting associations with rare variants for common diseases: application to analysis of sequence data. Am J Hum Genet 83:311–321
Manolio TA, Collins FS, Cox NJ, Goldstein DB, Hindorff LA, Hunter DJ, McCarthy MI, Ramos EM, Cardon LR, Chakravarti A, Cho JH, Guttmacher AE, Kong A, Kruglyak L, Mardis E, Rotimi CN, Slatkin M, Valle D, Whittemore AS, Boehnke M, Clark AG, Eichler EE, Gibson G, Haines JL, Mackay TF, McCarroll SA, Visscher PM (2009) Finding the missing heritability of complex diseases. Nature 461:747–753
Martin HC, Kim GE, Pagnamenta AT, Murakami Y, Carvill GL, Meyer E, Copley RR, Rimmer A, Barcia G, Fleming MR, Kronengold J, Brown MR, Hudspith KA, Broxholme J, Kanapin A, Cazier JB, Kinoshita T, Nabbout R; The WGS500 Consortium, Bentley D, McVean G, Heavin S, Zaiwalla Z, McShane T, Mefford HC, Shears D, Stewart H, Kurian MA, Scheffer IE, Blair E, Donnelly P, Kaczmarek LK, Taylor JC (2014) Clinical in severe early-onset epilepsy reveals new genes and improves molecular diagnosis. Hum Mol Genet Feb 11 [Epub ahead of print]
Ng SB, Turner EH, Robertson PD, Flygare SD, Bigham AW, Lee C, Shaffer T, Wong M, Bhattacharjee A, Eichler EE, Bamshad M, Nickerson DA, Shendure J (2009) Targeted capture and massively parallel sequencing of 12 human exomes. Nature 461:272–276
Ng SB, Buckingham KJ, Lee C, Bigham AW, Tabor HK, Dent KM, Huff CD, Shannon PT, Jabs EW, Nickerson DA, Shendure J, Bamshad MJ (2010) Exome sequencing identifies the cause of a mendelian disorder. Natl Genet 42:30–35
Pelak K, Shianna KV, Ge D, Maia JM, Zhu M, Smith JP, Cirulli ET, Fellay J,Dickson SP, Gumbs CE, Heinzen EL, Need AC, Ruzzo EK, Singh A, Campbell CR, HongLK, Lornsen KA, McKenzie AM, Sobreira NL, Hoover-Fong JE, Milner JD, Ottman R, Haynes BF, Goedert JJ, Goldstein DB (2010) The characterization of twenty sequenced human genomes. PLoS Genet 6: e1001111
Pomraning KR, Smith KM, Freitag M (2009) Genome-wide high throughput analysis of DNA methylation in eukaryotes. Methods 47:142–150
Rockman MV, Wray GA (2002) Abundant raw material for cis-regulatory evolution in humans. Mol Biol Evol 19:1991–2004
Sinnett D, Beaulieu P, Belanger H, Lefebvre JF, Langlois S, Theberge MC, Drouin S, Zotti C, Hudson TJ, Labuda D (2006) Detection and characterization of DNA variants in the promoter regions of hundreds of human disease candidate genes. Genomics 87:704–710
Wallace DC (2005) A Mitochondrial paradigm of metabolic and degenerative diseases, aging and cancer. Annu Rev Genet 39:359–407
Weissig V, Cheng SM, D'Souza GG (2004) Mitochondrial pharmaceutics. Mitochondrion 3:229–244

Index

A

β3-adrenergic receptor, 165
Abacavir, 23
Acculturation, 158
Additive model, 57
Admixture, 109, 135
Age-related macular degeneration, 93
Aggression, 281
Alaskan Eskimos, 155
Alcohol consumption, 279
Alcohol use, 158
Aleut form of subsistence, 131
Allele-specific expression analysis, 79
Allelic heterogeneity, 145
Allelic odds ratio, 54
Alternative, 78
American Indians, 155
Amish, 141, 142
Amish microcephaly, 146
Amish of Lancaster County, PA: History and Background, 142
Anabaptists, 142
Anabaptist genealogy database, 143
Anthropometric measurements, 159
Anxiety, 281, 283
Apes, 103
APOC3, 148
Archeological sites in the central Islands, 131
Arizona, 157
Assembly, 90
Association, 290
Association analysis, 67, 160
Autocorrelation indices, 135
Autosomal, 34
Autosomal dominant, 40
Autosomal recessive, 40

B

Baboon, 46, 103
BACPAC resources, 124
Basicranium, 260, 261, 265, 266
Bayesian quantitative trait nucleotide (BQTN) analysis, 167
Behavioral genetics, 277
Bioinformatic and transcriptomic analyses, 168
Bioinformatics, 106, 113
Biomedical research, 103
Bisulfite sequencing, 295
BLAST, 117
BLAT, 118
Bmp4, 259, 264
Body mass index (BMI), 187
Bone mineral density, 148
Breakpoint, 119

C

Calmodulin, 259
Cancer, 96
Candidate gene, 168, 187, 293
Cardiometabolic, 184
Cardiovascular disease (CVD), 155
Carotid artery wall thickness, 156
Case-control design, 54
Causal gene, 4
CCL3L1, 93
CDNA libraries, 107
CentiMorgans, 39
Charcot-Marie-Tooth neuropathy type 1A, 86
Chimpanzee, 17
Cholesterol, 159
Chromosomal rearrangements, 118
Chromosomal region expression array, 121
Cingular remnant, 268
Circular binary segmentation, 89
Cis, 73
Clinic examination, 156
CNPs/CNVs, 14
Cohen syndrome, 145
COL1A2
  G610C, 149
Colony-managed, 2
Common variant, 182

R. Duggirala et al. (eds.), *Genome Mapping and Genomics in Human and Non-Human Primates*,
Genome Mapping and Genomics in Animals 5, DOI 10.1007/978-3-662-46306-2,

Comparative genomic, 15, 113
Complex disease, 19, 70
Complex disease genomics, 289
Complex traits, 69, 189
Conditioned learning, 280
Conservation, 17, 105
Copy number polymorphism, 13
Coronary artery calcification, 148
Correlation analysis, 69
Co-segregation, 33
Craniosynostotic disorders
Apert syndrome, 264
Crouzon syndrome, 264
Crossover, 37
CVD risk factors, 155

D

Dakotas, 157
Data sharing, 170
De novo assembly, 106
Deep phenotyping, 5
Default-Mode Network, 253
Degree of relationship, 42
Delaunay triangulation methods, 133
Deletions, 85
Demic expansion of the Aleuts, 135
Demographic, 104
Demographic characteristics, 158
Diabetes, 156
Dichotomous, 254
Diet, 158
Discordant relative pairs, 43
Disease epidemics, 132
Disease susceptibility, 289
Disruption during world war II, 132
Distribution of mtDNA haplogroups, 134
DNA, 67, 158
DNA collection, 133
DNA sequence alignment, 115
Dopamine, 280
Dopaminergic signal, 281
Drosophila melanogaster, 15
Duplication, 85

E

Earliest evidence for human habitation in the Aleutian Islands, 131
Enamel knots, 267
Endophenotypes, 281
Environmental correlation, 160
Environmental risk factors, 158
EPIC markers, 107
Epigenetics, 256
Etiology, 70
Europeans, 182
Evolution, 1
Exclusion mapping, 39
Exome sequencing, 293
Exon, 69
Exon capture, 107
Experiment-wide statistical significance, 61
Exposome, 80
Expressed sequence tag, 9, 106
Expression QTLs, 167
Extended families, 155

F

Familial defective apolipoprotein B-100 R3500Q, 147
Family-based association, 60
Family history form, 157
Family study, 295
Fels longitudinal study, 268
FGFR2, 264
FGFR3, 264
Fibroblast growth factors, 263
Field studies, 103
Fine-mapping, 62, 161
Fitness, 103
Food frequency questionnaires, 159
Founder, 141, 143
Founder effect, 137
Fractional anisotropy, 252
Fragment length polymorphisms, 34
Freezing duration, 279
Functional allele, 55, 62
Functional craniology, 262
Functional magnetic resonance imaging, 247
Functional variant, 3, 290

G

GeneCards, 123
Gene coding, 113
Gene duplication, 106
Gene expansions, 16
Gene expression, 62, 95
Gene expression analysis, 167
Gene flow, 109, 137
Gene networks, 113
Gene x environment interactions, 113
Genetic, 181
Genetic architecture of the old order amish (OOA), 141
Genetic barriers, 136
Genetic correlation, 160, 250
Genetic discontinuity, 136
Genetic distance, 39
Genetic drift, 141, 143
Genetic epidemiology, 72
Genetic population structure, 129
Genetic variation, 2, 103, 189, 277, 286
Genome-enabled species, 105
Genome-wide arrays, 57
Genome-wide association, 155, 249
Genome-wide association studies (GWAS), 69, 129, 147

Genome-wide significance, 61
Genome annotation, 9
Genome complexity reduction, 107
Genome scanning, 290
Genomic control, 59
Genomics, 1
Genotype-by-environment (G x E) interaction, 213
Genotyping, 159
Geospiza, 259
Gly574Arg, 146
GOCADAN, 155
Gray matter, 248
GWAS, 291

H
HapMap Project, 19
Hardy-Weinberg-Castle equilibrium, 129
Haseman–Elston method, 44
Hepatic lipase gene, 167
Hereditary neuropathy with liability to pressure palsies, 86
Heredity and Phenotype Intervention (HAPI) Heart Study, 148
Heritability, 3, 160, 213, 252, 278, 289
Hidden Markov models, 89
HIV/AIDS, 93
Holoprosencephaly, 264
Hominini, 260
Hominoidea, 270
Homozygosity by descent mapping, 145
Howler monkey, 109
Human genome, 86
Human genome project, 7
HVS-I sequences, 133
Hybridization, 104, 109

I
11 island populations, 133
Identical twin, 42
Identity-by-descent allele sharing, 160
IHGSC, 8
Image-based, 255
Imaging genetics, 247
Imputation, 58, 91
Inbreeding, 46
INDELs, 85
Independent assortment, 37
Indian heritage, 159
Individual admixture, 60
Inflammation, 159
Informed consent, 158
Insulin resistance, 184
Integration, 270
Interference, 40
Intermatch distances, 138
International HapMap project, 56
Introgression, 109
Inupiat Eskimos, 157
Inversion, 118
Island populations, 129
Isolation-by-distance model, 135

J
J. Craig Venter, 8
Jakob Ammann, 142

K
KANSL1, 94
KEGG pathway, 126
Kinship, 108
Kinship coefficients, 45
Koolen syndrome, 94

L
Lemur, 106
Life history traits, 3
Likelihood ratio test, 38
Linkage, 290
Linkage analysis, 33, 67, 155
Linkage disequilibrium, 46, 55, 107, 161, 182
Linkage map, 35, 108, 115
Locus-specific heritability, 53
LOD, 39
LOD scores, 170
Longitudinal data, 156

M
Macaca mulatta
  Rhesus macaque, 266
Macaques, 103
Manhattan plot, 58
Mantel tests of matrices, 135
Mapping functions, 39
Maternal, 281
Maturity onset diabetes of the young 5, 94
Maximum likelihood variance decomposition methods, 160
Medical history, 158
Medication use, 159
MeDIP-Seq, 295
Meioses, 38
Mendelian and pedigree errors., 160
Meta-analysis, 58, 62, 191, 256
Metabolic syndrome (MS), 181
Metagenomics, 108
Methylation, 294
Mexican Americans, 155, 181
Micro-homology, 91
Microarray, 69, 88
Microarray expression profiles, 159
Microbes, 108
Microsatellite markers, 116, 159

Microsatellites, 37, 108
Minority populations, 155
Missing heritability, 63, 182
Mitochondrial DNA haplotype, 133
Mitochondrial sequencing, 295
Model-based linkage, 40
Model species, 103
Modularity, 270
Monmonier's algorithm, 133
Monozygotic twins, 250
Morbidity, 157
Morphological integration, 266
Mortality, 157
Multidimensional scaling, 136
Multifactorial, 181
Multiple comparisons, 60
Multipoint, 42

N
Natural, 103
NCBI trace archive, 117
Neanderthals, 109
Neurocranial, 266
Neurocranium, 260, 261, 265
Neuroreceptor proteins, 281
Neurotransmitters, 280
Neutral markers, 104
Next-generation sequencing, 105, 215, 292
Non-allelic homologous recombination, 91
Non-synonymous variants, 293
Noncoding RNAs, 10
Nonhuman primate, 1, 45, 281
Nonrecombining Y-chromosome markers, 130
Novelty-seeking, 278
NPL (nonparametric linkage) score, 43
Nutrigenetics, 24
Nutrigenomics, 24

O
Obesity, 156, 181
Oklahoma, 157
Oligo, 124
OMIM, 123
Ontological pathway, 126
Osteogenesis imperfecta, 149
Outlier loci, 104
Oxidative stress, 159

P
Paired-end read, 90
Papio hamadryas, 266
Paternity testing, 46
Pathophysiology, 71
PedCut, 146
PedHunter, 143
Pedigree, 108, 156, 290
Pedigree relationships, 158
Pedigreed colonies, 278
PedSys, 159
Penetrance, 40
Penetrance function, 35
Pharmacogenetics, 23
Phase, 37
Phenocopy, 41
Phenylalanine hydroxylase deficiency, 145
Phenylketonuria, 24
Phylogenetic, 103
Phylogenetic reconstruction, 270
Physical activity, 159
Physical examination, 158
Pleiotropy, 160
Polk directory, 156
Polydactyly, 264
Polymorphic markers, 105
Polyunsaturated/saturated fat ratio, 159
Population stratification, 59
Population subdivision, 129
Positional cloning, 188
Positional cloning approaches, 191
Prevalence, 54
Principal components, 60
Proband, 156
Promoter, 292
Pyrosequencing, 106

Q
17q12, 94
17q21.31, 94
Q or Q3 haplogroups, 137
QTL, 115, 290
Quality control, 57
Quantile–quantile plot, 58
Quantitative phenotypes, 249
Quantitative traits, 2, 36, 44
Quantitative trait locus (QTL), 36, 155

R
RADSeq, 107
Rare variants, 5, 181, 289
Read-depth, 90
Recombination, 33
Recombination hotspots, 40
Reference genome, 105
Regenerative medicine, 271
Regulatory elements, 113
Relative pairs, 162
Relative risk, 44
Relocations to the commander islands, 132
Replication, 61, 181
Reproductive history, 158
Resequencing, 106, 293
Resequencing technologies, 161
Resting-state, 256

Reward-seeking, 280
RFLP analyses, 133
Rhesus macaque, 17, 46
RNA, 67, 106
RNA sequencing, 69, 121, 294
Russian contact, 132

S
SAMOVA, 136
SAMOVA analysis, 133
Sample size, 54
Sampling, 133
San Antonio Family Heart Study (SAFHS), 155
Sanger sequencing, 293
Saquinus fuscicollis
  Saddle-back Tamarin, 266
Saturated fat, 159
Segmental duplications, 12, 86
Sequence variation, 289
Sequencher, 124
Sequencing, 1, 62, 89
Serotonin, 280
Serotonin transporter, 282
Shared environments, 278
Shh, 263
Shotgun, 105
Sib-pair linkage, 35
Signaling molecules, 280
Single nucleotide polymorphisms (SNPs), 155, 160, 194
SiteSeer, 123
Sitosterolemia
  Gly574Arg, 146, 147
Smoking, 158
SNP distributions, 130
SNPs, 107
Social behaviors, 279
Social dominance, 283
Socioeconomic status, 158
SOLAR, 160
Somatic mosaicism, 96
Spatial analysis of molecular variance, 133
Splanchnocranial, 266
Splanchnocranium, 260, 261, 265
Split-read, 89
Statistical, 70
Statistical genetic analysis, 160
Statistical power, 54
Stress, 280
Strong Heart Family Study (SHFS), 155
STRs, 130
Structural variation, 106
Study design, 289, 295
Susceptibility loci, 181
Syndactyly, 264
Synthetic associations, 215
Syntenic quantitative trait loci, 113
Systems biology, 262

T
α-thalassemias, 86
Tag SNPs, 56
Taxonomic, 104
Temperament, 279
Tissue specificity, 71
Total calories, 159
Trans, 71
Transcript, 68
Transcript abundance, 70
Transcription, 78, 292
Transcriptome, 67, 106, 294
Transcriptomic analysis, 155
Transcripts, 106
Transmission, 37
Transporter Proteins, 282
Tribal affiliation, 157
Tribal enrollment, 159
Triglycerides, 145
Type 2 diabetes (T2D), 181

U
UCSC genome bioinformatics browser, 118
UCSC table browser, 121
Unique, 278
Unrelated individuals, 289, 295

V
Variance component, 35, 45
Variance component analyses, 278
Variance component model, 290
Variation in complex disease traits, 113
Vervet, 46, 277
Victor McKusick, 150
Vista genome browser, 116

W
Warfarin, 23
White matter, 252
Whole genome expression profiling, 126
Whole genome sequence, 46, 182, 248
Whole genome sequence data, 160
Whole genome SNP typing, 167
Wild populations, 103
Winner's curse, 62
Wnt, 263

X
X-linked, 34, 40

Y
Y-chromosome haplogroups, 133, 134
Y-linked, 40

Printed by Printforce, the Netherlands